An Introduction to

Bayesian Inference in Econometrics

An Introduction to Bayesian Inference in Econometrics

ARNOLD ZELLNER

Wiley Classics Library Edition Published 1996

A WILEY-INTERSCIENCE PUBLICATION
JOHN WILEY & SONS, INC.
New York • Chichester • Brisbane • Toronto • Singapore

Wiley Classics Library Edition Published 1996.

To order books or for customer service please, call 1(800)-CALL-WILEY (225-5945).

Library of Congress Catalogue Card Number: 70–156329
ISBN 0-471-98165-6
ISBN 0-471-16937-4 Wiley Classics Library Edition

Printed in the United States of America

10 9 8 7 6 5 4

to Agnes and our sons

Preface

The purpose of this book is to provide readers with an introduction to Bayesian inference in econometrics. An effort has been made to relate the problems of inference in econometrics to the general problems of inference in science and to indicate how the Bayesian approach relates to general problems of scientific inference in Chapter I.

In Chapter II some fundamental concepts and operations employed in the Bayesian approach to inference are presented, discussed, and applied in analyses of several simple and important problems. Chapters III through IX are devoted to Bayesian analyses of models often encountered in econometric work, with the main emphasis on estimation. Many comparisons of Bayesian results with sampling theory results are made. In Chapter X I treat the problems of testing and comparing hypotheses, and in Chapter XI I analyze several control problems relating to regression and other processes. In Chapter XII I present some concluding remarks. Appendices A and B provide a resumé of the properties of a number of important univariate and multivariate distributions. In Appendix C univariate and bivariate numerical integration techniques are briefly described.

I have tried to keep the analysis and notation in this book as simple as possible. However, readers are assumed to be familiar with basic concepts and operations of probability theory, differential and integral calculus, and matrix algebra. A background in econometrics and statistics at about the level of A. S. Goldberger's text, *Econometric Theory*,[1] is desirable for an appreciation of the econometric relevance of the stochastic models considered and for an assessment of comparisons of Bayesian and sampling theory results.

For several years the material in this book has been presented to graduate students in economics and business at the University of Chicago in a course entitled Bayesian Inference in Econometrics. From experience in this course, I found that students not only mastered technical elements of the Bayesian approach but also gained considerable understanding of the properties of both sampling theory and Bayesian approaches to inference and the criteria used to appraise alternative systems of inference. Thus this teaching

[1] A. S. Goldberger, *Econometric Theory*, New York: Wiley 1964.

experience lends substantial support to Lindley's statement that the Bayesian and orthodox "approaches complement each other and together provide a substantially better understanding of statistics than either approach on its own."[2]

In teaching Bayesian Inference in Econometrics the material in Chapter I can serve as a basis for introducing students to some basic issues in the philosophy and methodology of science. It is desirable for an instructor to relate this material to philosophical and methodological issues in economics and econometrics. The text of the chapter and the questions at its end, are designed to encourage students to think about the nature and foundations of science and scientific method so that they may have a better understanding of research in economics and econometrics.

Chapter II provides a resumé of basic concepts and principles of Bayesian analysis along with some simple, but important, applications. Since most of the remaining chapters involve applications of the concepts and principles set forth in Chapter II, it is clearly vital that students master the subject matter of Chapter II. Perhaps the most difficult topics in Chapter II are the role and nature of prior information in analyses of data and the use of probability density functions to represent prior information. These topics require careful and thorough discussion.

Chapters III through IX are in essence technical chapters wherein the principles of Chapter II are applied in analyses of a number of models encountered in economics and econometrics. While the Bayesian principles involved in these analyses are the same, each problem has its peculiar technical features. In mastering technical features, the student is introduced to a number of distributions and operations which are thought to be valuable in a number of different contexts. Also, problems and applications are included which give the analyses relevance to current econometric research. In working some of the problems students will require use of numerical integration computer programs which are generally available at computation centers or which can easily be programmed.[3] Experience in using numerical integration programs will be valuable in approaching the analysis of a broad range of applied problems.

Chapter X takes up problems of comparing and testing hypotheses and models. The material in this chapter is in the nature of an introduction and may suggest fruitful areas for further methodological as well as applied work. In Chapter XI some control problems are analyzed, and again there is an opportunity for additional work on methods and on applications.

[2] D. V. Lindley, *Introduction to Probability and Statistics from a Bayesian Viewpoint, Part 2. Inference.* Cambridge: Cambridge University Press, 1965, p. 70.

[3] See, for example, B. Noble, *Numerical Methods, II: Differences, Integration and Differential Equations*, New York: Interscience Publishers, Inc., 1964 and Appendix C.

Finally, Chapter XII provides one person's, my own, summary and concluding remarks. Since systems of inference are extremely controversial, it is not to be expected that all will agree with what is presented. Thus, in teaching, Chapter XII can be utilized as a point of departure for developing each student's assessment of the Bayesian approach.

The content of this volume reflects intellectual interaction with a number of individuals. The work of H. Jeffreys has contributed importantly to advancing my understanding of the problems of inference and of elements of the Bayesian approach. In addition, papers and books by G. A. Barnard, G. E. P. Box, I. J. Good, D. V. Lindley, H. Raiffa and R. Schlaifer, and L. J. Savage have had an important influence on me, both with respect to fundamental conceptual problems and technical matters. In regard to current and former colleagues, I have benefited considerably from association with and the writings of G. E. P. Box, A. S. Goldberger, I. Guttman, M. Stone, and G. C. Tiao at the University of Wisconsin, Madison, Wisconsin, during the period 1960–1966. At the University of Chicago J. Drèze, S. J. Press, H. V. Roberts, D. Sharma, and H. Thornber have been stimulating colleagues in connection with the development of the present volume. Among past and current students who have contributed to the current work in the classroom and as research assistants, the following have been particularly helpful: V. K. Chetty, R. V. Cooper, M. S. Geisel, P. M. Laub, C. J. Park, J. F. Richard, N. S. Revankar, U. Sankar, P. A. V. B. Swamy, and H. Thornber.

Much of the research reported in this volume was supported by grants from the National Science Foundation for which I am grateful. Under the initial grant, GS-151, research was carried forward in association with G. C. Tiao. During periods of the first and second renewals of the original grant S. J. Press and H. Thornber participated in the research. Thanks are due to Tiao, Press, and Thornber for their valuable contributions and splendid cooperation in achieving the objectives of the NFS grants. The H. G. B. Alexander Endowment Fund gave salary support to the author since 1966 when he was appointed H. G. B. Alexander Professor of Economics and Statistics in the Graduate School of Business, University of Chicago.

Mrs. Shirley Black provided expert assistance in typing and proofing the manuscript for which I express my sincere thanks.

Chicago, Illinois
April 1971

ARNOLD ZELLNER

Contents

CHAPTER I

Remarks on Inference in Economics

It is a mistake for a sculptor or a painter to speak or write very often about his job. It releases tension needed for his work. By trying to express his aims with rounded-off logical exactness, he can easily become a theorist whose actual work is only a caged-in exposition of conceptions evolved in terms of logic and words.

Henry Moore[1]

What Moore has said about discussing work in art is undoubtedly applicable to discussions of methodology in economics. The importance, however, of knowing what it is we are doing in economic research makes it worthwhile on occasion to reflect on the general foundations of our work.

1.1 THE UNITY OF SCIENCE

The point of view taken herein is that scientific inferences made about economic phenomena are not fundamentally different from inferences made about phenomena in other areas of science. This stress on the unity of science has been elegantly expressed by Karl Pearson in the following words[2]:

> Now this is the peculiarity of scientific method, that when once it has become a habit of mind, that mind converts all facts whatsoever into science. The field of science is unlimited; its material endless, every group of natural phenomena, every phase of social life, every stage of past or present development is material for science. *The unity of all science consists alone in its method, not in its material.* The man who classifies facts of any kind whatever, who sees their mutual relation and describes their sequences, is applying the scientific method and is a man of science. The facts may belong to the past history of mankind, to the social statistics of our great cities, to the atmosphere of the most distant stars, to the digestive organs of a worm, or to the life of a scarcely visible bacillus. It is not the facts themselves which form science, but the methods by which they are dealt with.

To see that inferences in economic research are not fundamentally different from inferences in other areas of science it is relevant to review the kinds of

[1] Henry Moore, "Notes on Sculpture," in B. Ghiselin, ed., *The Creative Process*. New York: Mentor Books, 1952, p. 73.
[2] Karl Pearson, *The Grammar of Science*. London: Everyman, 1938, p. 16.

inferences that are relied on in scientific work. Aristotle lists three types of inference, namely, deductive, inductive, and reductive (also translated from Greek as "abductive" or "retroductive"). It is important that the nature of these types of inference be set forth clearly to appreciate their role in economic research.

1.2 DEDUCTIVE INFERENCE

On the nature of deductive inference, Reichenbach writes as follows[3]:

> Logical proof is called *deduction*; the conclusion is obtained by deducing it from other statements, called the premises of the argument. The argument is so constructed that if the premises are true the conclusion must also be true. . . . It unwraps, so to speak, the conclusion that was wrapped up in the premises.

There can be no doubt that deductive inference plays an important role in economics. It must be appreciated, however, that deductive inference *alone* is inadequate to serve as a basis for inference in economics. Primarily, this is the case because, as Jeffreys points out,[4]

> Traditional or deductive logic admits only three attitudes to any proposition: definite proof, disproof, or blank ignorance. But no number of previous instances of a rule provide a deductive proof that the rule will hold in a new instance. There is always the formal possibility of an exception.

This will be recognized as a restatement of Hume's views on the impossibility of complete certainty in knowledge; for example, we cannot be completely certain (probability equal to one) on purely deductive or inductive grounds that the sun will rise tomorrow. The fact that an exception to a rule or law is always possible means that deductive logic with its extreme attitudes of definite proof, disproof, or blank ignorance is inadequate to deal with the usual situations faced by researchers, situations in which the researcher requires and produces statements less extreme than those yielded by deductive logic.

Another point made by Jeffreys, which illustrates the inadequacies of deduction to serve as the sole process in research, is the fact that for any given set of data there is usually an infinite number of possible laws that will "explain" the data precisely; for example, suppose we observe the consumption and income of N households and suppose, merely for argument's sake, that the plot of consumption c against income y is *exactly* linear. We recognize that the same data will also be *exactly* described by the following infinity of laws:

$$c = \alpha + \beta y + f(y)(y_1 - y)(y_2 - y) \cdots (y_n - y),$$

[3] Hans Reichenbach, *The Rise of Scientific Philosophy*. Berkeley: University of California Press, 1958, p. 37.

[4] Harold Jeffreys, *Theory of Probability* (3rd ed.). Oxford: Clarendon, 1961, pp. 2–3.

where $f(y)$ is any arbitrary function that is not infinite at y_i, $i = 1, 2, \ldots, n$. Further, an infinity of these laws will be contradicted by one additional observation. Now deductive logic alone has nothing to say about which of these laws is the one for the researcher to choose. Some broader principles of choice are required, one of which is the principle of simplicity; that is, given a collection of models, which fit the facts equally well, the simplest is chosen. Abstracting from the obvious problem of defining simplicity, some choose the simplest model because they believe that it will predict best. Others assert that it is worthwhile to consider simple models because, although they may not be of greatest ultimate value, they are valuable, for they make strong statements about phenomena that are readily testable. This facilitates the primary activity of learning from experience. Although no definite conclusions are yet available on these two positions, both involve a prejudice for working with simple models.

With respect to the issue of simplicity, the following remarks of W. G. Cochran are relevant and interesting,

> About 20 years ago, when asked in a meeting what can be done in observational studies to clarify the step from association to causation, Sir Ronald Fisher replied: "Make your theories elaborate." The reply puzzled me at first, since by Occam's razor the advice usually given is to make theories as simple as is consistent with known data. What Sir Ronald meant, as the subsequent discussion showed, was that when constructing a causal hypothesis one should envisage as many *different* consequences of its truth as possible, and plan observational studies to discover whether each of these consequences is found to hold.[5]

Thus, although a theory may be simple, it is usually desirable that its implications be far reaching and investigated to evaluate their empirical validity.

In summary, our position is that deductive inference is an important ingredient in scientific inference but that, by itself, it is an inadequate foundation on which to base all inference. This view, of course, conflicts with the view that economics is a purely deductive science. It cannot be denied that there are those in economics and elsewhere who are active in deducing the logical implications of stated assumptions. A prime example of this type of work is Arrow's *Social Choice and Individual Values*. It must be recognized, however, that work of this type is just part of what is done in economic research. The empirical relevance of such deductive results is a primary issue. To assess this issue the wider process of induction is needed.

An apparently opposite view is strongly stated by Popper who writes,[6] "This appraisal of the hypothesis relies solely upon *deductive* consequences (predictions) which may be drawn from the hypothesis: *There is no need even*

[5] Quotation from W. G. Cochran, "The Planning of Observational Studies of Human Populations," *J. Roy. Statist. Soc.*, Series A, Part 2, 234–255 (1965).

[6] Karl R. Popper, *The Logic of Scientific Discovery*. New York; Science Editions, 1961, p. 315.

to mention 'induction.'" [His italics.] The major point to be appreciated in assessing Popper's position is that for him induction is viewed more narrowly than we are viewing it. (Cf. p. 27 ff.) Herein, we follow Jeffreys' point of view, quite at variance with Popper's, that inductive logic is such that deductive logic is encompassed within it. Its statements of proof and disproof are limiting cases of the types of statement yielded by inductive logic. Thus, in our view, inductive logic and deductive logic should not be regarded as mutually exclusive alternatives. In inductive logic deduction plays an important role; however, because inductive logic is broader than deductive logic, the rules for making inductive inferences must be stated and will differ in certain respects from those governing deductive inferences. Further, as regards Popper's statement of the infinite regress argument, namely, that to justify an inductive approach an inductive argument is required which itself needs inductive justification and so on, it seems clear that a deductive approach is open to the same kind of criticism. The best that can be done now, it seems, is to follow Jeffreys by adopting a pragmatic solution, that is, not to prove that induction is valid—since if this could be done deductively induction would be reduced to deduction which is impossible—nor to show that induction is valid by empirical generalizations—since in this case the argument would be circular—but to state a priori rules governing inductive logic, accepted independently of experience. Then "induction is the application of the rules to observational data."[7] Jeffreys further remarks,[8] "All that can be done is to state a set of hypotheses, as plausible as possible, and see where they lead us." We shall see that these hypotheses or rules for inductive inference encompass many elements of Popper's deductive approach, which is as it should be since herein induction is being viewed as a broader process than deduction, one in fact that includes deductive logic as a special limiting case.

1.3 INDUCTIVE INFERENCE

Jeffreys aptly remarks,[9]

> The fundamental problem of scientific progress, and a fundamental one of everyday life, is that of learning from experience. Knowledge obtained in this way is partly merely description of what we have already observed, but part consists of making inferences from past experience to predict future experience. This part may be called generalization or induction. It is the most important part; events that are merely described and have no apparent relation to others may as well be forgotten, and in fact usually are.

Note that for Jeffreys induction is *not* mere description and inductive generalizations are *not* economical modes of describing past observations. In

[7] Harold Jeffreys, *op. cit.*, p. 8.
[8] *Ibid.*, p. 8.
[9] *Ibid.*, p. 1.

fact, he is critical of Mach, who took this latter point of view, because "Mach missed the point that to describe an observation that has not been made yet is not the same thing as to describe one that has been made; consequently he missed the whole problem of induction."[10] Although Jeffreys emphasizes generalization, he is careful not to exclude description as part of the process of learning from experience and of generalization for prediction. In fact, we shall see that unusual facts play an important role in the process of reduction, the third kind of inference. Also, Jeffreys emphasizes

> . . . that inference from past observations to future ones is not deductive. The observations not yet made may concern events either in the future or simply at places not yet inspected. It is technically called induction. . . . There is an element of uncertainty in all inferences of the kind considered.[11]

1.4 REDUCTIVE INFERENCE

This type of inference, also referred to as "abductive" or "retroductive," is the most illusive to define and discuss. Pierce states that induction is the experimental testing of a finished theory; it can never originate any idea whatever.[12] This is a somewhat narrower view of induction than taken by Jeffreys, since Jeffreys includes generalization as part of the inductive process. Jeffreys, however, is vague on the process of generalization. According to Pierce, *abduction or reduction suggests that something may be*; that is, it involves studying facts and devising theories to explain them. For Pierce and others the link of reduction with the unusual fact is emphasized. Examples in economics are not hard to find. Kuznets' finding that the long-run saving ratio was constant led to reductive activities that resulted in several well-known theoretical explanations.

Although we recognize that unusual and surprising facts often trigger the reductive process to produce new concepts and generalizations, it is still pertinent to probe more deeply into the nature of the process. Here Hadamard's work on discovery in the mathematical field seems particularly relevant. He writes,[13] "Indeed, it is obvious that invention or discovery, be it in mathematics or anywhere else, takes place by combining ideas." In agreement with Poincaré, Hadamard views the problem of discovery or invention as one of choice among the many possible combinations of ideas. These combinations are formed in both the conscious and unconscious minds.

[10] Harold Jeffreys, *Scientific Inference* (2nd ed.). Cambridge: Cambridge University Press, 1957, p. 15.
[11] *Ibid.*, p. 13.
[12] See N. R. Hanson, *Patterns of Discovery*. Cambridge: Cambridge University Press, 1958, p. 85.
[13] Jacques Hadamard, *The Psychology of Invention in the Mathematical Field*. New York: Dover, 1945, p. 29.

It is the task of the researcher to avoid choosing useless combinations and to select only those that are useful, usually only a small fraction of the total. In this process choice is "imperatively governed by the sense of scientific beauty."[14] Further, most of this work is done in the unconscious mind. To make this process work, however, the conscious mind plays an important role; it "starts its action and defines, to a greater or lesser extent, the general direction in which that unconscious has to work."[15] It mobilizes ideas initially which lead to a stirring up of ideas previously held and this leads to new combinations. In this respect it is important that the conscious mind not be restricted to narrow lines of thought or held too closely to previous lines of thought. "Thinking aside" leads to a richer variety of ideas from which combinations can be formed. Being in touch with developments in several fields serves the same purpose, and above all, a good deal of hard preparatory work seems to be required for the reductive process to work—"sudden inspirations . . . never happen except after some days of voluntary effort which has appeared absolutely fruitless and whence nothing good seems to have come, where the way taken seems totally astray."[16] This, for Hadamard, represents an answer to those who adopt the "chance," "rest," and "forgetting" hypotheses of discovery. The preparatory work period is accompanied and followed by an incubation period and finally by illumination. After that the conscious mind precises the illumination and proceeds to verify it.

There are at least two conceptions of this process: (a) a goal being given and how to reach it and (b) discovering a fact and then imagining how it could be useful. "Now, paradoxical as it seems, that second kind of invention is the more general one and becomes more and more so as science advances. Practical application is found by not looking for it, and one can say that the whole progress of civilization rests on that principle."[17]

It is clear from the above discussion that many aspects of reduction are not fully understood. Some significant features of the process do emerge, however. First, there is recognition that the process is one of choosing particular combinations of ideas that seem fruitful. In this choice both the unconscious and conscious minds play a role importantly guided by an esthetic sense of scientific beauty. This esthetic sense is indeed a subjective element bearing some relation to the concept of simplicity, albeit not necessarily a one-to-one correspondence. As was stated above in the discussion of the concept of simplicity, there are those who argue that simplicity is to be desired in the choice of combinations of ideas on grounds other than esthetics. Second, in the process of reduction, the conscious mind takes an important part in

[14] *Ibid.*, p. 39.
[15] *Ibid.*, p. 46.
[16] *Ibid.*, p. 45.
[17] *Ibid.*, p. 124.

choosing the general area of investigation and in the period of intensive preparatory work. The preparatory work usually involves observation and experimentation. Observation, that is, a strong interaction with the data of a problem, oftentimes is the key factor in reduction. During this preparatory phase old combinations of ideas are being disturbed, new combinations are formed, and the problem of choice is forced on the investigator. Once a choice is made, the problems of precising and verification must be faced.

Since reductive inference is far from being completely understood, fruitful rules governing this kind of inference have not yet been formulated. Ideally, we should like to have useful rules covering *both* reductive and inductive inference. Lacking them, we review a set of rules just for inductive inference.

1.5 JEFFREYS' RULES FOR A THEORY OF INDUCTIVE INFERENCE[18]

Remembering that the most important part of induction is generalization from past experience and data to predict still unobserved phenomena, we now review a set of rules put forward by Jeffreys to govern the process of induction.

Rule 1. All hypotheses used must be explicitly stated and the conclusions must follow from the hypotheses.

Rule 2. A theory of induction must be self-consistent; that is, it must not be possible to derive contradictory conclusions from the postulates and any given set of observational data.

Rule 3. Any rule given must be applicable in practice. A definition is useless unless the thing defined can be recognized in terms of the definition when it occurs. The existence of a thing or the estimate of a quantity must not involve an impossible experiment.

Rule 4. A theory of induction must provide explicitly for the possibility that inferences made by it may turn out to be wrong.

Rule 5. A theory of induction must not deny any empirical proposition a priori; any precisely stated empirical proposition must be formally capable of being accepted in the sense of the last rule, given a moderate amount of relevant evidence.

Jeffreys regards these five rules as "essential." Rules 1 and 2 impose on inductive logic criteria already required in pure mathematics. We may add that they are usually required in economics as well. The third and fifth rules bring to the fore the distinction between a priori and empirical propositions. Note that the third rule incorporates elements of Bridgeman's operationalism and, very importantly, rules out impossible experiments. Finally, Rule 4

[18] Cf., Jeffreys, *Theory of Probability, loc. cit.*

makes explicit the distinction between induction and deduction; that is, it imposes on us recognition of the fact that scientific laws may have to be modified or even replaced as new evidence accumulates. However, "we do accept inductive inference in some sense; we have a certain amount of confidence that it will be right in any particular case, though this confidence does not amount to logical certainty."[19]

In addition to the five rules stated above, Jeffreys states three more which are "useful guides."

Rule 6. The number of postulates should be reduced to a minimum.

Rule 7. Although we do not regard the human mind as a perfect reasoner, we must accept it as a useful one and the only one available. The theory need not represent actual thought processes in detail but should agree with them in outline.

Rule 8. In view of the greater complexity of induction, we cannot hope to develop it more thoroughly than deduction. We therefore take it as a rule that an objection carries no weight if an analogous objection invalidates part of generally accepted pure mathematics.

Rule 6 is essentially a restatement of Ockham's rule and, as such, is likely to be regarded as acceptable. Rule 7 is indeed important. It means that a theory of induction must agree in outline with thought processes, in particular those that play a role in thinking about and evaluating generalizations or propositions about empirical phenomena. Finally, Rule 8 does not appear objectionable even though it must be recognized that there are disputes about the foundations of pure mathematics.

1.6 IMPLICATIONS OF THE RULES

The eight rules listed in Section 1.5 have important implications for theories of the inductive process. As Jeffreys remarks,[20]

> They rule out . . . any definition of probability that attempts to define probability in terms of infinite sets of possible observations, for we cannot in practice make an infinite number of observations. The Venn limit, the hypothetical infinite population of Fisher, and the ensemble of Willard Gibbs are useless to us by Rule 3. . . . In fact, no "objective" definition of probability in terms of actual or possible observations, or possible properties of the world, is admissible. For, if we made anything in our fundamental principles depend on observations or on the structure of the world, we should have to say either (1) that the observations we can make, and the structure of the world, are initially unknown; then we cannot know our fundamental principles, and we have no possible starting-point; or (2) that we know a priori something about observations on the structure of the world, and this is illegitimate by Rule 5.

[19] *Ibid.*, p. 9.
[20] *Ibid.*, p. 11.

He goes on to explain that "*the essence of the present theory* is that no probability . . . is simply a frequency. The fundamental idea is that of a reasonable degree of belief, which satisfies certain rules of consistency and can in consequence of these rules be formally expressed by numbers. . . ."[21] Thus, in terms of de Finetti's classification of probability theories, Jeffreys' is a subjective theory which attempts to provide consistent procedures for behavior under uncertainty as contrasted with those subjective theories that try to characterize psychological and rational behavior under uncertainty.[22]

With probability regarded as representing a degree of reasonable belief rather than a frequency, numerical probabilities can be associated with degrees of confidence that we have in propositions about empirical phenomena, a distinctive feature of the Bayesian approach to inference. As Jeffreys puts it, ". . . there is a valid primitive idea expressing the degree of confidence that we may reasonably have in a proposition, even though we may not be able to give either a deductive proof or a disproof of it"[23]; for example, when considering a particular explanation of an observed event, a researcher may remark that the explanation is "probably true." What he means by the phrase "probably true" is that, based on previous information, studies, and experience, he has a high degree of confidence in the explanation. The Bayesian approach, and Jeffreys' theory in particular, involves a quantification of such phrases as "probably true" or "probably false" by utilizing numerical probabilities to represent degrees of confidence or belief that individuals have in propositions. By using probabilities in this connection, we automatically allow for the possibility that a proposition may not be valid in accord with Rule 4. Also, to the extent that normal thought processes involve associating probabilities with uncertain propositions, we may also state that the formalization of this procedure in the Bayesian approach is in accord with Jeffreys' Rule 7.

Of course, the degree of reasonable belief that we have in a proposition, say a proposition about economic behavior deduced from the permanent income hypothesis, depends on the state of our current information. Therefore, in general, a probability representing a degree of reasonable belief that we have in a proposition is always a conditional probability, conditional on our present state of information. As our information relating to a particular proposition changes, we revise its probability or our belief in it. This process of revising probabilities associated with propositions in the face of new

[21] *Ibid.*, p. 401.

[22] Some classify Jeffreys' theory of probability as "necessary" or even "objective," since he provides procedures which, if adopted, produce identical results for different investigators, given the same data and model. Although this is so, his view of probability as an individual's degree of reasonable belief is a subjective one.

[23] Jeffreys, *Theory of Probability*, *op. cit.*, p. 15.

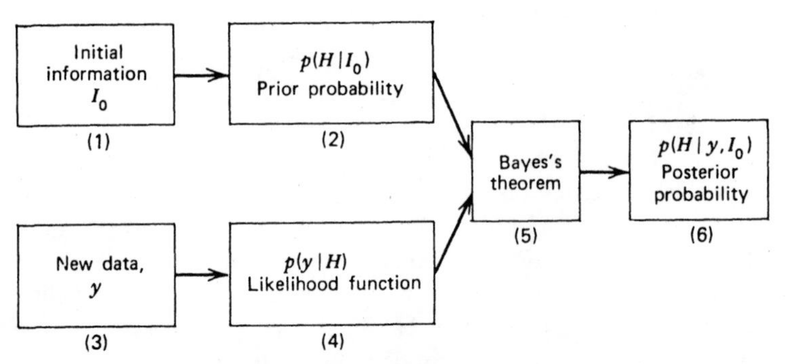

Figure 1.1 The process of revising probabilities, given new data.

information is the essence of learning from experience. We shall see in what follows that the process of revising probabilities representing degrees of belief in propositions to incorporate new information can be made operational and quantitative, in accord with Rule 3, by use of a simple rule of probability theory, namely Bayes' theorem. Schematically the process of revising probabilities, given new data denoted by y, is represented as in Figure 1.1. In the upper left-hand boxes, (1) and (2), we indicate that our initial or prior probability associated with a particular proposition H, $p(H|I_0)$, is based on our initial information I_0. This information is generally of various types, usually a combination of information from previous data and studies, theoretical considerations, and casual observation.[24] In the lower left-hand boxes, (3) and (4), we show that the probability density function (pdf), $p(y|H)$, for the new observation, y, given H, the proposition. This pdf is the well-known likelihood function. Then the prior probability $p(H|I_0)$ is combined with the likelihood function $p(y|H)$ by means of Bayes' theorem to yield the posterior probability $p(H|y, I_0)$. The posterior probability $p(H|y, I_0)$ is seen to depend on both the prior information I_0 and the sample information y. In this way we obtain a revision of our initial prior probability $p(H|I_0)$ to reflect the information in our new data y; that is, $p(H|I_0)$ is transformed via Bayes' theorem into $p(H|y, I_0)$.

If we were concerned about a parameter θ, we would use the approach shown in Figure 1.1 but with θ replacing H; that is, in box (2) we would have $p(\theta|I_0)$ rather than $p(H|I_0)$, where $p(\theta|I_0)$ is a prior pdf for the parameter θ, given our initial information. This prior pdf represents our initial beliefs about the parameter θ based on our initial information I_0. In box (4) we would have $p(y|\theta)$, the likelihood function. Then, on combining $p(\theta|I_0)$ and $p(y|\theta)$ by use of Bayes' theorem, we would obtain the posterior pdf, $p(\theta|y, I_0)$ in box (6). This latter pdf incorporates our initial information, as represented in

[24] In Chapter 2 we consider prior information in more detail.

the prior pdf, $p(\theta|I_0)$, and our sample information y. The posterior pdf $p(\theta|y, I_0)$, can be employed to make probability statements about θ; for example, to compute the probability that $a < \theta < b$, where a and b are given numbers. This and other uses of posterior pdf's are illustrated in subsequent chapters. At present it is pertinent to emphasize that the posterior pdf represents our beliefs about the parameter θ and incorporates both prior and sample information. As the sample information grows, under very general conditions it will more and more dominate the posterior pdf which will become more concentrated about the true value of the parameter. In addition, if two individuals have different prior pdf's, perhaps because of their having different initial information, under rather nonrestrictive conditions their posterior pdf's will become similar as they combine additional common data with their respective prior pdf's, for as the data base grows the information it provides will swamp the initial prior information.

It is of the utmost importance to realize that the procedure given graphically in Figure 1.1 and described in the text is operational and applicable in the analysis of a wide range of models and problems in econometrics and other areas of science. This is as it should be, since the procedure outlined is central to the inductive process as Jeffreys and others view it. That there is a unified and operational approach to problems of inference in econometrics and other areas of science is a fundamental point that should be appreciated. Whether we analyze, for example, time series, regression, or "simultaneous equation" models, the approach and principles will be the same. This stands in contrast to other approaches to inference that involve special techniques and principles for different problems.[25]

Since, in the past, most econometricians have employed non-Bayesian techniques in their work, it is useful and interesting to compare Bayesian and non-Bayesian analyses of a range of models and problems. In the following chapters this comparative approach is pursued, since, as Anscombe remarked some years ago about the state of statistics, "A just appreciation of the situation can only be had by studying the orthodox and the Bayesian approaches to a variety of statistical problems, noting what each one does and how well it does it."[26]

[25] B. de Finetti, in a lecture at Frascati in June 1968, used I. J. Good's phrase, "ad hockeries," to describe this aspect of non-Bayesian approaches to inference.

[26] F. J. Anscombe, "Bayesian Statistics," *Am. Statisn.*, **15**, 21–24 (1961), p. 21. The Reverend Thomas Bayes was an English Presbyterian minister who was born in 1702 and died in 1761. His paper, "An Essay Toward Solving a Problem in the Doctrine of Chances," was published posthumously by Richard Price in *Phil. Trans. Roy. Soc. (London)*, **53** (1763), 370–418, and is reprinted in *Biometrika*, **45**, 293–315 (1958). See also *Facsimiles of Two Papers by Bayes* (with commentary by W. Edwards Deming). New York: Hafner, 1963.

QUESTIONS AND PROBLEMS

1. What are some examples of simple economic theories or models that have been particularly fruitful?
2. Is it the case that simple theories have been more useful than complicated theories in economics?
3. Are there any problems associated with representing degrees of belief about hypotheses on a unidimensional probability scale?
4. Are economic theories expressed in terms of a few mathematical equations and parameters always simpler than those expressed in terms of a larger number of equations and parameters? (See H. Jeffreys, *Theory of Probability* (1966 ed.), pp. 47–49, for an interesting discussion of this issue.)
5. In economics what are some instances wherein recognition of an unusual fact or statistical regularity led to the formulation of a new economic theory?
6. What are examples of economic theories that are combinations of ideas and concepts from several different areas of knowledge?
7. With respect to any competing economic theories, for example, the absolute and permanent-income theories of consumer behavior, provide numerical probabilities to represent your degrees of belief in alternative theories. On what kinds of evidence do you base your degrees of belief in competing theories?
8. If, in the process of reduction, a strong interaction with the data of a problem is considered important in formulating a theory, how can it be established that the theory so derived is more than a description of the facts?
9. What accounts for the usually observed fact that different researchers have differing degrees of belief in a particular economic theory or proposition?
10. If different researchers have widely divergent beliefs about the validity of a theory, is this to be construed as evidence against the theory?
11. In what ways can prior beliefs about the nature of economic phenomena condition the design and formulation of a research project, say one to determine the economic effects of public regulation of private utility companies?
12. Can the use of prior knowledge and beliefs in formulating a research project condition and even vitiate the final research results?
13. In the area of economic policy making, what are some examples of policy recommendations that are critically dependent on degree of belief in one or another economic theory?
14. What are some specific cases in which those concerned with economic policy appear to have differing degrees of belief about particular economic theories?

CHAPTER II

Principles of Bayesian Analysis with Selected Applications

In this chapter we view some basic principles and concepts of Bayesian analysis and provide analyses of some relatively simple but important models and problems.

2.1 BAYES'S THEOREM

An essential element of the Bayesian approach is Bayes's theorem, also referred to in the literature as the *principle of inverse probability*.[1] Here we state the theorem for continuous random variables. Let $p(\mathbf{y}, \boldsymbol{\theta})$ denote the joint probability density function (pdf)[2] for a random observation vector $\mathbf{y}$ and a parameter vector $\boldsymbol{\theta}$, also considered random. The parameter vector $\boldsymbol{\theta}$ may have as its elements coefficients of a model, variances and covariances of disturbance terms, and so on. Then, according to usual operations with pdf's, we have

$$\begin{aligned} p(\mathbf{y},\boldsymbol{\theta}) &= p(\mathbf{y}|\boldsymbol{\theta})\, p(\boldsymbol{\theta}) \\ &= p(\boldsymbol{\theta}|\mathbf{y})\, p(\mathbf{y}) \end{aligned} \tag{2.1}$$

and thus

$$p(\boldsymbol{\theta}|\mathbf{y}) = \frac{p(\boldsymbol{\theta})\, p(\mathbf{y}|\boldsymbol{\theta})}{p(\mathbf{y})}, \tag{2.2}$$

[1] In problems involving "inverse probability" we have given data and from the information in the data try to infer what random process generated them. On the other hand, in problems of "direct probability" we know the random process, including values of its parameters, and from this knowledge make probability statements about outcomes or data produced by the known random process. Problems of statistical estimation are thus seen to be problems in "inverse probability," whereas many gambling problems are problems in "direct probability."

[2] Here and below we use the symbol p to denote pdf's generally and not one specific pdf. The argument of the function p as well as the context in which it is used will identify the particular pdf being considered.

with $p(\mathbf{y}) \neq 0$.[3] We can write this last expression as follows:

$$
\begin{aligned}
p(\boldsymbol{\theta}|\mathbf{y}) &\propto p(\boldsymbol{\theta})\, p(\mathbf{y}|\boldsymbol{\theta}) \\
&\propto \text{prior pdf} \times \text{likelihood function},
\end{aligned}
\tag{2.3}
$$

where $\propto$ denotes proportionality, $p(\boldsymbol{\theta}|\mathbf{y})$ is the *posterior pdf* for the parameter vector $\boldsymbol{\theta}$, given the sample information $\mathbf{y}$, $p(\boldsymbol{\theta})$ is the *prior pdf*[4] for the parameter vector $\boldsymbol{\theta}$, and $p(\mathbf{y}|\boldsymbol{\theta})$, viewed as a function of $\boldsymbol{\theta}$, is the well-known *likelihood function.*[5] Equation 2.3 is a statement of Bayes's theorem, a simple mathematical result in the theory of probability. Note that the joint posterior pdf, $p(\boldsymbol{\theta}|\mathbf{y})$, has all the prior and sample information incorporated in it. The prior information enters the posterior pdf via the prior pdf, whereas all the sample information enters via the likelihood function. In this latter connection the "likelihood principle" states that $p(\mathbf{y}|\boldsymbol{\theta})$, considered as a function of $\boldsymbol{\theta}$ ". . . constitutes the entire evidence of the experiment, that is, it tells all that the experiment has to tell."[6] The posterior pdf is employed in the Bayesian approach to make inferences about parameters.

Example 2.1. Assume that we have n independent observations, $\mathbf{y}' = (y_1, y_2, \ldots, y_n)$, drawn from a normal population with unknown mean μ and *known* variance $\sigma^2 = \sigma_0^2$. We wish to obtain the posterior pdf for μ. Applying (2.3) to this particular problem, we have

$$
p(\mu|\mathbf{y}, \sigma_0^2) \propto p(\mu)\, p(\mathbf{y}|\mu, \sigma_0^2),
\tag{2.4}
$$

where $p(\mu|\mathbf{y}, \sigma_0^2)$ is the posterior pdf for the parameter μ, given the sample information $\mathbf{y}$ and the assumed known value σ_0^2, $p(\mu)$ is the prior pdf for μ, and $p(\mathbf{y}|\mu, \sigma_0^2)$, viewed as a function of the unknown parameter μ is the likelihood function. The likelihood function is given by $\prod_{i=1}^{n} p(y_i|\mu, \sigma_0^2)$, or

$$
\begin{aligned}
p(\mathbf{y}|\mu, \sigma_0^2) &= (2\pi\sigma_0^2)^{-n/2} \exp\left[-\frac{1}{2\sigma_0^2}\sum_{i=1}^{n}(y_i - \mu)^2\right] \\
&= (2\pi\sigma_0^2)^{-n/2} \exp\left[-\frac{1}{2\sigma_0^2}[\nu s^2 + n(\mu - \hat{\mu})^2]\right],
\end{aligned}
\tag{2.5}
$$

[3] The quantity $p(\mathbf{y})$, the reciprocal of the normalizing constant for the pdf in (2.2), can be written as $p(\mathbf{y}) = \int p(\boldsymbol{\theta})\, p(\mathbf{y}|\boldsymbol{\theta})\, d\boldsymbol{\theta}$.

[4] As noted in Chapter 1, the prior pdf depends on the state of our initial information denoted by I_0. Here, to simplify the notation, we do not show this dependence explicitly; that is, we write $p(\boldsymbol{\theta})$ rather than $p(\boldsymbol{\theta}|I_0)$.

[5] The likelihood function is often written as $l(\boldsymbol{\theta}|\mathbf{y})$ to emphasize that it is *not* a pdf, whereas $p(\mathbf{y}|\boldsymbol{\theta})$ is a pdf for the observations given the parameters.

[6] L. J. Savage, "Subjective Probability and Statistical Practice," in L. J. Savage et al., *The Foundations of Statistical Inference.* London and New York: Methuen and Wiley, 1962, pp. 9–35, p. 17. Savage presents a discussion of the likelihood principle and provides references to earlier literature.

where $\nu = n - 1$, $\hat{\mu} = (1/n) \sum_{i=1}^{n} y_i$, the sample mean, and

$$s^2 = (1/\nu) \sum_{i=1}^{n} (y_i - \hat{\mu})^2,$$

the sample variance.[7]

As regards a prior pdf for μ, we assume that our prior information regarding this parameter can be represented by the following univariate normal pdf:

$$p(\mu) = \frac{1}{\sqrt{2\pi}\,\sigma_a} \exp\left[-\frac{1}{2\sigma_a^2}(\mu - \mu_a)^2\right], \tag{2.6}$$

where μ_a is the prior mean and σ_a^2 is the prior variance, parameters whose values are assigned by the investigator on the basis of his initial information. Then, on using Bayes's theorem to combine the likelihood function in (2.5) and the prior pdf in (2.6), we obtain the following posterior pdf for μ:

$$\begin{aligned} p(\mu|\mathbf{y}, \sigma_0^2) &\propto p(\mu)\, p(\mathbf{y}|\mu, \sigma_0^2) \\ &\propto \exp\left\{-\frac{1}{2}\left[\frac{(\mu - \mu_a)^2}{\sigma_a^2} + \frac{n}{\sigma_0^2}(\mu - \hat{\mu})^2\right]\right\} \\ &\propto \exp\left[-\left(\frac{\sigma_a^2 + \sigma_0^2/n}{2\sigma_a^2\sigma_0^2/n}\right)\left(\mu - \frac{\hat{\mu}\sigma_a^2 + \mu_a\sigma_0^2/n}{\sigma_a^2 + \sigma_0^2/n}\right)^2\right], \end{aligned} \tag{2.7}$$

from which it is seen that μ is normally distributed, a posteriori, with mean

$$E\mu = \frac{\hat{\mu}\sigma_a^2 + \mu_a\sigma_0^2/n}{\sigma_a^2 + \sigma_0^2/n} = \frac{\hat{\mu}(\sigma_0^2/n)^{-1} + \mu_a(\sigma_a^2)^{-1}}{(\sigma_0^2/n)^{-1} + (\sigma_a^2)^{-1}} \tag{2.8}$$

and variance given by

$$\operatorname{Var}(\mu) = \frac{\sigma_a^2\sigma_0^2/n}{\sigma_a^2 + \sigma_0^2/n} = \frac{1}{(\sigma_0^2/n)^{-1} + (\sigma_a^2)^{-1}}. \tag{2.9}$$

Note that the posterior mean in (2.8) is a weighted average of the sample mean $\hat{\mu}$ and the prior mean μ_a, with the weights being the reciprocals of σ_0^2/n and σ_a^2. If we let $h_0 = (\sigma_0^2/n)^{-1}$ and $h_a = (\sigma_a^2)^{-1}$, then $E\mu = (\hat{\mu}h_0 + \mu_a h_a)/(h_0 + h_a)$, where the h's are often referred to as "precision" parameters. Also we have $\operatorname{Var}(\mu) = 1/(h_0 + h_a)$ from (2.9), and thus the precision parameter associated with the posterior mean is just $[\operatorname{Var}(\mu)]^{-1} = h_0 + h_a$, the sum of the sample and prior precision parameters.

To provide some illustrative numerical results suppose that in Example 2.1 our sample of $n = 10$ observations is

[7] The expression in the exponent in the second line of (2.5) is obtained by using the following result: $\sum_{i=1}^{n} (y_i - \mu)^2 = \sum_{i=1}^{n} [(y_i - \hat{\mu}) - (\mu - \hat{\mu})]^2 = \sum_{i=1}^{n} (y_i - \hat{\mu})^2 + n(\mu - \hat{\mu})^2$ with the cross product term $\sum_{i=1}^{n} (y_i - \hat{\mu})(\mu - \hat{\mu})$ disappearing, since $\sum_{i=1}^{n} (y_i - \hat{\mu}) = 0$.

Observation Number	Observations y_i
1	0.699
2	0.320
3	−0.799
4	−0.927
5	0.373
6	−0.648
7	1.572
8	−0.319
9	2.049
10	−3.077

$$\text{Sample mean:} \quad \hat{\mu} = \tfrac{1}{10} \sum_{i=1}^{10} y_i = -0.0757$$

where the y_i's are independently drawn from a normal population with unknown mean μ and known variance $\sigma^2 = \sigma_0^2 = 1.00$. Assume that our prior information is suitably represented by a normal pdf with prior mean $\mu_a = -0.0200$ and prior variance, $\sigma_a^2 = 2.00$. This prior pdf, which is plotted in Figure 2.1, represents our initial beliefs about the unknown parameter μ. On combining this prior pdf with the likelihood function, the posterior pdf is given by the expression in (2.7). For the particular sample shown above, with mean $\hat{\mu} = -0.0757$ and the values of the prior parameters $\mu_a = -0.02$ and $\sigma_a^2 = 2.00$, the mean of the posterior pdf from (2.8) is

$$E\mu = \frac{-0.0757/0.100 - 0.0200/2.00}{1/0.100 + 1/2.00} = -0.0730$$

and its variance from (2.9) is

$$\text{Var}(\mu) = \frac{1}{1/0.100 + 1/2.00} = 0.0952.$$

For comparison with the prior pdf the posterior pdf is plotted in Figure 2.1. It is seen that combining the information contained in just 10 independent observations with our prior information has resulted in a considerable reduction in our uncertainty about the parameter μ; that is, our prior variance is $\sigma_a^2 = 2.00$, whereas the variance of our posterior pdf is 0.0952. In addition, our posterior mean $E\mu = -0.0730$ is not very different from $\hat{\mu} = -0.0757$, the sample mean, but is quite a bit larger in absolute value than our prior mean, $\mu_a = -0.0200$. Note, however, that our prior pdf has a substantial variance, $\sigma_a^2 = 2.00$, and thus initially there is substantial

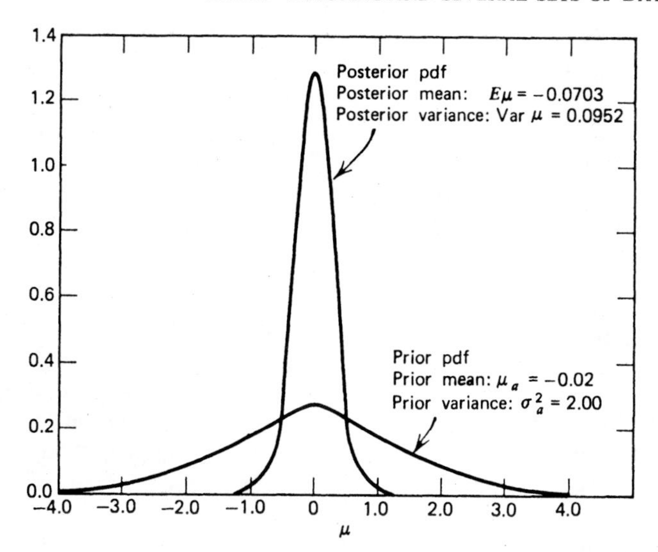

Figure 2.1 Plots of prior and posterior pdf's for μ. The prior and posterior pdf's are shown in (2.6) and (2.7), respectively.

probability density in the vicinity of -0.0730; that is, in this case our prior information is somewhat "vague" or "diffuse" in relation to the information in the sample.

2.2 BAYES'S THEOREM AND SEVERAL SETS OF DATA

If initially our prior pdf for a parameter vector $\boldsymbol{\theta}$ is $p(\boldsymbol{\theta})$ and we obtain a set of data $\mathbf{y}_1$ with pdf $p(\mathbf{y}_1|\boldsymbol{\theta})$, then from (2.3) the posterior pdf is

$$p(\boldsymbol{\theta}|\mathbf{y}_1) \propto p(\boldsymbol{\theta})\, p(\mathbf{y}_1|\boldsymbol{\theta}). \tag{2.10}$$

If we now obtain a new set of data, $\mathbf{y}_2$, generated independently of the first set, with pdf $p(\mathbf{y}_2|\boldsymbol{\theta})$, we can form the posterior pdf for $\boldsymbol{\theta}$ as follows. Use the posterior pdf in (2.10) as the prior pdf in the analysis of the new set of data $\mathbf{y}_2$ to obtain by means of Bayes's theorem

$$p(\boldsymbol{\theta}|\mathbf{y}_1, \mathbf{y}_2) \propto p(\boldsymbol{\theta}|\mathbf{y}_1)\, p(\mathbf{y}_2|\boldsymbol{\theta}), \tag{2.11}$$

where $p(\boldsymbol{\theta}|\mathbf{y}_1, \mathbf{y}_2)$ is the posterior pdf based on the information in $p(\boldsymbol{\theta})$ and the two samples of data $\mathbf{y}_1$ and $\mathbf{y}_2$. It is interesting to note that, since $p(\boldsymbol{\theta}|\mathbf{y}_1) \propto p(\boldsymbol{\theta})\, p(\mathbf{y}_1|\boldsymbol{\theta})$ from (2.10), (2.11) may be written as

$$p(\boldsymbol{\theta}|\mathbf{y}_1, \mathbf{y}_2) \propto p(\boldsymbol{\theta})\, p(\mathbf{y}_1|\boldsymbol{\theta})\, p(\mathbf{y}_2|\boldsymbol{\theta}). \tag{2.12}$$

In (2.12) $p(\mathbf{y}_1|\boldsymbol{\theta})\, p(\mathbf{y}_2|\boldsymbol{\theta})$ is the likelihood function for $\boldsymbol{\theta}$ based on the combined samples $\mathbf{y}_1$ and $\mathbf{y}_2$. Therefore it is the case that we obtain the same posterior pdf for $\boldsymbol{\theta}$, whether we proceed sequentially from $p(\boldsymbol{\theta})$ to $p(\boldsymbol{\theta}|\mathbf{y}_1)$ and then to $p(\boldsymbol{\theta}|\mathbf{y}_1, \mathbf{y}_2)$ or whether we use the likelihood function for the combined samples $p(\mathbf{y}_1, \mathbf{y}_2|\boldsymbol{\theta})$ in conjunction with the prior pdf $p(\boldsymbol{\theta})$. This general feature of the process of combining information in a prior pdf with information in successive samples can easily be shown to hold for cases involving more than two independent samples of data.

2.3 PRIOR PROBABILITY DENSITY FUNCTIONS

The prior pdf, denoted $p(\boldsymbol{\theta})$ in (2.3),[8] represents our prior information about parameters of a model; that is, in the Bayesian approach prior information about parameters of models is usually represented by an appropriately chosen pdf. In Example 2.1, for example, prior information about a mean μ is represented in (2.6) by a normal pdf with prior mean μ_a and variance σ_a^2. The prior mean and variance μ_a and σ_a^2 are assigned values by the investigator in accord with his prior information about the parameter μ. If this normal prior pdf is judged an adequate representation of the available prior information, it can be used, as demonstrated above, to obtain the posterior pdf for μ. On the other hand, if the prior information is not adequately represented by a normal prior pdf, another prior pdf that does so will be used by the investigator. To take a specific example, if we have a scalar parameter θ, say a proportion, that by its very nature is limited to the closed interval 0 to 1, it would not be appropriate to employ a normal prior pdf for θ, since a normal pdf does not limit the range of θ to the closed interval 0 to 1. The pdf chosen for θ should be one, say possibly a beta pdf, that can incorporate the available information on the range of θ. Considerations of this sort point up the importance of exercising care and thought in choosing a prior pdf to represent prior information.

As regards the nature of prior information, we recognize that it may be information contained in samples of past data which have been generated in a reasonably scientific manner and which are available for further analysis. When a prior pdf represents information of this kind, we shall term such a prior pdf a "data-based" (DB) prior. In other cases prior information may arise from introspection, casual observation, or theoretical considerations; that is, from sources other than currently available samples of past data of the kind described above. When a prior pdf represents information of this kind, we refer to it as a "nondata-based" (NDB) prior. Although in many

[8] In general, the pdf $p(\boldsymbol{\theta})$ will involve some prior parameters that we have not shown explicitly in order to simplify the notation.

situations prior pdf's represent both DB and NDB information, we think that the distinction between these two kinds of information is worth making, since they obviously have somewhat different characteristics.

It is extremely difficult to formulate general precepts regarding the appropriate uses of the two kinds of prior information mentioned above, since much depends on the objectives of analyses; for example, if an individual wishes to determine how new sample information modifies his own beliefs about parameters of a model and his initial information is NDB, he will, of course, use a NDB prior pdf in conjunction with a likelihood function to obtain a posterior pdf. Then, on comparing his posterior pdf with his NDB prior pdf, he can determine how the information in his sample data has modified his initial NDB beliefs, a fundamental operation in much scientific work. Again, if an economist is carrying through an analysis of sample data in order to make a policy decision, he may indeed incorporate NDB as well as DB prior information in his analysis to ensure that his final decision will be based on all his available information, prior and sample.

Although the above uses of NDB prior information are extremely valuable, it must be noted that one person's NDB prior information can differ from that of another's. In a research situation this is just another way of stating that different investigators may have different views, a not unusual state of affairs; for example, in the early days of Keynesian employment theory there were some old line quantity theorists who argued that the investment multiplier could be negative, zero, or positive. These views conflicted with those of the Keynesians, who argued, on the basis of theoretical considerations and casual observation, that the multiplier is strictly positive. Given a model for observations involving the multiplier and data, it is possible to compute a posterior pdf for the investment multiplier to determine what the information in the data has to say about the value of the multiplier. An analysis of this sort might yield the conclusion that the probability that the multiplier will be negative is negligibly small. Thus information in data can be employed to make comparisons of alternative prior beliefs or hypotheses. Specific techniques for making such comparisons are provided in Chapter 10. In addition, a framework for making a choice among alternative conflicting beliefs or hypotheses which utilizes information in sample data is described and applied.

It is possible that two investigators working with the same model and DB prior information can arrive at different posterior beliefs if they base their prior information on different bodies of past data. These investigators can be brought into agreement by pooling their past samples of data and thereby providing them with the same DB prior information.

Whether prior information is DB or NDB, it is conceivable that there is little prior information; for example, there may be no past sample data information available. A situation involving NDB prior information may be one in which an investigator has vague ideas about the phenomenon under study and in which case we refer to our prior information as "vague" or "diffuse." If our prior information relates to parameters of a model and is vague or diffuse, we employ a "diffuse" prior pdf in the analysis of our data. Various considerations and principles used in obtaining "diffuse" prior pdf's are discussed in the appendix to this chapter. To illustrate the use of a diffuse prior pdf consider the following example.

Example 2.2. Consider n independent observations $\mathbf{y}' = (y_1, y_2, \ldots, y_n)$ drawn from a normal population with unknown mean μ and known standard deviation $\sigma = \sigma_0$. Assume that our prior information regarding the value of μ is vague or diffuse. To represent knowing little about the value of μ, we follow Jeffreys (see the appendix to this chapter) by taking

$$p(\mu) \propto \text{constant} \qquad -\infty < \mu < \infty \tag{2.13}$$

as our prior pdf.[9] Then the posterior pdf for μ, $p(\mu|\mathbf{y}, \sigma = \sigma_0)$ is given by

$$\begin{aligned} p(\mu|\mathbf{y}, \sigma = \sigma_0) &\propto p(\mu)l(\mu|\mathbf{y}, \sigma = \sigma_0) \qquad -\infty < \mu < \infty \\ &\propto \exp\left[-\frac{n}{2\sigma_0^2}(\mu - \hat{\mu})^2\right], \end{aligned} \tag{2.14}$$

where $l(\mu|\mathbf{y}, \sigma = \sigma_0)$ is the likelihood function and $\hat{\mu} = \sum_{i=1}^{n} y_i/n$, the sample mean. It is seen that the posterior pdf is normal with mean $\hat{\mu}$ and variance σ_0^2/n. This same result would be obtained in Example 2.1 if there we spread out the normal prior pdf for μ (i.e., allowed $\sigma_a \to \infty$).

When we have NDB prior information that we wish to incorporate in an analysis, the problem of choosing a prior pdf to represent the available prior information must be faced. Ideally, we should like to have a prior pdf represent our prior information as accurately as possible and yet be relatively simple so that mathematical operations can be performed conveniently; for example, in Example 2.1 our prior information about a mean was assumed to be adequately represented by the normal pdf in (2.6), which is relatively simple and mathematically convenient. We shall see from what follows that

[9] This prior pdf is improper; that is $\int_{-\infty}^{\infty} p(\mu)\, d\mu$ is not finite. Jeffreys and others make extensive use of improper prior pdf's to represent "knowing little." Jeffreys remarks in his *Theory of Probability* (3rd ed.). Oxford: Clarendon, 1961, p. 119, that use of improper pdf's poses no difficulty, and in fact Renyi's axioms and his accompanying definition of conditional probability can be used to state Bayes' theorem when improper prior pdf's are employed. See D. V. Lindley, *Introduction to Probability and Statistics from a Bayesian Viewpoint. Part 1. Probability*. Cambridge: Cambridge University Press, 1965, pp. 11, 13, for a brief discussion of this point.

(2.6) is an example of a "natural conjugate" prior pdf.[10] Such prior pdf's are often useful in representing prior information, relatively simple, and mathematically tractable.

We now explain the definition of a natural conjugate prior pdf. Let $p(\mathbf{y}|\boldsymbol{\theta}, n)$ be the pdf for an $n \times 1$ vector of observations $\mathbf{y}$, where $\boldsymbol{\theta}$ is a parameter vector. If $p(\mathbf{y}|\boldsymbol{\theta}, n) = p_1(\mathbf{t}|\boldsymbol{\theta}, n)\, p_2(\mathbf{y})$, where $\mathbf{t}' = (t_1, t_2, \ldots, t_k)$, with $t_i = t_i(\mathbf{y})$ a function of the observations and $p_2(\mathbf{y})$ does not depend on $\boldsymbol{\theta}$, then the t_i's are defined as sufficient statistics.[11] A natural conjugate prior pdf for $\boldsymbol{\theta}$, say $f(\boldsymbol{\theta}|\cdot)$, is given by $f(\boldsymbol{\theta}|\cdot) \propto p_1(\mathbf{t}|\boldsymbol{\theta}, n)$, with the factor of proportionality depending on $\mathbf{t}$ and n but not $\boldsymbol{\theta}$. It is seen that $f(\boldsymbol{\theta}|\cdot)$, defined in this way, has a functional form precisely the same as that as $p_1(\mathbf{t}|\boldsymbol{\theta}, n)$; however, the argument of f is $\boldsymbol{\theta}$ and its $k + 1$ parameters are the elements of $\mathbf{t}$ and n. To represent prior information an investigator assigns values to $\mathbf{t}$ and n, say $\mathbf{t}_0$ and n_0, to obtain $f(\boldsymbol{\theta}|\mathbf{t}_0, n_0) \propto p_1(\mathbf{t}_0|\boldsymbol{\theta}, n_0)$ as his informative prior pdf.[12]

As an example of a natural conjugate prior pdf, consider the data in Example 2.2 with $\sigma = 1$. We have $p(\mathbf{y}|\mu, n) = (\sqrt{2\pi})^{-n} \exp[-\frac{1}{2}\sum_{i=1}^{n}(y_i - \mu)^2]$ $= (\sqrt{2\pi})^{-n} \exp\{-\frac{1}{2}[(n - 1)s^2 + n(\mu - \hat{\mu})^2]\}$, where $\hat{\mu} = \sum_{i=1}^{n} y_i/n$ and $(n - 1)s^2 = \sum_{i=1}^{n}(y_i - \hat{\mu})^2$. Then we can write $p(\mathbf{y}|\mu, n) = p_1(\hat{\mu}|\mu, n)\, p_2(\mathbf{y})$, with $p_1(\hat{\mu}|\mu, n) = \exp[-(n/2)(\mu - \hat{\mu})^2]$ and $p_2(\mathbf{y}) = (\sqrt{2\pi})^{-n} \exp[-(n-1)s^2/2]$. Clearly $\hat{\mu}$ is a sufficient statistic for μ, and the natural conjugate prior pdf for μ, $f(\mu|\cdot)$, is: $f(\mu|\hat{\mu}_0, n_0) = c \exp[-(n_0/2)(\mu - \hat{\mu}_0)^2]$, with $c = \sqrt{n_0/2\pi}$, a normal pdf with prior mean $\hat{\mu}_0$ and prior variance $1/n_0$. To use this prior pdf an investigator should check that it adequately represents his prior information and, if it does, supply values for its parameters $\hat{\mu}_0$ and n_0.

2.4 MARGINAL AND CONDITIONAL POSTERIOR DISTRIBUTIONS FOR PARAMETERS

As with usual joint pdf's, marginal and conditional pdf's can be obtained from a joint posterior pdf; for example, let $\boldsymbol{\theta}$ be partitioned as $\boldsymbol{\theta}' = (\boldsymbol{\theta}_1' \vdots \boldsymbol{\theta}_2')$ and suppose that we want the marginal posterior pdf for $\boldsymbol{\theta}_1$ which may

[10] See H. Raiffa and R. Schlaifer, *Applied Statistical Decision Theory*. Boston: Graduate School of Business Administration, Harvard University, 1961, Chapter 3, for a detailed discussion of natural conjugate prior pdf's.

[11] See, for example, Lindley, *Introduction to Probability and Statistics from a Bayesian Viewpoint. Part 2. Inference*. Cambridge: Cambridge University Press, 1965, pp. 46 ff. for further discussion of sufficient statistics.

[12] Note that when $p(\mathbf{y}|\boldsymbol{\theta}, n) = p_1(\mathbf{t}|\boldsymbol{\theta}, n)\, p_2(\mathbf{y})$ and $p(\boldsymbol{\theta})$ is a prior pdf for $\boldsymbol{\theta}$, the posterior pdf, $p(\boldsymbol{\theta}|\mathbf{y}, n)$ is $p(\boldsymbol{\theta}|\mathbf{y}, n) \propto p(\boldsymbol{\theta})\, p(\mathbf{y}|\boldsymbol{\theta}, n) \propto p(\boldsymbol{\theta})\, p_1(\mathbf{t}|\boldsymbol{\theta}, n)$. As explained in Section 2.11, for large n, $p(\boldsymbol{\theta}|\mathbf{y}, n)$ is approximately proportional to $p_1(\mathbf{t}|\boldsymbol{\theta}, n)$ under rather general conditions. Thus in large samples the posterior pdf assumes the form of $p_1(\mathbf{t}|\boldsymbol{\theta}, n)$, which is also the form of the natural conjugate prior pdf.

contain one or several elements of $\boldsymbol{\theta}$. This marginal posterior pdf, $p(\boldsymbol{\theta}_1|\mathbf{y})$, is readily obtained as follows:

$$p(\boldsymbol{\theta}_1|\mathbf{y}) = \int_{R_{\theta_2}} p(\boldsymbol{\theta}_1, \boldsymbol{\theta}_2|\mathbf{y})\, d\boldsymbol{\theta}_2 \tag{2.15a}$$

$$= \int_{R_{\theta_2}} p(\boldsymbol{\theta}_1|\boldsymbol{\theta}_2, \mathbf{y})\, p(\boldsymbol{\theta}_2|\mathbf{y})\, d\boldsymbol{\theta}_2, \tag{2.15b}$$

where R_{θ_2} denotes the region of $\boldsymbol{\theta}_2$ and $p(\boldsymbol{\theta}_1|\boldsymbol{\theta}_2, \mathbf{y})$ is the conditional posterior pdf for $\boldsymbol{\theta}_1$, given $\boldsymbol{\theta}_2$ and the sample information $\mathbf{y}$. Equation 2.15*b* illustrates the fact that the marginal posterior pdf for $\boldsymbol{\theta}_1$ may be viewed as an averaging of conditional posterior pdf's, $p(\boldsymbol{\theta}_1|\boldsymbol{\theta}_2, \mathbf{y})$, with the weight function being the marginal posterior pdf for $\boldsymbol{\theta}_2$, $p(\boldsymbol{\theta}_2|\mathbf{y})$. The integration shown in (2.15) provides an extremely useful way of getting rid of "nuisance" parameters, that is parameters that are not of special interest.

Example 2.3. Assume that we have n independent observations, $\mathbf{y}' = (y_1, y_2, \ldots, y_n)$, from a normal population with unknown mean μ and unknown standard deviation σ. If our prior information about values of the mean and standard deviation is vague or diffuse, we can represent this state of our initial information by taking our prior pdf as

$$p(\mu, \sigma)\, d\mu\, d\sigma \propto \frac{1}{\sigma}\, d\mu\, d\sigma, \qquad \begin{array}{c} -\infty < \mu < \infty. \\ 0 < \sigma < \infty. \end{array} \tag{2.16}$$

In (2.16) we have assumed μ and σ to be independently distributed, a priori, with μ and $\log \sigma$ each uniformly distributed [see the appendix at the end of this chapter for further discussion of (2.16)]. Then the joint posterior pdf for μ and σ is

$$\begin{aligned} p(\mu, \sigma|\mathbf{y}) &\propto p(\mu, \sigma) l(\mu, \sigma|\mathbf{y}), \qquad \begin{array}{c} -\infty < \mu < \infty, \\ 0 < \sigma < \infty. \end{array} \\ &\propto \sigma^{-(n+1)} \exp\left\{-\frac{1}{2\sigma^2}\left[\nu s^2 + n(\mu - \hat{\mu})^2\right]\right\}, \end{aligned} \tag{2.17}$$

where $l(\mu, \sigma|\mathbf{y}) \propto \sigma^{-n} \exp\left[-1/2\sigma^2 \sum_{i=1}^n (y_i - \mu)^2\right]$ is the likelihood function, $\nu = n - 1$, $\hat{\mu} = \sum_{i=1}^n y_i/n$, and $\nu s^2 = \sum_{i=1}^n (y_i - \hat{\mu})^2$. From the form of (2.17) it is clear that the conditional posterior pdf for μ given σ and the sample information, that is, $p(\mu|\sigma, \mathbf{y})$, is in the univariate normal form with conditional posterior mean and variance $E(\mu|\sigma, \mathbf{y}) = \hat{\mu}$ and $\operatorname{Var}(\mu|\sigma, \mathbf{y}) = \sigma^2/n$, respectively. Although these conditional results are of interest, it is clear that the conditional pdf for μ given σ and $\mathbf{y}$ depends critically on σ whose value is unknown. If we are mainly interested in μ, σ is a nuisance parameter and, as stated above, such a parameter can generally be integrated out of the posterior pdf. In the present instance we have[13]

$$
\begin{aligned}
p(\mu|\mathbf{y}) &= \int_0^\infty p(\mu, \sigma|\mathbf{y})\, d\sigma \\
&\propto \int_0^\infty \sigma^{-(n+1)} \exp\left\{-\frac{1}{2\sigma^2}[\nu s^2 + n(\mu - \hat{\mu})^2]\right\} d\sigma \\
&\propto \{\nu s^2 + n(\mu - \hat{\mu})^2\}^{-(\nu+1)/2}.
\end{aligned}
\tag{2.18}
$$

From (2.18) it is seen that the marginal posterior pdf for μ is in the form of a univariate Student t pdf[14] with mean $\hat{\mu}$; that is, the random variable

$$t = \frac{\mu - \hat{\mu}}{s/\sqrt{n}}$$

has a Student t pdf with $\nu = n - 1$ degrees of freedom.

If the parameter σ is of interest, we can integrate $p(\mu, \sigma|\mathbf{y})$ in (2.17) with respect to μ to obtain the marginal posterior pdf for σ, namely,[15]

$$
\begin{aligned}
p(\sigma|\mathbf{y}) &= \int_{-\infty}^\infty p(\mu, \sigma|\mathbf{y})\, d\mu \\
&\propto \sigma^{-(\nu+1)} \exp\left(-\frac{\nu s^2}{2\sigma^2}\right), \qquad 0 < \sigma < \infty.
\end{aligned}
\tag{2.19}
$$

This posterior pdf for σ is in the "inverted gamma" form (see Appendix A) and will be proper for $\nu > 0$. Further, from properties of (2.19) we have

$$E(\sigma|\mathbf{y}) = \frac{\sqrt{\nu/2}\,\Gamma[(\nu - 1)/2]}{\Gamma(\nu/2)} s \qquad \text{for } \nu > 1$$

and

$$\text{Var}(\sigma|\mathbf{y}) = \frac{\nu s^2}{\nu - 2} - [E(\sigma|\mathbf{y})]^2 \qquad \text{for } \nu > 2.$$

The modal value of the posterior pdf in (2.19) is $s\sqrt{\nu/(\nu + 1)}$.

[13] Note that $\int_0^\infty \sigma^{-(n+1)} \exp(-a/2\sigma^2)\, d\sigma = 2^{(n-2)/2}\Gamma(n/2)/a^{n/2}$. This result is easily obtained by letting $x = a/2\sigma^2$. Then the integral becomes $(2^{(n-2)/2}/a^{n/2})\int_0^\infty x^{(n-2)/2}e^{-x}\, dx = 2^{(n-2)/2}\Gamma(n/2)/a^{n/2}$, where Γ denotes the gamma function. In using this result in (2.18), $a = \nu s^2 + n(\mu - \hat{\mu})^2$ and the factor $2^{(n-2)/2}\Gamma(n/2)$ is absorbed in the factor of proportionality.

[14] See Appendix A at the end of the book for properties of this pdf.

[15] In integrating (2.17) with respect to μ, note that, for given σ, (2.17) is in the univariate normal form. Letting $z = \sqrt{n}(\mu - \hat{\mu})/\sigma$, $dz \propto d\mu/\sigma$, and thus (2.17) becomes $1/\sigma^n \exp(-\nu s^2/2\sigma^2) \exp(-z^2/2)\, d\sigma\, dz$ from which (2.19) follows.

2.5 POINT ESTIMATES FOR PARAMETERS

From Section 2.3 above it is seen that the Bayesian approach yields the complete posterior pdf for the parameter vector $\boldsymbol{\theta}$. If we wish, we may characterize this distribution in terms of a small number of measures, say measures of central tendency, dispersion, and skewness, with a measure of central tendency to serve as a point estimate. The problem of choosing a single measure of central tendency is a well-known problem in descriptive statistics. In some circumstances we may have a loss function, say $L = L(\boldsymbol{\theta}, \hat{\boldsymbol{\theta}})$, where $\hat{\boldsymbol{\theta}} = \hat{\boldsymbol{\theta}}(\mathbf{y})$ is a point estimate depending on the given sample observations $\mathbf{y}' = (y_1, \ldots, y_n)$. Since $\boldsymbol{\theta}$ is considered random, L is random. One generally used principle, which generates point estimates and which is in accord with the expected utility hypothesis, is to find the value of $\hat{\boldsymbol{\theta}}$ that minimizes the mathematical expectation of the loss function; that is

$$\min_{\hat{\boldsymbol{\theta}}} EL(\boldsymbol{\theta}, \hat{\boldsymbol{\theta}}) = \min_{\hat{\boldsymbol{\theta}}} \int_{R_\theta} L(\boldsymbol{\theta}, \hat{\boldsymbol{\theta}})\, p(\boldsymbol{\theta}|\mathbf{y})\, d\boldsymbol{\theta}, \tag{2.20}$$

which assumes that $EL(\boldsymbol{\theta}, \hat{\boldsymbol{\theta}})$ is finite and that a minimum exists.

As an important illustration of (2.20), consider the case of a quadratic loss function, $L = (\boldsymbol{\theta} - \hat{\boldsymbol{\theta}})'C(\boldsymbol{\theta} - \hat{\boldsymbol{\theta}})$, where C is a known nonstochastic positive definite symmetric matrix. Then the posterior expectation of the quadratic loss function is[16]

$$\begin{aligned} EL &= E(\boldsymbol{\theta} - \hat{\boldsymbol{\theta}})'C(\boldsymbol{\theta} - \hat{\boldsymbol{\theta}}) \\ &= E[(\boldsymbol{\theta} - E\boldsymbol{\theta}) - (\hat{\boldsymbol{\theta}} - E\boldsymbol{\theta})]'C[(\boldsymbol{\theta} - E\boldsymbol{\theta}) - (\hat{\boldsymbol{\theta}} - E\boldsymbol{\theta})] \\ &= E(\boldsymbol{\theta} - E\boldsymbol{\theta})'C(\boldsymbol{\theta} - E\boldsymbol{\theta}) + (\hat{\boldsymbol{\theta}} - E\boldsymbol{\theta})'C(\hat{\boldsymbol{\theta}} - E\boldsymbol{\theta}). \end{aligned} \tag{2.21}$$

The first term of this last expression does not involve $\hat{\boldsymbol{\theta}}$. The remaining term $(\hat{\boldsymbol{\theta}} - E\boldsymbol{\theta})'C(\hat{\boldsymbol{\theta}} - E\boldsymbol{\theta})$ is nonstochastic and will be minimized if we take $\hat{\boldsymbol{\theta}} = E\boldsymbol{\theta}$, given that C is positive definite. Thus for positive definite quadratic loss functions the mean $E\boldsymbol{\theta}$ of the posterior pdf $p(\boldsymbol{\theta}|\mathbf{y})$, if it exists, is an optimal point estimate. For other loss functions similar analysis can be performed to yield optimal point estimates.

Example 2.4. Consider Example 2.1 when our loss function is $L(\mu, \check{\mu}) = c(\mu - \check{\mu})^2$, where $\check{\mu}$ is a point estimate, and c is a positive constant. Then, taking $\check{\mu} = E\mu = (h_o\hat{\mu} + h_a\mu_a)/(h_o + h_a)$, the mean of the posterior pdf for μ will minimize $EL = cE(\mu - \check{\mu})^2$.

[16] In the second line of (2.21) the posterior mean $E\boldsymbol{\theta}$ has been subtracted from $\boldsymbol{\theta}$ and added to $\hat{\boldsymbol{\theta}}$ which does not affect the value of EL. In going from the second line of (2.21) to the third, the cross terms disappear; that is $E(\boldsymbol{\theta} - E\boldsymbol{\theta})'C(\hat{\boldsymbol{\theta}} - E\boldsymbol{\theta}) = 0$ since $E(\boldsymbol{\theta} - E\boldsymbol{\theta}) = \mathbf{0}$.

Example 2.5. Suppose that our loss function is $L = |\theta - \hat{\theta}|$ and the posterior pdf for θ is a proper continuous pdf, $p(\theta|\mathbf{y})$, with $a \leq \theta \leq b$ where a and b are known. Then the point estimate $\hat{\theta}$ which minimizes expected loss can be found as follows:

$$\begin{aligned} EL &= \int_a^b |\theta - \hat{\theta}|\, p(\theta|\mathbf{y})\, d\theta \\ &= \int_a^{\hat{\theta}} (\hat{\theta} - \theta)\, p(\theta|\mathbf{y})\, d\theta + \int_{\hat{\theta}}^b (\theta - \hat{\theta})\, p(\theta|\mathbf{y})\, d\theta \\ &= \hat{\theta}\, P(\hat{\theta}|\mathbf{y}) - \int_a^{\hat{\theta}} \theta\, p(\theta|\mathbf{y})\, d\theta + \int_{\hat{\theta}}^b \theta p\,(\theta|\mathbf{y})\, d\theta - \hat{\theta}[1 - P(\hat{\theta}|\mathbf{y})], \end{aligned}$$

where $P(\hat{\theta}|\mathbf{y}) = \int_a^{\hat{\theta}} p(\theta|\mathbf{y})\, d\theta$ is the cumulative posterior distribution function. Then, on differentiation[17] with respect to $\hat{\theta}$ and setting the derivative equal to zero, we have

$$\frac{dEL}{d\hat{\theta}} = P(\hat{\theta}|\mathbf{y}) - 1 + P(\hat{\theta}|\mathbf{y}) = 0$$

or

$$P(\hat{\theta}|\mathbf{y}) = \tfrac{1}{2}.$$

The $\hat{\theta}$ which satisfies this necessary condition for a minimum is the median of the posterior pdf. That this value for $\hat{\theta}$ produces a minimum of EL can be established by noting that $d^2EL/d\hat{\theta}^2$ is strictly positive for $\hat{\theta}$ = median of the posterior pdf. Thus for the absolute error function $L = |\theta - \hat{\theta}|$ the median of the posterior pdf is an optimal point estimate.

Next, we review a relationship between Bayesian and sampling theory approaches to point estimation. Let $\tilde{\boldsymbol{\theta}} = \tilde{\boldsymbol{\theta}}(\mathbf{y})$ be a sampling theory estimator.[18] The risk function associated with the estimator $\tilde{\boldsymbol{\theta}}$ is given by

$$r(\boldsymbol{\theta}) = \int_{R_{\mathbf{y}}} L(\boldsymbol{\theta}, \tilde{\boldsymbol{\theta}})\, p(\mathbf{y}|\boldsymbol{\theta})\, d\mathbf{y}, \tag{2.22}$$

where $L(\boldsymbol{\theta}, \tilde{\boldsymbol{\theta}})$ is a loss function, $p(\mathbf{y}|\boldsymbol{\theta})$ is a proper pdf for $\mathbf{y}$, given $\boldsymbol{\theta}$, and the integral in (2.22) is assumed to converge. As indicated explicitly in (2.22), the risk function depends on the value of the unknown parameter vector $\boldsymbol{\theta}$. Since it is impossible to find a $\tilde{\boldsymbol{\theta}}$ which minimizes $r(\boldsymbol{\theta})$ for all possible values of $\boldsymbol{\theta}$,[19]

[17] It is assumed that the needed derivatives exist for $a \leq \theta \leq b$.

[18] As is well known, the term "estimator" indicates that $\tilde{\boldsymbol{\theta}} = \tilde{\boldsymbol{\theta}}(\mathbf{y})$ is regarded as a random quantity.

[19] For example, if we take $\tilde{\boldsymbol{\theta}} = \mathbf{b}$, a vector of constants, this "estimator" will have smaller risk when $\boldsymbol{\theta} = \mathbf{b}$ than any other estimator and thus no single estimator can minimize $r(\boldsymbol{\theta})$ for all $\boldsymbol{\theta}$.

we shall seek the estimator that minimizes average risk when average risk is defined by

$$Er(\boldsymbol{\theta}) = \int_{R_\theta} p(\boldsymbol{\theta})\, r(\boldsymbol{\theta})\, d\boldsymbol{\theta}. \tag{2.23}$$

In (2.23) $p(\boldsymbol{\theta})$ is a "weighting function" used to weight the performance of $\tilde{\boldsymbol{\theta}}$, an estimator, in regions of the parameter space. Then our problem is to find the estimator that minimizes average risk, that is, that solves the following problem:

$$\min_{\tilde{\boldsymbol{\theta}}} Er(\boldsymbol{\theta}) = \min_{\tilde{\boldsymbol{\theta}}} \int_{R_\theta} \int_{R_{\mathbf{y}}} p(\boldsymbol{\theta})\, L(\boldsymbol{\theta}, \tilde{\boldsymbol{\theta}})\, p(\mathbf{y}|\boldsymbol{\theta})\, d\mathbf{y}\, d\boldsymbol{\theta}. \tag{2.24}$$

Given that the integrand of (2.24) is non-negative, we can interchange the order of integration and, using $p(\boldsymbol{\theta})\, p(\mathbf{y}|\boldsymbol{\theta}) = p(\mathbf{y})\, p(\boldsymbol{\theta}|\mathbf{y})$, write (2.24) as

$$\min_{\tilde{\boldsymbol{\theta}}} Er(\boldsymbol{\theta}) = \min_{\tilde{\boldsymbol{\theta}}} \int_{R_{\mathbf{y}}} \left[\int_{R_\theta} L(\boldsymbol{\theta}, \tilde{\boldsymbol{\theta}})\, p(\boldsymbol{\theta}|\mathbf{y})\, d\boldsymbol{\theta}\right] p(\mathbf{y})\, d\mathbf{y}. \tag{2.25}$$

The $\tilde{\boldsymbol{\theta}}$ that minimizes the expression in square brackets will minimize expected risk, provided that $Er(\boldsymbol{\theta})$ is finite, and this estimator is, by definition, the Bayes estimator.[20] Therefore, if a specification is made for the seriousness of estimation errors in the form of a loss function, $L(\boldsymbol{\theta}, \tilde{\boldsymbol{\theta}})$, and for the weighting of parameter values over which good performance is sought by a choice of $p(\boldsymbol{\theta})$, then on an average risk criterion the Bayesian estimator gives the best performance in repeated sampling.[21]

[20] When the double integral in (2.25) converges, and thus $Er(\boldsymbol{\theta})$ is finite, the $\tilde{\boldsymbol{\theta}}$ solving the minimization problem in (2.25) will also be a solution to the minimization problem in (2.20). If the double integral in (2.25) diverges, however, the minimization problem in (2.25) will have no solution, but still a solution to the problem in (2.20) often exists. When this is the situation, the solution to the minimization problem in (2.20) has been called a quasi-Bayesian estimator. Quasi-Bayesian estimators often arise when improper diffuse prior pdf's are employed along with usual loss functions, for example, quadr[illegible] loss functions. For further discussion of this point see H. Thornber, "Applications [illegible] Decision Theory to Econometrics," unpublished doctoral dissertation, University of Chicago, 1966, and M. Stone, "Generalized Bayes Decision Functions, Admissibility and the Exponential Family," *Ann. Math. Statist.*, **38**, 818–822 (1967).

[21] The relevance of the criterion of performance in *repeated samples* is questioned by some. They want an estimate that is appropriate for the given sample data and thus will solve the problem in (2.20) which involves no averaging over the sample space $R_{\mathbf{y}}$. When the solution to (2.20) is identical to the solution of (2.24), as it often is, this consideration makes no practical difference. On the other hand, many sampling theorists object to the introduction of the "weighting function" (prior pdf) $p(\boldsymbol{\theta})$ and therefore do not attach much importance to the minimal average risk property of Bayesian estimators.

2.6 BAYESIAN INTERVALS AND REGIONS FOR PARAMETERS

Given that the posterior pdf $p(\boldsymbol{\theta}|\mathbf{y})$ has been obtained, it is generally possible to compute the probability that the parameter vector $\boldsymbol{\theta}$ lies in a particular subregion, $\bar{R}$, of the parameter space as follows:

$$\Pr(\boldsymbol{\theta} \in \bar{R}|\mathbf{y}) = \int_{\bar{R}} p(\boldsymbol{\theta}|\mathbf{y})\, d\boldsymbol{\theta}. \tag{2.26}$$

The probability in (2.26) measures the degree of belief that $\boldsymbol{\theta} \in \bar{R}$ given the sample and prior information.

If we fix the probability in (2.26), say at 0.95, it is generally possible to find a region (or interval) $\bar{R}$, not necessarily unique, such that (2.26) holds. In many important problems with unimodal posterior pdf's, it is possible to obtain a unique region (or interval) $\bar{R}$ by imposing the conditions that its probability content be β, say $\beta = 0.95$, and that the posterior pdf's values over the region or interval be not less than those relating to any other region with the same probability content; for example, for unimodal symmetric posterior pdf's the region or interval with given probability content β, which is centered at the modal value of the posterior pdf is the Bayesian "highest posterior density" region or interval.[22]

Example 2.6. Consider Example 2.3 in which it was found that the posterior pdf of $(\mu - \hat{\mu})/s'$, where $s' = s/\sqrt{n}$ is a Student t pdf with $\nu = n - 1$ degrees of freedom. Thus the probability that μ will lie in a particular interval, say $\hat{\mu} \pm ks'$, with k given, can easily be evaluated by using tables of the Student t distribution.[23] Alternatively, k can be determined so that the posterior probability that $\hat{\mu} - ks' < \mu < \hat{\mu} + ks'$ is a given value, say $\beta = 0.90$. The interval so obtained, $\hat{\mu} \pm ks'$, is numerically exactly the same as a sampling theory confidence interval but is given an entirely different interpretation in

[22] See G. E. P. Box and G. C. Tiao, "Multiparameter Problems from a Bayesian Point of View," *Ann. Math. Statist.*, **36**, 1468–1482 (1965), for further discussion of "highest posterior density" Bayesian regions. In general, if we seek a "highest" interval with probability content β for a unimodal pdf, $p(x)$, it can be obtained by solving the following problem: minimize $(b - a)$ subject to $\int_a^b p(x)\,dx = \beta$. On differentiating $b - a + \lambda[\int_a^b p(x)\,dx - \beta]$, where λ is a Lagrange multiplier, partially with respect to a and b and setting these derivatives equal to zero, yields $1 + \lambda p(a) = 0$ and $1 + \lambda p(b) = 0$, and thus a and b must be such that $p(a) = p(b)$ for these necessary conditions to be satisfied. Determining a and b such that $\int_a^b p(x)\,dx = \beta$ with $p(a) = p(b)$ leads to a shortest interval with probability content β, and this interval will be a "highest" interval given that $p(x)$ is unimodal. In the example above in which z is a standardized normal variable $p(z)$ is unimodal and symmetric about zero. Thus taking $a = -z_\beta$ and $b = z_\beta$ satisfies the condition $p(a) = p(b)$.

[23] See, for example, N. V. Smirnov, *Tables for the Distribution and Density Function of t-Distribution.* New York: Pergamon, 1961.

the Bayesian approach. As is well known, the sampling theorist regards his interval as random and having probability $\beta = 0.90$ of covering the true value of the parameter. For the Bayesian whose work is conditional on the sample observations the interval $\hat{\mu} \pm ks'$ is regarded as given and his statement is that the posterior probability that μ will lie in the interval is $\beta = 0.90$. Note that the probability statements being made by the sampling theorist and the Bayesian are not identical.

2.7 MARGINAL DISTRIBUTION OF THE OBSERVATIONS

In certain instances it is of interest to obtain the marginal pdf for the observations, denoted by $p(\mathbf{y})$. This pdf can be obtained as follows:

$$\begin{aligned} p(\mathbf{y}) &= \int_{R_\theta} p(\boldsymbol{\theta}, \mathbf{y})\, d\boldsymbol{\theta} \\ &= \int_{R_\theta} p(\mathbf{y}|\boldsymbol{\theta})\, p(\boldsymbol{\theta})\, d\boldsymbol{\theta}. \end{aligned} \tag{2.27}$$

The second line of (2.27) indicates that the marginal pdf of the observations is an average of the conditional pdf $p(\mathbf{y}|\boldsymbol{\theta})$ with the prior pdf $p(\boldsymbol{\theta})$ serving as the weighting function.

Example 2.7. Let y_1 be an observation from a normal distribution with unknown mean μ and known standard deviation $\sigma = \sigma_0$. Then

$$p(y_1|\mu, \sigma = \sigma_0) = \frac{1}{\sqrt{2\pi}\,\sigma_0} \exp\left[-\frac{1}{2\sigma_0^2}(y_1 - \mu)^2\right].$$

If the prior pdf for μ is $p(\mu) = (\sqrt{2\pi}\,\sigma_a)^{-1} \exp[-(2\sigma_a^2)^{-1}(\mu - \mu_a)^2]$, $-\infty < \mu < \infty$, where μ_a and σ_a are the prior mean and standard deviation, respectively, the marginal pdf for y_1 is

$$\begin{aligned} p(y_1) &= \int_{-\infty}^{\infty} p(y_1|\mu, \sigma = \sigma_0)\, p(\mu)\, d\mu \\ &= (2\pi\sigma_0\sigma_a)^{-1} \int_{-\infty}^{\infty} \exp\left\{-\frac{1}{2}\left[\frac{(y_1 - \mu)^2}{\sigma_0^2} + \frac{(\mu - \mu_a)^2}{\sigma_a^2}\right]\right\} d\mu. \end{aligned}$$

On completing the square for μ in the exponent and performing the integration,[24] the result is

$$p(y_1) = \frac{1}{\sqrt{2\pi(\sigma_a^2 + \sigma_0^2)}} \exp\left[-\frac{(y_1 - \mu_a)^2}{2(\sigma_a^2 + \sigma_0^2)}\right].$$

[24] A less tedious way to derive $p(y_1)$ is to write $y_1 = \mu + \epsilon$ with ϵ, a scalar random variable normally distributed and independent of μ, with zero mean and variance σ_0^2. Then the mean of y_1 is μ_a, the mean of μ, and the variance of y_1 is $\sigma_a^2 + \sigma_0^2$. Since y_1 is linearly related to μ and ϵ, it will have a normal pdf.

Thus the marginal pdf for y_1 is normal with mean μ_a, the prior mean for μ, and variance $\sigma_a^2 + \sigma_0^2$. Since μ_a, σ_a^2, and σ_0^2 are assumed known, it is possible to use $p(y_1)$ to make probability statements about y_1, a fact that is often useful before y_1 is observed.

2.8 PREDICTIVE PROBABILITY DENSITY FUNCTIONS

On many occasions, given our sample information $\mathbf{y}$, we are interested in making inferences about other observations that are still unobserved, one part of the problem of prediction. In the Bayesian approach the pdf for the as yet unobserved observations, given our sample information, can be obtained and is known as the predictive pdf; for example, let $\tilde{\mathbf{y}}$ represent a vector of as yet unobserved observations. Write

$$p(\tilde{\mathbf{y}}, \boldsymbol{\theta}|\mathbf{y}) = p(\tilde{\mathbf{y}}|\boldsymbol{\theta}, \mathbf{y})\, p(\boldsymbol{\theta}|\mathbf{y}) \tag{2.28}$$

as the joint pdf for $\tilde{\mathbf{y}}$ and a parameter vector $\boldsymbol{\theta}$, given the sample information $\mathbf{y}$. On the right of (2.28) $p(\tilde{\mathbf{y}}|\boldsymbol{\theta}, \mathbf{y})$ is the conditional pdf for $\tilde{\mathbf{y}}$, given $\boldsymbol{\theta}$ and $\mathbf{y}$, whereas $p(\boldsymbol{\theta}|\mathbf{y})$ is the conditional pdf for $\boldsymbol{\theta}$ given $\mathbf{y}$, that is, the posterior pdf for $\boldsymbol{\theta}$. To obtain the predictive pdf, $p(\tilde{\mathbf{y}}|\mathbf{y})$, we merely integrate (2.28) with respect to $\boldsymbol{\theta}$; that is

$$\begin{aligned} p(\tilde{\mathbf{y}}|\mathbf{y}) &= \int_{R_{\boldsymbol{\theta}}} p(\tilde{\mathbf{y}}, \boldsymbol{\theta}|\mathbf{y})\, d\boldsymbol{\theta} \\ &= \int_{R_{\boldsymbol{\theta}}} p(\tilde{\mathbf{y}}|\boldsymbol{\theta}, \mathbf{y})\, p(\boldsymbol{\theta}|\mathbf{y})\, d\boldsymbol{\theta}. \end{aligned} \tag{2.29}$$

The second line of (2.29) indicates that the predictive pdf can be viewed as an average of conditional predictive pdf's, $p(\tilde{\mathbf{y}}|\boldsymbol{\theta}, \mathbf{y})$, with the posterior pdf for $\boldsymbol{\theta}$, $p(\boldsymbol{\theta}|\mathbf{y})$ serving as the weighting function.

Example 2.8. In Example 2.2 we had n independent observations $\mathbf{y}' = (y_1, y_2, \ldots, y_n)$ from a normal population with unknown mean μ and known standard deviation $\sigma = \sigma_0$. With diffuse prior information about μ, the posterior pdf [see (2.14)] was found to be normal with mean $\hat{\mu}$, the sample mean, and variance σ_0^2/n. We now wish to obtain the predictive pdf for a new observation, say $\tilde{y}_{n+1}$ which has not yet been observed. The two factors in the integrand of the second line of (2.29) are

$$p(\tilde{y}_{n+1}|\mu, \sigma = \sigma_0, \mathbf{y}) \propto \exp\left[-\frac{1}{2\sigma_0^2}(\tilde{y}_{n+1} - \mu)^2\right]$$

and from (2.14)

$$p(\mu|\sigma = \sigma_0, \mathbf{y}) \propto \exp\left[-\frac{n}{2\sigma_0^2}(\mu - \hat{\mu})^2\right], \qquad -\infty < \mu < \infty.$$

Then from (2.29)

$$(2.30)\quad \begin{aligned} p(\tilde{y}_{n+1}|\mathbf{y}) &= \int_{-\infty}^{\infty} p(\tilde{y}_{n+1}|\mu, \sigma = \sigma_0, \mathbf{y})\, p(\mu|\sigma = \sigma_0, \mathbf{y})\, d\mu \\ &\propto \int_{-\infty}^{\infty} \exp\left[-\frac{1}{2\sigma_0^2}[(\tilde{y}_{n+1} - \mu)^2 + n(\mu - \hat{\mu})^2]\right] d\mu. \end{aligned}$$

On completing the square on μ in this last expression[25] and integrating (2.30) with respect to μ, the predictive pdf for $\tilde{y}_{n+1}$ is

$$(2.31)\qquad p(\tilde{y}_{n+1}|\mathbf{y}) \propto \exp\left[-\frac{n}{2(n+1)\sigma_0^2}(\tilde{y}_{n+1} - \hat{\mu})^2\right].$$

It is seen that $\tilde{y}_{n+1}$ is normally distributed with mean $E(\tilde{y}_{n+1}|\mathbf{y}) = \hat{\mu}$, the sample mean, and variance $\mathrm{Var}(\tilde{y}_{n+1}|\mathbf{y}) = \sigma_0^2(n+1)/n$. The pdf in (2.31) can, of course, be employed to make probability statements about $\tilde{y}_{n+1}$ given $\mathbf{y}$.

2.9 POINT PREDICTION

The predictive pdf, $p(\tilde{\mathbf{y}}|\mathbf{y})$, can be used to obtain a point prediction; for example, we can use a measure of central tendency, say the mean or modal value, as a point prediction, or, if we have a loss function $L = L(\tilde{\mathbf{y}}, \hat{\mathbf{y}})$, where $\hat{\mathbf{y}}$ is a point prediction for $\tilde{\mathbf{y}}$, we can seek the vector $\hat{\mathbf{y}}$ that minimizes the mathematical expectation of the loss function; that is

$$(2.32)\qquad \min_{\hat{\mathbf{y}}} \int_{R\tilde{y}} L(\tilde{\mathbf{y}}, \hat{\mathbf{y}})\, p(\tilde{\mathbf{y}}|\mathbf{y})\, d\tilde{\mathbf{y}}.$$

If a solution to the problem in (2.32) exists, it is an optimal point prediction in the sense of minimizing expected loss. Analysis similar to that presented in Section 2.5 on point estimation provides the result that the mean of the predictive pdf is optimal if our loss function is quadratic; that is, if $L(\tilde{\mathbf{y}}, \hat{\mathbf{y}}) = (\tilde{\mathbf{y}} - \hat{\mathbf{y}})'Q(\tilde{\mathbf{y}} - \hat{\mathbf{y}})$, with Q a positive definite symmetric matrix, then taking $\hat{\mathbf{y}} = E(\tilde{\mathbf{y}}|\mathbf{y})$ as our point prediction provides minimal expected loss; for example, in Example 2.8 the mean of the predictive pdf is the sample mean $\hat{\mu}$, and this is an optimal point prediction for $\tilde{y}_{n+1}$, given that our loss function is of the form $L(\tilde{y}_{n+1}, \hat{y}_{n+1}) = c(\tilde{y}_{n+1} - \hat{y}_{n+1})^2$, $c > 0$. For other loss functions

[25] That is, $(\tilde{y}_{n+1} - \mu)^2 + n(\mu - \hat{\mu})^2 = \tilde{y}_{n+1}^2 + (n+1)\mu^2 - 2\mu(\tilde{y}_{n+1} + n\hat{\mu}) + n\hat{\mu}^2 = (n+1)[\mu^2 - 2\mu(\tilde{y}_{n+1} + n\hat{\mu})/(n+1)] + n\hat{\mu}^2 + \tilde{y}_{n+1}^2 = (n+1)[\mu - (\tilde{y}_{n+1} + n\hat{\mu})/(n+1)]^2 + n(\tilde{y}_{n+1} - \hat{\mu})^2/(n+1)$. On substituting this last expression in the second line of (2.30), the integration with respect to μ can be done readily to yield (2.31). An alternative, simpler derivation of (2.31) is obtained from noting that $\tilde{y}_{n+1} = \mu + \epsilon_{n+1}$ with the normal random error ϵ_{n+1} independent of μ, given $\mathbf{y}$, with mean zero and known variance σ_0^2. Since both $\mu|\mathbf{y}$ and ϵ_{n+1} are normal, $\tilde{y}_{n+1}$ has a normal pdf with mean $E\tilde{y}_{n+1}|\mathbf{y} = E\mu|\mathbf{y} = \hat{\mu}$, since $E\mu|\mathbf{y} = \hat{\mu}$ from (2.14), and $\mathrm{Var}(\tilde{y}_{n+1}|\mathbf{y}) = \mathrm{Var}(\mu|\mathbf{y}) + \mathrm{Var}\,\epsilon_{n+1} = \sigma_0^2/n + \sigma_0^2 = \sigma_0^2(n+1)/n$.

similar analysis can be performed to obtain optimal point predictions, of course under the assumption that a solution to the problem in (2.32) exists.

2.10 PREDICTION REGIONS AND INTERVALS

Given that we have the predictive pdf, $p(\tilde{\mathbf{y}}|\mathbf{y})$, we can, for a given region (or interval) $\bar{R}$, generally evaluate

$$\Pr(\tilde{\mathbf{y}} \in \bar{R}|\mathbf{y}) = \int_{\bar{R}} p(\tilde{\mathbf{y}}|\mathbf{y})\, d\tilde{\mathbf{y}}, \tag{2.33}$$

where $\bar{R}$ is a subspace of $R_{\tilde{\mathbf{y}}}$, the space of the elements of $\tilde{\mathbf{y}}$. In (2.33) we have the probability that the future observation vector $\tilde{\mathbf{y}}$ will lie in the region $\bar{R}$. Alternatively, given a stated probability in (2.33), we can seek a region $\bar{R}$ such that (2.33) is satisfied. As with regions for parameters in Section 2.6, this region can be made unique for unimodal pdf's if we require it to be a "highest predictive density" region; that is, a region with the given probability content and such that the predictive pdf's values over the region are not less than those relating to any other region with the same probability content.

Example 2.9. In Example 2.8 the predictive pdf for $\tilde{y}_{n+1}$ in (2.31) is normal with mean $\hat{\mu}$ and variance $\sigma_0^2(n+1)/n$. Then $z = (\tilde{y}_{n+1} - \hat{\mu})/\bar{\sigma}_0$, with $\bar{\sigma}_0 = \sigma_0\sqrt{(n+1)/n}$, has a normal pdf with zero mean and unit variance. From tables of the standardized normal distribution we can find the $\Pr\{a < z < b\}$, where a and b are given constants. The statement $a < z < b$ is equivalent to $\hat{\mu} + a\bar{\sigma}_0 < \tilde{y}_{n+1} < \hat{\mu} + b\bar{\sigma}_0$ and thus the probability that $\tilde{y}_{n+1}$ will satisfy these inequalities is the same as $\Pr\{a < z < b\}$. On the other hand, if we are required to find a and b such that $\Pr\{a < z < b\} = \beta$, where β is given, it is clear that there are many possible values for a and b such that $\Pr\{a < z < b\} = \beta$. The requirement that the interval be a "highest" interval leads to a unique a and b, namely, $a = -z_\beta$ and $b = z_\beta$, where the area over the interval $-z_\beta$ to z_β is just β.

2.11 SOME LARGE SAMPLE PROPERTIES OF BAYESIAN POSTERIOR PDF'S

In this section we discuss briefly some large sample properties of posterior pdf's.[26] First, let us consider the posterior pdf for a scalar parameter θ:

$$\begin{aligned} p(\theta|\mathbf{y}) &\propto p(\theta)l(\theta|\mathbf{y}) \\ &\propto p(\theta)e^{\log l(\theta|y)}, \end{aligned} \tag{2.34}$$

[26] For other discussions of this topic see Jeffreys, *op. cit.*, p. 193 ff.; Lindley, *op. cit.*, p. 128 ff., and "The Use of Prior Probability Distributions in Statistical Inference and Decisions," in J. Neyman (Ed.) *Proc. Fourth Berkeley Symp. Math. Statist. Probab.* Berkeley: University of California Press, **1**, 453–468 (1961); L. LeCam, "On Some

where $p(\theta)$ is our prior pdf and $l(\theta|\mathbf{y})$ denotes the likelihood function based on n independent sample observations, $\mathbf{y}' = (y_1, y_2, \ldots, y_n)$. We assume that both $p(\theta)$ and $l(\theta|\mathbf{y})$ are nonzero in the parameter space and have continuous derivatives and that $l(\theta|\mathbf{y})$ has a unique maximum at $\theta = \hat{\theta}$, the maximum likelihood estimate.

In general, as Jeffreys points out, $\log l(\theta|\mathbf{y})$ will be of order n, whereas $p(\theta)$ does not depend on n, the sample size. Thus, heuristically, in large-sized samples the likelihood factor in (2.34) will dominate the posterior pdf. Since under general conditions the likelihood function assumes a normal shape as n gets large, with center at the maximum likelihood estimate $\hat{\theta}$, the posterior pdf will be normal in large samples with mean equal to the maximum likelihood estimate $\hat{\theta}$.

To put these considerations in more explicit terms, we can expand both factors of (2.34) around the maximum likelihood estimate $\hat{\theta}$ as follows:

$$(2.35)\qquad \begin{aligned} p(\theta) &= p(\hat{\theta}) + (\theta - \hat{\theta})p'(\hat{\theta}) + \tfrac{1}{2}(\theta - \hat{\theta})^2 p''(\hat{\theta}) + \cdots \\ &= p(\hat{\theta})\left[1 + \frac{(\theta - \hat{\theta})p'(\hat{\theta})}{p(\hat{\theta})} + \frac{\tfrac{1}{2}(\theta - \hat{\theta})^2 p''(\hat{\theta})}{p(\hat{\theta})} + \cdots\right] \end{aligned}$$

and, with $g(\theta) = \log l(\theta|\mathbf{y})$,

$$(2.36)\qquad \begin{aligned} \exp\{g(\theta)\} &= \exp\{g(\hat{\theta}) + \tfrac{1}{2}(\theta - \hat{\theta})^2 g''(\hat{\theta}) + \tfrac{1}{6}(\theta - \hat{\theta})^3 g'''(\hat{\theta}) + \cdots\} \\ &\propto \exp\{\tfrac{1}{2}(\theta - \hat{\theta})^2 g''(\hat{\theta})\}[1 + \tfrac{1}{6}(\theta - \hat{\theta})^3 g'''(\hat{\theta}) + \cdots], \end{aligned}$$

where the fact that $g'(\hat{\theta}) = 0$ (since $\hat{\theta}$ is the maximum likelihood estimate) has been employed and where the expansion $e^x = 1 + x + \cdots$ has been utilized. Then, on multiplying (2.35) and (2.36), we have

$$(2.37)\qquad p(\theta|\mathbf{y}) \propto e^{\frac{1}{2}(\theta - \hat{\theta})^2 g''(\hat{\theta})}\left[1 + \frac{(\theta - \hat{\theta})p'(\hat{\theta})}{p(\hat{\theta})} + \frac{\tfrac{1}{2}(\theta - \hat{\theta})^2 p''(\hat{\theta})}{p(\hat{\theta})} + \tfrac{1}{6}(\theta - \hat{\theta})^3 g'''(\hat{\theta}) + \cdots\right].$$

The leading term in (2.37), $e^{\frac{1}{2}(\theta - \hat{\theta})^2 g''(\hat{\theta})}$, is in the normal form, centered at the maximum likelihood estimate $\hat{\theta}$ with variance[27]

$$\operatorname{Var}(\theta|\mathbf{y}) \doteq [-g''(\hat{\theta})]^{-1} = \left[-\frac{d^2 \log l(\theta|\mathbf{y})}{d\theta^2}\right]^{-1}_{\theta=\hat{\theta}}.$$

Thus, if we use just the leading term of (2.37), the approximate large sample posterior pdf for θ is

---

Asymptotic Properties of Maximum Likelihood and Related Bayes Estimates," *Univ. Calif. Publ. Statist.*, **1**, 277–330 (1953); and "Les Propriétés Asymptotiques des Solutions de Bayes," *Publ. Inst. Statist.*, University of Paris, Vol. 7, 1958, pp. 17–35; R. A. Johnson, "An Asymptotic Expansion for Posterior Distributions," *Ann. Math. Statist.* **38**, 1899–1906 (1967).

[27] Note that, since $g(\theta)$ has a maximum at $\theta = \hat{\theta}$, $g''(\hat{\theta}) < 0$.

$$p(\theta|\mathbf{y}) \doteq \frac{|g''(\hat{\theta})|^{1/2}}{\sqrt{2\pi}} e^{-1/2(\theta-\hat{\theta})^2|g''(\hat{\theta})|}. \tag{2.38}$$

Since $|g''(\hat{\theta})|$ is usually of order n, as n gets large the posterior pdf becomes sharply centered around $\hat{\theta}$, that is, $|g''(\hat{\theta})|^{-1}$, the variance, becomes smaller as n grows larger.

With respect to the quality of the approximation in (2.38), Jeffreys points out that $\theta - \hat{\theta}$ is of order $n^{-1/2}$, and thus in (2.37) the terms $(\theta - \hat{\theta})p'(\hat{\theta})/p(\hat{\theta})$ and $\frac{1}{6}(\theta - \hat{\theta})^3 g'''(\hat{\theta})$ are of order $n^{-1/2}$,[28] whereas $\frac{1}{2}(\theta - \hat{\theta})^2 p''(\hat{\theta})/p(\hat{\theta})$ is of order n^{-1}. Thus the approximation in (2.38)[29] involves an error of order $n^{-1/2}$.

Example 2.10. Assume that we have n independent observations from a normal population with unknown mean μ and known standard deviation $\sigma = \sigma_0$. It is well known that the sample mean $\hat{\mu} = \sum_{i=1}^n y_i/n$ is the maximum likelihood estimate for μ. Then, employing (2.38) for *any* prior pdf satisfying the assumptions set forth above, the posterior pdf, $p(\mu|\mathbf{y}, \sigma^2)$, can be approximated as follows in large samples:

$$\begin{aligned} p(\mu|\mathbf{y}, \sigma^2) &\doteq \frac{|g''(\hat{\mu})|^{1/2}}{\sqrt{2\pi}} e^{-1/2(\mu-\hat{\mu})^2|g''(\hat{\mu})|} \\ &\doteq \frac{\sqrt{n}}{\sqrt{2\pi}\,\sigma_0} e^{-(n/2\sigma_0^2)(\mu-\hat{\mu})^2} \end{aligned}$$

where

$$g(\mu) = \log l(\mu|\mathbf{y}, \sigma_0) = -\log\sqrt{2\pi} - n\log\sigma_0 - \frac{1}{2\sigma_0^2}\sum_{i=1}^n (y_i - \mu)^2$$

and

$$g''(\hat{\mu}) = \frac{-n}{\sigma_0^2}.$$

Thus the large sample posterior pdf for μ is a normal pdf with mean $\hat{\mu}$ and variance $|g''(\hat{\mu})|^{-1} = \sigma_0^2/n$.

The above argument generalizes easily to the case in which we have a vector of parameters, say $\boldsymbol{\theta}$, rather than a scalar parameter; that is, in large samples the posterior pdf for $\boldsymbol{\theta}$ will be approximately normal with mean $\hat{\boldsymbol{\theta}}$, the maximum likelihood estimate, and covariance matrix

$$\left[-\frac{\partial^2 \log l(\boldsymbol{\theta}|\mathbf{y})}{\partial\theta_i\,\partial\theta_j}\right]^{-1}_{\boldsymbol{\theta}=\hat{\boldsymbol{\theta}}}. \tag{2.39}$$

[28] $g'''(\hat{\theta})$ is usually of order n if it is nonzero.

[29] It is possible to improve the approximation in (2.38) by retaining additional terms appearing in the square brackets in (2.37). See, for example, Lindley, "The Use of Prior Probability Distributions in Statistical Inference and Decisions," *op. cit.*, p. 457 ff.

In this case, proceeding as above, we can expand the two factors in the posterior pdf for $\boldsymbol{\theta}$, $p(\boldsymbol{\theta}|y) \propto p(\boldsymbol{\theta})\, l(\boldsymbol{\theta}|\mathbf{y}) = p(\boldsymbol{\theta})e^{g(\boldsymbol{\theta}|\mathbf{y})}$, where $g(\boldsymbol{\theta}|\mathbf{y}) = \log l(\boldsymbol{\theta}|\mathbf{y})$. Then, if we retain just the leading term in the expansion, we have, with $\dot{\propto}$ denoting "approximately proportional to,"

$$p(\boldsymbol{\theta}|\mathbf{y}) \mathrel{\dot{\propto}} \exp\left[-\tfrac{1}{2}(\boldsymbol{\theta} - \hat{\boldsymbol{\theta}})'C(\boldsymbol{\theta} - \hat{\boldsymbol{\theta}})\right], \tag{2.40}$$

which is in the multivariate normal form with mean $\hat{\boldsymbol{\theta}}$, the maximum likelihood estimate, and covariance matrix C^{-1}, which is just the matrix in (2.39).[30]

It is indeed interesting to observe the close agreement of Bayesian results in large samples with those flowing from the maximum likelihood approach. Of course, a moot problem is how large a sample size is required for these large sample approximate results to be reasonably accurate. Fortunately there is usually no need to rely on large-sample approximate results, since finite sample posterior pdf's are available, given the elements appearing in Bayes' theorem. In certain instances, however, in which computational problems arise in the analysis of complicated posterior pdf's the above large-sample results are useful.

2.12. APPLICATION OF PRINCIPLES TO ANALYSIS OF THE PARETO DISTRIBUTION

Assume that we have n independent observations $\mathbf{y}' = (y_1, y_2, \ldots, y_n)$, each with the Pareto pdf given by

$$p(y_i|A, \alpha) = \frac{\alpha A^\alpha}{y_i^{\alpha+1}} \qquad \begin{array}{l} 0 < \alpha < \infty, \\ 0 < A < y_i < \infty. \end{array} \tag{2.41}$$

Such a pdf has been frequently assumed to represent the distribution of incomes above a known value A. Given that A is known, the only unknown parameter in (2.41) is α. We shall obtain the posterior pdf for α. From (2.41) the likelihood function is

$$l(\alpha|\mathbf{y}, A) = \prod_{i=1}^{n} p(y_i|A, \alpha),$$

or

$$\begin{aligned} l(\alpha|\mathbf{y}, A) &= \frac{\alpha^n A^{n\alpha}}{(y_1 y_2 \cdots y_n)^{\alpha+1}} \\ &= \frac{\alpha^n A^{n\alpha}}{G^{n(\alpha+1)}}, \end{aligned} \tag{2.42}$$

where $G = (y_1 y_2 \cdots y_n)^{1/n}$ is the geometric mean of the observations.

[30] Note that since $\hat{\boldsymbol{\theta}}$ is assumed to be a maximizing value, C will be a positive definite matrix.

As regards prior pdf for α, we assume that our information about the value of this parameter is diffuse or vague and represent this state of our prior information by assuming $\log \alpha$ uniformly distributed[31] which implies

$$p(\alpha) \propto \frac{1}{\alpha}, \qquad 0 < \alpha < \infty. \tag{2.43}$$

On combining this prior pdf with the likelihood function in (2.42), the posterior pdf for α is

$$\begin{aligned} p(\alpha|A, \mathbf{y}) &\propto \frac{\alpha^{n-1}A^{n\alpha}}{G^{n\alpha}} \\ &\propto \alpha^{n-1}e^{-an\alpha} \end{aligned} \tag{2.44}$$

where $a = lnG/A$. The posterior pdf in (2.44) is seen to be in the form of a gamma pdf. The normalized posterior pdf is thus

$$p(\alpha|A, \mathbf{y}) = \frac{(an)^n}{\Gamma(n)} \alpha^{n-1}e^{-an\alpha}, \qquad 0 < \alpha < \infty, \tag{2.45}$$

which will be proper for $n > 0$. This posterior pdf for α represents our knowledge about α based on the information in our sample $\mathbf{y}$ and the prior pdf in (2.43). If we wish, we can easily compute the posterior probability that $c_1 < \alpha < c_2$, where c_1 and c_2 are given numbers.[32] Also, since the posterior moments of α are given by $E\alpha^r = (an)^{-r}\,\Gamma(n + r)/\Gamma(n)$, $r = 1, 2, \ldots$, we have[33]

$$E\alpha = \frac{1}{a} = \frac{1}{ln(G/A)}, \tag{2.46}$$

which is an optimal point estimate for a quadratic loss function in the sense of minimizing posterior expected loss.

If we have a new sample of q independent observations, each with a pdf in the Pareto form (2.41), we can use the posterior pdf in (2.45) as a prior pdf in the analysis of the new sample; that is, the likelihood function for the new sample, denoted by $\mathbf{y}_*$, is

$$l(\alpha|A, \mathbf{y}_*) \propto \frac{\alpha^q A^{q\alpha}}{G_*^{\,q(\alpha+1)}}, \tag{2.47}$$

[31] This form for the diffuse prior is in accord with Jeffreys' rule for a parameter that can assume values from 0 to ∞; see the appendix to this chapter. Also, the prior pdf for α in (2.43) is a Jeffreys' invariant prior pdf, since $|\text{Inf}_\alpha|^{1/2} = |-E(d^2 \log l/d\alpha^2)|^{1/2} \propto 1/\alpha$, where Inf_α is Fisher's information matrix.

[32] The computation of $\int_{c1}^{c2} p(\alpha|A, \mathbf{y})\, d\alpha$ can be done by using a numerical integration program; see Appendix C.

[33] It is interesting to note that $1/ln(G/A)$ is the maximum likelihood estimate for α.

where G_* is the geometric average of the q new observations. On combining (2.44) and (2.47), the posterior pdf, based on both samples, is

$$
\begin{aligned}
p(\alpha|A, \mathbf{y}, \mathbf{y}_*) &\propto \frac{\alpha^{n+q-1}A^{(n+q)\alpha}}{(G^n G_*{}^q)^{\alpha+1}} \\
&\propto \frac{\alpha^{n+q-1}A^{(n+q)\alpha}}{(G_2)^{(n+q)\alpha}} \qquad (2.48)\\
&\propto \alpha^{n+q-1}e^{-a_2(n+q)\alpha},
\end{aligned}
$$

where G_2 is the geometric mean of the $n + q$ observations in the two samples and $a_2 = ln(G_2/A)$. It is seen that (2.48) is in the same gamma form as (2.44) and thus can be easily analyzed.

Often in analyses of the Pareto distribution the available data are not the individual observations $y_1, y_2, \ldots, y_n$ but are in the form of frequencies, $n_0, n_1, \ldots, n_T$, where n_t is the number of individuals whose y values, say incomes, fall in a particular income interval, say x_t to x_{t+1}, where $x_{t+1} > x_t$, $x_0 = A$, $x_{T+1} = \infty$, and $t = 0, 1, 2, \ldots, T-1, T$. From the Pareto pdf in (2.41) the probability that an individual chosen at random will have a y value such that $x_t < y < x_{t+1}$ is

$$
\Pr\{x_t < y < x_{t+1}\} = \int_{x_t}^{x_{t+1}} \frac{\alpha A^\alpha}{y^{\alpha+1}}\,dy = A^\alpha\left(\frac{1}{x_t^\alpha} - \frac{1}{x_{t+1}^\alpha}\right)
$$

for $t = 0, 1, 2, \ldots, T-1$. For the interval $x_T < y < \infty$ the $\Pr\{x_T < y < \infty\} = A^\alpha/x_T{}^\alpha$. Then, given N individuals selected at random, the probability that n_t individuals have y values in x_t to x_{t+1} for $t = 0, 1, 2, \ldots, T-1$ and n_T have y values in the interval x_T to ∞ is[34]

$$
\frac{N!}{\prod_{t=0}^T n_t!}\,\frac{A^{\alpha n_T}}{x_T^{\alpha n_T}} \prod_{t=0}^{T-1} A^{\alpha n_t}\left(\frac{1}{x_t^\alpha} - \frac{1}{x_{t+1}^\alpha}\right)^{n_t},
$$

where $N = \sum_{t=0}^T n_t$. This is a pdf for the random n_t's, which, when viewed as a function of the unknown parameter α, is the likelihood function. It can be expressed more compactly as

$$
\begin{aligned}
l(\alpha|A, \mathbf{n}, N) &\propto \frac{A^{\alpha N}}{(\prod_{t=0}^T x_t^{n_t})^\alpha} \prod_{t=0}^{T-1}\left[1 - \left(\frac{x_t}{x_{t+1}}\right)^\alpha\right]^{n_t} \qquad (2.49)\\
&\propto e^{-aN\alpha} \prod_{t=0}^{T-1}\left[1 - \left(\frac{x_t}{x_{t+1}}\right)^\alpha\right]^{n_t}
\end{aligned}
$$

where $a = lnG/A$ with $G = (\prod_{t=0}^T x_t^{n_t})^{1/N}$ and $\mathbf{n}' = (n_0, n_1, \ldots, n_T)$.

[34] This expression, based on the multinomial distribution, is given in D. J. Aigner and A. S. Goldberger, "On the Estimation of Pareto's Law," Workshop Paper 6818, Social Systems Research Institute, University of Wisconsin, Madison.

Given a prior pdf for α, say $p(\alpha)$, it can be combined with (2.49) to yield the following posterior pdf

$$p(\alpha|A, \mathbf{n}, N) \propto p(\alpha)e^{-aN\alpha} \prod_{t=0}^{T-1} \left[1 - \left(\frac{x_t}{x_{t+1}}\right)^{\alpha}\right]^{n_t}. \tag{2.50}$$

If little prior information is available, the prior pdf could be taken as shown in (2.43). If more prior information about the value of α is available, $p(\alpha)$ can be taken in a form to represent it. In either case the posterior pdf in (2.50) can be normalized and analyzed by using numerical integration techniques[35]; for example, given a loss function $L(\alpha, \hat{\alpha})$, the value of $\hat{\alpha}$ which minimizes the posterior expectation of the loss function can be obtained numerically by evaluating $EL(\alpha, \hat{\alpha})$ for different values of $\hat{\alpha}$. Also, posterior intervals can be obtained by using numerical integration techniques.

To provide an application of these results for grouped data, we employ the following U.S. data[36] for 1961 which relate to $N = 1004$ households with incomes of $A = \$10{,}000$ or more.

As regards prior assumptions about the parameter α in (2.50), we assume that we know little about this parameter and represent it by taking $\log \alpha$ uniformly distributed which implies $p(\alpha) \propto 1/\alpha$, $0 < \alpha < \infty$. With this prior pdf inserted in (2.50) and using the data numerical integration procedures were employed to obtain the following normalized posterior pdf for α:

$$p(\alpha|A, \mathbf{n}, N) = k\alpha^{-1}e^{-aN\alpha} \prod_{t=0}^{T-1} \left[1 - \left(\frac{x_t}{x_{t+1}}\right)^{\alpha}\right]^{n_t}$$

Table 2.1 FREQUENCY DISTRIBUTION OF HOUSEHOLDS WITH INCOMES OF $10,000 OR GREATER, UNITED STATES, 1961

Income Interval (dollars)	Relative Frequency n_t/N	Absolute Frequency n_t	t	X_t (10^4 dollars)
10,000–14,999	0.170319	171	0	1
15,000–24,999	0.221116	222	1	1.5
25,000–49,999	0.159363	160	2	2.5
50,000–99,999	0.219124	220	3	5.0
100,000–149,999	0.0478088	48	4	10.0
150,000–499,999	0.137450	138	5	15.0
500,000–	0.0448207	45	6	50.0
Totals	1.000	$N = 1004$		

[35] See Appendix C.

[36] These data are presented in R. Barlow, H. Brazer and J. N. Morgan, *Economic Behavior of the Affluent*. Washington, D.C.: Brookings Institution, 1966, p. 193, Table D-1. These same data have been analyzed with various sampling theory approaches by Aigner and Goldberger, *loc. cit.*

with k being the normalizing constant and $a = lnG/A$ with $G = (\prod_{t=0}^{T} x_t^{n_t})^{1/N}$. In Figure 2.2 a plot of this posterior pdf is provided. The posterior mean and

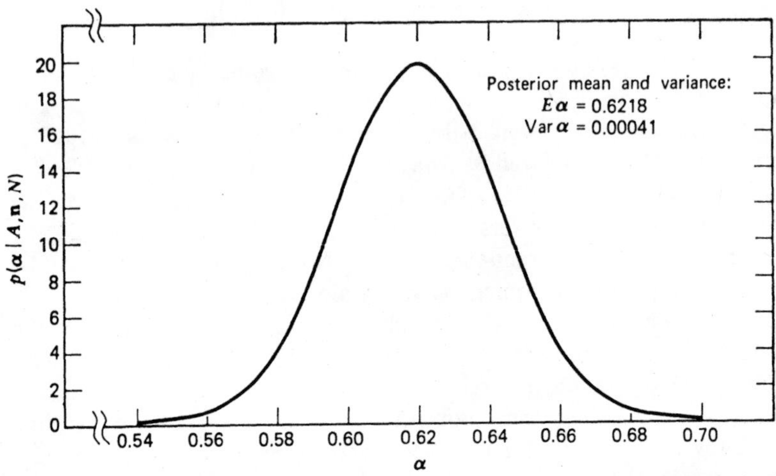

Figure 2.2 Posterior distribution for parameter of Pareto distribution derived from grouped data.

variance, computed by numerical integration, are $E\alpha = 0.6218$ and $\text{Var } \alpha = 0.00041$. Of course, for a quadratic loss function the optimal point estimate is the posterior mean.[37] If loss functions other than the quadratic are deemed more appropriate by an investigator, he can readily compute estimates which minimize expected loss for any particular loss function provided that the mathematical expectation of the loss function and a minimum exist. Further in this instance in which the sample size is rather large, the form of the posterior pdf resembles that of a normal pdf.

2.13 APPLICATION OF PRINCIPLES TO ANALYSIS OF THE BINOMIAL DISTRIBUTION

Consider the outcomes of n independent events, each of which can occur in one of two mutually exclusive ways, say A and B; for example, the n events may be n independent tosses of a two-sided coin. The outcome on each toss is either a head (A) or a tail (B). Let θ be the probability, assumed to be the same for all events, that A will occur and $1 - \theta$, the probability that B will occur. Then the probability of observing n_1 A's and $n - n_1$ B's in n independent events is given by the discrete binomial pdf

[37] For this set of data the maximum likelihood estimate is 0.6218, numerically the same as the posterior mean.

$$p(n_1|\theta, n) = \binom{n}{n_1}\theta^{n_1}(1 - \theta)^{n-n_1}, \tag{2.51}$$

where

$$\binom{n}{n_1} = \frac{n!}{n_1!\,(n - n_1)!}$$

with $0! \equiv 1$. The function in (2.51), viewed as a function of the unknown parameter θ, is the likelihood function.

Suppose that we have some prior information about θ and can represent it by the following beta pdf[38]

$$p(\theta) = k\theta^{a-1}(1 - \theta)^{b-1}, \qquad \begin{array}{l} a, b > 0, \\ 0 \leq \theta \leq 1, \end{array} \tag{2.52}$$

where $k = \Gamma(a + b)/\Gamma(a)\,\Gamma(b)$ is the normalizing constant and a and b are prior parameters whose values represent our prior information about θ. In assigning values to a and b, note that, for the beta pdf, $E\theta = a/(a + b)$ and $\text{Var } \theta = ab/(a + b)^2(a + b + 1)$. Given that a and b have been assigned suitable values to represent the available prior information about θ, (2.52) can be combined with the likelihood function in (2.51) to yield the posterior pdf for θ,

$$p(\theta|n_1, n) \propto \theta^{n_1+a-1}(1 - \theta)^{n-n_1+b-1}, \tag{2.53}$$

which is in the beta form (2.52) with parameters $a' = n_1 + a$ and $b' = n - n_1 + b$. The normalizing constant for (2.53) is $\Gamma(a' + b')/\Gamma(a')\,\Gamma(b')$, whereas the posterior mean is $a'/(a' + b')$ and the posterior modal value is $\theta_{\text{mod}} = (a' - 1)/(a' + b' - 2)$. Posterior probabilities, for example $\Pr(c < \theta < d)$, where c and d are given numbers, can be easily evaluated by numerical integration or by use of available tables of the incomplete beta function.[39] Also, as is well known, a random variable with a beta pdf can be transformed to a variable with an F-distribution. In the present instance, if we let $x = (b'/a')\theta/(1 - \theta)$, a posteriori x has an F distribution with $2a'$ and $2b'$ degrees of freedom.[40] This is a useful fact[41] for making posterior probability statements about $\theta/(1 - \theta)$, the odds relating to the events A and B.

[38] See Appendix A for properties of the beta pdf.

[39] See, for example, K. Pearson, *Tables of the Incomplete Beta Function*, Cambridge: Cambridge University Press, 1948.

[40] In general, if θ has a beta pdf, $p(\theta)\,d\theta \propto \theta^{a-1}(1 - \theta)^{b-1}\,d\theta$, then the pdf for $x = (b/a)\theta/(1 - \theta)$ is $p(x) \propto x^{a-1}/(b + ax)^{a+b} \propto x^{2a/2-1}/[1 + (2a/2b)x]^{(2a+2b)/2}$, which is in the $F_{2a,2b}$ form; see Appendix A. In deriving this expression for $p(x)$, note that $\theta = x/(x + b/a)$, $(1 - \theta) = (b/a)(x + b/a)^{-1}$ and $d\theta \propto dx/(x + b/a)^2$.

[41] That is, tables of the F-distribution can be consulted to obtain probabilities. Alternatively, it is easy to use numerical integration techniques to compute probabilities; for example the probability that x will lie in a given interval.

When our prior information about θ is diffuse or vague, there has been considerable discussion in the literature about the form of the prior pdf to represent knowing little about the value of θ.[42] We shall adopt the following improper pdf to represent vague knowledge about the value of θ:

$$p(\theta) \propto \frac{1}{\theta(1-\theta)}, \qquad 0 \leq \theta \leq 1. \tag{2.54}$$

This prior can be viewed as a limiting form of the prior in (2.52) as both a and b approach zero. Alternatively, as Jeffreys and Lindley[43] point out, $\eta = \theta/(1 - \theta)$ ranges from 0 to ∞, and if we take $v = \log \eta$ uniformly distributed over the entire real line this implies (2.54) as the prior for θ.[44]

On combining the prior pdf in (2.54) with the likelihood function in (2.51), the following is the posterior pdf for θ:[45]

$$p(\theta|n_1, n) \propto \theta^{n_1-1}(1-\theta)^{n-n_1-1}, \qquad 0 \leq \theta \leq 1, \tag{2.55}$$

which is in the beta form with parameters n_1 and $n - n_1$. The posterior pdf will be proper for $n_1 > 0$ and $n - n_1 > 0$; that is, if our sample information includes at least one occurrence of the event A and one of the event B. With this condition satisfied, the posterior mean of θ is $E(\theta|n_1, n) = n_1/n$, the sample proportion, and the posterior variance is $\text{Var}(\theta|n_1, n) = n_1(n - n_1)/n^2(n + 1)$. Further, given that we have the posterior pdf in (2.55), the posterior probability that θ will lie in any given interval can be readily computed.

2.14 REPORTING THE RESULTS OF BAYESIAN ANALYSES

In reporting the results of Bayesian analyses involving estimation of parameters in scientific journals, it is important to provide at least (a) a detailed discussion of the stochastic model assumed to generate the observations, (b) a full discussion of prior assumptions about parameter values, (c) the sample information, and (d) information about posterior pdf's for parameters of interest.

With respect to the stochastic model for the observations, subject matter considerations should be reviewed to justify its form and stochastic assumptions. Given that this has been done satisfactorily, the likelihood function $p(\mathbf{y}|\boldsymbol{\theta})$ should be shown explicitly, where $\mathbf{y}$ is an observation vector and $\boldsymbol{\theta}$ is a parameter vector.

[42] See, for example, Jeffreys, *Theory of Probability*, *op. cit.*, pp. 123–125.

[43] Lindley, *op. cit.*, p. 145.

[44] That is, from $v = \log \eta = \log \theta/(1 - \theta)$, $dv = d \log \theta - d \log (1 - \theta) = d\theta/\theta(1 - \theta)$ and thus $p(v)\, dv \propto dv$ implies $p(\theta)\, d\theta \propto [\theta(1 - \theta)]^{-1}\, d\theta$.

[45] If, instead of (2.54), we had used the Bayes-Laplace uniform prior, $p(\theta) \propto$ constant, $0 \leq \theta \leq 1$, the exponents in (2.55) would each be changed by one, the equivalent of two sample observations, which will not be important in moderate-sized samples.

As regards prior assumptions about $\boldsymbol{\theta}$, that is, a choice of prior pdf for $\boldsymbol{\theta}$, all information used to make such a choice should be explicitly stated. If data-based prior information is being employed, this fact should be noted and references provided to the sources of such prior information. If nondata-based prior information is employed, it should be carefully examined and explicated. In this way the reader will understand what information is being added to the sample information in performing an analysis. Of course, if little prior information is available or if the investigator wishes to show what results from an analysis assuming little prior information, he will use a vague or diffuse prior pdf.

With respect to reporting the data employed in an analysis, it is good procedure to describe in detail how they were obtained and have them available for any interested party by including them in the report or by making it known that they can be obtained on request. By having the data available other parties can perform analyses using whatever prior pdf's they choose to use. Also, should there be any controversy about the form of the likelihood function, the data can be employed to explore alternative formulations.[46]

With respect to reporting information about posterior pdf's for parameters of interest, it is good practice to report the complete posterior pdf and to provide summary characteristics, say measures of central tendency and dispersion. Also posterior intervals (or regions) often help readers to appreciate what the prior and sample information implies about the values of parameters.

By paying special attention to the above points, readers will understand how the reporting investigator learned from the information in his sample[47]; that is, he will have information regarding the investigator's initial beliefs about the parameters and model and can then see how they are altered by the data. This change in beliefs is indeed an essential part of the process of learning from experience.

APPENDIX PRIOR DISTRIBUTIONS TO REPRESENT "KNOWING LITTLE"

As stated in the body of this chapter, there are situations in which investigators know little, or wish to proceed as if they knew little, about parameters

[46] In some cases $p(\mathbf{y}|\boldsymbol{\theta}) = g(t_1, t_2, \ldots, t_k|\theta)\, h(\mathbf{y})$, where $t_i = t_i(\mathbf{y})$, $i = 1, 2, \ldots, k$, are functions of the observations called sufficient statistics. Reporting just the t_i's allows others to perform their own analyses, provided that they accept the form of the likelihood function as being satisfactory. To investigate alternative likelihood functions the complete sample $\mathbf{y}$ is usually required.

[47] See C. Hildreth, "Bayesian Statisticians and Remote Clients," *Econometrica*, **31**, 422–438 (1963), for further consideration of the problem of reporting.

of a model. Thus there is a need for explicit rules for selecting prior distributions to represent "knowing little" or ignorance. Perhaps, surprisingly, filling this need has been a difficult and controversial aspect of the Bayesian approach to inference. In this appendix we consider some approaches that have been put forward to deal with this problem.

When a parameter's value is completely unknown, Jeffreys[48] suggests two rules for choosing a prior distribution, which, according to him, ". . . cover the commonest cases." He states that, "If the parameter may have any value in a finite range, or from $-\infty$ to $+\infty$, its prior probability should be taken as uniformly distributed. If it arises in such a way that it may conceivably have any value from 0 to ∞, the prior probability of its logarithm should be taken as uniformly distributed."[49]

Let us consider application of Jeffreys' first rule to the case of an unknown parameter, μ, say a mean, which can conceivably assume values from $-\infty$ to $+\infty$. Jeffreys' prescription for representing ignorance about the value of μ is to take

$$(1) \qquad p(\mu)\, d\mu \propto d\mu, \qquad -\infty < \mu < \infty;$$

that is, $p(\mu) \propto$ constant. This rectangular pdf is obviously improper since $\int_{-\infty}^{\infty} p(\mu)\, d\mu = \infty$. Given that we know $-\infty < \mu < \infty$ to be a certain statement, this means that Jeffreys is using ∞ to represent the probability of the certain event, $-\infty < \mu < \infty$, rather than 1.[50] The fact that (1) integrates to ∞ is a virtue from Jeffreys' point of view since then $\Pr\{a < \mu < b\}/\Pr\{c < \mu < d\} = 0/0$ is indeterminate, where a, b, c, and d are any finite numbers.[51] Since this ratio of probabilities is indeterminate, we can make no statement about the odds that μ lies in any particular pair of finite intervals. Jeffreys views this property of (1) as a formal representation of ignorance.

[48] Jeffreys, *op. cit.*, p. 117. Plackett regards Jeffreys' work as "an authoritative modern version of the Bayes-Laplace procedure" for representing ignorance. See R. L. Plackett, "Current Trends in Statistical Inference," *J. Roy. Statist. Soc.*, Series A, Part 2, 249–267, p. 251 (1966).

[49] *Ibid.*, p. 117.

[50] Jeffreys, *ibid.*, p. 21, remarks, ". . . there are cases where we wish to express ignorance over an infinite range of values of a quantity, and it may be convenient to express certainty that the quantity lies in that range by ∞ (rather than 1). . . ."

[51] It may appear that $\Pr\{a < \mu < b\} = 0$ is an informative statement about μ. This statement, however, does not logically imply that μ is outside the interval a to b with certainty given that μ is uniformly distributed, $-\infty$ to ∞. Jeffreys (p. 21) provides an example to illustrate this point. If x is a continuous random variable with a uniform pdf from 0 to 1, the fact that $\Pr(x = \frac{1}{2}) = 0$ does not logically imply that $x \neq \frac{1}{2}$ with certainty, since $\frac{1}{2}$ is a possible value of x.

If, in place of (1), we had taken, say,

$$p(\mu)\,d\mu = \frac{1}{2M}\,d\mu, \qquad -M \le \mu \le M, \tag{2}$$

a proper pdf, we would have introduced prior information about the range of μ and thus would not be completely ignorant about the value of μ. With (2) we have for finite a and b in the closed interval $-M$ to M, $\Pr\{a < \mu < b\} = (b - a)/2M \neq 0$, and thus

$$\frac{\Pr\{a < \mu < b\}}{\Pr\{c < \mu < d\}} = \frac{(b - a)/2M}{(d - c)/2M} = \frac{b - a}{d - c}, \tag{3}$$

where c and d are finite and in the closed interval $-M$ to M. The ratio of probabilities in (3) is determinate in contrast to what follows from (1). However, if we consider $\lim_{M\to\infty} \Pr\{a < \mu < b\}/\lim_{M\to\infty} \Pr\{c < \mu < d\}$, this ratio of limiting probabilities is in the form 0/0. In this sense we can regard (2) as an approximation to (1) as M grows large.

It may be asked if we are incorporating information about μ by the choice of the rectangular form of the pdf in (1) or (2). By Jeffreys' line of reasoning the indeterminancy of the ratios $\Pr\{a < \mu < b\}/\Pr\{c < \mu < d\}$ seems adequate to justify the use of the rectangular pdf. Other improper pdf's, however, have this property. There appears to be no way of answering this question about the form of the pdf to express ignorance without introducing a measure of information. If we agree to measure information in a pdf, for example $p(\mu)$, by

$$H = \int_{-M}^{M} p(\mu) \log p(\mu)\,d\mu, \tag{4}$$

a measure employed by many, including Shannon,[52] the proper pdf which minimizes H is the one shown in (2).[53] Thus the rectangular pdf is a "minimal information" prior pdf. By letting M get large we get an approximation to (1).

[52] See, for example, C. E. Shannon, "The Mathematical Theory of Communication," *Bell System Tech. J.* (July–October 1948), reprinted in C. E. Shannon and W. Weaver, *The Mathematical Theory of Communication*. Urbana: University of Illinois Press, 1949, pp. 3–91. Shannon defines $W = -H$ as the entropy or uncertainty associated with a pdf, say $p(\mu)$.

[53] That is, minimize $H = \int_{-M}^{M} p(\mu) \log p(\mu)\,d\mu$ subject to $\int_{-M}^{M} p(\mu)\,d\mu = 1$. Form the Lagrangian expression $H + \lambda[\int_{-M}^{M} p(\mu)\,d\mu - 1]$, where λ is a Lagrange multiplier. Then, on varying $p(\mu)$, we have as the condition for H to be minimized subject to the condition, $[1 + \log p(\mu) + \lambda]\,\delta p(\mu) = 0$. Thus $p(\mu) = e^{-(\lambda+1)}$. On taking $\lambda + 1 = \log 2M$, $p(\mu) = 1/2M$ is the proper pdf which minimizes H.

Example 2.3 in the text of this chapter illustrates that on combining the improper pdf in (1) with a likelihood function by use of Bayes' theorem the resulting posterior pdf is proper. The sample information in the likelihood function has in this case taken us from an improper, "noninformative" pdf for μ to a proper, "informative" posterior pdf. In this way we have moved from ignorance, as represented by (1), to a more informed position, as represented by our proper posterior pdf; for example, a posteriori, $\Pr\{a < \mu < b|\mathbf{y}\}/\Pr\{c < \mu < d|\mathbf{y}\}$ is not indeterminate.

The second rule that Jeffreys gives pertains to parameters, which by their nature, can assume values from 0 to ∞; for example a standard deviation σ. For such a parameter Jeffreys suggests taking its logarithm uniform; that is, if we let $\theta = \log \sigma$, the prior pdf for θ is taken as follows:

$$p(\theta)\, d\theta \propto d\theta, \qquad -\infty < \theta < \infty. \tag{5}$$

With $\theta = \log \sigma$, note that θ's range is $-\infty$ to ∞ and thus (5) is consistent with Jeffreys' first rule. Since $d\theta = d\sigma/\sigma$, (5) implies

$$p(\sigma)\, d\sigma \propto \frac{d\sigma}{\sigma}, \qquad 0 < \sigma < \infty, \tag{6}$$

as the improper pdf to represent ignorance about σ.

Jeffreys presents several interesting observations about (6) and possible alternatives to it. First, (6) is invariant to transformations of the form $\phi = \sigma^n$; that is, $d\phi = n\sigma^{n-1}\, d\sigma$ and thus $d\phi/\phi \propto d\sigma/\sigma$. For Jeffreys this invariance property is important because, for example, some parameterize a model in terms of the standard deviation σ and others, in terms of the variance σ^2, or a precision parameter, $h = 1/\sigma^2$. As can easily be checked, if we take $d\sigma/\sigma$ as our prior for σ, this logically implies $d\sigma/\sigma \propto d\sigma^2/\sigma^2 \propto dh/h$. Thus by applying Jeffreys' rule to σ, $0 < \sigma < \infty$, to σ^2, $0 < \sigma^2 < \infty$, or to h, $0 < h < \infty$, provides prior pdf's in the same form and consistent with one another in the sense that $d\sigma/\sigma \propto d\sigma^2/\sigma^2 \propto dh/h$ is satisfied. Further, posterior probability statements, based on the alternative parameters, will be consistent.

Next, Jeffreys points out that (6) has the following properties: (a) $\int_0^\infty d\sigma/\sigma = \infty$; (b) $\int_0^a d\sigma/\sigma = \infty$; and (c) $\int_a^\infty d\sigma/\sigma = \infty$. Property (a) indicates that again ∞ is being used to represent certainty. Then (b) and (c) together imply that $\Pr\{0 < \sigma < a\}/\Pr\{a < \sigma < \infty\}$ is indeterminate and thus nothing can be said about the ratio of these two probabilities; that is, the odds pertaining to the propositions $0 < \sigma < a$ and $a < \sigma < \infty$. Again this indeterminacy is regarded as a formal representation of ignorance.

As alternatives to (6), Jeffreys considers $p(\sigma) \propto$ constant and $p(\sigma) \propto e^{-k\sigma}$ with $0 < \sigma < \infty$. The first of these pdf's has the property that the probability that $\sigma > c$, where c is positive and finite, is ∞, which is certainty on Jeffreys' scale. Thus the probability that $0 < \sigma < c$ is zero, which implies that we know something about σ. Therefore Jeffreys considers $p(\sigma) \propto$ constant, $0 < \sigma < \infty$, as an unacceptable representation of ignorance about the value of σ. With respect to $p(\sigma) \propto e^{-k\sigma}$ Jeffreys notes that a factor k must be introduced in the exponent, since σ has the dimensions of length and the exponent of e must be a pure number. Also, with a pdf in the form $p(\sigma) \propto e^{-k\sigma}$, $\Pr\{0 < \sigma < c\}/\Pr\{c < \sigma < \infty\}$, for finite positive c, has a finite determinate value that contradicts the premise that we know nothing about the value of σ. Further, if k were unknown, we should have to introduce a prior pdf for it so that no progress would have been made.

We have noted that $\theta = \log \sigma$ is a parameter such that $-\infty < \theta < \infty$, given that $0 < \sigma < \infty$. Then the information measure in (4) will be minimized by taking $p(\theta) \propto$ constant, and this is an information theoretic justification for taking $\theta = \log \sigma$ uniformly distributed, which implies (6).[54] In general, if $\eta = f(\sigma)$, where f is differentiable, $f(0) = -\infty$, and $f(\infty) = \infty$, we could take $p(\eta)\, d\eta \propto d\eta$, in accord with Jeffreys' first rule. This implies $p(\sigma)\, d\sigma \propto f'(\sigma)\, d\sigma$. If $f(\sigma) = \log \sigma$, we get Jeffreys' prior pdf, shown in (6). Jeffreys wants $f(\sigma)$ to be such that it involves no new parameters. This condition would rule out, for example, $f(\sigma) = e^{k\sigma}(1 - 1/\sigma)$. Jeffreys states that $f'(\sigma)$ must be of the form $A\sigma^n$, where A and n are constants, if we want to express ignorance about σ, given only the knowledge that $0 < \sigma < \infty$. He further argues that only by taking $n = -1$ will the ratio $\Pr\{0 < \sigma < a\}/\Pr\{a < \sigma < \infty\}$ be indeterminate. This result, in addition to the property of "invariance with respect to powers," is for him a compelling reason to take (6) as the prior pdf for σ representing ignorance.

In Example 2.3 in the body of this chapter we have seen how the improper prior pdf, $p(\sigma) \propto 1/\sigma$, combines with a likelihood function to yield a proper posterior pdf for σ. As mentioned above, $p(\sigma) \propto 1/\sigma$ implies $p(\sigma^2) \propto 1/\sigma^2$, and thus we use the same form for the pdf to express ignorance about σ^2 (or any power of σ). Thus, if one investigator uses the parameter σ while another uses $\phi = \sigma^2$, given that their priors to represent ignorance are $p(\sigma) \propto 1/\sigma$ and $p(\phi) \propto 1/\phi$, respectively, their posterior probabilities relating to σ and ϕ will be consistent.

It seems that some are hesitant to employ the improper pdf's recommended by Jeffreys. Rather they introduce "locally uniform" or "gentle"

[54] The pdf for σ which minimizes H is assumed to be proper. Thus, if $v_1 < \sigma < v_2$, the proper pdf minimizing H is $p(\sigma) = (\log v_2/v_1)^{-1}(d\sigma/\sigma)$. As $v_2 \to \infty$ and $v_1 \to 0$, we get an approximation to Jeffreys' improper pdf. On this point Jeffreys, *op. cit.*, p. 122, comments ". . . (1) an intermediate range contributes most of the values of the various integrals in any case, and the termini would make a negligible contribution; (2) in the mathematical definition of an infinite integral a finite range is always considered in the first place and then allowed to tend to infinity. Thus the results derived by using infinite integrals are just equivalent to making v_1 [in $p(\sigma) = (\log v_2/v_1)^{-1}(d\sigma/\sigma)$] tend to 0 and v_2 to infinity."

prior pdf's for unknown parameters.[55] These terms are employed for prior pdf's which are "reasonably flat" or "gentle" over the range in which the likelihood function assumes appreciable values. Outside this range it matters little what shape the prior pdf has, since, in deriving the posterior pdf, the prior pdf gets multiplied by small likelihood values. The situation is as shown in Figure 2.3.

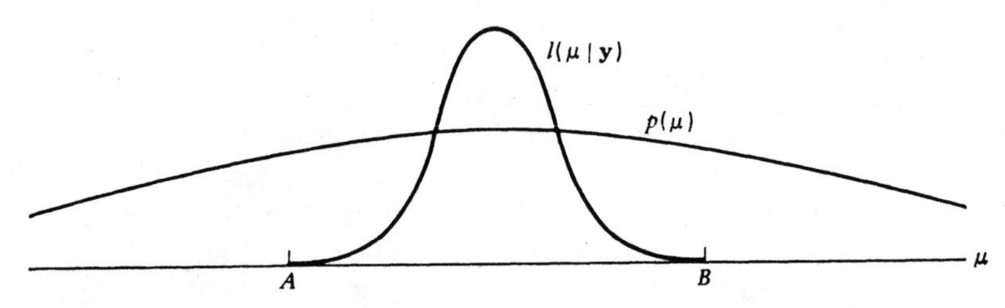

Figure 2.3 Example of a "locally uniform" prior pdf, $p(\mu)$.

Since the posterior pdf $\propto$ prior pdf $\times$ likelihood function, it is clear that the shape of the prior pdf to the left of the point A and to the right of the point B will have little influence on the shape of the posterior pdf. Analytically, for a parameter μ, the posterior pdf $p(\mu|\mathbf{y})$ is given by

$$(7) \qquad p(\mu|\mathbf{y}) \propto p(\mu) l(\mu|\mathbf{y}).$$

Let μ_0 be a value of μ located in the region in which $l(\mu|\mathbf{y})$ assumes appreciable values. In many examples μ_0 can be taken as the modal value of $l(\mu|\mathbf{y})$. Then expand $p(\mu)$ as follows

$$(8) \qquad p(\mu) = p(\mu_0) + (\mu - \mu_0)\, p'(\mu_0) + \tfrac{1}{2}(\mu - \mu_0)^2 p''(\mu_0) + \cdots.$$

If the first-order and higher order terms in this expansion are negligible in the region in which the likelihood function assumes appreciable values, as would be the case if $p(\mu)$ were flat or "gentle" in this region, we have

$$p(\mu|\mathbf{y}) \mathrel{\dot{\propto}} l(\mu|\mathbf{y}),$$

where $\dot{\propto}$ denotes "approximately proportional to."

By taking $p(\mu)$ "locally uniform" and *proper* it is clear that Jeffreys' condition for complete ignorance is not being satisfied [see the discussion in connection with (2)]. Also, this procedure for choosing a prior pdf implies that we know something about the likelihood function, which may or may not

[55] See, for example, L. J. Savage, "Bayesian Statistics," in *Decision and Information Processes*. New York: Macmillan, 1962. G. E. P. Box and G. C. Tiao, "A Further Look at Robustness via Bayes Theorem," *Biometrika*, **49**, 419–433 (1962).

be the situation in practical cases. If this information about the range of μ and the likelihood function is available, and some claim that it usually is, it can be used to good advantage; for example, if it is known that $-M < \mu < M$ and that the design of an experiment is such that the likelihood function can assume appreciable values for $-2M < \mu < 2M$, it is apparent that use of a prior $p(\mu)\,d\mu \propto d\mu$ for $-\infty < \mu < \infty$ can lead to inferences at variance with the prior information $-M < \mu < M$. Thus, when an investigator knows something about μ and the experimental design, it is, of course, important that this information be considered in making inferences about μ. When such information is not available, it usually makes very little practical difference whether we use a "locally uniform" prior pdf or Jeffreys' improper pdf for μ.

In the discussion above we have encountered several examples of invariance properties of prior pdf's; for example, it was noted that $p(\sigma)\,d\sigma \propto d\sigma/\sigma$ is invariant with respect to powers of σ. Jeffreys has provided a remarkable generalization of this invariance property. He points out[56] that if our prior pdf for the parameter vector $\boldsymbol{\theta}$ is taken as

$$p(\boldsymbol{\theta}) \propto |\text{Inf}_{\boldsymbol{\theta}}|^{1/2}, \tag{9}$$

where $\text{Inf}_{\boldsymbol{\theta}}$ is Fisher's information matrix for the parameter vector $\boldsymbol{\theta}$, that is,

$$\text{Inf}_{\boldsymbol{\theta}} = -E\left[\frac{\partial^2 \log p(\mathbf{y}|\boldsymbol{\theta})}{\partial\theta_i\,\partial\theta_j}\right], \tag{10}$$

where the expectation E is with respect to the pdf for $\mathbf{y}$, $p(\mathbf{y}|\boldsymbol{\theta})$, the prior pdf in (9) will be invariant in the following sense. If an investigator parametrizes his model in terms of the elements of $\boldsymbol{\eta}$, where $\boldsymbol{\eta} = F(\boldsymbol{\theta})$ with F a one-to-one differentiable transformation of the elements of $\boldsymbol{\theta}$, and takes his prior pdf for $\boldsymbol{\eta}$ as

$$p(\boldsymbol{\eta}) \propto |\text{Inf}_{\boldsymbol{\eta}}|^{1/2}, \tag{11}$$

his posterior probability statements will be consistent with those of a person using the parameter vector $\boldsymbol{\theta}$ in conjunction with the prior pdf in (9). A proof of this property follows.

Let $p(\mathbf{y}|\boldsymbol{\theta})$, where $\boldsymbol{\theta}' = (\theta_1, \theta_2, \ldots, \theta_k)$, be a pdf for the observation vector $\mathbf{y}$. The information matrix for $\boldsymbol{\theta}$, given in (10), can alternatively be expressed as

$$\text{Inf}_{\boldsymbol{\theta}} = E\left(\frac{\partial \log p}{\partial\theta_i}\frac{\partial \log p}{\partial\theta_j}\right), \qquad i, j = 1, 2, \ldots, k, \tag{12}$$

where we have written p for $p(\mathbf{y}|\boldsymbol{\theta})$ and the expectation E is taken with respect to $\mathbf{y}$. Let $\boldsymbol{\eta} = F(\boldsymbol{\theta})$ be any one-to-one differentiable transformation

[56] Jeffreys, *op. cit.*, p. 179 ff.

of the elements of $\boldsymbol{\theta}$; that is, $\eta_1 = f_1(\boldsymbol{\theta})$, $\eta_2 = f_2(\boldsymbol{\theta}), \ldots, \eta_k = f_k(\boldsymbol{\theta})$. Then we have

$$|\text{Inf}_{\boldsymbol{\theta}}|^{1/2}\, d\boldsymbol{\theta} = |\text{Inf}_{\boldsymbol{\eta}}|^{1/2}\, d\boldsymbol{\eta}. \tag{13}$$

Proof.[57] Since

$$\partial \log p/\partial\theta_i = \sum_{j=1}^{k} \frac{\partial \log p}{\partial\eta_j}\frac{\partial\eta_j}{\partial\theta_i},$$

the (i, j)th element of $\text{Inf}_{\boldsymbol{\theta}}$, $\text{Inf}_{\boldsymbol{\theta},i,j}$, can be expressed as

$$\text{Inf}_{\boldsymbol{\theta},i,j} = \sum_{r=1}^{k}\sum_{s=1}^{k} \frac{\partial\eta_r}{\partial\theta_i}(\text{Inf}_{\boldsymbol{\eta},r,s})\frac{\partial\eta_s}{\partial\theta_j},$$

where

$$I_{\boldsymbol{\eta},r,s} = E\left(\frac{\partial \log p}{\partial\eta_r}\frac{\partial \log p}{\partial\eta_s}\right).$$

Then

$$\text{Inf}_{\boldsymbol{\theta}} = (\text{Inf}_{\boldsymbol{\theta},i,j}) = J\,\text{Inf}_{\boldsymbol{\eta}}J' \tag{14a}$$

and

$$|\text{Inf}_{\boldsymbol{\theta}}|^{1/2} = |J|\,|\text{Inf}_{\boldsymbol{\eta}}|^{1/2}, \tag{14b}$$

where J is the Jacobian matrix, associated with the transformation $\boldsymbol{\eta} = F(\boldsymbol{\theta})$, which has the typical element $\partial\eta_j/\partial\theta_i$. Noting that $|J|\, d\boldsymbol{\theta} = d\boldsymbol{\eta}$, we have from (14*b*)

$$|\text{Inf}_{\boldsymbol{\theta}}|^{1/2}\, d\boldsymbol{\theta} = |\text{Inf}_{\boldsymbol{\eta}}|^{1/2}\, d\boldsymbol{\eta},$$

which was to be shown.

The importance of this result is that if investigator A parameterizes a model in terms of $\boldsymbol{\theta}$ and uses $|\text{Inf}_{\boldsymbol{\theta}}|^{1/2}\, d\boldsymbol{\theta}$ as his prior pdf his posterior pdf is $p(\boldsymbol{\theta}|y)\, d\boldsymbol{\theta} \propto |\text{Inf}_{\boldsymbol{\theta}}|^{1/2}p(\mathbf{y}|\boldsymbol{\theta})\, d\boldsymbol{\theta}$, whereas if investigator B parameterizes the model in terms of $\boldsymbol{\eta} = F(\boldsymbol{\theta})$ his posterior pdf is $p(\boldsymbol{\eta}|\mathbf{y})\, d\boldsymbol{\eta} \propto |\text{Inf}_{\boldsymbol{\eta}}|^{1/2}p(\mathbf{y}|\boldsymbol{\eta})\, d\boldsymbol{\eta}$. Since (13) is true, B can employ $\boldsymbol{\eta} = F(\boldsymbol{\theta})$ to transform his posterior pdf to relate to $\boldsymbol{\theta}$ and he gets exactly the posterior pdf that A has obtained. Alternatively, A can use $\boldsymbol{\eta} = F(\boldsymbol{\theta})$ to express his posterior pdf in terms of $\boldsymbol{\eta}$ and, given (13), this posterior for $\boldsymbol{\eta}$ will be precisely B's. Thus, when investigators take their prior pdf's proportional to the square root of the information matrix, they are led to consistent posterior pdf's in the sense explained above.

To consider invariance properties more generally, it is useful to note that the Bayesian "transformation" of a given prior pdf, $p(\boldsymbol{\theta})$,

$$p(\boldsymbol{\theta}|\mathbf{y}) \propto p(\boldsymbol{\theta})\, p(\mathbf{y}|\boldsymbol{\theta}), \tag{15}$$

[57] This proof, essentially the same as Jeffreys', was presented in a lecture by M. Stone in 1965.

involves several components, namely, a pdf for $\mathbf{y}$, $p(\mathbf{y}|\boldsymbol{\theta})$, a parameter space Ω, and a sample space S. We shall write

$$\mathscr{F} = \{p(\mathbf{y}|\boldsymbol{\theta}),\ \theta \in \Omega \subset R^k, \text{ and } \mathbf{y} \in S \subset R^n\} \tag{16}$$

to represent collections of these quantities where Ω denotes an open subset in the k-dimensional Euclidean space R^k and S, an open subset in R^n. We assume that $\mathscr{F}$ includes just pdf's for $\mathbf{y}$, given $\boldsymbol{\theta}$, which have a continuous $\boldsymbol{\theta}$ derivative for all $\mathbf{y} \in S$. Hartigan[58] considers properties of the Bayesian transformation in (15) for various $\mathscr{F}$'s. He establishes that (15) has the following invariance properties when $p(\boldsymbol{\theta}) \propto |\text{Inf}_{\boldsymbol{\theta}}|^{1/2}$, that is, when the prior pdf is of the form suggested by Jeffreys:

I. *S-labeling invariance*: If $\mathbf{z} = G(\mathbf{y})$ is a differentiable one-to-one transformation which takes the sample space S for $\mathbf{y}$ into S^*, the sample space for $\mathbf{z}$, then

$$p(\boldsymbol{\theta}|\mathbf{z}) \propto p(\boldsymbol{\theta}|\mathbf{y}). \tag{17}$$

This property is particularly important, for example, when the transformation $\mathbf{z} = G(\mathbf{y})$ involves a change in the units of measurement.

II. *Ω-labeling invariance*: If $\boldsymbol{\eta} = F(\boldsymbol{\theta})$ is a differentiable one-to-one transformation of $\boldsymbol{\theta}$, then there exists $p^*(\mathbf{y}|\boldsymbol{\eta}) \equiv p(\mathbf{y}|\boldsymbol{\theta})$ and

$$p(\boldsymbol{\eta}|\mathbf{y})\, d\boldsymbol{\eta} \propto p(\boldsymbol{\theta}|\mathbf{y})\, d\boldsymbol{\theta}. \tag{18}$$

We have provided a proof of this property and commented on its importance.

III. *Ω-restriction invariance*: Assume that $\boldsymbol{\theta} \in \Omega^* \subset \Omega$. Then the posterior pdf based on $p^*(\mathbf{y}|\boldsymbol{\theta})$ with $\boldsymbol{\theta} \in \Omega^*$, that is, $p^*(\boldsymbol{\theta}|\mathbf{y}) \propto p(\boldsymbol{\theta})\, p^*(\mathbf{y}|\boldsymbol{\theta})$ will be proportional to $p(\boldsymbol{\theta}|\mathbf{y}) \propto p(\boldsymbol{\theta})\, p(\mathbf{y}|\boldsymbol{\theta})$ with $\boldsymbol{\theta} \in \Omega^*$.

This property implies that when we use Jeffreys' prior we get the same posterior if we work with the likelihood function defined for $\boldsymbol{\theta} \in \Omega^*$ that we would get if we worked with the likelihood function defined for $\boldsymbol{\theta} \in \Omega$ and then restricted the resulting posterior pdf $p(\boldsymbol{\theta}|\mathbf{y})$ to be zero outside the region $\boldsymbol{\theta} \in \Omega^*$.

IV. *Sufficiency invariance*: If $\mathbf{t}' = (t_1, t_2, \ldots, t_m)$ is sufficient for $\boldsymbol{\theta}$ and $p^*(\mathbf{t}|\boldsymbol{\theta})$ is the pdf for the sufficient statistics, then $p^*(\boldsymbol{\theta}|\mathbf{t}) \propto p(\boldsymbol{\theta}|\mathbf{y})$, where

$$p^*(\boldsymbol{\theta}|\mathbf{t}) \propto p^*(\boldsymbol{\theta})\, p^*(\mathbf{t}|\boldsymbol{\theta})$$

and $p(\boldsymbol{\theta}|\mathbf{y})$ is given by (15) when the priors are of Jeffreys' form.

[58] J. Hartigan, "Invariant Prior Distributions," *Ann. Math. Statist.*, **35**, 836–845 (1964).

V. *Direct product invariance*: If we have two independent samples, $\mathbf{y}_1$ and $\mathbf{y}_2$, with pdf's $p_1(\mathbf{y}_1|\boldsymbol{\theta}_1)$ and $p_2(\mathbf{y}_2|\boldsymbol{\theta}_2)$, respectively, $\boldsymbol{\theta}_1 \in \Omega_1$ and $\boldsymbol{\theta}_2 \in \Omega_2$, and $p(\mathbf{y}|\boldsymbol{\theta}) \propto p_1(\mathbf{y}_1|\boldsymbol{\theta}_1)\,p(\mathbf{y}_2|\boldsymbol{\theta}_2)$, where $\boldsymbol{\theta} \in \Omega = \Omega_1 \times \Omega_2$, then

$$p(\boldsymbol{\theta}|\mathbf{y}) \propto p_1(\boldsymbol{\theta}_1|\mathbf{y}_1)\,p_2(\boldsymbol{\theta}_2|\mathbf{y}_2),$$

where $p_i(\boldsymbol{\theta}_i|\mathbf{y}_i) \propto p_i(\boldsymbol{\theta}_i)\,p_i(\mathbf{y}_i|\boldsymbol{\theta}_i)$ for $i = 1, 2$, and $p(\boldsymbol{\theta}|\mathbf{y}) \propto p(\boldsymbol{\theta})\,p(\mathbf{y}|\boldsymbol{\theta})$ when the prior pdf's $p_1(\boldsymbol{\theta}_1)$, $p_2(\boldsymbol{\theta}_2)$, and $p(\boldsymbol{\theta})$ are each taken in Jeffreys' form.

VI. *Repeated product invariance*: Suppose that $\mathbf{y}_1, \mathbf{y}_2, \ldots, \mathbf{y}_m$ are each $n \times 1$ independent observation vectors from $p(\mathbf{y}|\boldsymbol{\theta})$. Then

$$p(\mathbf{y}_1, \mathbf{y}_2, \ldots, \mathbf{y}_m|\boldsymbol{\theta}) = \prod_{i=1}^{m} p(\mathbf{y}_i|\boldsymbol{\theta})$$

and

$$p^*(\boldsymbol{\theta}|\mathbf{y}_1, \mathbf{y}_2, \ldots, \mathbf{y}_m) \propto p^*(\boldsymbol{\theta}) \prod_{i=1}^{m} p(\mathbf{y}_i|\boldsymbol{\theta}).$$

If

$$p^*(\boldsymbol{\theta}|\mathbf{y}_1, \mathbf{y}_2, \ldots, \mathbf{y}_m) \propto p(\boldsymbol{\theta}|\mathbf{y}_1) \prod_{i=2}^{m} p(\mathbf{y}_i|\boldsymbol{\theta}),$$

where $p(\boldsymbol{\theta}|\mathbf{y}_1) \propto p(\boldsymbol{\theta})\,p(\mathbf{y}_1|\boldsymbol{\theta})$, we have "repeated product invariance." Taking $p^*(\boldsymbol{\theta})$ and $p(\boldsymbol{\theta})$ in Jeffreys' form will provide this property.[59]

These six invariance properties are indeed important properties relating to prior pdf's, and the fact that Jeffreys' prior, $p(\boldsymbol{\theta}) \propto |\text{Inf}_{\boldsymbol{\theta}}|^{1/2}$, has them is fortunate. However, in what sense can Jeffreys' prior be considered as a representation of "knowing little" or "ignorance"? As in the discussion of the uniform distribution as representing "knowing little," it appears necessary to consider this problem in information theoretic terms. To do so, let $p(y|\boldsymbol{\theta})$ be the pdf for y given $\boldsymbol{\theta}$. Then define

$$I_y(\boldsymbol{\theta}) = \int p(y|\boldsymbol{\theta}) \log p(y|\boldsymbol{\theta})\, dy \tag{19}$$

as measuring the information in $p(y|\boldsymbol{\theta})$. A priori, the prior average information will be defined as

$$\bar{I}_y = \int I_y(\boldsymbol{\theta})\, p(\boldsymbol{\theta})\, d\boldsymbol{\theta}, \tag{20}$$

where $p(\boldsymbol{\theta})$ is a prior pdf, here proper. Then we introduce

$$\begin{aligned} G &= \bar{I}_y - \int p(\boldsymbol{\theta}) \log p(\boldsymbol{\theta})\, d\boldsymbol{\theta} \\ &= \int I_y(\boldsymbol{\theta})\, p(\boldsymbol{\theta})\, d\boldsymbol{\theta} - \int p(\boldsymbol{\theta}) \log p(\boldsymbol{\theta})\, d\boldsymbol{\theta} \end{aligned} \tag{21}$$

[59] Hartigan, *loc. cit.*, notes that if $p(\mathbf{y}|\boldsymbol{\theta}) = p_1(\mathbf{y}_1|\boldsymbol{\theta})\,p_2(\mathbf{y}_2|\boldsymbol{\theta})$, where p_1 and p_2 are *different* pdf's, requiring the "repeated product invariance" property implies that other invariance properties, including the Ω-labeling invariance property, will be violated.

to measure the gain in information; that is, the prior average information associated with an observation y, denoted $\bar{I}_y$, minus the information in our prior pdf $p(\boldsymbol{\theta})$, as measured by $\int p(\boldsymbol{\theta}) \log p(\boldsymbol{\theta})\, d\boldsymbol{\theta}$.

We now *define* a "minimal information" prior pdf to be one that maximizes G for given $p(y|\boldsymbol{\theta})$. Although this is not the only possible definition of a "minimal information" prior pdf (e.g., other measures of information could be employed), it is of interest to apply this definition to a few particular cases to illustrate prior pdf's yielded by it and compare them with Jeffreys' prior pdf's.

Consider first

$$p(y|\theta) = \frac{1}{\sqrt{2\pi}} \exp\left[-\tfrac{1}{2}(y - \theta)^2\right], \qquad -\infty < y < \infty;$$

that is, y is normally distributed with unknown mean θ and known variance equal to one. Then

$$\begin{aligned} I_y(\theta) &= \int_{-\infty}^{\infty} p(y|\theta) \log p(y|\theta)\, dy \\ &= \int_{-\infty}^{\infty} \left[-\tfrac{1}{2}\log 2\pi - \tfrac{1}{2}(y - \theta)^2\right] p(y|\theta)\, dy \\ &= -\tfrac{1}{2}(\log 2\pi + 1), \end{aligned}$$

which is independent of θ. Therefore, for proper $p(\theta)$,

$$\int I_y(\theta)\, p(\theta)\, d\theta = -\tfrac{1}{2}(\log 2\pi + 1)$$

and

$$G = -\tfrac{1}{2}(\log 2\pi + 1) - \int p(\theta) \log p(\theta)\, d\theta$$

will be maximized if we minimize $\int p(\theta) \log p(\theta)\, d\theta$. As shown above, the solution to this problem is a rectangular or uniform pdf for θ; that is, $p(\theta) \propto$ constant. This is also the form of the prior yielded by Jeffreys' rule; that is, $p(\theta) \propto |\text{Inf}_\theta|^{1/2} \propto$ constant.

As a second example, consider

$$p(y|\sigma) = \frac{1}{\sqrt{2\pi}\,\sigma} \exp\left[-\frac{y^2}{2\sigma^2}\right], \qquad -\infty < y < \infty.$$

Here we have assumed y normal with known mean equal to zero and unknown standard deviation σ. Then

$$\begin{aligned} I_y(\sigma) &= \int_{-\infty}^{\infty} p(y|\sigma) \log p(y|\sigma)\, dy \\ &= -\tfrac{1}{2}(\log 2\pi + 1) - \log \sigma, \end{aligned}$$

and for proper $p(\sigma)$

$$G = -\tfrac{1}{2}\log(2\pi + 1) - \int \log \sigma \, p(\sigma)\, d\sigma - \int p(\sigma) \log p(\sigma)\, d\sigma.$$

The necessary condition for G to be a maximum subject to $\int p(\sigma)\, d\sigma = 1$ is $-\log \sigma - 1 - \log p(\sigma) + \lambda = 0$ where λ is a Lagrange multiplier. This implies that

$$p(\sigma) \propto \frac{1}{\sigma}, \tag{22}$$

a result also in accord with $p(\sigma) \propto |\mathrm{Inf}_\sigma|^{1/2} \propto 1/\sigma$.

As a third example, consider

$$p(y|\theta, \sigma) = \frac{1}{\sqrt{2\pi}\,\sigma} \exp\left[-\frac{1}{2\sigma^2}(y - \theta)^2\right], \qquad -\infty < y < \infty,$$

wherein both θ and σ are unknown. Then

$$\begin{aligned} I_y(\theta, \sigma) &= \int_{-\infty}^{\infty} p(y|\theta, \sigma) \log p(y|\theta, \sigma)\, dy \\ &= -\tfrac{1}{2}(\log 2\pi + 1) - \log \sigma, \end{aligned}$$

and for proper $p(\theta, \sigma)$

$$G = -\tfrac{1}{2}(\log 2\pi + 1) - \iint \log \sigma \, p(\theta, \sigma)\, d\theta\, d\sigma - \iint p(\theta, \sigma) \log p(\theta, \sigma)\, d\theta\, d\sigma;$$

G will be maximized, subject to $\iint p(\theta, \sigma)\, d\theta\, d\sigma = 1$ for

$$p(\theta, \sigma) \propto \frac{1}{\sigma}. \tag{23}$$

Thus the "minimal information" prior pdf is one in which θ and σ are independent, with θ and $\log \sigma$ uniformly distributed.

A prior pdf in the form of (23) is the one used extensively by Jeffreys, even though[60]

$$\begin{aligned} p(\theta, \sigma) &\propto |\mathrm{Inf}_{\theta,\sigma}|^{1/2} \\ &\propto \frac{1}{\sigma^2}. \end{aligned} \tag{24}$$

Jeffreys explains his departure from the use of the general rule in terms of a

[60] Note that the information matrix is given by

$$\mathrm{Inf}_{\theta,\sigma} = \begin{pmatrix} \dfrac{1}{\sigma^2} & 0 \\ 0 & \dfrac{2n}{\sigma^2} \end{pmatrix}$$

and thus (24) follows.

prior judgment that θ and σ are independent.[61] Then he applies his rule separately to θ and σ to obtain a prior pdf in the form of (23). He notes that the prior pdf in (23) will be invariant to transformations of the following kind, $\eta = \theta + k\sigma$.

If we consider an asymptotic form of the quantity G in (21),

$$G_a = \int p(\boldsymbol{\theta}) \log \sqrt{n|\text{Inf}_{\boldsymbol{\theta}}|}\, d\boldsymbol{\theta} - \int p(\boldsymbol{\theta}) \log p(\boldsymbol{\theta})\, d\boldsymbol{\theta}, \tag{25}$$

where n = number of independent drawings from $p(y|\boldsymbol{\theta})$, and seek the prior pdf, $p(\boldsymbol{\theta})$, which maximizes G_a subject to $\int p(\boldsymbol{\theta})\, d\boldsymbol{\theta} = 1$, the result is just[62]

$$p(\boldsymbol{\theta}) \propto |\text{Inf}_{\boldsymbol{\theta}}|^{1/2}, \tag{26}$$

which is in the form of Jeffreys' invariant prior pdf. Thus for the *asymptotic* form of G given in (25) Jeffreys' prior is a minimal information prior. For the nonasymptotic form of G in (21), however, it must be recognized, as shown above, that Jeffreys' invariant priors are not always minimal information prior pdf's since they do not always maximize G.[63] When a Jeffreys' prior pdf is not the prior pdf that maximizes G, it is the case that use of Jeffreys' prior involves introduction of more prior information in an analysis than associated with the use of a prior pdf that maximizes G. When the number of parameters is large, this difference can be important; for example, in k normal populations with unknown means, $\boldsymbol{\theta}' = (\theta_1, \theta_2, \ldots, \theta_k)$, and the common variance, σ^2, as mentioned above, Jeffreys' prior pdf is $p(\boldsymbol{\theta}, \sigma) \propto 1/\sigma^{k+1}$ which, for large k, contrasts sharply with the minimal information prior $p(\boldsymbol{\theta}, \sigma) \propto 1/\sigma$. It is thus important, as Jeffreys himself emphasizes, to examine the form of vague or diffuse prior pdf's carefully in order to avoid putting some unwanted prior information into an analysis, a consideration that is particularly relevant for small sample situations.[64]

[61] Jeffreys, *op. cit.*, p. 182, recognizes that if we have k normal populations with unknown means $\theta_1, \theta_2, \ldots, \theta_k$ and common unknown variance σ^2 his invariant prior pdf is $|\text{Inf}|^{1/2} \propto 1/\sigma^{k+1}$. He deems this unsatisfactory since, for example, if we have n_i observations from population i, with $n = \sum_{i=1}^{k} n_i$, the marginal posterior pdf for θ_1 will be in the univariate Student t form with n degrees of freedom for any k. That there is no loss of degrees of freedom associated with integrating out the $k - 1$ means other than θ_1 is due to the appearance of a factor $1/\sigma$ for each mean in Jeffreys' prior and is considered unreasonable by Jeffreys.

[62] This result appears in Lindley, *op. cit.*, p. 467.

[63] Prior pdf's that maximize G do not in general have the Ω-labeling invariance property of Jeffreys' prior pdf. However, investigators using different parameterizations can get compatible results if they adopt the convention of using minimal information prior pdf's (i.e., priors that maximize G) for any given parameterization when they know little about the values of the parameters.

[64] A small sample situation can be roughly defined as one in which the ratio of the number of parameters to the number of observations is not small.

QUESTIONS AND PROBLEMS

1. Assume that the following data represent experimental measures of yields of a new variety of rice: 10.40, 10.36, 9.16, 10.03, 9.31, 9.75, 8.69, 9.89. If these data are regarded as having been generated by independent drawings from a normal distribution, derive the posterior distributions for the population mean yield and the population standard deviation using a diffuse prior pdf for the mean μ and standard deviation σ, namely, $p(\mu, \sigma) \propto 1/\sigma$, with $-\infty < \mu < \infty$ and $0 < \sigma < \infty$. What are the first two moments of the marginal posterior pdf's for μ and for σ?
2. From (2.19), which is the posterior pdf for a standard deviation σ, derive the posterior pdf for the variance of $\phi = \sigma^2$ and expressions for the posterior mean and variance of ϕ. Then show that the posterior pdf for $x = \nu s^2/\phi$ is the χ^2 pdf with ν degrees of freedom shown in (A.35) of Appendix A.
3. Suppose that another set of experiments, independent of those referred to in Problem 1, produced the following yields for a variety of rice currently in use: 8.47, 7.35, 12.08, 7.83, 8.43, 10.29, 11.34, 8.40. If these data are regarded as having been generated by independent drawings from a normal distribution with mean θ and standard deviation τ, what is the likelihood function for the parameters μ, θ, σ, and τ, given the data of Problems 1 and 3? If the μ, θ, $\log \sigma$, and $\log \tau$ are assumed to be independently and uniformly distributed a priori, each with a range of $-\infty$ to $+\infty$, derive the joint marginal posterior pdf for σ^2 and τ^2. Then, using the analysis in Appendix A, Section 6, show that the posterior pdf for $\phi = \sigma^2/\tau^2$ is in the form of an F pdf. Compute and plot this pdf. Also compare the posterior pdf's for θ and for μ. How does the new variety of rice compare with the one currently in use?
4. If price, p, is related to yield per acre, y, by $p = \alpha_1 - \alpha_2 N y$, with α_1 and α_2 positive parameters and N a given number of acres under cultivation, sales revenue is $SR = Nyp = Ny\alpha_1 - \alpha_2(Ny)^2$. If y is assumed to be random with mean μ and variance σ^2, obtain an expression for the mathematical expectation of SR, given N, α_1 and α_2. How does ESR depend on μ and σ^2? Use the posterior pdf for μ and σ^2, obtained in Problem 1, to obtain the posterior mean of ESR.
5. As an alternative prior pdf for the analysis of Example 2.3 consider the following natural-conjugate, "normal-gamma" pdf for μ and σ:

$$p(\mu, \sigma | \mu_a, h_a, \sigma_a, \nu_a) = p_1(\sigma | \sigma_a, \nu_a)\, p_2(\mu | \sigma, \mu_a, h_a), \qquad \begin{array}{c} -\infty < \mu < \infty, \\ 0 < \sigma < \infty, \end{array}$$

with

$$p_1(\sigma | \sigma_a, \nu_a) \propto (\sigma^{\nu_a + 1})^{-1} \exp\left(-\frac{\nu_a \sigma_a^2}{2\sigma^2}\right),$$

a pdf in the inverted gamma form (see Appendix A, Section 4), and

$$p_2(\mu | \sigma, \mu_a, h_a) \propto \frac{1}{\sigma} \exp\left[-\frac{h_a}{2\sigma^2}(\mu - \mu_a)^2\right],$$

a pdf in the univariate normal form, where μ_a, h_a, σ_a, and ν_a are given parameters of the prior pdf with $h_a, \sigma_a, \nu_a > 0$.

(a) What is the modal value of the marginal prior pdf for σ, $p_1(\sigma|\sigma_a, \nu_a)$? What condition on ν_a is required for the mean and the variance of this pdf to exist? (See Appendix A, Section 4.)

(b) What are the mean and variance of the normal conditional prior pdf for μ, given σ, μ_a, and h_a, $p_2(\mu|\sigma, \mu_a, h_a)$?

(c) After integrating the joint prior pdf for μ and σ with respect to σ, show that the marginal prior pdf for μ is in the form of a univariate Student t pdf. What condition on ν_a is required for the mean and variance of this pdf to exist? (See Appendix A, Section 2.) Given that this condition is satisfied, derive the prior mean and variance of μ.

(d) Comment on other properties of the marginal prior pdf's for μ and σ, particularly how their properties depend on values of the prior parameters.

6. Combine the natural-conjugate prior pdf in Problem 5 with the following normal likelihood function:

$$l(\mu, \sigma|\mathbf{y}) \propto \frac{1}{\sigma^n} \exp\left[-\frac{1}{2\sigma^2} \sum_{i=1}^{n} (y_i - \mu)^2\right]$$

$$\propto \frac{1}{\sigma^n} \exp\left[-\frac{1}{2\sigma^2} [\nu s^2 + n(\mu - \hat{\mu})^2]\right],$$

where

$$\mathbf{y}' = (y_1, y_2, \ldots, y_n), \qquad \nu = n - 1, \qquad \hat{\mu} = \sum_{i=1}^{n} y_i/n$$

and

$$\nu s^2 = \sum_{i=1}^{n} (y_i - \hat{\mu})^2,$$

to obtain the joint posterior pdf for μ and σ.

(a) What are the form and properties of the marginal posterior pdf for σ?

(b) What are the form and properties of the conditional posterior pdf for μ given σ, the prior parameters' values, and the sample observations $\mathbf{y}$?

(c) What are the form and properties of the marginal posterior pdf for μ?

7. In Problem 2 the posterior pdf for $\phi = \sigma^2$ was obtained. Consider and compare the following two loss functions:

$$L_1 = k_1(\phi - \hat{\phi}_1)^2 \quad \text{and} \quad L_2 = k_2\left(\frac{\phi - \hat{\phi}_2}{\phi}\right)^2,$$

where the k's are positive numerical constants and the $\hat{\phi}$'s are point estimates. What values for $\hat{\phi}_1$ and for $\hat{\phi}_2$ minimize expected loss?

8. Contrast the point estimates for ϕ obtained in Problem 7 with the maximum likelihood and other sampling theory estimators for $\phi = \sigma^2$ in the normal mean problem with likelihood function, as shown in Problem 6. In particular, consider the mean-square error of an estimator of the following form $\hat{\phi} = cs^2$, where c is a constant. Show that $E(\hat{\phi} - \phi)^2$, where the expectation is with respect to the pdf for s^2 with ϕ fixed, has a minimal value for $c = \nu/(\nu + 2)$, where $\nu = n - 1$. In taking the expectation, note that $\nu s^2/\phi$ has a χ^2 pdf with $\nu = n - 1$ degrees of freedom. Compare the resulting estimator $\nu s^2/(\nu + 2)$ with what was obtained as point estimates in Problem 7.

9. If z is a strictly positive random variable and if $y = lnz$ is normally distributed with mean μ and standard deviation σ, then z is said to have a log-normal distribution. What is the pdf for z? Show that e^{μ} and $e^{\mu + \frac{1}{2}\sigma^2}$ are the median and mean, respectively, of the pdf for z.

10. Let y_i denote the natural logarithm of the ith individual's annual labor income, z_i; that is, $y_i = lnz_i$, $i = 1, 2, \ldots, n$. Further assume that the y_i's are normally and independently distributed, each with mean μ and variance σ^2. Then under the assumption that the prior pdf for μ and σ is $p(\mu, \sigma) \propto 1/\sigma$, with $-\infty < \mu < \infty$ and $0 < \sigma < \infty$, show that the posterior pdf for $ln\theta = \mu$, where θ is the median of the log-normal pdf for annual labor income, has a univariate Student t form and thus θ has a "log-Student t" posterior pdf.

11. In Problem 10 show that the natural logarithm of the mean, η, of the log-normal pdf, $\ln \eta = \mu + \frac{1}{2}\sigma^2$, has a normal conditional posterior pdf, given σ, with conditional mean $\hat{\mu} + \frac{1}{2}\sigma^2$ and conditional variance σ^2/n, where $\hat{\mu} = \sum_{i=1}^{n} y_i/n$. Then provide the joint posterior pdf for $\ln \eta$ and σ and explain how this bivariate posterior pdf can be transformed to one for η and σ; indicate how the latter bivariate pdf can be normalized and analyzed by employing bivariate numerical integration techniques.

12. Suppose that a two-sided coin is fairly tossed and comes down with the head side upward. Given this outcome for a single toss, what is the likelihood function and the maximum likelihood estimate of the probability θ of obtaining a head on a single toss? If our prior pdf for θ is uniform, $0 \leq \theta \leq 1$, what is the mean of the posterior pdf for θ? On the other hand, if our prior pdf for θ is given by (2.54), what is the posterior mean? How do you interpret the fact that the maximum likelihood estimate and the posterior means are numerically quite different?

13. In Problem 12 plot the posterior pdf's associated with the two different prior pdf's. What can be said about the precision with which θ can be estimated from a sample of size $n = 1$? What are the posterior variances and the variance of the maximum likelihood estimator?

14. Assume that of 10 randomly selected consumers from a homogeneous large population, four responded that they bought product A and six responded that they did not. Under the assumption that consumers' choices of products are independent and that there is a common probability θ of buying product A for all members of the population, what is the likelihood function and maximum likelihood estimate for θ? What is the posterior pdf for θ employing the prior pdf in (2.54)? Find its mean and modal value and provide a justification for the selection of the posterior mean as a point estimate of the probability of buying Brand A, given a quadratic loss function.

15. In Problem 14 obtain the posterior pdf for θ by using an informative prior pdf in the form of a beta pdf with prior mean equal to 0.5 and prior variance equal to 0.024. Compare prior and posterior moments.

16. What might have been the source or sources of the prior information described in Problem 15?

CHAPTER III

The Univariate Normal Linear Regression Model

In this chapter we first take up the analysis of the simple univariate normal linear regression model and then turn to the normal linear multiple regression model.[1] Throughout we adopt assumptions of normality, independence, linearity, homoscedasticity, and absence of measurement errors. Certain departures from these specifying assumptions and their analysis are treated in subsequent chapters.

3.1 THE SIMPLE UNIVARIATE NORMAL LINEAR REGRESSION MODEL

3.1.1 Model and Likelihood Function

In the simple univariate normal linear regression model we have *one* random variable (hence the term "univariate"), the "dependent" variable, whose variation is to be explained, at least in part, by the variation of another variable, the "independent" variable. That part of the variation of the dependent variable unexplained by variation in the independent variable is assumed to be produced by an unobserved random "error" or "disturbance" variable which may be viewed as representing the collective action of a number of minor factors that produce variation in the dependent variable. Formally, with the dependent variable, denoted by y, and the independent variable denoted by x, we have the following relationship:

$$y_i = \beta_1 + \beta_2 x_i + u_i, \qquad i = 1, 2, \ldots, n, \tag{3.1}$$

[1] Bayesian analyses of the univariate normal linear regression model appear in H. Jeffreys, *Theory of Probability* (3rd rev. ed.). Oxford: Clarendon, 1966, pp. 147–161; D. V. Lindley, *Introduction to Probability and Statistics from a Bayesian Viewpoint. Part 2. Inference*. Cambridge: Cambridge University Press, 1965, Chapter 8; and H. Raiffa and R. Schlaifer, *Applied Statistical Decision Theory*. Boston: Graduate School of Business Administration, Harvard University, 1961, Chapter 13.

where y_i = ith observation on the dependent variable,
x_i = ith observation on the independent variable,
u_i = ith unobserved value of the random disturbance or error variable, and
β_1 and β_2 = regression parameters, namely, the "intercept" and "slope coefficient," respectively.

Note that the relation in (3.1) is linear in β_1, β_2, and u_i, hence the term "linear" regression.[2]

Assumption 1. The u_i, $i = 1, 2, \ldots, n$, are normally and independently distributed, each with zero mean and common variance σ^2.

Regarding the independent variable, we make the following assumption:

Assumption 2. The x_i, $i = 1, 2, \ldots, n$, are fixed nonstochastic variables.

Alternatively, we can make the following assumption about x_i:

Assumption 3. The x_i, $i = 1, 2, \ldots, n$, are random variables distributed independently of the u_i, with a pdf *not* involving the parameters β_1, β_2, and σ.

To form the likelihood function under assumptions 1 and 3, we write the joint pdf for $\mathbf{y}' = (y_1, y_2, \ldots, y_n)$ and $\mathbf{x}' = (x_1, x_2, \ldots, x_n)$, namely

$$p(\mathbf{y}, \mathbf{x} | \beta_1, \beta_2, \sigma^2, \boldsymbol{\theta}) = p(\mathbf{y} | \mathbf{x}, \beta_1, \beta_2, \sigma^2)\, g(\mathbf{x} | \boldsymbol{\theta}), \tag{3.2}$$

when $\boldsymbol{\theta}$ denotes the parameters of the marginal pdf for $\mathbf{x}$. Since by assumption (3) $\boldsymbol{\theta}$ does not involve β_1, β_2, or σ, the likelihood function for β_1, β_2, and σ can be formed from the first factor on the right-hand side of (3.2). Note from (3.1) that for given $\mathbf{x}$, β_1, β_2, and σ^2, $\mathbf{y}$ will be normally distributed with $E(y_i | x_i, \beta_1, \beta_2, \sigma^2) = \beta_1 + \beta_2 x_i$ and $\text{Var}(y_i | x_i, \beta_1, \beta_2, \sigma^2) = \sigma^2$, $i = 1, 2, \ldots, n$. Further, the y_i, given the x_i, β_1, β_2, and σ^2, will be independently distributed. Thus we have

$$p(\mathbf{y} | \mathbf{x}, \beta_1, \beta_2, \sigma) \propto \frac{1}{\sigma^n} \exp\left[-\frac{1}{2\sigma^2} \sum (y_i - \beta_1 - \beta_2 x_i)^2\right], \tag{3.3}$$

with the summation extending from $i = 1$ to $i = n$. Also, (3.3) would have resulted had we adopted Assumption 2 about the x_i rather than Assumption 3. The expression in (3.3), viewed as a function of the parameters β_1, β_2, and σ, is the likelihood function to be combined with our prior pdf for the parameters.

[2] The relation in (3.1) need not be linear in the "underlying" variables; for example, it may be that $y_i = \log w_i$, where w_i, is the ith observation on an underlying variable, or x_i may represent z_i^2, where z_i is an observation on an underlying variable.

3.1.2 Posterior Pdf's for Parameters with a Diffuse Prior Pdf

For our prior pdf for β_1, β_2 and σ we assume that β_1, β_2 and $\log \sigma$ are uniformly and independently distributed, which implies

$$p(\beta_1, \beta_2, \sigma) \propto \frac{1}{\sigma} \qquad \begin{matrix} -\infty < \beta_1, \beta_2 < \infty, \\ 0 < \sigma < \infty. \end{matrix} \tag{3.4}$$

Then, on combining (3.3) and (3.4), the joint posterior pdf for β_1, β_2, and σ is given by

$$p(\beta_1, \beta_2, \sigma|\mathbf{y}, \mathbf{x}) \propto \frac{1}{\sigma^{n+1}} \exp\left[-\frac{1}{2\sigma^2} \sum (y_i - \beta_1 - \beta_2 x_i)^2\right]. \tag{3.5}$$

This joint posterior pdf, which serves as a basis for making inferences about β_1, β_2, and σ, can be analyzed conveniently by taking note of the following algebraic identity[3]:

$$\sum (y_i - \beta_1 - \beta_2 x_i)^2 = \nu s^2 + n(\beta_1 - \hat{\beta}_1)^2 + (\beta_2 - \hat{\beta}_2)^2 \sum x_i^2 + 2(\beta_1 - \hat{\beta}_1)(\beta_2 - \hat{\beta}_2) \sum x_i, \tag{3.6}$$

where $\nu = n - 2$,

$$\hat{\beta}_1 = \bar{y} - \hat{\beta}_2 \bar{x}, \qquad \hat{\beta}_2 = \frac{\sum (x_i - \bar{x})(y_i - \bar{y})}{\sum (x_i - \bar{x})^2}, \tag{3.7}$$

$$s^2 = \nu^{-1} \sum (y_i - \hat{\beta}_1 - \hat{\beta}_2 x_i)^2, \tag{3.8}$$

with $\bar{y} = n^{-1} \sum y_i$ and $\bar{x} = n^{-1} \sum x_i$. To establish (3.6) we write

$$\sum (y_i - \beta_1 - \beta_2 x_i)^2 = \sum \{(y_i - \hat{\beta}_1 - \hat{\beta}_2 x_i) - [(\beta_1 - \hat{\beta}_1) + (\beta_2 - \hat{\beta}_2)x_i]\}^2.$$

On expanding the rhs, note that the cross-product term vanishes and thus (3.6) results.

On substituting from (3.6) in (3.5), we have

$$p(\beta_1, \beta_2, \sigma|\mathbf{y}, \mathbf{x}) \propto \frac{1}{\sigma^{n+1}} \times \exp\left\{-\frac{1}{2\sigma^2} [\nu s^2 + n(\beta_1 - \hat{\beta}_1)^2 + (\beta_2 - \hat{\beta}_2)^2 \sum x_i^2 + 2(\beta_1 - \hat{\beta}_1)(\beta_2 - \hat{\beta}_2) \sum x_i]\right\}. \tag{3.9}$$

From (3.9) it is immediately seen that the conditional posterior pdf for β_1 and β_2, given σ, is in the bivariate normal form with mean $(\hat{\beta}_1, \hat{\beta}_2)$ and

[3] If the expression in (3.6) is substituted in (3.3), the likelihood function can be expressed in terms of s^2, $\hat{\beta}_1$, and $\hat{\beta}_2$, which are sufficient statistics.

covariance matrix

$$\sigma^2 \begin{bmatrix} n & \sum x_i \\ \sum x_i & \sum x_i^2 \end{bmatrix}^{-1} = \sigma^2 \begin{bmatrix} \dfrac{\sum x_i^2}{n \sum (x_i - \bar{x})^2} & \dfrac{-\bar{x}}{\sum (x_i - \bar{x})^2} \\ \dfrac{-\bar{x}}{\sum (x_i - \bar{x})^2} & \dfrac{1}{\sum (x_i - \bar{x})^2} \end{bmatrix}.$$

Of course, since σ^2 is rarely known in practice, this result is not very useful. To obtain the marginal posterior pdf for β_1 and β_2 we integrate (3.9) with respect to σ to obtain

$$\begin{aligned} p(\beta_1, \beta_2 | \mathbf{y}, \mathbf{x}) &= \int_0^\infty p(\beta_1, \beta_2, \sigma | \mathbf{y}, \mathbf{x})\, d\sigma \\ &\propto [\nu s^2 + n(\beta_1 - \hat{\beta}_1)^2 + (\beta_2 - \hat{\beta}_2)^2 \textstyle\sum x_i^2 \\ &\qquad + 2(\beta_1 - \hat{\beta}_1)(\beta_2 - \hat{\beta}_2) \textstyle\sum x_i]^{-n/2}, \end{aligned} \tag{3.10}$$

which is seen to be in the form of a bivariate Student t pdf (see Appendix B). From properties of the bivariate Student t pdf we have the following results:

$$p(\beta_1 | \mathbf{y}, \mathbf{x}) \propto \left[\nu + \frac{\sum (x_i - \bar{x})^2}{s^2 \sum x_i^2/n} (\beta_1 - \hat{\beta}_1)^2\right]^{-(\nu+1)/2}, \quad -\infty < \beta_1 < \infty, \tag{3.11}$$

$$p(\beta_2 | \mathbf{y}, \mathbf{x}) \propto \left[\nu + \frac{\sum (x_i - \bar{x})^2}{s^2} (\beta_2 - \hat{\beta}_2)^2\right]^{-(\nu+1)/2}, \quad -\infty < \beta_2 < \infty. \tag{3.12}$$

If we make the following transformations in (3.11) and (3.12),

$$\left[\frac{\sum (x_i - \bar{x})^2}{s^2 \sum x_i^2/n}\right]^{1/2} (\beta_1 - \hat{\beta}_1) = t_\nu \tag{3.13}$$

and

$$\frac{\beta_2 - \hat{\beta}_2}{s/[\sum (x_i - \bar{x})^2]^{1/2}} = t_\nu, \tag{3.14}$$

the random variable t_ν has the Student t pdf with ν degrees of freedom. These results enable us to make inferences about β_1 and β_2, using tables of the t-distribution.

As regards the posterior pdf for σ, it can be obtained by integrating (3.9) with respect to β_1 and β_2. This operation yields

$$p(\sigma | \mathbf{y}, \mathbf{x}) \propto \frac{1}{\sigma^{\nu+1}} \exp\left(-\frac{\nu s^2}{2\sigma^2}\right), \qquad 0 < \sigma < \infty. \tag{3.15}$$

From (3.15) σ is distributed in the form of an inverted gamma function (see Appendix A). Thus we have[4]

$$E\sigma = s\left(\sqrt{\frac{\nu}{2}}\right)^{1/2} \frac{\Gamma[(\nu - 1)/2]}{\Gamma(\nu/2)}; \qquad \text{Var}(\sigma) = \frac{s^2\nu}{\nu - 2} - (E\sigma)^2.$$

[4] For the pdf in (3.15) to be proper we must have $\nu > 0$; for the mean to exist we need $\nu > 1$; and for the variance to exist we need $\nu > 2$. See Appendix A.

Further, if we transform from σ to σ^2, the posterior pdf for the variance is

$$(3.16) \qquad p(\sigma^2|\mathbf{y}, \mathbf{x}) \propto [(\sigma^2)^{n/2}]^{-1} \exp\left(-\frac{\nu s^2}{2\sigma^2}\right), \qquad 0 < \sigma^2 < \infty.$$

Finally, the posterior pdf for the precision parameter $h = 1/\sigma^2$ is given by

$$(3.17) \qquad p(h|\mathbf{y}, \mathbf{x}) \propto h^{\nu/2-1} \exp\left(-\frac{\nu s^2 h}{2}\right), \qquad 0 < h < \infty.$$

It is seen from (3.17) that the variable $\nu s^2 h$ has the χ^2 pdf with ν degrees of freedom. From (3.16) and (3.17) we have, for example, $E\sigma^2 = \nu s^2/(\nu - 2)$ and $Eh = 1/s^2$. Further properties of these pdf's are discussed in Appendix A.

To make joint posterior inferences about β_1 and β_2 we shall show that the following quantity

$$(3.18) \qquad \psi = \frac{[n(\beta_1 - \hat{\beta}_1)^2 + (\beta_2 - \hat{\beta}_2)^2 \sum x_i^2 + 2(\beta_1 - \hat{\beta}_1)(\beta_2 - \hat{\beta}_2) \sum x_i]}{2s^2}$$

is distributed a posteriori as $F_{2,\nu}$. To show this let us write

$$\psi = \boldsymbol{\delta}' A \boldsymbol{\delta},$$

where $\boldsymbol{\delta}' = (\beta_1 - \hat{\beta}_1, \beta_2 - \hat{\beta}_2)$ and

$$(3.19) \qquad A \equiv \frac{1}{2s^2}\begin{bmatrix} 1 & \sum x_i \\ \sum x_i & \sum x_i^2 \end{bmatrix}.$$

Using this notation, the posterior pdf for $\boldsymbol{\delta}$ is given from (3.10) by

$$(3.20) \qquad p(\boldsymbol{\delta}|\mathbf{y}, \mathbf{x}) \propto \left(1 + \frac{2}{\nu}\boldsymbol{\delta}' A \boldsymbol{\delta}\right)^{-n/2}$$

Now, since A is positive definite, we can write $A = K'K$, where K is nonsingular and thus $\boldsymbol{\delta}' A \boldsymbol{\delta} = (K\boldsymbol{\delta})' K \boldsymbol{\delta} = \mathbf{V}'\mathbf{V}$, where $\mathbf{V} = K\boldsymbol{\delta}$ is a 2×1 vector and $\mathbf{V}' = (v_1, v_2)$. Then

$$(3.21) \qquad p(\mathbf{V}|\mathbf{y}, \mathbf{x}) \propto \left(1 + \frac{2}{\nu}\mathbf{V}'\mathbf{V}\right)^{-n/2}.$$

Now let

$$v_1 = \psi^{1/2} \cos \theta,$$

$$v_2 = \psi^{1/2} \sin \theta.$$

The Jacobian of this transformation is $\frac{1}{2}$. Note, too, that $\mathbf{V}'\mathbf{V} = v_1^2 + v_2^2 = \psi(\cos^2 \theta + \sin^2 \theta) = \psi$. Thus

$$(3.22) \qquad p(\psi|\mathbf{y}, \mathbf{x}) \propto \left(1 + \frac{2}{\nu}\psi\right)^{-(\nu+2)/2},$$

which is the $F_{2,\nu}$ pdf.[5] This result can be employed to construct posterior confidence regions for β_1 and β_2.

In (3.10) we noted that β_1 and β_2 are distributed in the bivariate Student t form. An important property of this bivariate pdf is that a single linear combination of variables so distributed has a pdf in the univariate Student t form. This result is illustrated below in obtaining the posterior pdf of η_0, defined by

$$E(y|x = x_0) = \eta_0 = \beta_1 + \beta_2 x_0. \tag{3.23}$$

It is seen that η_0 is a linear form in β_1 and β_2 and thus will be distributed in the univariate Student t form with mean $\hat{\eta}_0 = \hat{\beta}_1 + \hat{\beta}_2 x_0$; that is

$$\frac{\eta_0 - \hat{\eta}_0}{s[1/n + (x_0 - \bar{x})^2/\sum (x_i - \bar{x})^2]^{1/2}} \sim t_\nu. \tag{3.24}$$

The result in (3.24) can be derived by changing variables in (3.10) from β_1 and β_2 to η_0 and β_2 as follows:

$$\eta_0 - \hat{\eta}_0 = \beta_1 - \hat{\beta}_1 + (\beta_2 - \hat{\beta}_2)x_0,$$
$$\beta_2 - \hat{\beta}_2 = \beta_1 - \hat{\beta}_2.$$

The Jacobian of this transformation is 1. Then, on integrating out β_2, the marginal posterior pdf for η_0 is given by

(3.25)
$$p(\eta_0|\mathbf{y}, \mathbf{x}) \propto \left[\nu + \frac{n \sum (x_i - \bar{x})^2}{s^2 \sum (x_i - x_0)^2}(\eta_0 - \hat{\eta}_0)^2\right]^{-(\nu+1)/2}, \qquad -\infty < \eta_0 < \infty.$$

Then note that $\sum (x_i - x_0)^2 = \sum [(x_i - \bar{x}) - (x_0 - \bar{x})]^2 = \sum (x_i - \bar{x})^2 + n(x_0 - \bar{x})^2$ and thus (3.24) follows. The result in (3.25) provides the complete posterior pdf for η_0 and (3.24) can be utilized to construct posterior intervals for η_0.

3.1.3 Application to Analysis of the Investment Multiplier

To illustrate application of the results presented above, we interpret (3.1) as relating income, the dependent variable, to autonomous investment, the independent variable. The parameter β_2 is then termed the investment multiplier. If our prior information about β_1, β_2, and σ is vague, we can employ the diffuse pdf in (3.4) to represent it. Note that this involves assuming $-\infty < \beta_2 < \infty$; that is, our prior information is not precise enough to fix even the algebraic sign of the multiplier.[6] Our data, taken from a paper by

[5] In general $p(F) \propto F^{(m-2)/2}/(1 + m/qF)^{(m+q)/2}$ is the $F_{m,q}$ pdf with $0 < F < \infty$; see Appendix A.

[6] In early discussions some argued that the investment multiplier might be positive, negative, or possibly zero.

Haavelmo,[7] are given in Table 3.1. From these data, with $n = 20$, we compute the sample quantities shown in (3.7) and (3.8):

$$\hat{\beta}_1 = 345, \quad \hat{\beta}_2 = 3.05, \quad s^2 = 662.8.$$

Table 3.1 HAAVELMO'S DATA[a] ON INCOME AND INVESTMENT

Year	Income[b] ($ per capita, deflated)	Investment[c] ($ per capita, deflated)	Year	Income[b] ($ per capita, deflated)	Investment[c] ($ per capita, deflated)
1922	433	39	1932	372	22
1923	483	60	1933	381	17
1924	479	42	1934	419	27
1925	486	52	1935	449	33
1926	494	47	1936	511	48
1927	498	51	1937	520	51
1928	511	45	1938	477	33
1929	534	60	1939	517	46
1930	478	39	1940	548	54
1931	440	41	1941	629	100

[a] T. Haavelmo, "Methods of Measuring the Marginal Propensity to Consume," *J. Am. Statist. Assoc.*, **42**, p. 88 (1947).

[b] Income is per capita personal disposable income deflated by the BLS Cost of Living Index, 1935–1939 = 100.

[c] Haavelmo defines investment to be the per capita price-deflated difference between personal disposable income and personal consumption expenditures. He is aware of the fact that this empirical measure contains certain elements which are not strictly investment outlays.

Further we have

$$\bar{x} = 45.35 \quad \text{and} \quad \frac{1}{n}\sum (x_i - \bar{x})^2 = 285.55.$$

Then, utilizing the results in (3.13) and (3.14) along with tables of the Student t pdf,[8] we obtain the posterior pdf's for β_1 and β_2 shown in Figure 3.1.

From the bottom panel of Figure 3.1 we see that the posterior pdf for the multiplier β_2 is centered at $\hat{\beta}_2 = 3.05$, the posterior mean. Further, we see

[7] T. Haavelmo, "Methods of Measuring the Marginal Propensity to Consume," *J. Am. Statist. Assoc.*, **42**, 105–122 (1947), reprinted in William C. Hood and T. C. Koopmans, *Studies in Econometric Method.* New York: Wiley, 1953, pp. 75–91.

[8] See, for example, N. V. Smirnov, *Tables for the Distribution and Density Functions of t-Distribution.* New York: Pergamon, 1961. These tables give values of $p(t_\nu)$, that is, ordinates of the Student t pdf with ν degrees of freedom. In our problem $\nu = 18$. To obtain ordinates of the posterior pdf for β_2 we have from (3.14) $dt_\nu = k\, d\beta_2$ where $k = [\sum (x_i - \bar{x})^2]^{1/2}/s$. Thus $p(t_\nu)\, dt_\nu = p(t_\nu)k\, d\beta_2$, and the ordinates of the posterior pdf for β_2 are given by $p(t_\nu)k$, with the values of $p(t_\nu)$ obtained from the tables.

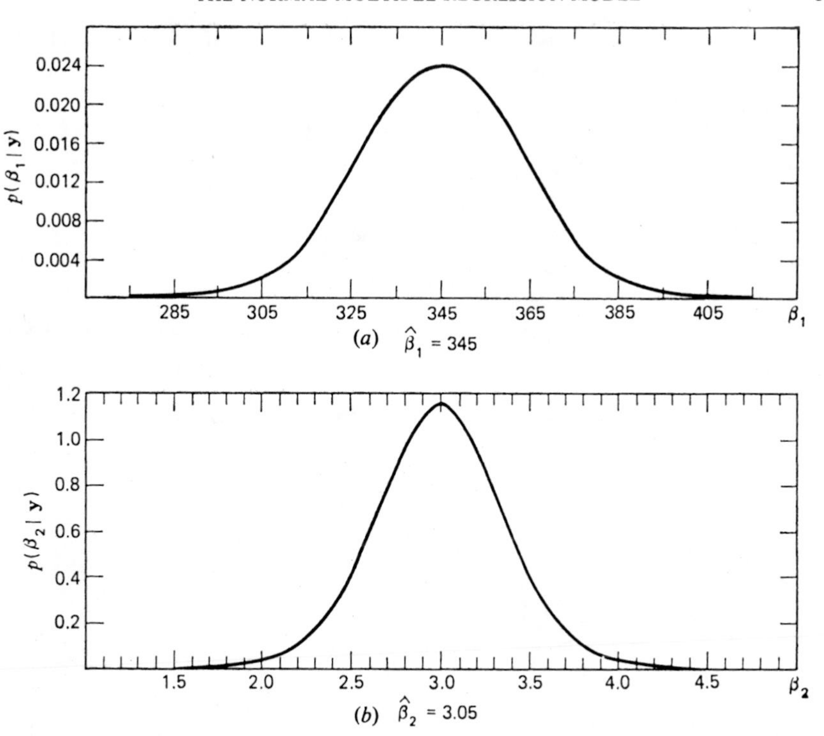

Figure 3.1 Posterior pdf's for the intercept (β_1) and investment multiplier (β_2) based on Haavelmo's model and data, and diffuse prior distributions for parameters. (*a*) Posterior pdf for intercept. (*b*) Posterior pdf for the multiplier.

that there is an almost negligible posterior probability that the multiplier is negative. Thus, although our prior beliefs did not preclude negative values for the multiplier, use of the present model and the information in the sample data has resulted in a posterior pdf that indicates that negative values for the multiplier are improbable.

3.2 THE NORMAL MULTIPLE REGRESSION MODEL

3.2.1 Model and Likelihood Function

With the normal multiple regression model, we assume that an $n \times 1$ vector of observations $\mathbf{y}$ on our dependent variable satisfies

$$\mathbf{y} = X\boldsymbol{\beta} + \mathbf{u}, \tag{3.26}$$

where X = an $n \times k$ matrix, with rank k, of observations on k independent variables,

$\boldsymbol{\beta}$ = a $k \times 1$ vector of regression coefficients,

$\mathbf{u}$ = an $n \times 1$ vector of disturbance or error terms.

We assume that the elements of $\mathbf{u}$ are normally and independently distributed, each with mean zero and common variance σ^2; that is $E\mathbf{u} = \mathbf{0}$ and $E\mathbf{uu}' = \sigma^2 I_n$, where I_n is an $n \times n$ unit matrix. With respect to the matrix X, if the regression equation is assumed to have a nonzero intercept, all elements in the first column of X will be ones; that is, the first column is $\boldsymbol{\iota}$ with $\boldsymbol{\iota}' = (1, 1, \ldots, 1)$. The remaining elements of X may be nonstochastic or stochastic, as in Section 3.1. If elements of X are stochastic, it is assumed that they are distributed independently of $\mathbf{u}$ with a distribution that does *not* involve the parameters $\boldsymbol{\beta}$ and σ.

Under the above assumptions the joint pdf for the elements of $\mathbf{y}$, given X, $\boldsymbol{\beta}$, and σ, is

$$
\begin{aligned}
p(\mathbf{y}|X, \boldsymbol{\beta}, \sigma) &\propto \frac{1}{\sigma^n} \exp\left[-\frac{1}{2\sigma^2}(\mathbf{y} - X\boldsymbol{\beta})'(\mathbf{y} - X\boldsymbol{\beta})\right] \\
&\propto \frac{1}{\sigma^n} \exp\left\{-\frac{1}{2\sigma^2}[\nu s^2 + (\boldsymbol{\beta} - \hat{\boldsymbol{\beta}})'X'X(\boldsymbol{\beta} - \hat{\boldsymbol{\beta}})]\right\},
\end{aligned}
\tag{3.27}
$$

where $\nu = n - k$,

$$\hat{\boldsymbol{\beta}} = (X'X)^{-1}X'\mathbf{y}. \tag{3.28}$$

and

$$s^2 = \frac{(\mathbf{y} - X\hat{\boldsymbol{\beta}})'(\mathbf{y} - X\hat{\boldsymbol{\beta}})}{\nu} \tag{3.29}$$

are sufficient statistics. The second line of (3.27) makes use of the following algebraic identity:

$$
\begin{aligned}
(\mathbf{y} - X\boldsymbol{\beta})'(\mathbf{y} - X\boldsymbol{\beta}) &= [\mathbf{y} - X\hat{\boldsymbol{\beta}} - X(\boldsymbol{\beta} - \hat{\boldsymbol{\beta}})]'[\mathbf{y} - X\hat{\boldsymbol{\beta}} - X(\boldsymbol{\beta} - \hat{\boldsymbol{\beta}})] \\
&= (\mathbf{y} - X\hat{\boldsymbol{\beta}})'(\mathbf{y} - X\hat{\boldsymbol{\beta}}) + (\boldsymbol{\beta} - \hat{\boldsymbol{\beta}})'X'X(\boldsymbol{\beta} - \hat{\boldsymbol{\beta}}),
\end{aligned}
$$

since the cross-product terms

$$(\boldsymbol{\beta} - \hat{\boldsymbol{\beta}})'X'(\mathbf{y} - X\hat{\boldsymbol{\beta}}) = (\boldsymbol{\beta} - \hat{\boldsymbol{\beta}})'[X'\mathbf{y} - X'X(X'X)^{-1}X'\mathbf{y}] = 0.$$

3.2.2 Posterior Pdf's for Parameters with a Diffuse Prior Pdf

As prior pdf in the analysis of the multiple regression model, we assume that our information is diffuse or vague and represent it by taking the elements of $\boldsymbol{\beta}$ and $\log \sigma$ independently and uniformly distributed; that is

$$p(\boldsymbol{\beta}, \sigma) \propto \frac{1}{\sigma}, \qquad \begin{matrix} -\infty < \beta_i < \infty, \\ 0 < \sigma < \infty, \end{matrix} \qquad \text{for } i = 1, 2, \ldots, k. \tag{3.30}$$

On combining (3.27) and (3.30), the joint posterior pdf for the parameters $\boldsymbol{\beta}$ and σ is

$$p(\boldsymbol{\beta}, \sigma|\mathbf{y}, X) \propto \frac{1}{\sigma^{n+1}} \exp\left\{-\frac{1}{2\sigma^2}[\nu s^2 + (\boldsymbol{\beta} - \hat{\boldsymbol{\beta}})'X'X(\boldsymbol{\beta} - \hat{\boldsymbol{\beta}})]\right\}. \tag{3.31}$$

From (3.31) it is seen immediately that the conditional posterior pdf for $\boldsymbol{\beta}$, given σ [i.e., $p(\boldsymbol{\beta}|\sigma, \mathbf{y}, X)$], is a k-dimensional multivariate normal pdf with mean $\hat{\boldsymbol{\beta}}$ and covariance matrix $(X'X)^{-1}\sigma^2$. Although this fact is interesting and useful in certain derivations, σ^2 is rarely known in practice and thus the conditional covariance matrix $(X'X)^{-1}\sigma^2$ cannot be evaluated. To get rid of the troublesome parameter σ, we integrate (3.31) with respect to σ to obtain the following marginal posterior pdf for the elements of $\boldsymbol{\beta}$:

$$\begin{aligned} p(\boldsymbol{\beta}|\mathbf{y}, X) &= \int_0^\infty p(\boldsymbol{\beta}, \sigma|\mathbf{y}, X)\, d\sigma \\ &\propto \{\nu s^2 + (\boldsymbol{\beta} - \hat{\boldsymbol{\beta}})'X'X(\boldsymbol{\beta} - \hat{\boldsymbol{\beta}})\}^{-n/2}, \end{aligned} \tag{3.32}$$

which is in the form of a multivariate Student t pdf. This posterior pdf serves as a basis for making inferences about $\boldsymbol{\beta}$. Before turning to further analysis of it, we note that the marginal posterior pdf for σ can be obtained from (3.31) by integrating with respect to the elements of $\boldsymbol{\beta}$; that is,

$$\begin{aligned} p(\sigma|\mathbf{y}, X) &= \int_{-\infty}^{\infty} \cdots \int_{-\infty}^{\infty} p(\boldsymbol{\beta}, \sigma|\mathbf{y}, X)\, d\boldsymbol{\beta} \\ &\propto \frac{1}{\sigma^{\nu+1}} \exp\left(-\frac{\nu s^2}{2\sigma^2}\right), \end{aligned} \tag{3.33}$$

which is in the form of an inverted gamma pdf and exactly in the same form as (3.15), except that here $\nu = n - k$. By simple changes of variable the posterior pdf's for σ^2 or $h = 1/\sigma^2$ can be obtained from (3.33) if they are wanted.

We now return to the analysis of (3.32), the marginal posterior pdf for $\boldsymbol{\beta}$. First, we derive the marginal posterior pdf for a single element of $\boldsymbol{\beta}$, say β_1. This can be done in two ways, namely, by integrating (3.31) with respect to $\beta_2, \beta_3, \ldots, \beta_k$ and then with respect to σ or by integrating (3.32) with respect to $\beta_2, \beta_3, \ldots, \beta_k$. We take the second route here. For convenience rewrite (3.32) as follows:

$$p(\boldsymbol{\delta}|\mathbf{y}, X) \propto (\nu + \boldsymbol{\delta}'H\boldsymbol{\delta})^{-n/2}, \tag{3.34}$$

with $\boldsymbol{\delta}' = (\boldsymbol{\beta} - \hat{\boldsymbol{\beta}})'$ and $H = X'X/s^2$. Now let $\delta_1 = \beta_1 - \hat{\beta}_1$, a scalar, and $\boldsymbol{\delta}_2' = (\beta_2 - \hat{\beta}_2, \beta_3 - \hat{\beta}_3, \ldots, \beta_k - \hat{\beta}_k)$. Then

$$\boldsymbol{\delta}'H\boldsymbol{\delta} = \delta_1^2 h_{11} + \boldsymbol{\delta}_2'H_{22}\boldsymbol{\delta}_2 + 2\delta_1 H_{12}\boldsymbol{\delta}_2, \tag{3.35}$$

where H has been partitioned to conform with the partitioning of $\boldsymbol{\delta}$; that is

$$H = \left(\begin{array}{c|c} h_{11} & H_{12} \\ \hline H_{21} & H_{22} \end{array}\right),$$

with h_{11}, a scalar, H_{12} a $1 \times (k-1)$ vector, H_{21} a $(k-1) \times 1$ vector and H_{22} a $(k-1) \times (k-1)$ matrix. Now complete the square on $\boldsymbol{\delta}_2$ in (3.35) as follows

$$
\begin{aligned}
\boldsymbol{\delta}'H\boldsymbol{\delta} &= \delta_1^2 h_{11} - \delta_1^2 H_{12}H_{22}^{-1}H_{21} \\
&\qquad + (\boldsymbol{\delta}_2 + \delta_1 H_{22}^{-1}H_{21})'H_{22}(\boldsymbol{\delta}_2 + \delta_1 H_{22}^{-1}H_{21}) \\
&= \delta_1^2(h_{11} - H_{12}H_{22}^{-1}H_{21}) \\
&\qquad + (\boldsymbol{\delta}_2 + \delta_1 H_{22}^{-1}H_{21})'H_{22}(\boldsymbol{\delta}_2 + \delta_1 H_{22}^{-1}H_{21}).
\end{aligned} \tag{3.36}
$$

Then substitute from (3.36) in (3.34) to obtain

$$
\begin{aligned}
p(\delta_1, \boldsymbol{\delta}_2 | \mathbf{y}, X) &\propto \left(\nu + \frac{\delta_1^2}{h^{11}}\right)^{-n/2} \\
&\qquad \times [1 + (\boldsymbol{\delta}_2 + \delta_1 H_{22}^{-1}H_{21})'C(\boldsymbol{\delta}_2 + \delta_1 H_{22}^{-1}H_{21})]^{-n/2},
\end{aligned} \tag{3.37}
$$

with $C = H_{22}/(\nu + \delta_1^2/h^{11})$, where $h^{11} = (h_{11} - H_{12}H_{22}^{-1}H_{21})^{-1}$, the (1, 1) element of the inverse of H. Now (3.37) can be integrated with respect to $\boldsymbol{\delta}_2$ by using the properties of the multivariate Student t pdf to yield[9]:

$$
\begin{aligned}
p(\delta_1 | \mathbf{y}, X) &\propto \left(\nu + \frac{\delta_1^2}{h^{11}}\right)^{-n/2} |C|^{-\frac{1}{2}}. \\
&\propto \left(\nu + \frac{\delta_1^2}{h^{11}}\right)^{-(n-k+1)/2}.
\end{aligned} \tag{3.38}
$$

Thus

$$
\frac{\delta_1}{(h^{11})^{\frac{1}{2}}} = \frac{\beta_1 - \hat{\beta}_1}{s(m^{11})^{\frac{1}{2}}} \sim t_\nu, \tag{3.39}
$$

where m^{11} is the (1, 1) element of $(X'X)^{-1}$. Since the choice of which element of $\boldsymbol{\beta}$ is labeled β_1 is open, we have for the ith element of $\boldsymbol{\beta}$:

$$
\frac{\delta_i}{(h^{ii})^{\frac{1}{2}}} = \frac{\beta_i - \hat{\beta}_i}{s(m^{ii})^{\frac{1}{2}}} \sim t_\nu. \tag{3.40}
$$

This result enables us to make inferences about β_i conveniently by consulting the t tables for $\nu = n - k$ degrees of freedom. Note also that a simple change of variable in (3.38), namely, $F = \delta_1^2/h^{11} = (\beta_1 - \hat{\beta}_1)^2/s^2m^{11}$, yields

$$
P(F) \propto F^{-\frac{1}{2}}(\nu + F)^{-(\nu+1)/2}, \qquad 0 < F < \infty, \tag{3.41}
$$

and thus $F = (\beta_1 - \hat{\beta}_1)^2/s^2m^{11}$ has the $F_{1,\nu}$ pdf.

If we are interested in the marginal posterior pdf of a subset of elements of $\boldsymbol{\beta}$ (or of $\boldsymbol{\delta} = \boldsymbol{\beta} - \hat{\boldsymbol{\beta}}$), we can partition $\boldsymbol{\delta}' = (\boldsymbol{\delta}_1' \vdots \boldsymbol{\delta}_2')$ and write the quadratic form $\boldsymbol{\delta}'H\boldsymbol{\delta}$ appearing in (3.34) as follows:

$$
\begin{aligned}
\boldsymbol{\delta}'H\boldsymbol{\delta} &= (\boldsymbol{\delta}_1' \vdots \boldsymbol{\delta}_2') \begin{pmatrix} H_{11} & H_{12} \\ H_{21} & H_{22} \end{pmatrix} \begin{pmatrix} \boldsymbol{\delta}_1 \\ \boldsymbol{\delta}_2 \end{pmatrix} \\
&= \boldsymbol{\delta}_1'H_{11}\boldsymbol{\delta}_1 + \boldsymbol{\delta}_2'H_{22}\boldsymbol{\delta}_2 + 2\boldsymbol{\delta}_1'H_{12}\boldsymbol{\delta}_2,
\end{aligned} \tag{3.42}
$$

[9] The integration of $|C|^{\frac{1}{2}}[1 + (\boldsymbol{\delta}_2 + \delta_1 H_{22}^{-1}H_{21})'C(\boldsymbol{\delta}_2 + \delta_1 H_{22}^{-1}H_{21})]^{-n/2}$ with respect to $\boldsymbol{\delta}_2$ yields a constant independent of δ_1. Since C involves δ_1, it must appear in (3.38).

where the partitioning of H has been done to conform with that of $\boldsymbol{\delta}$. On completing the square on $\boldsymbol{\delta}_2$ in (3.42) we have

$$\boldsymbol{\delta}'H\boldsymbol{\delta} = \boldsymbol{\delta}_1'H_{11}\boldsymbol{\delta}_1 - \boldsymbol{\delta}_1'H_{12}H_{22}^{-1}H_{21}\boldsymbol{\delta}_1 + (\boldsymbol{\delta}_2 + H_{22}^{-1}H_{21}\boldsymbol{\delta}_1)'H_{22}(\boldsymbol{\delta}_2 + H_{22}^{-1}H_{21}\boldsymbol{\delta}_1),$$

which, when substituted in (3.34), yields

$$(3.43) \quad p(\boldsymbol{\delta}_1, \boldsymbol{\delta}_2|\mathbf{y}, X) \propto [\nu + \boldsymbol{\delta}_1'(H_{11} - H_{12}H_{22}^{-1}H_{21})\boldsymbol{\delta}_1 + (\boldsymbol{\delta}_2 + H_{22}^{-1}H_{21}\boldsymbol{\delta}_1)'H_{22}(\boldsymbol{\delta}_2 + H_{22}^{-1}H_{21}\boldsymbol{\delta}_1)]^{-n/2}.$$

As an aside, we see from (3.43) that the conditional pdf for $\boldsymbol{\delta}_2$, given $\boldsymbol{\delta}_1$, $p(\boldsymbol{\delta}_2|\boldsymbol{\delta}_1, \mathbf{y}, X)$, is in the multivariate Student t form with conditional mean $-H_{22}^{-1}H_{21}\boldsymbol{\delta}_1$. Thus, since $\boldsymbol{\delta}_2 = \boldsymbol{\beta}_2 - \hat{\boldsymbol{\beta}}_2$, the conditional mean of $\boldsymbol{\beta}_2$, given $\boldsymbol{\beta}_1$, is $\hat{\boldsymbol{\beta}}_2 - H_{22}^{-1}H_{21}(\boldsymbol{\beta}_1 - \hat{\boldsymbol{\beta}}_1)$.

To obtain the marginal posterior pdf for $\boldsymbol{\delta}_1$, we have to integrate (3.43) with respect to $\boldsymbol{\delta}_2$ which can be done by using properties of the multivariate Student t pdf. This operation results in[10]

$$(3.44) \quad p(\boldsymbol{\delta}_1|\mathbf{y}, X) \propto [\nu + \boldsymbol{\delta}_1'(H_{11} - H_{12}H_{22}^{-1}H_{21})\boldsymbol{\delta}_1]^{-(n-k_2)/2},$$

where k_2 is the number of elements in $\boldsymbol{\delta}_2$. Note that $n - k_2 = n - k + k - k_2 = \nu + k_1$ so that the exponent of (3.44) is $-(\nu + k_1)/2$. Thus from (3.44) the marginal posterior pdf for $\boldsymbol{\delta}_1$ is in the multivariate Student t form with $E\boldsymbol{\delta}_1 = E(\boldsymbol{\beta}_1 - \hat{\boldsymbol{\beta}}_1) = \mathbf{0}$, that is $E\boldsymbol{\beta}_1 = \hat{\boldsymbol{\beta}}_1$, $\nu > 1$, and $E\boldsymbol{\delta}_1\boldsymbol{\delta}_1' = E(\boldsymbol{\beta}_1 - \hat{\boldsymbol{\beta}}_1)(\boldsymbol{\beta}_1 - \hat{\boldsymbol{\beta}}_1)' = [(\nu/(\nu - 2)](H_{11} - H_{12}H_{22}^{-1}H_{21})^{-1}$, $\nu > 2$, where it is to be remembered that the matrices $H_{\alpha l}$, $\alpha, l = 1, 2$, are submatrices of $H = X'X/s^2$.

Next, using the prior assumptions in (3.30), we turn to the problem of deriving the posterior pdf of a linear combination of the elements of $\boldsymbol{\beta}$, say $\alpha = \mathbf{l}'\boldsymbol{\beta}$, where α is a scalar parameter and $\mathbf{l}'$ is a $1 \times k$ vector of fixed numbers; for example, if the elements of $\boldsymbol{\beta}$ are Cobb-Douglas production function parameters, we might be interested in $\alpha = \beta_1 + \beta_2 + \cdots + \beta_k$, which is the "returns to scale" parameter. In this application $\mathbf{l}' = (1, 1, \ldots, 1)$. In other situations other linear combinations of the elements of $\boldsymbol{\beta}$ may be of interest. To obtain the posterior pdf of $\alpha = \mathbf{l}'\boldsymbol{\beta}$ we note that the joint posterior pdf for $\boldsymbol{\beta}$ and σ can be written as

$$(3.45) \quad p(\boldsymbol{\beta}, \sigma|\mathbf{y}, X) = p(\boldsymbol{\beta}|\sigma, \mathbf{y}, X)\, p(\sigma|\mathbf{y}, X),$$

[10] Write (3.43) as

$$p(\boldsymbol{\delta}_1, \boldsymbol{\delta}_2|\mathbf{y}, X) \propto \{[\nu + \boldsymbol{\delta}_1'(H_{11} - H_{12}H_{22}^{-1}H_{21})\boldsymbol{\delta}_1]^{-n/2}|C|^{-1/2}\} \times \{|C|^{1/2}[1 + (\boldsymbol{\delta}_2 + H_{22}^{-1}H_{21}\boldsymbol{\delta}_1)'C(\boldsymbol{\delta}_2 + H_{22}^{-1}H_{21}\boldsymbol{\delta}_1)]^{-n/2}\},$$

with $C \equiv H_{22}/[\nu + \boldsymbol{\delta}_1'(H_{11} - H_{12}H_{22}^{-1}H_{21})\boldsymbol{\delta}_1]$. By integrating over $\boldsymbol{\delta}_2$ the second factor yields a constant independent of $\boldsymbol{\delta}_1$ and the first factor is proportional to the expression shown in (3.44).

with $p(\boldsymbol{\beta}|\sigma, \mathbf{y}, X)$ in the multivariate normal form. Thus, conditional on σ, $\alpha = \mathbf{l}'\boldsymbol{\beta}$ will be normally distributed, since it is a linear combination of normally distributed variables, the elements of $\boldsymbol{\beta}$, with mean $\hat{\alpha} = \mathbf{l}'\hat{\boldsymbol{\beta}}$, and conditional posterior variance

$$E[(\alpha - \hat{\alpha})^2|\sigma, \mathbf{y}, X)] = E[\mathbf{l}'(\boldsymbol{\beta} - \hat{\boldsymbol{\beta}})(\boldsymbol{\beta} - \hat{\boldsymbol{\beta}})'\mathbf{l}|\sigma, \mathbf{y}, X] = \mathbf{l}'(X'X)^{-1}\mathbf{l}\sigma^2,$$

since the conditional covariance matrix for $\boldsymbol{\beta}$, given σ, is $(X'X)^{-1}\sigma^2$. Thus the marginal posterior pdf for α can be obtained by integrating $p(\alpha, \sigma|\mathbf{y}, X) = p(\alpha|\sigma, \mathbf{y}, X)p(\sigma|\mathbf{y}, X)$ with respect to σ. Letting $c = \mathbf{l}'(X'X)^{-1}\mathbf{l}$, we have

$$p(\alpha|\sigma, \mathbf{y}, X) \propto \frac{1}{\sigma} \exp\left[-\frac{(\alpha - \hat{\alpha})^2}{2\sigma^2 c}\right],$$

$$p(\sigma|\mathbf{y}, X) \propto \frac{1}{\sigma^{\nu+1}} \exp\left(-\frac{\nu s^2}{2\sigma^2}\right),$$

and

$$p(\alpha|\mathbf{y}, X) = \int_0^\infty p(\alpha|\sigma, \mathbf{y}, X)\, p(\sigma|\mathbf{y}, X)\, d\sigma$$

$$(3.46) \qquad \propto \int_0^\infty \frac{1}{\sigma^{\nu+2}} \exp\left\{-\frac{1}{2\sigma^2}\left[\nu s^2 + \frac{(\alpha - \hat{\alpha})^2}{c}\right]\right\} d\sigma$$

$$\propto \left[\nu + \frac{(\alpha - \hat{\alpha})^2}{s^2 c}\right]^{-(\nu+1)/2}.$$

Thus α has a posterior pdf in the univariate Student t form; that is,

$$(3.47) \qquad \frac{\alpha - \hat{\alpha}}{s c^{1/2}} \sim t_\nu,$$

with $\nu = n - k$, $\hat{\alpha} = \mathbf{l}'\hat{\boldsymbol{\beta}}$ and $c = \mathbf{l}'(X'X)^{-1}\mathbf{l}$. This fact can be utilized to make inferences about α.[11]

3.2.3 Posterior Pdf Based on an Informative Prior Pdf

We next take up the problem of using the posterior pdf in (3.31) as a prior pdf in the analysis of a new sample of data generated by the same regression process. To distinguish between the two samples subscripts 1 and 2 are employed. With this notation the posterior pdf in (3.31), which we use as a prior pdf for analysis of a new sample, is

$$(3.48) \qquad \begin{aligned} p(\boldsymbol{\beta}, \sigma|\mathbf{y}_1, X_1) &\propto \frac{1}{\sigma^{n_1+1}} \exp\left[-\frac{1}{2\sigma^2}(\mathbf{y}_1 - X_1\boldsymbol{\beta})'(\mathbf{y}_1 - X_1\boldsymbol{\beta})\right] \\ &\propto \frac{1}{\sigma^{n_1+1}} \exp\left\{-\frac{1}{2\sigma^2}[\nu_1 s_1^2 + (\boldsymbol{\beta} - \hat{\boldsymbol{\beta}}_1)'X_1'X_1(\boldsymbol{\beta} - \hat{\boldsymbol{\beta}}_1)]\right\}, \end{aligned}$$

[11] See, Appendix B for the derivation of the joint distribution of several linear combinations of variables having a multivariate Student pdf.

with $\nu_1 = n_1 - k$, $\hat{\boldsymbol{\beta}}_1 = (X_1'X_1)^{-1}X_1'\mathbf{y}_1$ and $\nu_1 s_1^2 = (\mathbf{y}_1 - X_1\hat{\boldsymbol{\beta}}_1)'(\mathbf{y}_1 - X_1\hat{\boldsymbol{\beta}}_1)$. Viewing (3.48) as a prior pdf, we see that it factors into a normal part for $\boldsymbol{\beta}$, given σ, with mean $\hat{\boldsymbol{\beta}}_1$ and covariance matrix $(X_1'X_1)^{-1}\sigma^2$ and a marginal pdf for σ in the inverted gamma form with parameters ν_1 and s_1^2; that is, from the second line of (3.48)

$$p(\boldsymbol{\beta}|\sigma, \hat{\boldsymbol{\beta}}_1, s_1^2) \propto \frac{1}{\sigma^k} \exp\left[-\frac{1}{2\sigma^2}(\boldsymbol{\beta} - \hat{\boldsymbol{\beta}}_1)'X_1'X_1(\boldsymbol{\beta} - \hat{\boldsymbol{\beta}}_1)\right]$$

and

$$p(\sigma|\hat{\boldsymbol{\beta}}_1, s_1^2) \propto \frac{1}{\sigma^{\nu_1+1}} \exp\left(-\frac{\nu_1 s_1^2}{2\sigma^2}\right).$$

Thus the prior pdf's parameters are just the quantities $\hat{\boldsymbol{\beta}}_1$, s_1^2, $X_1'X_1$, and ν_1.

The likelihood function for the second sample $(\mathbf{y}_2, X_2)$, where $\mathbf{y}_2$ is an $n_2 \times 1$ vector of observations on the dependent variable in the second sample and X_2 is an $n_2 \times k$ matrix with rank k of observations on the k independent variables in the second sample, is assumed to be given by

$$l(\boldsymbol{\beta}, \sigma|\mathbf{y}_2, X_2) \propto \frac{1}{\sigma^{n_2}} \exp\left[-\frac{1}{2\sigma^2}(\mathbf{y}_2 - X_2\boldsymbol{\beta})'(\mathbf{y}_2 - X_2\boldsymbol{\beta})\right]. \tag{3.49}$$

Note that $\boldsymbol{\beta}$ and σ for the second sample are assumed to be the same parameters as those for the first sample.

On combining the prior pdf in (3.48) with the likelihood function in (3.49), we obtain the posterior pdf:

$$\begin{aligned} p(\boldsymbol{\beta}, \sigma|\mathbf{y}_1, \mathbf{y}_2, X_1, X_2) \propto {} & \frac{1}{\sigma^{n_1+n_2+1}} \\ & \times \exp\left\{-\frac{1}{2\sigma^2}[(\mathbf{y}_1 - X_1\boldsymbol{\beta})'(\mathbf{y}_1 - X_1\boldsymbol{\beta}) \right. \\ & \left. + (\mathbf{y}_2 - X_2\boldsymbol{\beta})'(\mathbf{y}_2 - X_2\boldsymbol{\beta})]\right\}. \end{aligned} \tag{3.50}$$

This expression can be brought into a more convenient form on completing the square in the exponent; that is,

$$(\mathbf{y}_1 - X_1\boldsymbol{\beta})'(\mathbf{y}_1 - X_1\boldsymbol{\beta}) + (\mathbf{y}_2 - X_2\boldsymbol{\beta})'(\mathbf{y}_2 - X_2\boldsymbol{\beta}) = \nu s^2 + (\boldsymbol{\beta} - \tilde{\boldsymbol{\beta}})'M(\boldsymbol{\beta} - \tilde{\boldsymbol{\beta}}),$$

where

$$\begin{aligned} M &= X_1'X_1 + X_2'X_2, \\ \tilde{\boldsymbol{\beta}} &= M^{-1}(X_1'\mathbf{y}_1 + X_2'\mathbf{y}_2), \\ \nu s^2 &= (\mathbf{y}_1 - X_1\tilde{\boldsymbol{\beta}})'(\mathbf{y}_1 - X_1'\tilde{\boldsymbol{\beta}}) + (\mathbf{y}_2 - X_2\tilde{\boldsymbol{\beta}})'(\mathbf{y}_2 - X_2\tilde{\boldsymbol{\beta}}), \\ \nu &= n_1 + n_2 - k. \end{aligned}$$

Thus (3.50) can be written as

$$p(\boldsymbol{\beta}, \sigma|\mathbf{y}_1, \mathbf{y}_2, X_1, X_2) \propto \frac{1}{\sigma^{n_1+n_2+1}} \exp\left\{-\frac{1}{2\sigma^2}[\nu s^2 + (\boldsymbol{\beta} - \tilde{\boldsymbol{\beta}})'M(\boldsymbol{\beta} - \tilde{\boldsymbol{\beta}})]\right\}. \tag{3.51}$$

It is seen that (3.51) is in exactly the same form as (3.31) and thus it can be analyzed by using exactly the same techniques. Further, if we had pooled the two samples, based our likelihood function on both samples, and used the diffuse prior pdf in (3.30), the resulting posterior pdf would be exactly (3.50), from which (3.51) can be derived.

3.2.4 Predictive Pdf

In this section we derive the predictive pdf for a vector of q future observations, $\tilde{\mathbf{y}}' = (y_{n+1}, y_{n+2}, \ldots, y_{n+q})$, which is assumed to be generated by the multiple regression process specified at the beginning of this section; that is, with n sample observations $\mathbf{y}$, given X, and with the diffuse prior assumptions in (3.30), we wish to derive the pdf for $\tilde{\mathbf{y}}$ which is assumed to be generated by

$$\tilde{\mathbf{y}} = \tilde{X}\boldsymbol{\beta} + \tilde{\mathbf{u}}, \tag{3.52}$$

where $\tilde{X}$ is a $q \times k$ matrix of given values for the independent variables in the q future periods and $\tilde{\mathbf{u}}$ is a $q \times 1$ vector of future disturbance terms normally and independently distributed, each with mean zero and common variance σ^2.

As mentioned in Chapter 2, one way of deriving the predictive pdf is to write down the joint pdf $p(\tilde{\mathbf{y}}, \boldsymbol{\beta}, \sigma | X, \tilde{X}, \mathbf{y})$ and integrate with respect to $\boldsymbol{\beta}$ and σ to obtain the marginal pdf for $\tilde{\mathbf{y}}$, which is the predictive pdf. In the present problem this joint pdf factors as follows:

$$p(\tilde{\mathbf{y}}, \boldsymbol{\beta}, \sigma | X, \tilde{X}, \mathbf{y}) = p(\tilde{\mathbf{y}} | \boldsymbol{\beta}, \sigma, \tilde{X})\, p(\boldsymbol{\beta}, \sigma | \mathbf{y}, X) \tag{3.53}$$

with $p(\boldsymbol{\beta}, \sigma | \mathbf{y}, X)$ being the posterior pdf for $\boldsymbol{\beta}$ and σ, shown in (3.31) and

$$p(\tilde{\mathbf{y}} | \boldsymbol{\beta}, \sigma, \tilde{X}) \propto \frac{1}{\sigma^q} \exp\left[-\frac{1}{2\sigma^2}(\tilde{\mathbf{y}} - \tilde{X}\boldsymbol{\beta})'(\tilde{\mathbf{y}} - \tilde{X}\boldsymbol{\beta})\right]. \tag{3.54}$$

With this noted, (3.53) is proportional to

$$\frac{1}{\sigma^{n+q+1}} \exp\left\{-\frac{1}{2\sigma^2}[(\mathbf{y} - X\boldsymbol{\beta})'(\mathbf{y} - X\boldsymbol{\beta}) + (\tilde{\mathbf{y}} - \tilde{X}\boldsymbol{\beta})'(\tilde{\mathbf{y}} - \tilde{X}\boldsymbol{\beta})]\right\} \tag{3.55}$$

and our problem is to integrate (3.35) with respect to σ and $\boldsymbol{\beta}$. On integrating with respect to σ we obtain

$$p(\boldsymbol{\beta}, \tilde{\mathbf{y}} | \mathbf{y}, X, \tilde{X}) \propto [(\mathbf{y} - X\boldsymbol{\beta})'(\mathbf{y} - X\boldsymbol{\beta}) + (\tilde{\mathbf{y}} - \tilde{X}\boldsymbol{\beta})'(\tilde{\mathbf{y}} - \tilde{X}\boldsymbol{\beta})]^{-(n+q)/2}. \tag{3.56}$$

On completing the square on $\boldsymbol{\beta}$ we have

$$\begin{aligned}
(\mathbf{y} - X\boldsymbol{\beta})'(\mathbf{y} - X\boldsymbol{\beta}) &+ (\tilde{\mathbf{y}} - \tilde{X}\boldsymbol{\beta})'(\tilde{\mathbf{y}} - \tilde{X}\boldsymbol{\beta}) \\
&= \mathbf{y}'\mathbf{y} + \tilde{\mathbf{y}}'\tilde{\mathbf{y}} + \boldsymbol{\beta}'M\boldsymbol{\beta} - 2\boldsymbol{\beta}'(X'\mathbf{y} + \tilde{X}'\tilde{\mathbf{y}}) \\
&= \mathbf{y}'\mathbf{y} + \tilde{\mathbf{y}}'\tilde{\mathbf{y}} - (\mathbf{y}'X + \tilde{\mathbf{y}}'\tilde{X})M^{-1}(X'\mathbf{y} + \tilde{X}'\tilde{\mathbf{y}}) \\
&\quad + [\boldsymbol{\beta} - M^{-1}(X'\mathbf{y} + \tilde{X}'\tilde{\mathbf{y}})]'M[\boldsymbol{\beta} - M^{-1}(X'\mathbf{y} + \tilde{X}'\tilde{\mathbf{y}})],
\end{aligned}$$

where $M = X'X + \tilde{X}'\tilde{X}$. On substituting in (3.56) and integrating with respect to the k elements of $\boldsymbol{\beta}$, we obtain

$$(3.57)\quad p(\tilde{\mathbf{y}}|\mathbf{y}, X, \tilde{X}) \propto [\mathbf{y}'\mathbf{y} + \tilde{\mathbf{y}}'\tilde{\mathbf{y}} - (\mathbf{y}'X + \tilde{\mathbf{y}}'\tilde{X})M^{-1}(X'\mathbf{y} + \tilde{X}'\tilde{\mathbf{y}})]^{-(\nu+q)/2},$$

where $\nu = n - k$. To put (3.57) in a more intelligible form we write the quantity within brackets on the rhs of (3.57) as

$$(3.58)\quad \begin{aligned} &\mathbf{y}'(I - XM^{-1}X')\mathbf{y} + \tilde{\mathbf{y}}'(I - \tilde{X}M^{-1}\tilde{X}')\tilde{\mathbf{y}} - 2\tilde{\mathbf{y}}'\tilde{X}M^{-1}X'\mathbf{y} \\ &\quad = \mathbf{y}'(I - XM^{-1}X')\mathbf{y} - \mathbf{y}'XM^{-1}\tilde{X}'(I - \tilde{X}M^{-1}\tilde{X}')^{-1}\tilde{X}M^{-1}X'\mathbf{y} \\ &\qquad + [\tilde{\mathbf{y}} - (I - \tilde{X}M^{-1}\tilde{X}')^{-1}\tilde{X}M^{-1}X'\mathbf{y}]'(I - \tilde{X}M^{-1}\tilde{X}') \\ &\qquad \times [\tilde{\mathbf{y}} - (I - \tilde{X}M^{-1}\tilde{X}')^{-1}\tilde{X}M^{-1}X'\mathbf{y}]. \end{aligned}$$

Now note the following result which can be verified by direct matrix multiplication:

$$(3.59)\quad (I - \tilde{X}M^{-1}\tilde{X}')^{-1} = I + \tilde{X}(X'X)^{-1}\tilde{X}'.$$

Using this result, we have

$$(3.60)\quad \begin{aligned} (I - \tilde{X}M^{-1}\tilde{X}')^{-1}\tilde{X}M^{-1} &= [I + \tilde{X}(X'X)^{-1}\tilde{X}']\tilde{X}M^{-1} \\ &= \tilde{X}[I + (X'X)^{-1}\tilde{X}'\tilde{X}](X'X + \tilde{X}'\tilde{X})^{-1} \\ &= \tilde{X}(X'X)^{-1}. \end{aligned}$$

Substitution of this result in (3.58) leads to

$$\begin{aligned} &\mathbf{y}'(I - XM^{-1}X')\mathbf{y} - \mathbf{y}'XM^{-1}\tilde{X}'\tilde{X}(X'X)^{-1}X'\mathbf{y} \\ &\qquad\qquad + (\tilde{\mathbf{y}} - \tilde{X}\hat{\boldsymbol{\beta}})'(I - \tilde{X}M^{-1}\tilde{X}')(\tilde{\mathbf{y}} - \tilde{X}\hat{\boldsymbol{\beta}}) \\ &\quad = \mathbf{y}'\{I - X[M^{-1} + M^{-1}\tilde{X}'\tilde{X}(X'X)^{-1}]X'\}\mathbf{y} \\ &\qquad + (\tilde{\mathbf{y}} - \tilde{X}\hat{\boldsymbol{\beta}})'(I - \tilde{X}M^{-1}\tilde{X}')(\tilde{\mathbf{y}} - \tilde{X}\hat{\boldsymbol{\beta}}) \\ &\quad = \mathbf{y}'[I - XM^{-1}(X'X + \tilde{X}'\tilde{X})(X'X)^{-1}X']\mathbf{y} \\ &\qquad + (\tilde{\mathbf{y}} - \tilde{X}\hat{\boldsymbol{\beta}})'(I - \tilde{X}M^{-1}\tilde{X}')(\tilde{\mathbf{y}} - \tilde{X}\hat{\boldsymbol{\beta}}) \\ &\quad = \mathbf{y}'[I - X(X'X)^{-1}X']\mathbf{y} + (\tilde{\mathbf{y}} - \tilde{X}\hat{\boldsymbol{\beta}})'(I - \tilde{X}M^{-1}\tilde{X}')(\tilde{\mathbf{y}} - \tilde{X}\hat{\boldsymbol{\beta}}), \end{aligned}$$

where $\hat{\boldsymbol{\beta}} = (X'X)^{-1}X'\mathbf{y}$. Utilizing this result, and since $\mathbf{y}'[I - X(X'X)^{-1}X']\mathbf{y} = (\mathbf{y} - X\hat{\boldsymbol{\beta}})'(\mathbf{y} - X\hat{\boldsymbol{\beta}}) = \nu s^2$, we can write (3.57) as

$$(3.61)\quad p(\tilde{\mathbf{y}}|\mathbf{y}, X, \tilde{X}) \propto [\nu + (\tilde{\mathbf{y}} - \tilde{X}\hat{\boldsymbol{\beta}})'H(\tilde{\mathbf{y}} - \tilde{X}\hat{\boldsymbol{\beta}}]^{-(\nu+q)/2},$$

where $H = (1/s^2)(I - \tilde{X}M^{-1}\tilde{X}')$. It is seen from (3.61) that $\tilde{\mathbf{y}}$ is distributed in the multivariate Student t form. Thus we have for the mean of $\tilde{\mathbf{y}}$

$$(3.62)\quad E\tilde{\mathbf{y}} = \tilde{X}\hat{\boldsymbol{\beta}} \qquad \nu > 1$$

and for the covariance matrix

$$E(\tilde{\mathbf{y}} - E\tilde{\mathbf{y}})(\tilde{\mathbf{y}} - E\tilde{\mathbf{y}})' = \frac{\nu}{\nu - 2} H^{-1} \qquad \nu > 2$$

$$(3.63) \qquad = \frac{\nu s^2}{\nu - 2}(I - \tilde{X}M^{-1}\tilde{X}')^{-1}$$

$$= \frac{\nu s^2}{\nu - 2}[I + \tilde{X}(X'X)^{-1}\tilde{X}'].$$

Also, of course, the properties of the multivariate Student t pdf, established in connection with (3.32), apply here as well; for example, the marginal pdf of a single element of $\tilde{\mathbf{y}}$, say $\tilde{y}_i$, will be in the univariate t form:

$$(3.64) \qquad \frac{\tilde{y}_i - \tilde{X}(i)\hat{\boldsymbol{\beta}}}{\sqrt{h^{ii}}} \sim t_\nu,$$

where $\tilde{X}(i)$ is the ith row of $\tilde{X}$ and h^{ii} is the (i, i)th element of the inverse of H.

Last, as in (3.32) and (3.47), a linear combination of the elements of $\tilde{\mathbf{y}}$ will be distributed in the univariate Student t form; that is, let $\mathbf{l}$ be an $q \times 1$ nonstochastic vector with given elements. Then $V = \mathbf{l}'\tilde{\mathbf{y}}$ will be distributed in the univariate Student t form:

$$(3.65) \qquad \frac{V - \hat{V}}{(\mathbf{l}'H^{-1}\mathbf{l})^{1/2}} \sim t_\nu,$$

where $\hat{V} = \mathbf{l}'\tilde{X}\hat{\boldsymbol{\beta}}$.

A particular linear combination of future observations often encountered in economic work is the following:

$$(3.66) \qquad V = \frac{\tilde{y}_{n+1}}{1 + r} + \frac{\tilde{y}_{n+2}}{(1 + r)^2} + \cdots + \frac{\tilde{y}_{n+q}}{(1 + r)^q},$$

where r is a given discount rate.[12] For (3.66)

$$\mathbf{l}' = [(1 + r)^{-1}, (1 + r)^{-2}, \ldots, (1 + r)^{-q}].$$

With the result in (3.65) available, the distribution of the quantity V in (3.66) is known. Further, if we have a utility function depending on V, say $U(V)$, its expectation can be evaluated:

$$(3.67) \qquad EU(V) = \int_{-\infty}^{\infty} U(V)\, p(V|\mathbf{y})\, dV,$$

since from (3.65) we know the pdf for V, $p(V|\mathbf{y})$. Computation of (3.67) provides a means of comparing the expected utility associated with various

[12] We could easily modify (3.66) to incorporate different discount rates for different future periods if that were thought appropriate; that is

$$V = \frac{\tilde{y}_{n+1}}{1 + r_1} + \frac{\tilde{y}_{n+2}}{(1 + r_2)^2} + \cdots + \frac{\tilde{y}_{n+q}}{(1 + r_q)^q}.$$

V's, provided that each V is a linear combination of the future observations generated by a normal regression model.

3.2.5 Analysis of Model when $X'X$ is Singular

The moment matrix $X'X$ will be singular when the $n \times k$ matrix X is of rank q with $0 \leq q < k$. This occurs, for example, when the observations on the independent variables satisfy an exact linear relation and thus the columns of X are not linearly independent; that is, for $k = 2$, $X = (\mathbf{x}_1, \mathbf{x}_2)$, if $\mathbf{x}_1$ and $\mathbf{x}_2$ satisfy an exact linear relation it can easily be shown that $|X'X| = 0$ and thus $X'X$ is *not* of rank $k = 2$. This situation is commonly termed "multicollinearity." Another example in which X cannot be of rank k is when $n < k$; that is, when the number of observations is less than the number (k) of independent variables or coefficients to be estimated. This problem often arises in connection with analyses of reduced-form equations associated with large simultaneous equation econometric models.[13] Also, design matrices in the area of experimental design are often not of full rank.[14]

When $X'X$ is singular, for whatever reason, it is generally appreciated that prior information, in some form, must be added to the sample information in order to estimate *all* k regression coefficients. Below we analyze the model with a natural conjugate prior pdf. Then we review a sampling theory approach that utilizes generalized inverses and give it a Bayesian interpretation.[15]

In the approach to the analysis of the regression model when $X'X$ is singular, employed by Raiffa and Schlaifer[16] and Ando and Kaufman,[17] it is assumed that we have prior information about $\boldsymbol{\beta}$ and σ which can be represented by the following natural conjugate prior pdf:

$$p(\boldsymbol{\beta}, \sigma) = p(\boldsymbol{\beta}|\sigma)\, p(\sigma), \tag{3.68}$$

[13] See, for example, F. M. Fisher, "Dynamic Structure and Estimation in Economy-wide Econometric Models," in J. S. Duesenberry et al., *The Brookings Quarterly Econometric Model of the United States*. Chicago: Rand-McNally, 1965, pp. 589–635, especially p. 622 ff.

[14] See, for example, F. A. Graybill, *An Introduction to Linear Statistical Models*, New York: McGraw-Hill, 1961.

[15] The material in this section appears in S. J. Press and A. Zellner, "On Generalized Inverses and Prior Information in Regression Analysis," manuscript, September 1968.

[16] H. Raiffa and R. Schlaifer, *Applied Statistical Decision Theory*. Boston: Division of Research, Graduate School of Business Administration, Harvard University, 1961.

[17] A. Ando and G. M. Kaufman, "Bayesian Analysis of the Independent Multinormal Process—Neither Mean Nor Precision Known," *J. Amer. Statistical Association*, **60**, 347–358 (1965).

where

$$p(\boldsymbol{\beta}|\sigma) \propto \frac{|A|^{1/2}}{\sigma^k} \exp\left[-\frac{1}{2\sigma^2}(\boldsymbol{\beta} - \bar{\boldsymbol{\beta}})'A(\boldsymbol{\beta} - \bar{\boldsymbol{\beta}})\right] \tag{3.69}$$

and

$$p(\sigma) \propto \frac{1}{\sigma^{\nu_0+1}} \exp\left(-\frac{\nu_0 c_0^2}{2\sigma^2}\right), \qquad \nu_0 > 0. \tag{3.70}$$

In (3.69) we have a normal prior pdf for $\boldsymbol{\beta}$, given σ, with prior mean $\bar{\boldsymbol{\beta}}$ and prior covariance matrix $\sigma^2 A^{-1}$, which is assumed to be nonsingular. In (3.70) the prior pdf for σ is in the inverted gamma form with prior parameters ν_0 and c_0^2. The prior parameters $\bar{\boldsymbol{\beta}}$, A, ν_0, and c_0^2 must be assigned appropriate values to represent the prior information that is assumed to be available.

The prior pdf's in (3.69) and (3.70) can easily be combined with the likelihood function in the first line of (3.27) to yield the posterior pdf for $\boldsymbol{\beta}$ and σ[18]:

$$\begin{aligned} p(\boldsymbol{\beta}, \sigma|\mathbf{y}) &\propto \frac{1}{\sigma^{n+k+\nu_0+1}} \exp\left\{-\frac{1}{2\sigma^2}[\nu_0 c_0^2 + (\boldsymbol{\beta} - \bar{\boldsymbol{\beta}})'A(\boldsymbol{\beta} - \bar{\boldsymbol{\beta}}) \right. \\ &\qquad \left. + (\mathbf{y} - X\boldsymbol{\beta})'(\mathbf{y} - X\boldsymbol{\beta})]\right\} \\ &\propto \frac{1}{\sigma^{n'+k+1}} \exp\left\{-\frac{1}{2\sigma^2}[n'c^2 + (\boldsymbol{\beta} - \check{\boldsymbol{\beta}})'(A + X'X)(\boldsymbol{\beta} - \check{\boldsymbol{\beta}})]\right\}, \end{aligned} \tag{3.71}$$

where $n' = n + \nu_0$, $n'c^2 = \nu_0 c_0^2 + \mathbf{y}'\mathbf{y} + \bar{\boldsymbol{\beta}}'A\bar{\boldsymbol{\beta}} - \check{\boldsymbol{\beta}}'(A + X'X)\check{\boldsymbol{\beta}}$, and

$$\check{\boldsymbol{\beta}} = (A + X'X)^{-1}(A\bar{\boldsymbol{\beta}} + X'\mathbf{y}). \tag{3.72}$$

On integrating (3.71) with respect to σ, the marginal posterior pdf for $\boldsymbol{\beta}$ is

$$p(\boldsymbol{\beta}|\mathbf{y}) \propto [n'c^2 + (\boldsymbol{\beta} - \check{\boldsymbol{\beta}})'(A + X'X)(\boldsymbol{\beta} - \check{\boldsymbol{\beta}})]^{-(n'+k)/2}, \tag{3.73}$$

which is a proper posterior pdf in the multivariate Student t form with mean $\check{\boldsymbol{\beta}}$, shown in (3.72). Using (3.73), posterior inferences about all the elements of $\boldsymbol{\beta}$ can be made. Thus, given that we have prior information which can be adequately represented by the prior pdf's in (3.69) and (3.70), the Bayesian analysis of the model is quite straightforward, even though $X'X$ is assumed to be singular.

We next review a sampling theory approach to the analysis of the model

[18] In going from the first line of (3.71) to the second, we complete the square on $\boldsymbol{\beta}$ as follows: $(\boldsymbol{\beta} - \bar{\boldsymbol{\beta}})'A(\boldsymbol{\beta} - \bar{\boldsymbol{\beta}}) + (\mathbf{y} - X\boldsymbol{\beta})'(\mathbf{y} - X\boldsymbol{\beta}) = \boldsymbol{\beta}'(A + X'X)\boldsymbol{\beta} - 2\boldsymbol{\beta}'(A\bar{\boldsymbol{\beta}} + X'\mathbf{y}) + \mathbf{y}'\mathbf{y} + \bar{\boldsymbol{\beta}}'A\bar{\boldsymbol{\beta}} = (\boldsymbol{\beta} - \check{\boldsymbol{\beta}})'(A + X'X)(\boldsymbol{\beta} - \check{\boldsymbol{\beta}}) + \mathbf{y}'\mathbf{y} + \bar{\boldsymbol{\beta}}'A\bar{\boldsymbol{\beta}} - \check{\boldsymbol{\beta}}'(A + X'X)\check{\boldsymbol{\beta}}$, with $\check{\boldsymbol{\beta}} = (A + X'X)^{-1}(A\bar{\boldsymbol{\beta}} + X'\mathbf{y})$.

when $X'X$ is singular and give it a Bayesian interpretation. It is convenient and illuminating to reparameterize the model:

$$\mathbf{y} = XP\boldsymbol{\gamma} + \mathbf{u} \tag{3.74}$$

where $\boldsymbol{\gamma}$, a $k \times 1$ vector of parameters, is given by

$$\boldsymbol{\gamma} = P'\boldsymbol{\beta} \tag{3.75}$$

and P is a $k \times k$ orthogonal matrix[19] such that

$$P'X'XP = \begin{pmatrix} D & 0 \\ 0 & 0 \end{pmatrix}, \tag{3.76}$$

where D is a $q \times q$ nonsingular diagonal matrix with the nonzero characteristic roots of $X'X$ on the diagonal. Then the "normal equations" for $\boldsymbol{\gamma}$ are

$$P'X'XP\boldsymbol{\gamma} = P'X'\mathbf{y} \tag{3.77}$$

or

$$\begin{pmatrix} D & 0 \\ 0 & 0 \end{pmatrix}\begin{pmatrix} \boldsymbol{\gamma}_1 \\ \boldsymbol{\gamma}_2 \end{pmatrix} = \begin{pmatrix} P_1'X'\mathbf{y} \\ P_2'X'\mathbf{y} \end{pmatrix} = \begin{pmatrix} P_1'X'\mathbf{y} \\ \mathbf{0} \end{pmatrix}, \tag{3.78}$$

where $\boldsymbol{\gamma}' = (\boldsymbol{\gamma}_1' \vdots \boldsymbol{\gamma}_2')$, with $\boldsymbol{\gamma}_1$ and $\boldsymbol{\gamma}_2$ $q \times 1$ and $(k - q) \times 1$ vectors, respectively, and P_1 is a $k \times q$ submatrix of P given by $P = (P_1 \vdots P_2)$. From (3.76) note that $P_2'X' = 0$, and thus the vector on the rhs of (3.78) has a zero subvector.

The complete solution of the normal equations in (3.78) is given by[20]

$$\tilde{\boldsymbol{\gamma}} = (P'X'XP)^*P'X'\mathbf{y} + [I - (P'X'XP)^*P'X'XP]\mathbf{z}, \tag{3.79}$$

where $(P'X'XP)^*$ denotes a generalized inverse[21] (GI) of $P'X'XP$ and $\mathbf{z}$ is an

[19] Given that P is orthogonal, on substituting from (3.75) into (3.74) we have $\mathbf{y} = XPP'\boldsymbol{\beta} + \mathbf{u} = X\boldsymbol{\beta} + \mathbf{u}$, since $PP' = I$.

[20] See, for example, C. R. Rao, *Linear Statistical Inference and Its Applications*. New York: Wiley, 1965, p. 26.

[21] M^* is a GI of M if, and only if, $MM^*M = M$. This definition does not make M^* unique, as is well known. For further discussion of GI's, see C. R. Rao, *ibid.*, pp. 24–26, and "A Note on a Generalized Inverse of a Matrix with Applications to problems in Mathematical Statistics," *J. Roy. Statistical Soc.*, Series B, **24**, 152–158 (1962); T. N. E. Greville, "Some Applications of the Pseudoinverse of a Matrix," *SIAM Rev.*, **2**, 15–22 (1960), and "The Pseudoinverse of a Rectangular or Singular Matrix and Its Application to the Solution of Systems of Linear Equations," *ibid.*, **1**, 38–43 (1959); R. Penrose, "A Generalized Inverse for Matrices," *Proc. Cambridge Phil. Soc.*, **51**, 406–413 (1955); and E. H. Moore, *General Analysis*, Part I. Philadelphia: Memoirs of the American Philosophical Society, Vol. I, 1935.

arbitrary $k \times 1$ vector. On direct substitution of $\tilde{\boldsymbol{\gamma}}$, given in (3.79), into the normal equations it is seen that $\tilde{\boldsymbol{\gamma}}$ is a solution[22] to the normal equations for *any* GI of $P'X'XP$ and for any $\mathbf{z}$.

To show how the choice of a GI and a choice of $\mathbf{z}$ affect solutions to the normal equations note that

$$(P'X'XP)^* = \begin{pmatrix} D^{-1} & E \\ C & F \end{pmatrix} \tag{3.80}$$

for any selection of the matrices C, E, and F is a GI of $P'X'XP$. On substituting from (3.76) and (3.80) into (3.79) we have, with $\mathbf{z}' = (z_1' \vdots z_2')$,

$$\tilde{\boldsymbol{\gamma}} = \begin{pmatrix} \tilde{\boldsymbol{\gamma}}_1 \\ \tilde{\boldsymbol{\gamma}}_2 \end{pmatrix} = \begin{pmatrix} D^{-1}P_1'X'\mathbf{y} \\ CP_1'X'\mathbf{y} \end{pmatrix} + \begin{pmatrix} \mathbf{0} \\ \mathbf{z}_2 - CD\mathbf{z}_1 \end{pmatrix} \tag{3.81a}$$

or

$$\tilde{\boldsymbol{\gamma}}_1 = D^{-1}P_1'X'\mathbf{y} \tag{3.81b}$$

and

$$\begin{aligned} \tilde{\boldsymbol{\gamma}}_2 &= \mathbf{z}_2 + C(P_1'X'\mathbf{y} - D\mathbf{z}_1) \\ &= \mathbf{z}_2 + CD(\tilde{\boldsymbol{\gamma}}_1 - \mathbf{z}_1). \end{aligned} \tag{3.81c}$$

We see from (3.81*b*) that the estimator for $\boldsymbol{\gamma}_1$ is independent of the choice of GI and of $\mathbf{z}$. However, $\tilde{\boldsymbol{\gamma}}_2$ in (3.81*c*) obviously depends on C and thus on the choice of a GI and on the choice of $\mathbf{z}_1$ and $\mathbf{z}_2$.

If, for example, we use the Moore-Penrose GI for $P'X'XP$, namely,

$$(P'X'XP)^* = \begin{pmatrix} D^{-1} & 0 \\ 0 & 0 \end{pmatrix}, \tag{3.82}$$

or any GI with $C = 0$ [see (3.80)], the estimator for $\boldsymbol{\gamma}_1$ in (3.81*a*) is unaffected. However, $\tilde{\boldsymbol{\gamma}}_2$ in (3.81*c*) becomes

$$\tilde{\boldsymbol{\gamma}}_2 = \mathbf{z}_2, \tag{3.83}$$

where $\mathbf{z}_2$ is arbitrary.

To see what this analysis implies for the estimation of $\boldsymbol{\beta}$ we have $\boldsymbol{\beta} = P\boldsymbol{\gamma}$ from (3.75), and thus from (3.79)

$$\begin{aligned} \hat{\boldsymbol{\beta}} = P\tilde{\boldsymbol{\gamma}} &= P(P'X'XP)^*P'X'\mathbf{y} + [P - P(P'X'XP)^*P'X'XP]\mathbf{z} \\ &= (X'X)^*X'\mathbf{y} + [I - (X'X)^*X'X]P\mathbf{z}, \end{aligned} \tag{3.84}$$

[22] That is, with $N = P'X'XP$, the lhs of (3.77) is, with $\boldsymbol{\gamma} = \tilde{\boldsymbol{\gamma}}$, $N\tilde{\boldsymbol{\gamma}} = NN^*P'X'\mathbf{y} + N(I - N^*N)\mathbf{z} = NN^*P'X'\mathbf{y} + (N - NN^*N)\mathbf{z} = NN^*P'X'\mathbf{y} = P'X'\mathbf{y}$ since from the definition of N^*, $NN^*N = N$, and $NN^*P'X'\mathbf{y} = \begin{pmatrix} D & 0 \\ 0 & 0 \end{pmatrix}\begin{pmatrix} D^{-1} & E \\ C & F \end{pmatrix}\begin{pmatrix} P_1'X' \\ 0 \end{pmatrix}\mathbf{y} = P_1'X'\mathbf{y}$ $= P'X'\mathbf{y}$, given that $P_2'X' = 0$ from (3.76).

since $P(P'X'XP)^*P' = (X'X)^*$, a generalized inverse of $X'X$, the singular moment matrix.[23] Then from (3.81)

(3.85)

$$\tilde{\boldsymbol{\beta}} = \begin{pmatrix}\tilde{\boldsymbol{\beta}}_1\\ \tilde{\boldsymbol{\beta}}_2\end{pmatrix} = \begin{pmatrix}P_{11} & P_{12}\\ P_{21} & P_{22}\end{pmatrix}\begin{pmatrix}\tilde{\boldsymbol{\gamma}}_1\\ \tilde{\boldsymbol{\gamma}}_2\end{pmatrix} = \begin{bmatrix}P_{11}D^{-1}P_1'X'\mathbf{y} + P_{12}[\mathbf{z}_2 + CD(\tilde{\boldsymbol{\gamma}}_1 - \mathbf{z}_1)]\\ P_{21}D^{-1}P_1'X'\mathbf{y} + P_{22}[\mathbf{z}_2 + CD(\tilde{\boldsymbol{\gamma}}_1 - \mathbf{z}_1)]\end{bmatrix},$$

and it is obvious that the estimates $\tilde{\boldsymbol{\beta}}_1$ and $\tilde{\boldsymbol{\beta}}_2$ *both* depend on the choice of the GI and of $\mathbf{z}$.

We now wish to use the Bayesian approach to analyze the model, when $X'X$ is singular, with a prior pdf representing the prior information employed in the GI approach described above. The likelihood function, in terms of the parameters $\boldsymbol{\gamma}_1$, $\boldsymbol{\gamma}_2$, and σ, is given by

$$\text{(3.86)}\quad \begin{aligned} l(\boldsymbol{\gamma}_1, \boldsymbol{\gamma}_2, \sigma|\mathbf{y}) &\propto \frac{1}{\sigma^n}\exp\left[-\frac{1}{2\sigma^2}(\mathbf{y} - XP\boldsymbol{\gamma})'(y - XP\boldsymbol{\gamma})\right]\\ &\propto \frac{1}{\sigma^n}\exp\left\{-\frac{1}{2\sigma^2}[a + (\boldsymbol{\gamma}_1 - \tilde{\boldsymbol{\gamma}}_1)'D(\boldsymbol{\gamma}_1 - \tilde{\boldsymbol{\gamma}}_1)]\right\}, \end{aligned}$$

where $\tilde{\boldsymbol{\gamma}}_1 = D^{-1}P_1'X\mathbf{y}$ and $a = (\mathbf{y} - XP_1\tilde{\boldsymbol{\gamma}}_1)'(\mathbf{y} - XP_1\tilde{\boldsymbol{\gamma}}_1)$. It is of fundamental importance to observe that the likelihood function does not depend on $\boldsymbol{\gamma}_2$.[24]

In the GI approach based on a GI in the form of (3.80) with $C \neq 0$ we represent the prior information regarding $\boldsymbol{\gamma}_1$ and σ by the following improper diffuse pdf:

$$\text{(3.87)}\quad p(\boldsymbol{\gamma}_1, \sigma) \propto \frac{1}{\sigma}, \qquad -\infty < \gamma_{1i} < \infty, \qquad i = 1, 2, \ldots, q, \qquad 0 < \sigma < \infty.$$

The remaining prior information, corresponding to (3.81*b*), takes the form of the following linear relations or side conditions connecting $\boldsymbol{\gamma}_1$ and $\boldsymbol{\gamma}_2$:

$$\text{(3.88)}\qquad \boldsymbol{\gamma}_2 = \mathbf{z}_2 + CD(\boldsymbol{\gamma}_1 - \mathbf{z}_1).$$

These relations, for example, may be suggested by economic theory or other considerations. The matrix C and the vectors $\mathbf{z}_1$ and $\mathbf{z}_2$ must be assigned values in accord with the assumed available prior information.

Combination of the prior pdf in (3.87) with the likelihood function in (3.86) produces the posterior pdf for $\boldsymbol{\gamma}_1$ and σ:

$$\text{(3.89)}\quad p(\boldsymbol{\gamma}_1, \sigma|\mathbf{y}) \propto \frac{1}{\sigma^{n+1}}\exp\left\{-\frac{1}{2\sigma^2}[a + (\boldsymbol{\gamma}_1 - \tilde{\boldsymbol{\gamma}}_1)'D(\boldsymbol{\gamma}_1 - \tilde{\boldsymbol{\gamma}}_1)]\right\},$$

[23] From the definition of a GI, $P'X'XP(P'X'XP)^*P'X'XP = P'X'XP$. On premultiplying both sides by P and postmultiplying by P' we have $X'XP(P'X'XP)^*P'X'X = X'X$; that is, $P(P'X'XP)^*P'$ is a GI of $X'X$.

[24] This is perhaps more easily seen from $\mathbf{y} = XP\boldsymbol{\gamma} + \mathbf{u} = X(P_1 \vdots P_2)\begin{pmatrix}\boldsymbol{\gamma}_1\\ \boldsymbol{\gamma}_2\end{pmatrix} + \mathbf{u} = XP_1\boldsymbol{\gamma}_1 + \mathbf{u}$, since $XP_2 = 0$ from (3.76).

which will be proper if $n - q > 0$. The posterior pdf for $\boldsymbol{\gamma}_1$ has mean $\tilde{\boldsymbol{\gamma}}_1$. Further, from (3.88) the posterior expectation of $\boldsymbol{\gamma}_2$ is

$$E\boldsymbol{\gamma}_2 = \mathbf{z}_2 + CD(\tilde{\boldsymbol{\gamma}}_1 - \mathbf{z}_1), \tag{3.90}$$

which is identical to the point estimate in (3.81*b*) yielded by the sampling theory GI approach. Thus the improper prior pdf in (3.87), combined with the a priori relations in (3.88), yields posterior means for $\boldsymbol{\gamma}_1$ and $\boldsymbol{\gamma}_2$ which are identical to estimates provided by the GI approach.

When $C = 0$, as would be the case in the GI approach if we used the Moore-Penrose inverse (3.82), the a priori relations in (3.88) reduce to

$$\boldsymbol{\gamma}_2 = \mathbf{z}_2. \tag{3.91}$$

In Bayesian terms, if we assign a value to $\mathbf{z}_2$, and thus $\boldsymbol{\gamma}_2$, in (3.91), this represents "dogmatic" prior information about $\boldsymbol{\gamma}_2$; that is, the prior pdf for $\boldsymbol{\gamma}_2$ is degenerate with all its mass at the point $\mathbf{z}_2$. Further from (3.88) it is clear that taking $C = 0$ removes any dependence of $\boldsymbol{\gamma}_2$ on $\boldsymbol{\gamma}_1$ from the analysis.

The prior assumption in (3.91) is rather restrictive in many situations, given that the value of $\boldsymbol{\gamma}_2$ is not known precisely. One way we can loosen this prior assumption is by taking

$$p(\boldsymbol{\gamma}_2|\sigma) \propto \frac{1}{\sigma^{k-q}} \exp\left[-\frac{1}{2\sigma^2}(\boldsymbol{\gamma}_2 - \mathbf{z}_2)'Q(\boldsymbol{\gamma}_2 - \mathbf{z}_2)\right] \tag{3.92}$$

to be the prior pdf for the $k - q$ elements of $\boldsymbol{\gamma}_2$ and assuming $\boldsymbol{\gamma}_2$ independent of $\boldsymbol{\gamma}_1$. Given that Q is nonsingular, the mean of this pdf is $E\boldsymbol{\gamma}_2 = \mathbf{z}_2$ and $\text{Var}(\boldsymbol{\gamma}_2) = Q^{-1}\sigma^2$. Thus by using (3.92) we can relax the prior assumption in (3.91). When we combine the prior pdf's in (3.87) and (3.92)[25] with the likelihood function (3.86), the posterior pdf is

$$\begin{aligned} p(\boldsymbol{\gamma}_1, \boldsymbol{\gamma}_2, \sigma|\mathbf{y}) \propto & \frac{1}{\sigma^{n+k-q+1}} \\ & \times \exp\left\{-\frac{1}{2\sigma^2}[a + (\boldsymbol{\gamma}_2 - \mathbf{z}_2)'Q(\boldsymbol{\gamma}_2 - \mathbf{z}_2) \right. \\ & \left. + (\boldsymbol{\gamma}_1 - \tilde{\boldsymbol{\gamma}}_1)'D(\boldsymbol{\gamma}_1 - \tilde{\boldsymbol{\gamma}}_1)]\right\}. \end{aligned} \tag{3.93}$$

On transforming from $\boldsymbol{\gamma}$ to $\boldsymbol{\beta}$ this posterior pdf becomes[26]

[25] Here, as with $C = 0$, we are assuming $\boldsymbol{\gamma}_1$ and $\boldsymbol{\gamma}_2$ independent a priori.

[26] For comparison with (3.73) the following is the prior pdf for $\boldsymbol{\beta}$, given σ, implied by the prior assumptions regarding $\boldsymbol{\gamma}_1$ and $\boldsymbol{\gamma}_2$ in (3.87) and (3.92): $p(\boldsymbol{\beta}|\sigma) \propto \exp[-(1/2\sigma^2)(P_2'\boldsymbol{\beta} - \mathbf{z}_2)'Q(P_2'\boldsymbol{\beta} - \mathbf{z}_2)]$. Since the prior pdf for $\boldsymbol{\gamma}' = (\boldsymbol{\gamma}_1' \vdots \boldsymbol{\gamma}_2')$ is improper, this prior pdf for $\boldsymbol{\beta}$ is also improper.

(3.94) $$p(\boldsymbol{\beta}, \sigma|\mathbf{y}) \propto \frac{1}{\sigma^{n+k-q+1}} \exp\left[-\frac{1}{2\sigma^2}[a + (\boldsymbol{\beta} - \tilde{\boldsymbol{\beta}})'PFP'(\boldsymbol{\beta} - \tilde{\boldsymbol{\beta}})]\right],$$

with $\tilde{\boldsymbol{\beta}} = P\tilde{\boldsymbol{\gamma}}$, where $\tilde{\boldsymbol{\gamma}}' = (\tilde{\boldsymbol{\gamma}}_1' \vdots \mathbf{z}_2')$ and

(3.95) $$F = \begin{pmatrix} D & 0 \\ 0 & Q \end{pmatrix}.$$

Given that $n - q > 0$, the posterior pdf in (3.94) is proper, since $PFP' = P_1DP_1' + P_2QP_2'$ is a nonsingular matrix even though $P_1DP_1' = X'X$ and P_2QP_2' are each singular. The pdf in (3.94) can be used to make posterior inferences about $\boldsymbol{\beta}$ and σ.

Last, it is interesting to inquire what happens if we introduce a diffuse prior pdf for all elements of $\boldsymbol{\beta}$ when $X'X$ is singular. Given that we assume

(3.96) $$p(\boldsymbol{\beta}) \propto \text{constant}, \qquad -\infty < \beta_i < \infty, \qquad i = 1, 2, \ldots, k,$$

the posterior pdf for $\boldsymbol{\beta}$, given σ, is just the following improper pdf:

(3.97) $$\begin{aligned} p(\boldsymbol{\beta}|\sigma, \mathbf{y}) &\propto \exp\left[-\frac{1}{2\sigma^2}(\mathbf{y} - X\boldsymbol{\beta})'(\mathbf{y} - X\boldsymbol{\beta})\right] \\ &\propto \exp\left[-\frac{1}{2\sigma^2}(\boldsymbol{\beta} - \tilde{\boldsymbol{\beta}})'X'X(\boldsymbol{\beta} - \tilde{\boldsymbol{\beta}})\right], \end{aligned}$$

where $\tilde{\boldsymbol{\beta}}$ is *any* solution to the normal equations, that is, $X'X\tilde{\boldsymbol{\beta}} = X'\mathbf{y}$. As (3.85) indicates, $\tilde{\boldsymbol{\beta}}$ is not unique. Since $\boldsymbol{\beta} = P\boldsymbol{\gamma}$ from (3.75), we can express (3.97) in terms of $\boldsymbol{\gamma}$ as follows:

(3.98) $$p(\boldsymbol{\gamma}_1, \boldsymbol{\gamma}_2|\sigma, \mathbf{y}) \propto \exp\left[-\frac{1}{2\sigma^2}(\boldsymbol{\gamma}_1 - \tilde{\boldsymbol{\gamma}}_1)'D(\boldsymbol{\gamma}_1 - \tilde{\boldsymbol{\gamma}}_1)\right].$$

This posterior pdf is the product of a proper normal pdf for $\boldsymbol{\gamma}_1$, with unique mean $\tilde{\boldsymbol{\gamma}}_1$ and a diffuse posterior pdf for $\boldsymbol{\gamma}_2$, which, of course, is identical to the prior pdf for $\boldsymbol{\gamma}_2$, since the likelihood function in (3.86) does not involve $\boldsymbol{\gamma}_2$. Since the marginal posterior pdf for $\boldsymbol{\gamma}_1$ in (3.98) is a proper normal pdf, the distribution of the $q \times 1$ vector $\boldsymbol{\theta}_1 = H_1'\boldsymbol{\gamma}_1$, where H_1 is any $q \times q$ nonsingular matrix, will be proper and therefore inferences can be made about $\boldsymbol{\theta}_1$, given σ; for example, $E(\boldsymbol{\theta}_1|\sigma, \mathbf{y}) = H_1'\tilde{\boldsymbol{\gamma}}_1$ and $\text{Var}(\boldsymbol{\theta}_1|\sigma, \mathbf{y}) = H_1'D^{-1}H_1\sigma^2$. To relate these results to $\boldsymbol{\beta}$ note that $X\boldsymbol{\beta} = XP\boldsymbol{\gamma} = XP_1\boldsymbol{\gamma}_1$, since $XP_2 = 0$. If R' is any $q \times n$ matrix of rank q and $R'XP_1$ is a $q \times q$ nonsingular matrix, then $R'X\boldsymbol{\beta} = R'XP_1\boldsymbol{\gamma}_1$ is a $q \times 1$ vector with a proper normal pdf. Thus, even though $X'X$ is singular and we introduce no prior information, it is possible to make inferences about q linearly independent combinations of the elements of $\boldsymbol{\beta}$, $R'X\boldsymbol{\beta}$. In sampling theory terms, the q linear functions of the elements of $\boldsymbol{\beta}$, $R'X\boldsymbol{\beta}$, are called "estimable functions."[27]

[27] See, for example, Graybill, *op. cit.*, p. 227 ff.

QUESTIONS AND PROBLEMS

1. If the investment variable in the investment multiplier model in Section 3. is considered random, what assumptions about it are needed for the anal of Section 3.1.3 to be appropriate?
2. Using the prior assumptions employed in Section 3.1.3 and the data in Ta 3.1, derive and plot the posterior pdf for σ^2, the common variance of error terms in the investment multiplier model. Show that a posteri $\nu s^2/\sigma^2$ has a χ^2 pdf with $\nu = n - 2$ degrees of freedom and use this fact construct a 95% Bayesian confidence interval for the variance.
3. Using the data, assumptions, and investment multiplier model of Sect 3.1.3, derive the predictive pdf for income, given that investment assume value of 50. Compare the mean and variance of this predictive pdf with t of the predictive pdf for investment with a value equal to 100.
4. In Problem 3 compute a predictive interval with probability 0.80 of includ the unobserved value of income associated with investment equal to 50. the same for the value of income associated with investment equal to 1 Compute a predictive region with probability 0.80 of including the values income associated with investment equal to 50 and 100. Compare the t intervals with the region and interpret the comparison.
5. Suppose, in the analysis of the investment multiplier in Section 3.1.3, that had taken a Keynesian viewpoint by relating the investment multiplier β_2 the marginal propensity to consume α: $\beta_2 = 1/(1 - \alpha)$, with α satisfy $0 < \alpha < 1$ on a priori grounds.
   (a) What prior restriction on the range of β_2 is implied by the condit $0 < \alpha < 1$?
   (b) If α is assumed uniformly distributed over the interval zero to one, w is the implied pdf for β_2? Comment on its properties.
   (c) If α has a beta pdf with parameters a and b (see Appendix A, Section what is the implied pdf for $\beta_2 = 1/(1 - \alpha)$ with $0 < \alpha < 1$?
   (d) In the analysis of the investment multiplier model with the data of Ta 3.1, assume that the prior pdf for the parameters is given by $p(\beta_1, \beta_2, \sigma)$ $g(\beta_2)/\sigma$, with $-\infty < \beta_1 < \infty$, $0 < \sigma < \infty$, and $g(\beta_2)$, the prior obtained in (b) of this question. Derive the posterior pdf for β_2. U univariate numerical integration procedures (see Appendix C) to normal it. Then compare the results with the posterior pdf for β_2 plotted in S tion 3.1.3 to see how sensitive results are to changes in prior assun tions.
6. Consider the posterior pdf for parameters of the normal multiple regressi model shown in (3.31). What is the conditional posterior pdf for $\boldsymbol{\beta}$, given Give its mean vector and variance-covariance matrix.
7. Suppose that in a normal multiple regression model $\mathbf{y} = X\boldsymbol{\beta} + \mathbf{u}$ we p tition $X = (X_1 \vdots X_2)$ and $\boldsymbol{\beta}' = (\boldsymbol{\beta}_1' \vdots \boldsymbol{\beta}_2')$ and write $\mathbf{y} = X_1\boldsymbol{\beta}_1 + X_2\boldsymbol{\beta}_2 +$ How does the condition $X_1'X_2 = 0$ affect properties of the conditio

posterior pdf for $\boldsymbol{\beta}' = (\boldsymbol{\beta}_1' \vdots \boldsymbol{\beta}_2')$, given σ, which was derived in Problem 6? Also, what does the condition $X_1'X_2 = 0$ imply about the marginal posterior pdf for $\boldsymbol{\beta}' = (\boldsymbol{\beta}_1' \vdots \boldsymbol{\beta}_2')$ shown in (3.32) of the text? In particular, does the condition $X_1'X_2 = 0$ imply that elements of $\boldsymbol{\beta}_1$ will be uncorrelated and distributed independently of those of $\boldsymbol{\beta}_2$?

1957 U.S. ANNUAL SURVEY OF MANUFACTURES DATA FOR THE TRANSPORTATION EQUIPMENT INDUSTRY

State	Aggregate Value Added, V_a	Aggregate Capital Service Flow[a] K_a	Aggregate Man-Hours Worked,[b] L_a	No. of Establishments, N
	(millions of dollars)		(millions of man-hours)	
Alabama	126.148	3.804	31.551	68
California	3201.486	185.446	452.844	1372
Connecticut	690.670	39.712	124.074	154
Florida	56.296	6.547	19.181	292
Georgia	304.531	11.530	45.534	71
Illinois	723.028	58.987	88.391	275
Indiana	992.169	112.884	148.530	260
Iowa	35.796	2.698	8.017	75
Kansas	494.515	10.360	86.189	76
Kentucky	124.948	5.213	12.000	31
Louisiana	73.328	3.763	15.900	115
Maine	29.467	1.967	6.470	81
Maryland	415.262	17.546	69.342	129
Massachusetts	241.530	15.347	39.416	172
Michigan	4079.554	435.105	490.384	568
Missouri	652.085	32.840	84.831	125
New Jersey	667.113	33.292	83.033	247
New York	940.430	72.974	190.094	461
Ohio	1611.899	157.978	259.916	363
Pennsylvania	617.579	34.324	98.152	233
Texas	527.413	22.736	109.728	308
Virginia	174.394	7.173	31.301	85
Washington	636.948	30.807	87.963	179
West Virginia	22.700	1.543	4.063	15
Wisconsin	349.711	22.001	52.818	142

[a] Net capital stock is defined as "gross book value on December 31, 1957" minus "accumulated depreciation and depletion up to December 31, 1956," minus "depreciation and depletion charged in 1957." Capital service flow is defined as depreciation and depletion charged in 1957 plus 0.06 times the net capital stock plus the sum of insurance premiums, rental payments, and property taxes paid.

[b] These figures refer to production workers.

8. From (3.32) provide an explicit expression for the posterior mean $\hat{\boldsymbol{\beta}}_1$ of $\boldsymbol{\beta}_1$, a subvector of $\boldsymbol{\beta}$; that is, $\boldsymbol{\beta}' = (\boldsymbol{\beta}_1' \vdots \boldsymbol{\beta}_2')$. If $X_1'X_2 = 0$, show that the posterior mean of $\boldsymbol{\beta}_1$ is $\tilde{\boldsymbol{\beta}}_1 = (X_1'X_1)^{-1}X_1'\mathbf{y}$. (Note that $\tilde{\boldsymbol{\beta}}_1$ is the mean of the posterior pdf when the term $X_2\boldsymbol{\beta}_2$ is omitted from the regression model, given a diffuse prior pdf for $\boldsymbol{\beta}_1$ and σ.) If $X_1'X_2 \neq 0$, compare $\tilde{\boldsymbol{\beta}}_1$ with the posterior mean $\hat{\boldsymbol{\beta}}_1$ for $\boldsymbol{\beta}_1$. In particular, show that

$$\tilde{\boldsymbol{\beta}}_1 = \hat{\boldsymbol{\beta}}_1 + P\hat{\boldsymbol{\beta}}_2,$$

where $\hat{\boldsymbol{\beta}}' = (\hat{\boldsymbol{\beta}}_1' \vdots \hat{\boldsymbol{\beta}}_2')$, with $\hat{\boldsymbol{\beta}} = (X'X)^{-1}X'\mathbf{y}$, and $P = (X_1'X_1)^{-1}X_1'X_2$.

9. Shown in the table on p. 83 are data relating to the U.S. Transportation Equipment Industry for 1957.
Assume that the data are generated by a Cobb-Douglas production function; that is,

(a) $$\left(\frac{V_a}{N}\right)_i = A\left(\frac{L_a}{N}\right)_i^{\beta_2}\left(\frac{K_a}{N}\right)_i^{\beta_3} e^{u_i}$$

or

(b) $$\ln\left(\frac{V_a}{N}\right)_i = \beta_1 + \beta_2 \ln\left(\frac{L_a}{N}\right)_i + \beta_3 \ln\left(\frac{K_a}{N}\right)_i + u_i,$$

where $\beta_1 = \ln A$, β_2, and β_3 are parameters with unknown values; V_a, L_a, K_a, and N are defined in the table; the subscript i denotes values of the variables for the ith state, $i = 1, 2, \ldots, 25$, and u_i is a random disturbance term. We assume that the u_i's are normally and independently distributed, each with zero mean and common variance σ^2.

After reviewing alternative assumptions about independent variables in regression models, consider whether $\ln (L_a/N)_i$ and $\ln (K_a/N)_i$ can reasonably, on a priori grounds, be assumed to satisfy any or all of these alternative assumptions.

10. Under the assumption that it is appropriate to analyze the data in Problem 9 within a regression framework, what are the likelihood function and maximum likelihood estimates for β_1, β_2, β_3, σ^2, and A?

11. Derive and compute posterior pdf's for the parameters of (b) in Problem 9, using the data presented in that problem and the diffuse prior pdf $p(\boldsymbol{\beta}, \sigma) \propto 1/\sigma$ with $0 < \sigma < \infty$ and $-\infty < \beta_i < \infty$, $i = 1, 2$, and 3. How do the means of these posterior pdf's compare with the maximum likelihood estimates of corresponding parameters in Problem 10?

12. In Problem 11 derive and plot the marginal posterior pdf for $A = e^{\beta_1}$. Do the mean and higher posterior moments of A exist? Compare the modal value of the posterior pdf for A with the maximum likelihood estimate of A. What can be said about these two quantities in large samples?

13. In view of the economic theory of production functions, comment on the prior assumptions about the production function parameters that were introduced in Problem 11. Determine whether restricting β_2 and β_3 to be non-negative has a great effect on the numerical results in Problem 11.

14. Under the conditions of Problem 11, derive and compute the posterior pdf

for $\eta = \beta_2 + \beta_3$, the returns to scale parameter. What is an 85% Bayesian confidence interval for this parameter?

15. In connection with (b) in Problem 9, suppose that we assume constant returns to scale, that is, $\eta = \beta_2 + \beta_3 = 1$, and that $0 \leq \beta_2$ and $0 \leq \beta_3$. How would you formulate a prior pdf to reflect this information in the analysis of the data and (b) in Problem 9?

16. Let $\eta = \beta_2 + \beta_3$ be a returns to scale parameter and consider a prior pdf for η and β_2, namely, $p(\eta, \beta_2) = p_1(\eta)\, p_2(\beta_2|\eta)$. Can $p_1(\eta)$, the marginal prior pdf for η, and $p_2(\beta_2|\eta)$, the conditional pdf for β_2, given η, both be in form of beta pdf's with $0 \leq \eta \leq 2$ and $0 \leq \beta_2 < \eta$? Provide an example to illustrate your answer.

CHAPTER IV

Special Problems in Regression Analysis

The topics treated in this chapter provide examples of how some of the specifying assumptions of the regression model, considered in Chapter 3, can be relaxed; for example, we consider the regression model with autocorrelated errors. Since this and other departures from our "standard" assumptions are often encountered in practice, it is important to be able to deal with them. Failure to take account of possible departures from the standard assumptions can, of course, result in incorrect inferences. Thus it is imperative that users of the regression model evaluate critically the adequacy of specifying assumptions in applications.

4.1 THE REGRESSION MODEL WITH AUTOCORRELATED ERRORS[1]

Initially we analyze a simple regression model with a disturbance term generated by a first-order autoregressive process; that is,

$$\begin{aligned} &(4.1a) \qquad y_t = \beta x_t + u_t, \\ &(4.1b) \qquad u_t = \rho u_{t-1} + \epsilon_t \end{aligned} \qquad t = 1, 2, \ldots, T.$$

In (4.1*a*) y_t is the tth observation on the dependent variable, β is a scalar regression coefficient, x_t is the tth observation on an independent variable, assumed nonstochastic, and u_t is the tth error term. In (4.1*b*) the first-order autoregressive process, assumed to generate the error term u_t, is presented. It involves a scalar parameter ρ and an error term ϵ_t. It is assumed that the ϵ_t are normally and independently distributed with zero means and common variance σ^2. Note that if $\rho = 0$ (4.1*a*, *b*) would reduce to a simple regression model satisfying the standard assumptions of Chapter 3.

From (4.1*a*, *b*) we obtain

$$(4.1c) \qquad y_t = \rho y_{t-1} + \beta(x_t - \rho x_{t-1}) + \epsilon_t, \qquad t = 1, 2, \ldots, T.$$

[1] This section is based mainly on work in the following paper: A. Zellner and G. C. Tiao, "Bayesian Analysis of the Regression Model with Autocorrelated Errors," *J. Am. Statist. Assoc.*, **59**, 763–778 (1964).

Note that y_0 appears in (4.1*c*), and thus something must be said about initial conditions before we can proceed with the analysis of the model.

If we assume that the process represented by (4.1*a*, *b*) has been operative for $t = 0, -1, -2, \ldots, -T_0$, where T_0 is unknown, we can write $y_0 - \beta x_0 = M + \epsilon_0$, where $M = \rho(y_{-1} - \beta x_{-1})$; M is regarded as a parameter, since it depends on unobservable and unobserved quantities. Under these assumptions y_0 is normally distributed with mean $\beta x_0 + M$ and variance σ^2. These assumptions are broad enough to apply to explosive ($|\rho| \geq 1$) as well as nonexplosive ($|\rho| < 1$) schemes and to situations in which the process commences at any unknown point in the past.

On the other hand, it may be that y_0 is fixed and known; for example, if the observations relate to a price and $t = 0$ is the last period during which this price was fixed by a governmental body, it may be appropriate to take y_0 as fixed and known. This situation can also be represented in the framework introduced in the preceding paragraph by assuming that ϵ_0 has zero variance. Other assumptions which may be appropriate for other circumstances are that ϵ_0 is normal with known variance, σ_0^2 or that y_0 is distributed independently of $\mathbf{y}' = (y_1, y_2, \ldots, y_T)$ and has a distribution that does *not* involve any of the parameters of the model. From what follows it will be seen that any of the assumptions regarding y_0 lead to the same joint posterior pdf for the parameters β, ρ, and σ.

Under the assumptions introduced, the joint pdf for y_0 and $\mathbf{y}' = (y_1, y_2, \ldots, y_T)$ is given by

$$p(y_0, \mathbf{y}|\beta, \rho, \sigma, M) = p(y_0|\beta, \rho, \sigma, M)\, p(\mathbf{y}|y_0, \beta, \rho, \sigma, M)$$

$$\propto \frac{1}{\sigma^{T+1}} \exp\left\{-\frac{1}{2\sigma^2}(y_0 - \beta x_0 - M)^2 - \frac{1}{2\sigma^2}\sum_{t=1}^{T}[y_t - \rho y_{t-1} - \beta(x_t - \rho x_{t-1})]^2\right\}, \tag{4.2}$$

which, viewed as a function of the parameters, is the likelihood function, $l(\beta, \rho, M, \sigma|y_0, \mathbf{y})$, with $-\infty < \beta < \infty$, $-\infty < \rho < \infty$, $-\infty < M < \infty$, and $\sigma > 0$. Since $-\infty < \rho < \infty$, we are allowing the process in (4.1*b*) to be explosive or nonexplosive.

As regards prior assumptions, we assume that we have little prior information and represent it by assuming that β, ρ, $\log \sigma$, and M are uniformly and independently distributed; that is

$$p(\beta, \rho, M, \sigma) \propto \frac{1}{\sigma}. \tag{4.3}$$

On combining this prior pdf with the likelihood function, we obtain the following joint posterior for the parameters:

$$p(\beta, \rho, \sigma, M|\mathbf{y}) \propto \frac{1}{\sigma^{T+2}} \exp\left\{-\frac{1}{2\sigma^2}(y_0 - \beta x_0 - M)^2 - \frac{1}{2\sigma^2}\sum_{t=1}^{T}[y_t - \rho y_{t-1} - \beta(x_t - \rho x_{t-1})]^2\right\}. \tag{4.4}$$

If we are interested in investigating M, the initial level of the process in (4.1*c*), it is possible to obtain the posterior pdf for M by integrating (4.4) over β, ρ, and σ. If interest does not center on M, the influence of this parameter can be eliminated by integrating (4.4) with respect to M to yield

$$p(\beta, \rho, \sigma|\mathbf{y}) \propto \frac{1}{\sigma^{T+1}} \exp\left\{-\frac{1}{2\sigma^2}\sum_{t=1}^{T}[y_t - \rho y_{t-1} - \beta(x_t - \rho x_{t-1})]^2\right\}, \tag{4.5}$$

which is the joint posterior pdf for β, ρ, and σ. In deriving (4.5), y_0 was assumed normal with mean $M + \beta x_0$ and variance σ^2. It is straightforward to verify that by employing the other assumptions about y_0, discussed above, we would also obtain (4.5) as our posterior pdf.

On integrating (4.5) with respect to σ, we obtain the following bivariate posterior distribution:

$$\begin{aligned} p(\beta, \rho|\mathbf{y}) &\propto \{\sum [y_t - \rho y_{t-1} - \beta(x_t - \rho x_{t-1})]^2\}^{-T/2} \\ &\propto \{\sum [y_t - \beta x_t - \rho(y_{t-1} - \beta x_{t-1})]^2\}^{-T/2}, \end{aligned} \tag{4.6}$$

with the summations extending from $t = 1$ to $t = T$. The bivariate pdf in (4.6) enables us to make joint inferences about β and ρ; that is bivariate numerical integration procedures (see Appendix C) can be employed to evaluate the normalizing constant and to compute, for example, the posterior probability that $a_1 \leq \beta \leq a_2$ and $b_1 \leq \rho \leq b_2$, where a_1, a_2, b_1, and b_2 are given numbers. Also the contours of the posterior pdf can be readily computed to provide information about the shape of the bivariate posterior pdf.

To obtain the marginal posterior pdf for ρ complete the square on β in the first line of (4.6) and use properties of the univariate Student t pdf to integrate out β. Similarly, to obtain the marginal posterior pdf for β, complete the square on ρ in the second line of (4.6) and integrate out ρ again by using properties of the univariate Student t pdf. These operations yield

$$\begin{aligned} p(\beta|\mathbf{y}) &\propto [\sum (y_{t-1} - \beta x_{t-1})^2]^{-1/2} \\ &\quad \times \left\{\sum (y_t - x_t\beta)^2 - \frac{[\sum (y_{t-1} - \beta x_{t-1})(y_t - \beta x_t)]^2}{\sum (y_{t-1} - \beta x_{t-1})^2}\right\}^{-(T-1)/2}, \end{aligned} \tag{4.7}$$

$$\begin{aligned} p(\rho|\mathbf{y}) &\propto [\sum (x_t - \rho x_{t-1})^2]^{-1/2} \\ &\quad \times \left\{\sum (y_t - \rho y_{t-1})^2 - \frac{[\sum (x_t - \rho x_{t-1})(y_t - \rho y_{t-1})]^2}{\sum (x_t - \rho x_{t-1})^2}\right\}^{-(T-1)/2}. \end{aligned} \tag{4.8}$$

In order for the distribution in (4.8) to be proper, the quantity $\sum (x_t - \rho x_{t-1})^2$ must be positive. This implies that we must assume that for any ρ there exists some t such that $x_t \neq \rho x_{t-1}$, which is not very restrictive.[2]

The posterior pdf's in (4.7) and (4.8) can be analyzed by using univariate numerical integration procedures. To illustrate the results of these computations, we have computed these pdf's with data generated from the following model:

$$\begin{aligned} y_t &= 3x_t + u_t \\ u_t &= \rho u_{t-1} + \epsilon_t \end{aligned} \qquad t = 1, 2, \ldots, 15,$$

where the ϵ's, given in Table 4.1, were drawn from a table of standardized random normal deviates. The x's are rescaled investment expenditures taken from a paper by Haavelmo.[3] The first series of 15 observations was generated with $\rho = 0.5$ and the second set was generated with $\rho = 1.25$. We refer to the first set of y's as the "nonexplosive" series and the second set as the "explosive" series. Although we distinguish these two cases, it is important

Table 4.1

t	ϵ_t	x_t	y_t (for $\rho = 0.5$)	y_t (for $\rho = 1.25$)
0	...	3.0	9.500	9.500
1	0.699	3.9	12.649	13.024
2	0.320	6.0	18.794	19.975
3	−0.799	4.2	12.198	14.270
4	−0.927	5.2	14.372	16.760
5	0.373	4.7	13.909	15.923
6	−0.648	5.1	14.556	16.931
7	1.572	4.5	14.700	17.111
8	−0.319	6.0	18.281	22.195
9	2.049	3.9	13.890	18.992
10	−3.077	4.1	10.318	18.338
11	−0.136	2.2	5.473	14.012
12	−0.492	1.7	4.044	13.873
13	−1.211	2.7	6.361	17.855
14	−1.994	3.3	7.036	20.099
15	0.400	4.8	13.368	27.549

$u_0 = 0.5$

[2] However, if $x_t = 1$ for all t, the condition is violated for $\rho = 1$. With the x_t's all equal to 1, our prior pdf must assign a zero density to $\rho = 1$. From (4.1*c*) note that with $\rho = 1$ and the x_t's $= 1$, β does not appear in the model.

[3] T. Haavelmo, "Methods of Measuring the Marginal Propensity to Consume," *J. Am. Statist. Assoc.*, **42**, 105–122 (1947).

to realize that the results given in (4.6), (4.7), and (4.8) are appropriate in the analysis of both.

The marginal distributions of β and of ρ for these data are shown in Figure 4.1. It is seen that the posterior pdf for ρ, derived from the explosive series, is much sharper than that relating to the nonexplosive case.

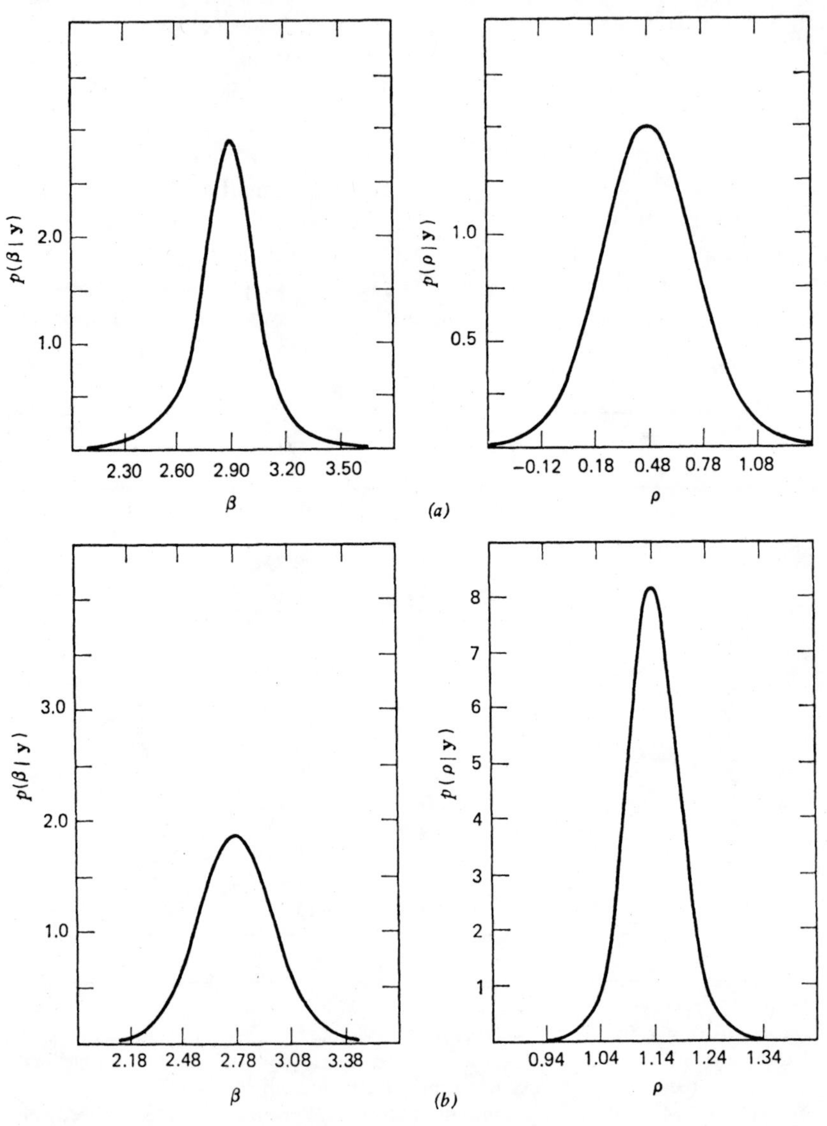

Figure 4.1 Marginal distributions of β and ρ. (*a*) Nonexplosives series ($T = 15$). (*b*) Explosive series ($T = 15$).

The posterior pdf's for β in Figure 4.1 enable us to make inferences about this parameter which incorporate an allowance for the departure from independence postulated in the model. That allowance be made for such a departure is extremely important because inferences would be markedly different if we analyzed these data under the assumption of independence. As shown in Chapter 3 under the assumption of independence ($\rho = 0$), we would have $p(\beta|\mathbf{y})$ in the univariate Student t form; that is,

$$\frac{\beta - \hat{\beta}}{s/(\sum x_t^2)^{\frac{1}{2}}} \sim t_\nu,$$

with $\nu = T - 1$, $\hat{\beta} = \sum x_t y_t / \sum x_t^2$ and $\nu s^2 = \sum (y_t - \hat{\beta} x_t)^2$. For our two sets of data the posterior pdf's for β under the independence assumption are shown in Figure 4.2 by the curves labeled $\rho = 0$. These distributions are far different from those shown in Figure 4.1.

To appreciate the situation fully it is instructive to write the marginal distribution of β as

$$p(\beta|\mathbf{y}) = \int_{-\infty}^{\infty} p(\beta|\rho, \mathbf{y})\, p(\rho|\mathbf{y})\, d\rho. \tag{4.9}$$

The integrand in (4.9) contains two factors, the conditional posterior pdf for β given ρ, $p(\beta|\rho, \mathbf{y})$, and the marginal posterior pdf for ρ, $p(\rho|\mathbf{y})$. Thus, as pointed out in Chapter 2, the marginal posterior pdf for β can be regarded as a suitably weighted average of the conditional pdf's $p(\beta|\rho, \mathbf{y})$, with $p(\rho|\mathbf{y})$ serving as the weight function; that is, the conditional pdf, $p(\beta|\rho, \mathbf{y})$, provides inferences about β for an assumed value of ρ. On the other hand, the marginal pdf, $p(\rho|\mathbf{y})$ reflects the plausibility of assertions about the value of ρ in the light of the data and our original assumptions. Unless the conditional pdf is insensitive to changes in ρ, it is clear that an assumption that ρ equals some fixed value, say $\rho = 0$ (observations independent) or $\rho = 1$ (first differences of the observations independent), could lead to a posterior pdf for β far different from that given in (4.7).

To analyze this point further, note that the conditional pdf for β, given ρ, which is easily obtained from (4.6), is

$$p(\beta|\rho, \mathbf{y}) \propto [s^2(\rho)]^{-\frac{1}{2}} \left\{1 + \frac{[\beta - \hat{\beta}(\rho)]^2}{\nu s^2(\rho)}\right\}^{-(\nu+1)/2}, \tag{4.10a}$$

where $\nu = T - 1$,

$$\hat{\beta}(\rho) = \frac{\sum (x_t - \rho x_{t-1})(y_t - \rho y_{t-1})}{\sum (x_t - \rho x_{t-1})^2}, \tag{4.10b}$$

and

$$\nu s^2(\rho) = \frac{\sum [y_t - \rho y_{t-1} - \hat{\beta}(\rho)(x_t - \rho x_{t-1})]^2}{\sum (x_t - \rho x_{t-1})^2}. \tag{4.10c}$$

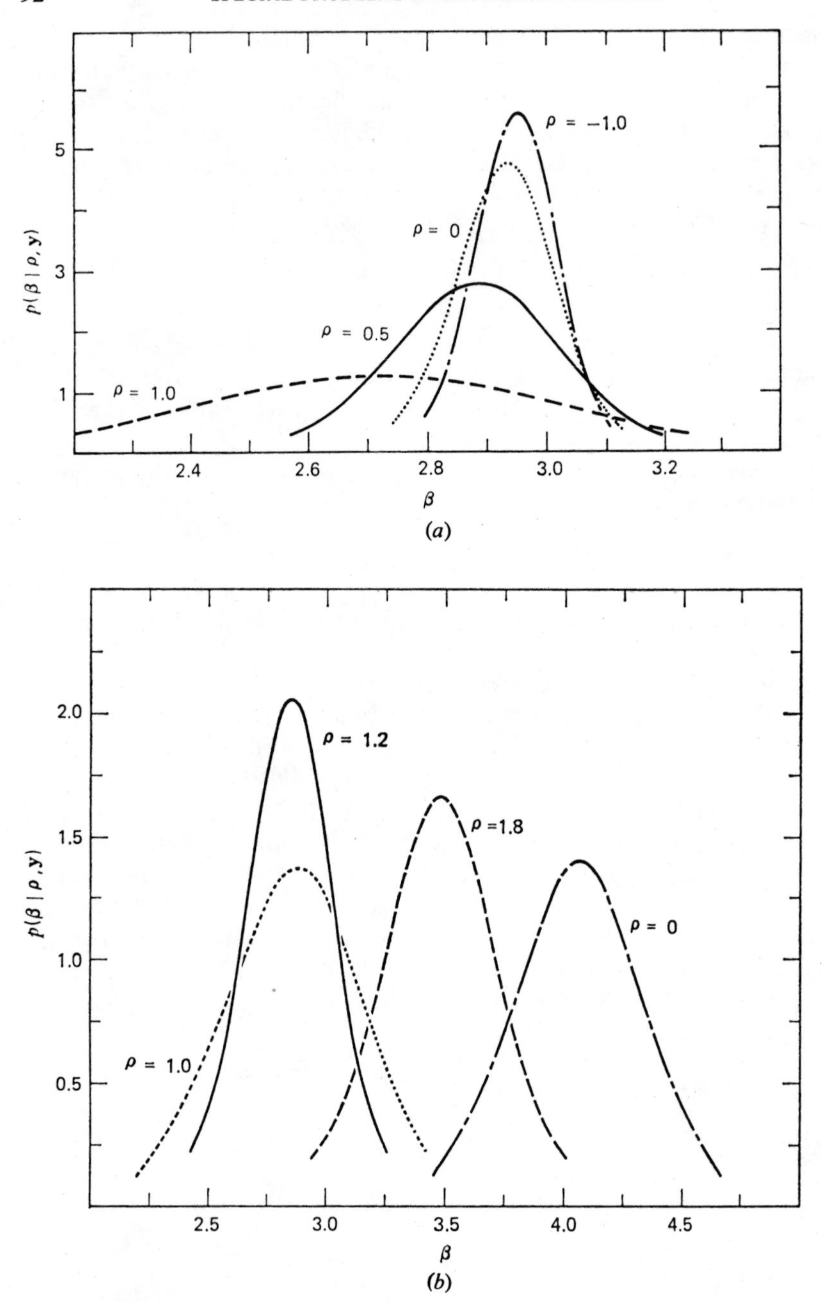

Figure 4.2 Conditional posterior distribution of β for various ρ. (*a*) Nonexplosive series ($T = 15$). (*b*) Explosive series ($T = 15$).

From (4.10)

$$\frac{\beta - \hat{\beta}(\rho)}{s(\rho)} \sim t_\nu; \tag{4.11}$$

that is, this quantity has the Student t pdf with $\nu = T - 1$ degrees of freedom. To show how sensitive inferences about β are to what is assumed about ρ, we have computed conditional posterior pdf's for β for various assumed values of ρ which are shown in Figure 4.2. The results indicate that for the nonexplosive series the center of the conditional pdf is relatively insensitive to changes in ρ, whereas the spread of the distribution is quite sensitive to such changes. On the other hand, both the center and spread in the explosive series change markedly as ρ is varied. Thus an inappropriate assumption about ρ can vitally affect an analysis. This fact underlines the importance of working with the marginal posterior pdf for β which incorporates a proper allowance for the role of ρ in the model.

We now generalize these methods to apply to the multiple regression model with errors generated by a first-order autoregressive process. Our model is[4]

$$\mathbf{y} = X\boldsymbol{\beta} + \mathbf{u}, \tag{4.12a}$$

$$\mathbf{u} = \rho\mathbf{u}_{-1} + \boldsymbol{\epsilon}, \tag{4.12b}$$

or, alternatively,

$$\mathbf{y} = \rho\mathbf{y}_{-1} + (X - \rho X_{-1})\boldsymbol{\beta} + \boldsymbol{\epsilon}, \tag{4.12c}$$

where $\mathbf{y}' = (y_1, y_2, \ldots, y_T)$ and $\mathbf{y}_{-1}' = (y_0, y_1, \ldots, y_{T-1})$ are $1 \times T$ vectors of observations, $\mathbf{u}' = (u_1, u_2, \ldots, u_T)$ and $\mathbf{u}_{-1}' = (u_0, u_1, \ldots, u_{T-1})$ are $(1 \times T)$ vectors of autocorrelated errors, $\boldsymbol{\beta}' = (\beta_1, \beta_2, \ldots, \beta_k)$ is a $(1 \times k)$ vector of regression coefficients, ρ is a scalar parameter, $\boldsymbol{\epsilon}' = (\epsilon_1, \epsilon_2, \ldots, \epsilon_T)$ is a $1 \times T$ vector of random errors, and

(4.13)

$$X = \begin{bmatrix} x_{11} & x_{12} & \cdots & x_{1k} \\ \vdots & \vdots & & \vdots \\ x_{T1} & x_{T2} & \cdots & x_{Tk} \end{bmatrix}, \qquad X_{-1} = \begin{bmatrix} x_{01} & x_{02} & \cdots & x_{0k} \\ \vdots & \vdots & & \vdots \\ x_{(T-1)1} & x_{(T-1)2} & \cdots & x_{(T-1)k} \end{bmatrix}$$

are $T \times k$ matrices with given elements.

As above, we assume that the elements of $\boldsymbol{\epsilon}$ are normally and independently distributed, each with mean zero and common variance σ^2. Further, we make the same assumptions about initial conditions and prior pdf's for ρ and σ as

[4] We assume that there is no intercept in the regression. If there is, our prior assumptions must preclude the value $\rho = 1$, for when $\rho = 1$ the intercept term disappears in (4.12c).

introduced. Last, we assume that the regression coefficients are a priori uniformly and independently distributed; that is

$$p(\boldsymbol{\beta}) \propto \text{const.} \quad -\infty < \beta_i < \infty, i = 1, 2, \ldots, k. \tag{4.14}$$

Under these assumptions the joint posterior pdf for $\boldsymbol{\beta}$, ρ, σ, and M is given by

$$\begin{aligned} p(\boldsymbol{\beta}, \rho, \sigma, M|\mathbf{y}) \propto \frac{1}{\sigma^{T+2}} \exp\bigg\{-\frac{1}{2\sigma^2}(y_0 - \mathbf{x}_0'\boldsymbol{\beta} - M)^2 \\ -\frac{1}{2\sigma^2}[\mathbf{y} - \rho\mathbf{y}_{-1} - (X - \rho X_{-1})\boldsymbol{\beta}]' \\ \times [\mathbf{y} - \rho\mathbf{y}_{-1} - (X - \rho X_{-1})\boldsymbol{\beta}]\bigg\}, \end{aligned} \tag{4.15}$$

where $\mathbf{x}_0' = (x_{01}, x_{02}, \ldots, x_{0k})$ is the first row of X_{-1}. On integrating (4.15) over M and σ, the joint posterior pdf for $\boldsymbol{\beta}$ and ρ is readily obtained as

$$\begin{aligned} p(\boldsymbol{\beta}, \rho|\mathbf{y}) &\propto \{[\mathbf{y} - \rho\mathbf{y}_{-1} - (X - \rho X_{-1})\boldsymbol{\beta}]'[\mathbf{y} - \rho\mathbf{y}_{-1} - (X - \rho X_{-1})\boldsymbol{\beta}]\}^{-T/2} \\ &\propto \{[\mathbf{y} - X\boldsymbol{\beta} - \rho(\mathbf{y}_{-1} - X_{-1}\boldsymbol{\beta})]'[\mathbf{y} - X\boldsymbol{\beta} - \rho(\mathbf{y}_{-1} - X_{-1}\boldsymbol{\beta})]\}^{-T/2}. \end{aligned} \tag{4.16}$$

For any fixed value of ρ it is seen from the first line of (4.16) that the conditional pdf for $\boldsymbol{\beta}$ is

$$p(\boldsymbol{\beta}|\rho, \mathbf{y}) \propto \left\{1 + \frac{[\boldsymbol{\beta} - \hat{\boldsymbol{\beta}}(\rho)]'H[\boldsymbol{\beta} - \hat{\boldsymbol{\beta}}(\rho)]}{\nu s^2(\rho)}\right\}^{-T/2}, \qquad \nu = T - k, \tag{4.17}$$

$$H = (X - \rho X_{-1})'(X - \rho X_{-1}),$$

$$\hat{\boldsymbol{\beta}}(\rho) = H^{-1}(X - \rho X_{-1})'(\mathbf{y} - \rho\mathbf{y}_{-1}),$$

$$\nu s^2(\rho) = [\mathbf{y} - \rho\mathbf{y}_{-1} - (X - \rho X_{-1})\hat{\boldsymbol{\beta}}(\rho)]'[\mathbf{y} - \rho\mathbf{y}_{-1} - (X - \rho X_{-1})\hat{\boldsymbol{\beta}}(\rho)].$$

The distribution in (4.17) is in the form of a multivariate Student t distribution. This result is not surprising, since, for given ρ, (4.12*c*) can be regarded as a usual regression model, with the results from Chapter 3 applicable.

We note that in deriving (4.17) it is implicitly assumed that the matrix H is positive definite for any fixed value of ρ. A necessary and sufficient condition for this to be so is given in Appendix 1 to this chapter. For the case $k = 1$ the condition reduces to that given in connection with (4.8); that is, for any ρ there exists some t such that $x_t \neq \rho x_{t-1}$. In the more general case, $k > 1$, the condition implies that any linear combination of the columns of the matrix of independent variables for periods $0, 1, \ldots, T$, must not satisfy an exact first-order autoregressive scheme. This is not a restrictive condition.

To obtain the marginal posterior pdf's for $\boldsymbol{\beta}$ and ρ[5] we merely integrate (4.16) with respect to these parameters. This can be done easily by completing squares and using properties of the univariate and multivariate Student t pdf's to yield

$$p(\boldsymbol{\beta}|\mathbf{y}) \propto [(\mathbf{y}_{-1} - X_{-1}\boldsymbol{\beta})'(\mathbf{y}_{-1} - X_{-1}\boldsymbol{\beta})]^{-1/2}$$

$$(4.18) \qquad \times \left\{(\mathbf{y} - X\boldsymbol{\beta})'(\mathbf{y} - X\boldsymbol{\beta}) - \frac{[(\mathbf{y} - X\boldsymbol{\beta})'(\mathbf{y}_{-1} - X_{-1}\boldsymbol{\beta})]^2}{(\mathbf{y}_{-1} - X_{-1}\boldsymbol{\beta})'(\mathbf{y}_{-1} - X_{-1}\boldsymbol{\beta})}\right\}^{-(T-1)/2}$$

and

$$p(\rho|\mathbf{y}) \propto \int \{\nu s^2(\rho) + [\boldsymbol{\beta} - \hat{\boldsymbol{\beta}}(\rho)]'H[\boldsymbol{\beta} - \hat{\boldsymbol{\beta}}(\rho)]\}^{-T/2}\, d\boldsymbol{\beta}$$

$$(4.19) \qquad \propto |H|^{-1/2}[\nu s^2(\rho)]^{-\nu/2}$$

$$\propto |H|^{-1/2}\{(\mathbf{y} - \rho\mathbf{y}_{-1})'[I - (X - \rho X_{-1})H^{-1}(X - \rho X_{-1})'](\mathbf{y} - \rho\mathbf{y}_{-1})\}^{-\nu/2},$$

where ν, H, $\hat{\beta}(\rho)$, and $s^2(\rho)$ have been defined in connection with (4.17).

If interest centers on the marginal posterior distribution of a single element of $\boldsymbol{\beta}$, say β_1, its posterior can be obtained in principle from (4.18) by integration. However, this integration, when viewed analytically or numerically, appears to be quite difficult, particularly when k is large. As an alternative, we have

$$(4.20) \qquad p(\beta_1, \rho|\mathbf{y}) = p(\rho|\mathbf{y})\, p(\beta_1|, \rho, \mathbf{y}),$$

with $p(\rho|\mathbf{y})$ given by (4.19) and $p(\beta_1|\rho, \mathbf{y})$ obtained from (4.17) by integration with respect to the elements of $\boldsymbol{\beta}$ other than β_1. From properties of the multivariate t distribution we have from (4.17)

$$(4.21) \qquad \frac{\beta_1 - \hat{\beta}_1(\rho)}{s(\rho)\sqrt{h^{11}}} \sim t_\nu,$$

where h^{11} denotes the (1, 1)th element of H^{-1}. The result in (4.21) gives us the form of the second factor on the rhs of (4.20). Then bivariate numerical integration procedures can be employed to integrate out ρ and thus to obtain the marginal posterior pdf for β_1.

As an alternative, we can obtain $p(\beta_1, \rho|\mathbf{y})$ in a different way by integrating (4.16) with respect to $\beta_2, \beta_3, \ldots, \beta_k$. To perform this integration we partition $\boldsymbol{\beta}' = (\beta_1 \vdots \bar{\boldsymbol{\beta}}')$, $X = (\mathbf{x} \vdots \bar{X})$ and $X_{-1} = (\mathbf{x}_{-1} \vdots \bar{X}_{-1})$, where $\mathbf{x}$ and $\mathbf{x}_{-1}$ denote the first column of X and X_{-1}, respectively. Then, with

$$(4.22) \qquad \mathbf{w} = \mathbf{y} - \rho\mathbf{y}_{-1} - (\mathbf{x} - \rho\mathbf{x}_{-1})\beta_1,$$

[5] The procedures described have been incorporated in a computer program. See H. Thornber, "Bayes Addendum to Technical Report 6603 'Manual for B34T—A Stepwise Regression Program'," Graduate School of Business, University of Chicago, September 1967.

we have

$$p(\beta_1, \bar{\boldsymbol{\beta}}, \rho|\mathbf{y}) \propto \{[\mathbf{w} - (\bar{X} - \rho\bar{X}_{-1})\bar{\boldsymbol{\beta}}]'[\mathbf{w} - (\bar{X} - \rho\bar{X}_{-1})\bar{\boldsymbol{\beta}}]\}^{-T/2}.$$

Integration with respect to $\bar{\boldsymbol{\beta}}$ yields

$$(4.23)\quad p(\beta_1, \rho|\mathbf{y}) \propto |\bar{H}|^{-1/2}\{\mathbf{w}'[I - (\bar{X} - \rho\bar{X}_{-1})\bar{H}^{-1}(\bar{X} - \rho\bar{X}_{-1})']\mathbf{w}\}^{-(T-k+1)/2},$$

where

$$\bar{H} = (\bar{X} - \rho\bar{X}_{-1})'(\bar{X} - \rho\bar{X}_{-1})$$

and $\mathbf{w}$ is defined in (4.22). The posterior pdf for β_1 can be obtained from (4.23) by numerical integration. The advantage of the form (4.23) is that its use involves inverting a $(k - 1) \times (k - 1)$ matrix $\bar{H}$, whereas use of (4.20) involves inverting a $k \times k$ matrix H. Note further that $\bar{H}$ is a λ-matrix[6] of second degree in ρ. Thus the inverse can be expressed as a λ-matrix of degree $2(k - 2)$ in ρ, divided by a scalar polynomial of degree $2(k - 1)$ in ρ. Putting the inverse of $\bar{H}$ in such a form is computationally convenient, since it will avoid the necessity for inverting a matrix for each value of ρ in the integration.

The above results and methods can be applied readily in practice to make inferences about $\boldsymbol{\beta}$ and ρ. Thus there seems little reason to develop *approximate* large-sample techniques. It is interesting, however, to compare the results of a large-sample approximate procedure with the results flowing from application of the results above. As shown in (4.12*c*), our model is

$$(4.24)\quad \mathbf{y} = \rho\mathbf{y}_{-1} + (X - \rho X_{-1})\boldsymbol{\beta} + \boldsymbol{\epsilon}.$$

Note that $\rho\boldsymbol{\beta}$ is a nonlinear combination of parameters. Let us linearize our model by expanding $\rho\boldsymbol{\beta}$ about maximum likelihood estimates,[7] say $\hat{\rho}$ and $\hat{\beta}$, and apply linear theory to the linearized model.[8] On expanding about the maximum likelihood estimates we obtain

$$(4.25)\quad \mathbf{y} \doteq \rho\mathbf{y}_{-1} + X\boldsymbol{\beta} - X_{-1}[\hat{\rho}\hat{\boldsymbol{\beta}} + (\rho - \hat{\rho})\hat{\boldsymbol{\beta}} + \hat{\rho}(\boldsymbol{\beta} - \hat{\boldsymbol{\beta}})] + \boldsymbol{\epsilon}$$

[6] See R. A. Frazer, W. J. Duncan, and A. R. Collar, *Elementary Matrices*, for a discussion of the properties of λ-matrices. A square λ-matrix of degree N takes the form $A_0\lambda^N + A_1\lambda^{N-1} + \cdots + A_{N-1}\lambda + A_N$, where the A_i, $i = 0, 1, 2, \ldots, N$, are square matrices whose elements are independent of λ.

[7] Various ways of computing maximum likelihood estimates of the parameters in (4.24) are reviewed in Zellner and Tiao, *op. cit.*, pp. 776–778.

[8] This appears to be the Bayesian analogue of the large-sample sampling theory approach suggested by W. A. Fuller and J. E. Martin, "The Effects of Autocorrelated Errors on the Statistical Estimation of Distributed Lag Models," *J. Farm Econ.*, **43**, 1961, 71–82. See also C. Hildreth and J. Y. Lu, "Demand Relations with Autocorrelated Disturbances," *Tech. Bull.* 276. East Lansing, Mich.: Michigan State University Agricultural Experiment Station, 1960.

or

$$\mathbf{y} - X_{-1}\hat{\rho}\hat{\boldsymbol{\beta}} \doteq \rho(\mathbf{y}_{-1} - X_{-1}\hat{\boldsymbol{\beta}}) + (X - \hat{\rho}X_{-1})\boldsymbol{\beta} + \boldsymbol{\epsilon}, \tag{4.26}$$

which is linear in the parameters ρ and $\boldsymbol{\beta}$. With the uniform prior pdf's with which we have been working, application of the linear theory of Chapter 3 to (4.26) leads to a posterior pdf for ρ and $\boldsymbol{\beta}$ in the multivariate t form. To illustrate results of this approach we have applied the linearization procedure to analyze the data, shown in Table 4.1, generated from our simple nonexplosive model. Then this sample of 15 observations was augmented to 20, 30, and 40 observations. In Figure 4.3 the resulting approximate posterior pdf's for our scalar parameter β are compared with the exact pdf's computed from (4.7).

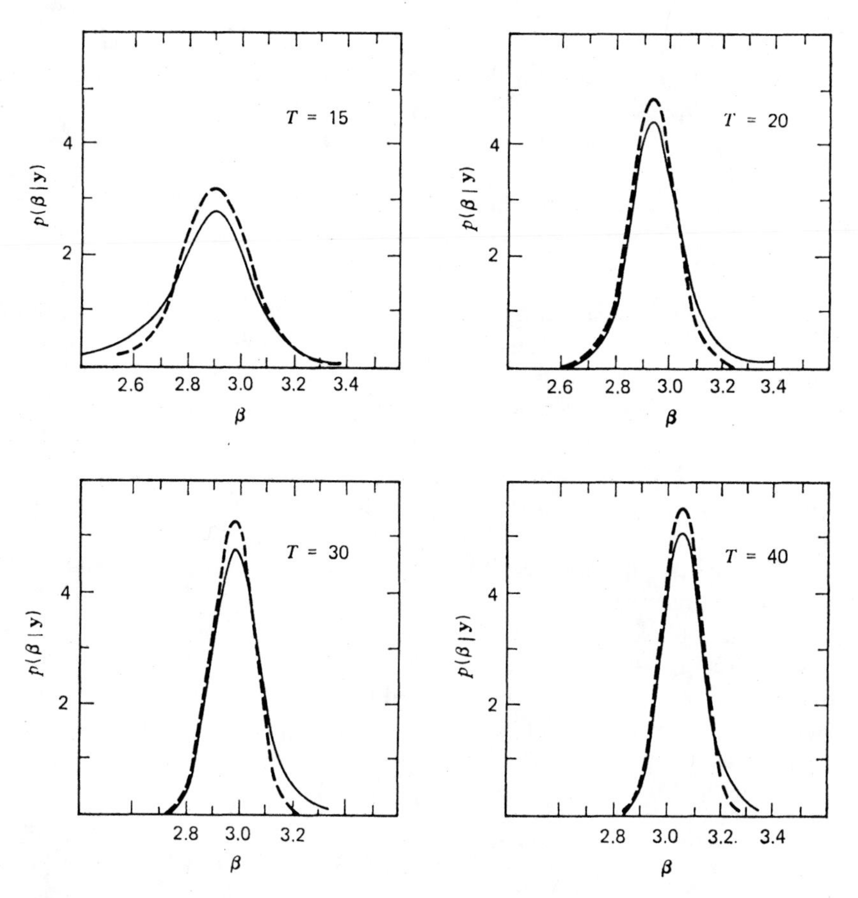

Figure 4.3 Exact (—) and approximate (- - -) marginal distribution of β for several sample sizes and nonexplosive series.

Although the modes of the approximate and exact pdf's occur at approximately the same values, it is seen that there are some rather large differences in the shapes of these pdf's. For $T = 40$ the approximate and exact pdf's are just in fair agreement. These results illustrate quite graphically how careful one must be in using large-sample approximate procedures.

4.2 REGRESSIONS WITH UNEQUAL VARIANCES[9]

Here we consider two normal linear regression equations:

$$\mathbf{y}_1 = X_1\boldsymbol{\beta} + \mathbf{u}_1, \tag{4.27}$$

$$\mathbf{y}_2 = X_2\boldsymbol{\beta} + \mathbf{u}_2, \tag{4.28}$$

where $\mathbf{y}_1$ = an $n_1 \times 1$ vector of observations on a dependent variable,
$\mathbf{y}_2$ = an $n_2 \times 1$ vector of observations on a dependent variable,
X_1 = an $n_1 \times k$ matrix, with rank k, of observations on k independent variables,
X_2 = an $n_2 \times k$ matrix, with rank k, of observations on k independent variables,
$\boldsymbol{\beta}$ = a $k \times 1$ vector of regression coefficients,
$\mathbf{u}_1$ = an $n_1 \times 1$ vector of error terms,
$\mathbf{u}_2$ = an $n_2 \times 1$ vector of error terms.

We assume that the elements of $\mathbf{u}_1$ and $\mathbf{u}_2$ are normally and independently distributed with zero means. The elements of $\mathbf{u}_1$ are assumed to have common variance σ_1^2, and the elements of $\mathbf{u}_2$ are assumed to have common variance σ_2^2. Note that if $\sigma_1^2 = \sigma_2^2 = \sigma^2$ or $\sigma_1^2 = c\sigma_2^2$, with c a *known* factor, we could use the methods of Chapter 3 to analyze our data; that is, we could write (4.27) and (4.28) as $y = X\boldsymbol{\beta} + \mathbf{u}$, with $\mathbf{y}' = (\mathbf{y}_1' \vdots \mathbf{y}_2')$, $X' = (X_1' \vdots X_2')$, and $\mathbf{u}' = (\mathbf{u}_1' \vdots \mathbf{u}_2')$, which is in the form of the standard model considered in Chapter 3. In this section we take up the situation in which $\sigma_1^2 \neq \sigma_2^2$. We analyze first the case in which σ_1^2 is known and σ_2^2 is unknown and then turn to the case in which both variances are unknown.

The problem posed in this section may be encountered in practice in the following circumstances. Suppose that the observations in (4.27) pertain to a particular historical period, say the period between World Wars I and II,[10] and that the observations in (4.28) pertain to the post-World War II period. We may be willing to assume that the regression coefficients are the same for the two periods,[11] that is, $\boldsymbol{\beta}$ is the same in (4.27) and (4.28), but that the error

[9] This section includes much of the material presented in G. C. Tiao and A. Zellner, "Bayes's Theorem and the Use of Prior Knowledge in Regression Analysis," *Biometrika*, **51**, 1 and 2, 219–230 (1964).

[10] With so many wars, we eschew use of the term "interwar."

[11] Below we show how this assumption can be relaxed.

terms' variances σ_1^2 and σ_2^2 are different in the two periods. Alternatively, (4.27) can be viewed as a regression model for a microunit, say a firm, and (4.28) as a regression model for a second microunit. Although we may be willing to assume that both microunits have the same regression coefficient vector $\boldsymbol{\beta}$, we may wish to assume that the units have disturbance terms that are independent but have different variances.

Let us turn to the case in which σ_1^2 is known, one that is not often encountered in practice but is considered here to bring out the relation between Bayesian and certain sampling theory results. The likelihood function is given by

$$
\begin{aligned}
l(\boldsymbol{\beta}, \sigma_2|\sigma_1, \mathbf{y}) \propto \frac{1}{\sigma_2^{n_2}} \exp\Big[-\frac{1}{2\sigma_1^2}(\mathbf{y}_1 - X_1\boldsymbol{\beta})'(\mathbf{y}_1 - X_1\boldsymbol{\beta}) \\
-\frac{1}{2\sigma_2^2}(\mathbf{y}_2 - X_2\boldsymbol{\beta})'(\mathbf{y}_2 - X_2\boldsymbol{\beta})\Big],
\end{aligned}
\tag{4.29}
$$

where $\mathbf{y}' = (\mathbf{y}_1'\mathbf{y}_2')$ and it is understood that we are given X_1 and X_2. As prior assumptions we assume that $\log \sigma_2$ and the elements of $\boldsymbol{\beta}$ are uniformly and independently distributed, which implies

$$
p(\boldsymbol{\beta}, \sigma_2) \propto \frac{1}{\sigma_2}, \qquad \begin{array}{ll} -\infty < \beta_i < \infty, & i = 1, 2, \ldots, k, \\ 0 < \sigma_2 < \infty. & \end{array}
\tag{4.30}
$$

On combining (4.29) and (4.30) and integrating with respect to σ_2, the posterior pdf for $\boldsymbol{\beta}$ is

$$
\begin{aligned}
p(\boldsymbol{\beta}|\sigma_1, \mathbf{y}) &\propto \exp\left[-\frac{1}{2\sigma_1^2}(\mathbf{y}_1 - X_1\boldsymbol{\beta})'(\mathbf{y}_1 - X_1\boldsymbol{\beta})\right] \\
&\quad \times [(\mathbf{y}_2 - X_2\boldsymbol{\beta})'(\mathbf{y}_2 - X_2\boldsymbol{\beta})]^{-\frac{1}{2}n_2} \\
&\propto \exp\left[-\frac{1}{2\sigma_1^2}(\boldsymbol{\beta} - \hat{\boldsymbol{\beta}}_1)'Z_1(\boldsymbol{\beta} - \hat{\boldsymbol{\beta}}_1)\right] \\
&\quad \times \left[1 + \frac{(\boldsymbol{\beta} - \hat{\boldsymbol{\beta}}_2)'Z_2(\boldsymbol{\beta} - \hat{\boldsymbol{\beta}}_2)}{\nu_2 s_2^2}\right]^{-\frac{1}{2}(\nu_2 + k)},
\end{aligned}
\tag{4.31}
$$

where $Z_1 = X_1'X_1$, $\hat{\boldsymbol{\beta}}_1 = Z_1^{-1}X_1'\mathbf{y}_1$, $Z_2 = X_2'X_2$, $\hat{\boldsymbol{\beta}}_2 = Z_2^{-1}X_2'\mathbf{y}_2$, $\nu_2 = n_2 - k$, and $\nu_2 s_2^2 = (\mathbf{y}_2 - X_2\hat{\boldsymbol{\beta}}_2)'(\mathbf{y}_2 - X_2\hat{\boldsymbol{\beta}}_2)$. It is seen that (4.31) is the product of two factors, the first in the normal form in $\boldsymbol{\beta}$ and the second in the multivariate Student t form. For this reason we refer to (4.31) as a "normal-t" pdf. On expanding the second factor in an asymptotic series (see Appendix 2 to this chapter), we have for the leading normal term in the expansion

$$
\begin{aligned}
p(\boldsymbol{\beta}|\sigma_1, \mathbf{y}) &\propto \exp\left[-\frac{1}{2\sigma_1^2}(\boldsymbol{\beta} - \hat{\boldsymbol{\beta}}_1)'Z_1(\boldsymbol{\beta} - \hat{\boldsymbol{\beta}}_1) - \frac{1}{2s_2^2}(\boldsymbol{\beta} - \hat{\boldsymbol{\beta}}_2)'Z_2(\boldsymbol{\beta} - \hat{\boldsymbol{\beta}}_2)\right] \\
&\propto \exp\left[-\tfrac{1}{2}(\boldsymbol{\beta} - \tilde{\boldsymbol{\beta}})'A(\boldsymbol{\beta} - \tilde{\boldsymbol{\beta}})\right],
\end{aligned}
\tag{4.32}
$$

with

$$\tilde{\boldsymbol{\beta}} = \left(\frac{1}{\sigma_1^2} Z_1 + \frac{1}{s_2^2} Z_2\right)^{-1}\left(\frac{1}{\sigma_1^2} X_1'\mathbf{y}_1 + \frac{1}{s_2^2} X_2'\mathbf{y}_2\right) \tag{4.33}$$

and

$$A = \left(\frac{1}{\sigma_1^2} Z_1 + \frac{1}{s_2^2} Z_2\right). \tag{4.34}$$

The second line of (4.32) is obtained simply by completing the square on $\boldsymbol{\beta}$ in the first line of (4.32).

It is interesting to note that (4.33) is the quantity that Theil[12] recommends as a *sampling theory* estimator incorporating prior stochastic information and for which he provides a large sample justification. Here the quantity in (4.33) appears as the mean of the leading normal term in an asymptotic expansion approximating the posterior pdf for $\boldsymbol{\beta}$. In Appendix 2 to this chapter methods are presented that enable us to take account of additional terms in the asymptotic expansion and thus to get a better approximation to the posterior pdf.

We now take up the case in which both σ_1 and σ_2 are unknown, the case most often encountered in practice. The likelihood function is given by

$$\begin{aligned} l(\boldsymbol{\beta}, \sigma_1, \sigma_2|\mathbf{y}) \propto \frac{1}{\sigma_1^{n_1}\sigma_2^{n_2}} \exp\Big[&-\frac{1}{2\sigma_1^2}(\mathbf{y}_1 - X_1\boldsymbol{\beta})'(\mathbf{y}_1 - X_1\boldsymbol{\beta}) \\ &- \frac{1}{2\sigma_2^2}(\mathbf{y}_2 - X_2\boldsymbol{\beta})'(\mathbf{y}_2 - X_2\boldsymbol{\beta})\Big]. \end{aligned} \tag{4.35}$$

As prior pdf, we assume that we have diffuse information about $\boldsymbol{\beta}$, σ_1, and σ_2 and represent this by

$$p(\boldsymbol{\beta}, \sigma_1, \sigma_2) \propto \frac{1}{\sigma_1\sigma_2} \qquad \begin{aligned} -\infty < \beta_i < \infty,& \\ 0 < \sigma_1 < \infty,& \qquad i = 1, 2, \ldots, k, \\ 0 < \sigma_2 < \infty.& \end{aligned} \tag{4.36}$$

Here we have assumed that the elements of $\boldsymbol{\beta}$, $\log \sigma_1$, and $\log \sigma_2$ are independently and uniformly distributed. The joint posterior pdf for the parameters is given by

$$\begin{aligned} p(\boldsymbol{\beta}, \sigma_1, \sigma_2|\mathbf{y}) \propto (\sigma_1^{n_1+1}\sigma_2^{n_2+1})^{-1} \exp\Big[&-\frac{1}{2\sigma_1^2}(\mathbf{y}_1 - X_1\boldsymbol{\beta})'(\mathbf{y}_1 - X_1\boldsymbol{\beta}) \\ &- \frac{1}{2\sigma_2^2}(\mathbf{y}_2 - X_2\boldsymbol{\beta})'(\mathbf{y}_2 - X_2\boldsymbol{\beta})\Big]. \end{aligned} \tag{4.37}$$

[12] H. Theil, "On the Use of Incomplete Prior Information in Regression Analysis," *J. Am. Statist. Assoc.*, **58**, 401–414 (1963).

The integrations over σ_1 and σ_2 are easily performed to yield the following joint posterior pdf for the elements of $\boldsymbol{\beta}$:

$$
\begin{aligned}
p(\boldsymbol{\beta}|\mathbf{y}) &\propto [(\mathbf{y}_1 - X_1\boldsymbol{\beta})'(\mathbf{y}_1 - X_1\boldsymbol{\beta})]^{-n_1/2}[(\mathbf{y}_2 - X_2\boldsymbol{\beta})'(\mathbf{y}_2 - X_2\boldsymbol{\beta})]^{-n_2/2} \\
&\propto \left[1 + \frac{(\boldsymbol{\beta} - \hat{\boldsymbol{\beta}}_1)'Z_1(\boldsymbol{\beta} - \hat{\boldsymbol{\beta}}_1)}{\nu_1 s_1^2}\right]^{-\frac{1}{2}(\nu_1 + k)} \\
&\quad \times \left[1 + \frac{(\boldsymbol{\beta} - \hat{\boldsymbol{\beta}}_2)'Z_2(\boldsymbol{\beta} - \hat{\boldsymbol{\beta}}_2)}{\nu_2 s_2^2}\right]^{-\frac{1}{2}(\nu_2 + k)},
\end{aligned}
\tag{4.38}
$$

where $\nu_i = n_i - k$, $Z_i = X_i'X_i$, $\hat{\boldsymbol{\beta}}_i = Z_i^{-1}X_i'\mathbf{y}_i$, and $\nu_i s_i^2 = (\mathbf{y}_i - X_i\hat{\boldsymbol{\beta}}_i)' \times (\mathbf{y}_i - X_i\hat{\boldsymbol{\beta}}_i)$, with $i = 1, 2$.

We see that (4.38) is the product of two factors, each in the multivariate Student t form. Thus we call this a multivariate "double-t" distribution. To analyze it we expand each of the factors in an asymptotic expansion (see Appendix 2) that yields the following as the leading normal term:

$$
\begin{aligned}
p(\boldsymbol{\beta}|\mathbf{y}) &\dot{\propto} \exp\left[-\frac{1}{2s_1^2}(\boldsymbol{\beta} - \hat{\boldsymbol{\beta}}_1)'Z_1(\boldsymbol{\beta} - \hat{\boldsymbol{\beta}}_1) - \frac{1}{2s_2^2}(\boldsymbol{\beta} - \hat{\boldsymbol{\beta}}_2)'Z_2(\boldsymbol{\beta} - \hat{\boldsymbol{\beta}}_2)\right] \\
&\dot{\propto} \exp\left[-\tfrac{1}{2}(\boldsymbol{\beta} - \bar{\boldsymbol{\beta}})'D(\boldsymbol{\beta} - \bar{\boldsymbol{\beta}})\right],
\end{aligned}
\tag{4.39}
$$

where $\dot{\propto}$ denotes "approximately proportional to,"

$$D = M_1 + M_2, \tag{4.40}$$

and

$$\bar{\boldsymbol{\beta}} = D^{-1}(M_1\hat{\boldsymbol{\beta}}_1 + M_2\hat{\boldsymbol{\beta}}_2), \tag{4.41}$$

where $M_1 = Z_1/s_1^2 = X_1'X_1/s_1^2$ and $M_2 = Z_2/s_2^2 = X_2'X_2/s_2^2$. Using these definitions, (4.41) can be written as

$$\bar{\boldsymbol{\beta}} = \left(\frac{1}{s_1^2}X_1'X_1 + \frac{1}{s_2^2}X_2'X_2\right)^{-1}\left(\frac{1}{s_1^2}X_1'\mathbf{y}_1 + \frac{1}{s_2^2}X_2'\mathbf{y}_2\right), \tag{4.42}$$

which is, of course, the mean of the leading normal term in the asymptotic expansion of the double t pdf in (4.38). The analysis required to take account of higher order terms in the asymptotic expansion and thus to get a better approximation to the posterior pdf is given in Appendix 2.

It is also interesting to observe that (4.42) would emerge in the sampling theory approach as an approximation to the generalized least squares estimator for the system in (4.27) and (4.28) if the unknown parameters σ_1^2 and σ_2^2 appearing in this estimator were replaced by s_1^2 and s_2^2, respectively; that is, the generalized least squares estimator is given by

$$(X'\Sigma^{-1}X)^{-1}X'\Sigma^{-1}\mathbf{y} = \left(\frac{1}{\sigma_1^2}X_1'X_1 + \frac{1}{\sigma_2^2}X_2'X_2\right)^{-1}\left(\frac{1}{\sigma_1^2}X_1'\mathbf{y}_1 + \frac{1}{\sigma_2^2}X_2'\mathbf{y}_2\right),$$

where $X' = (X_1' \vdots X_2')$, $\mathbf{y}' = (\mathbf{y}_1' \vdots \mathbf{y}_2')$, and

$$\Sigma = \begin{pmatrix} \sigma_1^2 I_{n_1} & 0 \\ 0 & \sigma_2^2 I_{n_2} \end{pmatrix}.$$

Thus, if we set $\sigma_1^2 = s_1^2$ and $\sigma_2^2 = s_2^2$, we obtain an approximation to the generalized least squares estimator which is usually given a large sample justification. In the Bayesian approach we see from (4.37) that the conditional posterior pdf for $\boldsymbol{\beta}$, given σ_1 and σ_2, is

$$\begin{aligned} p(\boldsymbol{\beta}|\sigma_1, \sigma_2, \mathbf{y}) &\propto \exp\left\{-\frac{1}{2}\left[\frac{1}{\sigma_1^2}(\mathbf{y}_1 - X_1\boldsymbol{\beta})'(\mathbf{y}_1 - X_1\boldsymbol{\beta}) + \frac{1}{\sigma_2^2}(\mathbf{y}_2 - X_2\boldsymbol{\beta})'(\mathbf{y}_2 - X_2\boldsymbol{\beta})\right]\right\} \\ &\propto \exp\left[-\tfrac{1}{2}(\boldsymbol{\beta} - \tilde{\boldsymbol{\beta}})'(X'\Sigma^{-1}X)(\boldsymbol{\beta} - \tilde{\boldsymbol{\beta}})\right], \end{aligned}$$

with

$$\begin{aligned} \tilde{\boldsymbol{\beta}} &= (X'\Sigma^{-1}X)^{-1}X'\Sigma^{-1}\mathbf{y} \\ &= \left(\frac{1}{\sigma_1^2}X_1'X_1 + \frac{1}{\sigma_2^2}X_2'X_2\right)^{-1}\left(\frac{1}{\sigma_1^2}X_1'\mathbf{y}_1 + \frac{1}{\sigma_2^2}X_2'\mathbf{y}_2\right). \end{aligned}$$

Thus in the Bayesian approach the generalized least squares quantity appears as the mean of the conditional posterior pdf $p(\boldsymbol{\beta}|\sigma_1, \sigma_2, \mathbf{y})$, which is in the multivariate normal form with covariance matrix

$$(X'\Sigma^{-1}X)^{-1} = \left(\frac{1}{\sigma_1^2}X_1'X_1 + \frac{1}{\sigma_2^2}X_2'X_2\right)^{-1}.$$

If in this conditional pdf we set $\sigma_1^2 = s_1^2$ and $\sigma_2^2 = s_2^2$, we obtain the approximation to the generalized least squares quantity as the mean of our conditional pdf. In large samples s_1^2 and s_2^2 will be close to the true values of σ_1^2 and σ_2^2 and thus the use of the conditional pdf may be satisfactory. In general, however, it is better to integrate with respect to σ_1 and σ_2 to obtain the marginal pdf for $\boldsymbol{\beta}$ and to base inferences on it rather than use the conditional pdf.

To illustrate application of these techniques we analyze a simple investment model with annual time series data, 1935–1954, relating to two corporations, General Electric and Westinghouse.[13] In this model price deflated gross investment is assumed to be a linear function of expected profitability and beginning-of-year real capital stock. Following Grunfeld,[14] the value of

[13] The data are taken from J. C. G. Boot and G. M. de Witt, "Investment Demand: An Empirical Contribution to the Aggregation Problem," *Intern. Econ. Rev.*, **1**, 3–30 (1960).

[14] Y. Grunfeld, "The Determinants of Corporate Investment," unpublished PhD thesis, University of Chicago, 1958.

outstanding shares at the beginning of the year is taken as a measure of a firm's expected profitability, an assumption that has received critical comment but which we use just for illustrative purposes. The two investment relations are

$$y_1(t) = \alpha_1 + \beta_1 x_{11}(t) + \beta_2 x_{12}(t) + u_1(t), \tag{4.43a}$$

$$y_2(t) = \alpha_2 + \beta_1 x_{21}(t) + \beta_2 x_{22}(t) + u_2(t), \tag{4.43b}$$

where t in parentheses denotes the value of a variable in year t ($t = 1, 2, \ldots, 20$), and

Variable	General Electric	Westinghouse
Annual real gross investment	$y_1(t)$	$y_2(t)$
Value of shares at beginning of year	$x_{11}(t)$	$x_{21}(t)$
Real capital at beginning of year	$x_{12}(t)$	$x_{22}(t)$
Error term	$u_1(t)$	$u_2(t)$

The parameters β_1 and β_2 in (4.43) are taken to be the same for the two firms in this illustrative example; however, α_1 and α_2, the intercepts, are assumed to be different to allow for possible differences in the investment behavior of the two firms. Further, $u_1(t)$ and $u_2(t)$ are assumed to be independently[15] and normally distributed for all t with zero means and variances σ_1^2 and σ_2^2, respectively.

If we employ a diffuse prior pdf for the parameters, namely,

$$p(\boldsymbol{\alpha}, \boldsymbol{\beta}, \sigma_1, \sigma_2) \propto \frac{1}{\sigma_1 \sigma_2}, \tag{4.44}$$

with $\boldsymbol{\alpha}' = (\alpha_1, \alpha_2)$ and $\boldsymbol{\beta}' = (\beta_1, \beta_2)$, we obtain the following joint posterior pdf:

$$\begin{aligned} p(\boldsymbol{\alpha}, \boldsymbol{\beta}, \sigma_1, \sigma_2 | \mathbf{y}) &\propto (\sigma_1^{n+1})^{-1} (\sigma_2^{n+1})^{-1} \\ &\times \exp\left\{-\frac{1}{2}\left[\frac{1}{\sigma_1^2}(\mathbf{y}_1 - \alpha_1 \boldsymbol{\iota} - X_1\boldsymbol{\beta})'(\mathbf{y}_1 - \alpha_1 \boldsymbol{\iota} - X_1\boldsymbol{\beta})\right.\right. \\ &\left.\left. + \frac{1}{\sigma_2^2}(\mathbf{y}_2 - \alpha_2 \boldsymbol{\iota} - X_2\boldsymbol{\beta})'(\mathbf{y}_2 - \alpha_2 \boldsymbol{\iota} - X_2\boldsymbol{\beta})\right]\right\}, \end{aligned} \tag{4.45}$$

where $n = 20$, $\boldsymbol{\iota}' = (1, 1, \ldots, 1)$, a $1 \times n$ vector, and X_1 and X_2 are $n \times 2$ matrices of observations on G.E.'s and Westinghouse's independent variables.

[15] Below we show how this assumption can be relaxed.

If interest does not center on α_1 and α_2, they can be integrated out of (4.45).[16] Then on integrating over σ_1 and σ_2, we have

$$p(\boldsymbol{\beta}|\mathbf{y}) \propto [\nu_1 + (\boldsymbol{\beta} - \hat{\boldsymbol{\beta}}_1)'M_1(\boldsymbol{\beta} - \hat{\boldsymbol{\beta}}_1)]^{-(\nu_1+2)/2} \times [\nu_2 + (\boldsymbol{\beta} - \hat{\boldsymbol{\beta}}_2)M_2(\boldsymbol{\beta} - \hat{\boldsymbol{\beta}}_2)]^{-(\nu_2+2)/2}, \tag{4.46}$$

where $\nu_1 = \nu_2 = 17$, $\hat{\boldsymbol{\beta}}_1$ is the least squares quantity, a 2×1 vector, obtained from G.E.'s data, $\hat{\boldsymbol{\beta}}_2$ is the least squares quantity based on Westinghouse's data,[17]

$$M_1 = \frac{1}{s_1^2}\left(X_1 - \frac{1}{n}\,\iota\iota'X_1\right)'\left(X_1 - \frac{1}{n}\,\iota\iota'X_1\right)$$

and

$$M_2 = \frac{1}{s_2^2}\left(X_2 - \frac{1}{n}\,\iota\iota'X_2\right)\left(X_2 - \frac{1}{n}\,\iota\iota'X_2\right).$$

The sample quantities are shown below:

General Electric	Westinghouse
$\hat{\boldsymbol{\beta}}_1 = \begin{pmatrix} 0.02655 \\ 0.1517 \end{pmatrix}$	$\hat{\boldsymbol{\beta}}_2 = \begin{pmatrix} 0.05289 \\ 0.09241 \end{pmatrix}$
$s_1^2 = 777.4463$	$s_2^2 = 104.3079$
$\nu_1 = 17$	$\nu_2 = 17$
$M_1 = \begin{bmatrix} 4185.1054 & 299.6748 \\ 299.6748 & 1535.0640 \end{bmatrix}$	$M_2 = \begin{bmatrix} 9010.5868 & 1871.1079 \\ 1871.1079 & 706.3320 \end{bmatrix}$

Further the quantity in (4.42),[18] $\bar{\boldsymbol{\beta}}' = (0.0373, 0.1446)$.

A plot of the contours of the joint posterior pdf for β_1 and β_2, given in (4.46), is shown in Figure 4.4. Also in this figure are lines showing the loci of conditional modes. We see that the posterior distribution is concentrated rather sharply in the region $0.0278 < \beta_1 < 0.0468$ and $0.1216 < \beta_2 < 0.1676$, with mode at approximately (0.0373, 0.1446). Further, β_1 and β_2 are negatively correlated and the contours are approximately elliptical. This is because the joint density function is nearly a bivariate normal distribution due to the fact that both ν_1 and ν_2 are fairly large in this example.

If interest centers on only one of the parameters, say β_1, we can obtain

[16] If interest does center on α_1 and α_2, (4.45) can be integrated with respect to $\boldsymbol{\beta}$, σ_1, and σ_2. Then α_1 and α_2 will be distributed in the form of two independent Student t variables. Further, the difference $\alpha_1 - \alpha_2$ has the Behrens-Fisher distribution.

[17] That is, let $\mathbf{z}_i = (I - \iota\iota'/n)\mathbf{y}_i$ and $W_i = (I - \iota\iota'/n)X_i$; then $\hat{\boldsymbol{\beta}}_i = (W_i'W_i)^{-1}W_i'\mathbf{z}_i$, $i = 1, 2$.

[18] Here, because we have integrated out α_1 and α_2, all moments appearing in (4.42) become moments about sample means.

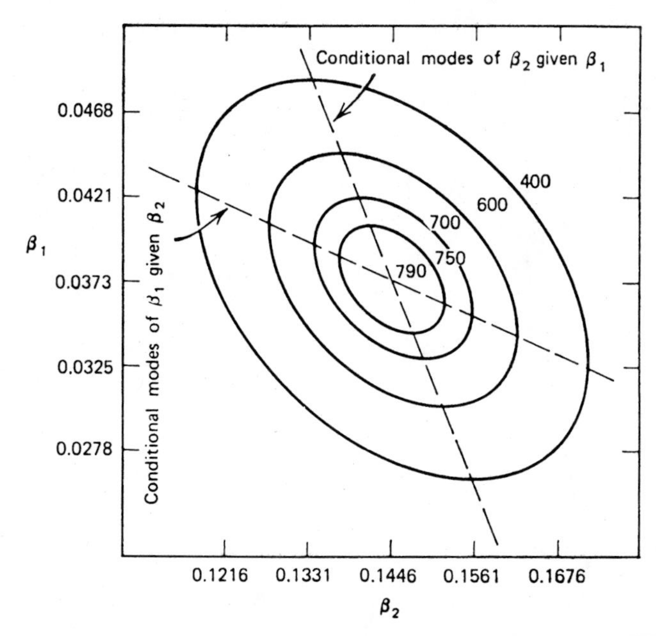

Figure 4.4 Contours of the joint posterior distribution of β_1 and β_2.

its marginal pdf by methods discussed in Appendix 2 and utilizing an asymptotic expansion of (4.46).[19] This marginal pdf is plotted as the solid line in Figure 4.5. Also shown in Figure 4.5 is an approximate posterior pdf for β_1, based on the leading normal term in the asymptotic expansion [see (4.39) and Appendix 2]. It is seen that the posterior pdf for β_1, given by the solid line, is somewhat flatter at the center and fatter in the tails than the approximating large-sample normal pdf, represented by the broken curve. A comparison of the first two moments of these pdf's is shown below:

	Large Sample Normal Approximation[a]	Finite Sample Approximation[b]
Mean	0.0373	0.03726
Variance	9.01445×10^{-5}	9.6158×10^{-5}

a. Broken curve in Figure 4.5.

b. Solid curve in Figure 4.5, based on asymptotic expansion of (4.46), disregarding terms for which $i + j > 2$ (see Appendix 2).

[19] In this particular example, with just two elements in $\boldsymbol{\beta}$, bivariate numerical integration techniques can be employed to obtain the marginal pdf for β_1. We use the asymptotic expansion here because it can be used when $\boldsymbol{\beta}$ contains more than two elements. Also, see footnote 20 below.

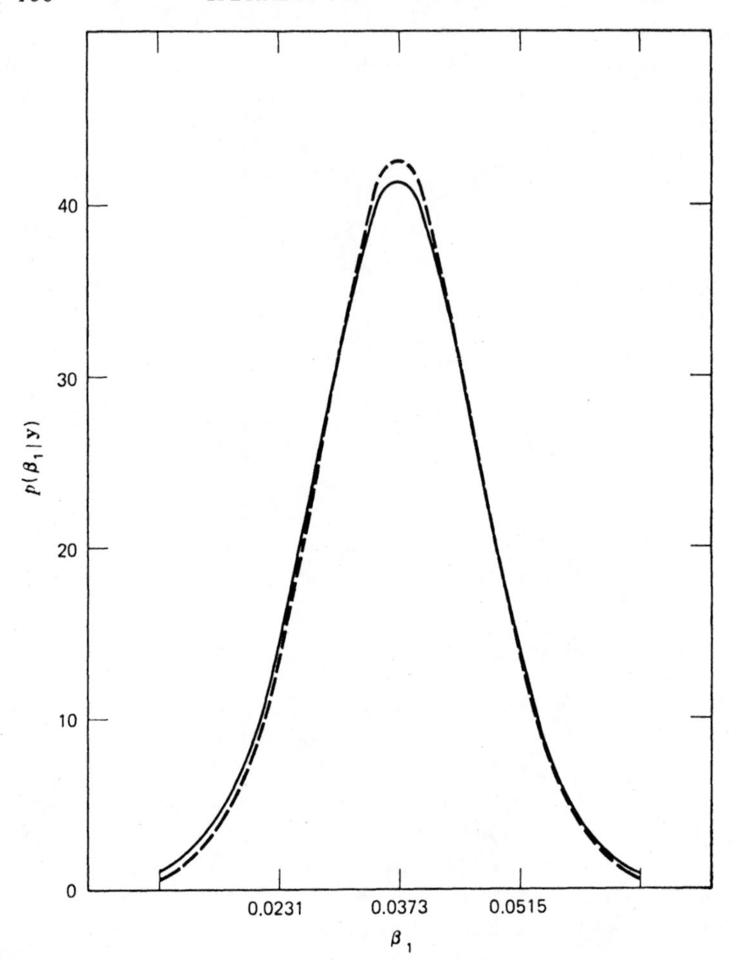

Figure 4.5 In this figure, the solid curve represents the posterior distribution of β_1 and the broken curve represents the limiting normal approximation.

The mean of β_1 is extremely close to its normal approximating value. On the other hand, the variance of β_1 is about 6% larger than that provided by the approximating normal distribution.

We have concentrated on making inferences about regression coefficients. In some circumstances we may be interested in making inferences about σ_1 and σ_2, given the model shown in (4.27) and (4.28), the likelihood function shown in (4.35), and prior assumptions shown in (4.36). In the joint posterior pdf in (4.37) we change variables from $\boldsymbol{\beta}$, σ_1, and σ_2 to $\boldsymbol{\beta}$, σ_1, and $\lambda = \sigma_1{}^2/\sigma_2{}^2$,

$0 < \lambda < \infty$. The Jacobian of this transformation is $J \propto \sigma_2^3/\sigma_1^2 = \sigma_1\lambda^{-3/2}$. Thus, in terms of $\boldsymbol{\beta}$, σ, and λ, (4.37) becomes[20]

$$p(\boldsymbol{\beta}, \sigma_1, \lambda|\mathbf{y}) \propto \frac{\lambda^{(n_2-2)/2}}{\sigma_1^{n_1+n_2+1}} \exp\left\{-\frac{1}{2\sigma_1^2}[(\mathbf{y}_1 - X_1\boldsymbol{\beta})'(\mathbf{y}_1 - X_1\boldsymbol{\beta}) + \lambda(\mathbf{y}_2 - X_2\boldsymbol{\beta})'(\mathbf{y}_2 - X_2\boldsymbol{\beta})]\right\}. \tag{4.47}$$

We now complete the square on $\boldsymbol{\beta}$ in the exponent:

$$\begin{aligned}
(\mathbf{y}_1 - X_1\boldsymbol{\beta})'(\mathbf{y}_1 - X_1\boldsymbol{\beta}) &+ \lambda(\mathbf{y}_2 - X_2\boldsymbol{\beta})'(\mathbf{y}_2 - X_2\boldsymbol{\beta}) \\
&= \boldsymbol{\beta}'(X_1'X_1 + \lambda X_2'X_2)\boldsymbol{\beta} - 2\boldsymbol{\beta}'(X_1'\mathbf{y}_1 + \lambda X_2'\mathbf{y}_2) + \mathbf{y}_1'\mathbf{y}_1 + \lambda\mathbf{y}_2'\mathbf{y}_2 \\
&= (\boldsymbol{\beta} - C_1^{-1}C_2)'C_1(\boldsymbol{\beta} - C_1^{-1}C_2) + \mathbf{y}_1'\mathbf{y}_1 + \lambda\mathbf{y}_2'\mathbf{y}_2 - C_2'C_1^{-1}C_2,
\end{aligned}$$

with $C_1 = X_1'X_1 + \lambda X_2'X_2$ and $C_2 = X_1'\mathbf{y}_1 + \lambda X_2'\mathbf{y}_2$. Then on substituting in (4.47) and integrating with respect to $\boldsymbol{\beta}$ we have

$$p(\sigma_1, \lambda|\mathbf{y}) \propto \frac{\lambda^{(n_2-2)/2}}{\sigma_1^{n_1+n_2-k+1}} |C_1|^{-1/2} \times \exp\left\{-\frac{1}{2\sigma_1^2}[\mathbf{y}_1'\mathbf{y}_1 + \lambda\mathbf{y}_2'\mathbf{y}_2 - C_2'C_1^{-1}C_2]\right\}, \tag{4.48}$$

which is the bivariate posterior pdf for σ_1 and λ. On integrating (4.48) with respect to σ_1 we have the following for the marginal posterior pdf for λ:

$$\begin{aligned}
p(\lambda|\mathbf{y}) &\propto \frac{\lambda^{(n_2-2)/2}|C_1|^{-1/2}}{(\mathbf{y}_1'\mathbf{y}_1 + \lambda\mathbf{y}_2'\mathbf{y}_2 - C_2'C_1^{-1}C_2)^{(n_1+n_2-k)/2}} \\
&\propto \frac{\lambda^{(n_2-2)/2}|X_1'X_1 + \lambda X_2'X_2|^{-1/2}}{[\mathbf{y}_1'\mathbf{y}_1 + \lambda\mathbf{y}_2'\mathbf{y}_2 - (X_1'\mathbf{y}_1 + \lambda X_2'\mathbf{y}_2)' \times (X_1'X_1 + \lambda X_2'X_2)^{-1}(X_1'\mathbf{y}_1 + \lambda X_2'\mathbf{y}_2)]^{(n_1+n_2-k)/2}}.
\end{aligned} \tag{4.49}$$

This posterior pdf can be analyzed by using univariate numerical integration techniques. It should be noted that if the regression coefficient vectors in (4.27) and (4.28) were not identical and we assumed a priori that all regression coefficients, $\log \sigma_1$ and $\log \sigma_2$, were uniformly and independently distributed, the posterior pdf for $\lambda = \sigma_1^2/\sigma_2^2$ would be in the form of an F distribution. The expression in (4.49) departs from being in the F form because it incorporates the information that the coefficient vectors in (4.27) and (4.28) are the same.

[20] Equation 4.47 can also be employed to obtain the marginal posterior pdf of a single element of $\boldsymbol{\beta}$, say β_1. Analytically integrate (4.47) with respect to σ_1 and the elements of $\boldsymbol{\beta}$ other than β_1. The result is a bivariate pdf for β_1 and λ which can be analyzed numerically.

4.3 TWO REGRESSIONS WITH SOME COMMON COEFFICIENTS[21]

Assume in connection with the system in (4.27) and (4.28) that the coefficient vectors appearing in these equations are not entirely the same; for example, in the numerical example of Section 4.2 we allowed the intercept terms to be different in the two investment relations. In general we may have

$$\mathbf{y}_1 = W_1\boldsymbol{\beta}_1 + W_2\boldsymbol{\beta}_2 + \mathbf{u}_1, \tag{4.50}$$

$$\mathbf{y}_2 = Z_1\boldsymbol{\beta}_1 + Z_2\boldsymbol{\gamma}_2 + \mathbf{u}_2, \tag{4.51}$$

where $\mathbf{y}_1$ and $\mathbf{y}_2$ are $n_1 \times 1$ and $n_2 \times 1$ vectors of observations on our dependent variables, $(W_1 \vdots W_2)$ is an $n_1 \times k_1$ matrix, with rank k_1 of given observations on k_1 independent variables, $(Z_1 \vdots Z_2)$ is an $n_2 \times k_2$ matrix, with rank k_2 of given observations on k_2 independent variables, $\boldsymbol{\beta}_1$ is an $m \times 1$ vector of coefficients appearing in both equations, $\boldsymbol{\beta}_2$ and $\boldsymbol{\gamma}_2$ are $m_1 \times 1$ and $m_2 \times 1$ vectors, respectively, of regression coefficients, and $\mathbf{u}_1$ and $\mathbf{u}_2$ are $n_1 \times 1$ and $n_2 \times 1$ vectors of error terms. Note that $k_1 = m + m_1$ and $k_2 = m + m_2$ and that W_1 has the dimension $n_1 \times m$ and Z_1, the dimension $n_2 \times m$. We assume that the disturbance terms in $\mathbf{u}_1$ and $\mathbf{u}_2$ are normally and independently distributed each with zero mean and common variance σ^2. For convenience we rewrite the system as follows:

$$\begin{pmatrix}\mathbf{y}_1\\ \mathbf{y}_2\end{pmatrix} = \begin{pmatrix}W_1 & W_2 & 0\\ Z_1 & 0 & Z_2\end{pmatrix}\begin{pmatrix}\boldsymbol{\beta}_1\\ \boldsymbol{\beta}_2\\ \boldsymbol{\gamma}_2\end{pmatrix} + \begin{pmatrix}\mathbf{u}_1\\ \mathbf{u}_2\end{pmatrix} \tag{4.52}$$

or

$$\mathbf{y} = X\boldsymbol{\beta} + \mathbf{u}, \tag{4.53}$$

with $\mathbf{y}' = (\mathbf{y}_1' \vdots \mathbf{y}_2')$, $\boldsymbol{\beta}' = (\boldsymbol{\beta}_1' \vdots \boldsymbol{\beta}_2' \vdots \boldsymbol{\gamma}_2')$, $\mathbf{u}' = (\mathbf{u}_1' \vdots \mathbf{u}_2')$, and X denoting the partitioned matrix on the rhs of (4.52). It is seen that (4.53) is in the form of a multiple regression model with $n_1 + n_2$ observations assumed to satisfy the standard assumptions. Thus with a diffuse prior on the elements of $\boldsymbol{\beta}$ and $\log \sigma$ the posterior pdf for $\boldsymbol{\beta}$ will be in the multivariate Student t form; that is

$$p(\boldsymbol{\beta}|\mathbf{y}) \propto \{\nu s^2 + (\boldsymbol{\beta} - \hat{\boldsymbol{\beta}})'X'X(\boldsymbol{\beta} - \hat{\boldsymbol{\beta}})\}^{-(n_1+n_2)/2} \tag{4.54}$$

with $\nu = n_1 + n_2 - m - m_1 - m_2$, $\hat{\boldsymbol{\beta}} = (X'X)^{-1}X'\mathbf{y}$ and $\nu s^2 = (\mathbf{y} - X\hat{\boldsymbol{\beta}})'(\mathbf{y} - X\hat{\boldsymbol{\beta}})$.

[21] This problem has been analyzed in V. K. Chetty, "On Pooling of Time Series and Cross-Section Data," *Econometrica*, **36**, 279–290 (1968). Here we take a somewhat different approach in the derivation of some of his results.

With (4.54) noted, the problem of obtaining the marginal posterior pdf for, say, $\boldsymbol{\beta}_1$ is just the problem of getting the marginal posterior pdf for a subset of a set of variables distributed in the multivariate Student t form, a problem considered in Chapter 3. Here we partition $\boldsymbol{\beta}' = (\boldsymbol{\beta}_1' \vdots \boldsymbol{\gamma}')$, $\hat{\boldsymbol{\beta}}' = (\hat{\boldsymbol{\beta}}_1' \vdots \hat{\boldsymbol{\gamma}}')$, and

$$X'X = \begin{pmatrix} M_{11} & M_{12} \\ M_{21} & M_{22} \end{pmatrix} = \begin{pmatrix} W_1'W_1 + Z_1'Z_1 & W_1'W_2 & Z_1'Z_2 \\ W_2'W_1 & W_2'W_2 & 0 \\ Z_2'Z_1 & 0 & Z_2'Z_2 \end{pmatrix},$$

with $\boldsymbol{\gamma}' = (\boldsymbol{\beta}_2' \vdots \boldsymbol{\gamma}_2')$ and $\hat{\boldsymbol{\gamma}}' = (\hat{\boldsymbol{\beta}}_2' \vdots \hat{\boldsymbol{\gamma}}_2')$. Then the marginal posterior pdf for $\boldsymbol{\beta}_1$ is given by

$$p(\boldsymbol{\beta}_1|\mathbf{y}) \propto \{\nu s^2 + (\boldsymbol{\beta}_1 - \hat{\boldsymbol{\beta}}_1)'H(\boldsymbol{\beta}_1 - \hat{\boldsymbol{\beta}}_1)\}^{-(n_1+n_2-m_1-m_2)/2} \tag{4.55}$$

with

$$\begin{aligned} H &= M_{11} - M_{12}M_{22}^{-1}M_{21} \\ &= W_1'W_1 - W_1'W_2(W_2'W_2)^{-1}W_2'W_1 + Z_1'Z_1 - Z_1'Z_2(Z_2'Z_2)^{-1}Z_2'Z_1. \end{aligned} \tag{4.56}$$

Note that $n_1 + n_2 - m_1 - m_2 = \nu + m$, and thus the exponent of (4.55) can be written as $-(\nu + m)/2$. The mean of (4.55), $\hat{\boldsymbol{\beta}}_1$, a subvector of $\hat{\boldsymbol{\beta}} = (X'X)^{-1}X'\mathbf{y}$, can be solved for explicitly and is given by

$$\hat{\boldsymbol{\beta}}_1 = H^{-1}(V_1\tilde{\boldsymbol{\beta}}_1 + V_2\check{\boldsymbol{\beta}}_1), \tag{4.57}$$

where H is defined in (4.56), $V_1 = W_1'W_1 - W_1'W_2(W_2'W_2)^{-1}W_2'W_1$, $V_2 = Z_1'Z_1 - Z_1'Z_2(Z_2'Z_2)^{-1}Z_2'Z_1$, $\tilde{\boldsymbol{\beta}}_1$ is the least squares quantity obtained from a least squares fit of $\mathbf{y}_1$ on W_1 and W_2, and $\check{\boldsymbol{\beta}}_1$ is the least squares quantity obtained from a least squares fit of $\mathbf{y}_2$ on Z_1 and Z_2. From (4.56) and the definitions of V_1 and V_2 we have $H = V_1 + V_2$. Thus $\hat{\boldsymbol{\beta}}_1$ in (4.57) is a "matrix weighted average" of $\tilde{\boldsymbol{\beta}}_1$ and $\check{\boldsymbol{\beta}}_1$. Similar analysis can be performed to obtain the posterior pdf's for $\boldsymbol{\beta}_2$ and $\boldsymbol{\gamma}_2$.

The above analysis will be useful in pooling data from two sources. Note that it is not equivalent to what would be obtained had we analyzed (4.51) conditional on $\boldsymbol{\beta}_1 = \tilde{\boldsymbol{\beta}}_1$, the least squares quantity obtained from (4.50). In this case, for example, we obtain a conditional rather than a marginal posterior pdf for $\boldsymbol{\gamma}_2$; the marginal pdf can be obtained from (4.54) as was done for $\boldsymbol{\beta}_1$ above. Last, should the elements of $\mathbf{u}_1$ and $\mathbf{u}_2$ in (4.50) and (4.51) have differing variances, say σ_1^2 and σ_2^2, the methods of Section 4.2 would have to be employed in the analysis of (4.50) and (4.51).[22]

[22] See V. K. Chetty, *loc. cit.*, for further details and an application of these methods.

APPENDIX 1

Here we provide the lemma needed to establish that the matrix H appearing in (4.17) is positive definite.

Lemma. Let X_*' be the $k \times (T+1)$ augmented matrix $X_*' = [\mathbf{x}_0' \vdots X']$, where $\mathbf{x}_0 = (x_{01}, x_{02}, \ldots, x_{0k})$ and let $\mathbf{z}' = (1, \rho, \rho^2, \ldots, \rho^T)$ be a $1 \times (T+1)$ vector. If $\mathbf{z}$ and X_* are linearly independent, H is positive definite.

Proof. It suffices to show that the matrix $X - \rho X_{-1}$ is of rank k. We can write

$$X - \rho X_{-1} = AX_*,$$

where

$$A = \begin{bmatrix} -\rho & 1 & & & \\ & -\rho & 1 & & \\ & & \ddots & \ddots & \\ & & & -\rho & 1 \end{bmatrix}$$

is a $T \times (T+1)$ matrix with all elements not shown being zero. It is easily seen that A is of rank T and $\mathbf{w} = \mathbf{z}$ is the only nontrivial solution of the system of equations $A\mathbf{w} = 0$. Since X_* and $\mathbf{z}$ are assumed to be linearly independent, there exists a $(T+1) \times (T-k)$ matrix C such that $B = [\mathbf{z} \vdots X_* \vdots C]$ is a $(T+1) \times (T+1)$ nonsingular matrix. Thus the rank of the product AB is T, but note that

$$AB = [0 \vdots AX_* \vdots AC]$$

has only T nonzero columns. Hence the rank of AX_* must be k and the lemma follows.[23]

APPENDIX 2

In this appendix we present asymptotic expansions of the multivariate "normal-t" pdf in (4.31) and the multivariate "double-t" pdf in (4.38).

With respect to (4.31), the factor in the multivariate Student t form can be expanded as follows. We can write

$$\left(1 + \frac{Q_2}{\nu_2}\right)^{-(\nu_2+k)/2} = \exp\left[-\frac{\nu_2 + k}{2}\log\left(1 + \frac{Q_2}{\nu_2}\right)\right], \tag{1}$$

[23] The foregoing condition is also necessary; that is, if $\mathbf{z}$ and X_* are not linearly independent, H will not be positive definite.

where $Q_2 = (\boldsymbol{\beta} - \hat{\boldsymbol{\beta}}_2)'Z_2(\boldsymbol{\beta} - \hat{\boldsymbol{\beta}}_2)/s_2^2$. Then, on employing

$$\log\left(1 + \frac{Q_2}{\nu_2}\right) = \frac{Q_2}{\nu_2} - \frac{1}{2}\left(\frac{Q_2}{\nu_2}\right)^2 + \frac{1}{3}\left(\frac{Q_2}{\nu_2}\right)^3 - \cdots$$

$$= \frac{Q_2}{\nu_2} + R,$$

where R is the remainder term, (1) can be written as:

$$\exp\left(-\frac{Q_2}{2}\right) \exp\left\{-\frac{1}{2}\left[k\frac{Q_2}{\nu_2} + (\nu_2 + k)R\right]\right\}.$$

Now expand the second exponential factor as $e^x = 1 + x + x^2/2! + x^3/3! + \cdots$ to obtain

$$\exp\left\{-\frac{Q_2}{2}\right\} \sum_{i=0}^{\infty} q_i \nu_2^{-i}, \tag{2}$$

where

$$q_0 = 1, \qquad q_1 = \tfrac{1}{4}[Q_2^2 - 2kQ_2],$$
$$q_2 = \tfrac{1}{96}[3Q_2^4 - 4(3k + 4)Q_2^3 + 12k(k + 2)Q_2^2], \quad \text{and so on.}$$

Thus (4.31) can be approximated by

$$\exp\left\{-\tfrac{1}{2}\left[(\boldsymbol{\beta} - \hat{\boldsymbol{\beta}}_1)'\frac{Z_1}{\sigma_1^2}(\boldsymbol{\beta} - \hat{\boldsymbol{\beta}}_1) + Q_2\right]\right\} \sum_{i=0}^{\infty} q_i \nu_2^{-i}$$
$$= \exp\left[-\tfrac{1}{2}(\boldsymbol{\beta} - \tilde{\boldsymbol{\beta}})'A(\boldsymbol{\beta} - \tilde{\boldsymbol{\beta}})\right] \sum_{i=0}^{\infty} q_i \nu_2^{-i},$$

with $\tilde{\boldsymbol{\beta}}$ and A given in (4.33) and (4.34), respectively. Thus $\tilde{\boldsymbol{\beta}}$ is the mean of the leading normal term in the asymptotic expansion of the multivariate "normal-t" pdf as stated in the text.

In the case of the multivariate "double-t" pdf in (4.38), namely

$$\left(1 + \frac{Q_1}{\nu_1}\right)^{-(\nu_2 + k)/2}\left(1 + \frac{Q_2}{\nu_2}\right)^{-(\nu_2 + k)/2},$$

with $Q_1 = (\boldsymbol{\beta} - \hat{\boldsymbol{\beta}}_1)'Z_1(\boldsymbol{\beta} - \hat{\boldsymbol{\beta}}_1)/s_1^2$, both factors are expanded exactly as described above to yield

$$\exp\left[-\tfrac{1}{2}(Q_1 + Q_2)\right] \sum_{i=0}^{\infty} p_i \nu_1^{-i} \sum_{i=0}^{\infty} q_i \nu_2^{-i}$$
$$= \exp\left[-\tfrac{1}{2}(\boldsymbol{\beta} - \bar{\boldsymbol{\beta}})'D(\boldsymbol{\beta} - \bar{\boldsymbol{\beta}})\right] \sum_{i,j=0}^{\infty} p_i q_j \nu_1^{-i} \nu_2^{-j},$$

where $\bar{\boldsymbol{\beta}}$ and D are shown in (4.40) and (4.41), respectively, the q_i's are as defined above, and the p_i's are given by $p_0 = 1$, $p_1 = \frac{1}{4}(Q_1^2 - 2kQ_1)$, $p_2 =$

$\frac{1}{96}[3Q_1^4 - 4(3k + 4)Q_1^3 + 12k(k + 2)Q_1^2]$, and so on. Thus $\bar{\boldsymbol{\beta}}$ is the mean of the leading normal term in the asymptotic expansion of the multivariate "double-t" pdf. Further, in the paper by Tiao and Zellner, *loc. cit.*, methods are described for taking higher order terms into account in analyzing this pdf; that is, integration of the above series is accomplished by noting that each term is a bivariate polynomial in Q_1 and Q_2. Thus integration of each term involves evaluating mixed moments of the quadratic forms Q_1 and Q_2, which is done by employing the bivariate moment-cumulant inversion formulas given by Cook.[24]

QUESTIONS AND PROBLEMS

1. Using the model shown in (4.1) and the prior assumptions shown in (4.3), derive the conditional predictive pdf for y_{T+1}, given that $\rho = \rho_0$, where ρ_0 is a given value. How do the mean and variance of y_{T+1}, given $\rho = \rho_0$, depend on ρ_0? Explain how the unconditional predictive pdf for y_{T+1} can be computed.
2. Suppose that the parameter ρ, appearing in (4.1*b*) is believed to satisfy $0 < \rho < 1$ and to have prior mean and variance equal to 0.5 and 0.04, respectively. How can this prior information be represented by a beta pdf?
3. Use the prior pdf in Problem 2 along with other prior assumptions in Section 4.1 to obtain the joint posterior pdf for the parameters β and ρ in the simple regression model in (4.1).
4. For each of the two sets of data in Table 4.1 use the result obtained in Problem 3 to compute a marginal posterior pdf for ρ by means of a bivariate numerical integration. Comment on the properties of the resulting posterior pdf's for ρ.
5. In Chapter 3, Table 3.1, Haavelmo's data on income (y) and investment (x) are presented. Use these data to compute the posterior pdf for ρ in the following model:

$$\begin{aligned} y_t - \bar{y} &= \beta(x_t - \bar{x}) + u_t, \\ u_t &= \rho u_{t-1} + \epsilon_t, \end{aligned} \qquad t = 1, 2, \ldots, 20,$$

   where $\bar{y}$ and $\bar{x}$ are sample means for income and investment, respectively. Employ the assumptions of Section 4.1. What could account for the posterior pdf for ρ being centered far from zero?
6. In Problem 5 compute the marginal posterior pdf for the investment multiplier β and compare it with the posterior pdf shown in Figure 3.1.
7. Assume that (4.1*a*) has an intercept term, β_0; that is, $y_t = \beta_0 + \beta x_t + u_t$. By combining this with (4.1*b*) we obtain $y_t = \beta_0(1 - \rho) + \rho y_{t-1} + \beta(x_t - \rho x_{t-1})$

[24] M. B. Cook, "Bivariate κ-statistics and Cumulants of their Joint Sampling Distribution," *Biometrika*, **38**, 179–195 (1951).

$+ \epsilon_t$. Is there a difficulty in estimating β_0 from this equation when $\rho = 1$? Will this difficulty be present if a priori we restrict the range of ρ by $0 < |\rho| < 1$ or use a prior pdf that assigns zero probability density to the value $\rho = 1$?

8. Using the likelihood function in (4.2), evaluate Fisher's information matrix, a typical element of which is given by $-E(\partial^2 \log l/\partial\theta_i\, \partial\theta_j)$, where θ_i and θ_j denote the ith and jth parameters, respectively, and the expectation is taken with respect to the pdf for the y's. In performing this derivation, note that (a) $y_t - \beta x_t = \rho^t M + \sum_{j=0}^{t} \rho^j \epsilon_{t-j}$, (b) $E(y_t - \beta x_t) = \rho^t M$, and (c) $E(y_t - \beta x_t)^2 = \rho^{2t} M^2 + \sigma^2(1 - \rho^{2(t+1)})/(1 - \rho^2)$. From an examination of $-E(\partial^2 \log l/\partial\rho^2)$ what can be said about the information regarding ρ when $|\rho| > 1$? If $0 < |\rho| < 1$ and T is large, show that the part of the information matrix pertaining to ρ and β is approximately diagonal. What does this imply?

9. For the model in (4.12), given $\rho = \rho_0$, derive the predictive pdf for a vector of future observations, say $\mathbf{z}' = (y_{T+1}, y_{T+2}, \ldots, y_{T+q})$, assumed to be generated by $\mathbf{z} = W\boldsymbol{\beta} + \mathbf{u}_*$ with W a given $q \times k$ matrix and u_* a $q \times 1$ vector of future error terms generated by the same process as the elements of $\mathbf{u}$ in (4.12). Use a diffuse prior pdf for the unknown parameters of the model.

10. Derive the marginal posterior pdf's for α_1 and α_2 in (4.43a,b) using a diffuse prior pdf for the parameters and assuming that $\sigma_1^2 \neq \sigma_2^2$. Compare these pdf's with those obtained using the same prior assumptions but with $\sigma_1^2 = \sigma_2^2 = \sigma^2$ and a diffuse prior pdf for σ.

11. Interpret the pdf's plotted in Figure 4.5.

12. In (4.43a,b), corresponding slope coefficients β_1 and β_2 are assumed to be the same in the two relationships. If this assumption is questioned, what can be computed to provide a check on this point?

13. For the system in (4.50) and (4.51), with a diffuse prior pdf for the parameters, derive the joint marginal posterior pdf for the elements of $\boldsymbol{\gamma}_2$ and provide its mean and variance-covariance matrix.

14. Provide an analysis of (4.50) and (4.51) when the elements of $\mathbf{u}_1$ have variance σ_1^2 and those of $\mathbf{u}_2$ have variance σ_2^2 with $\sigma_1^2 \neq \sigma_2^2$ and σ_1^2 and σ_2^2 are assumed to be independently distributed a priori.

CHAPTER V

On Errors in the Variables

It has been generally recognized that economic data often contain errors and that the presence of measurement errors can vitally affect the results of analyses. In view of these generally accepted propositions it is not surprising to observe that considerable effort has been expended on the development of methods for analyzing data which are contaminated with measurement errors. In this chapter we consider several models and problems related to measurement errors. After analysis of several preliminary problems, which illustrate problems associated with certain basic "errors-in-the-variables" models (EVM's), we take up the analysis of the classical EVM. This model can be viewed as a generalization of the simple regression model, considered in Chapter 3, which takes account of random measurement errors in both the dependent and independent variables. Two forms of the classical EVM, namely the functional and the structural, are analyzed by maximum likelihood and Bayesian techniques. This comparative approach is particularly revealing in the present instance because, as will be seen, prior information plays a vital role in both sampling theory and Bayesian approaches.

After analyzing the classical EVM, we consider a form of this model which incorporates special assumptions about the systematic parts of observed variables, namely that they can be represented by the systematic parts of regression equations. As will be seen, this analysis is closely related to "instrumental variable" estimation techniques for the EVM. Although the analyses of this chapter cover only a subset of those bearing on EVM's, this subset is of considerable importance in econometric work.

5.1 THE CLASSICAL EVM: PRELIMINARY PROBLEMS

Before turning to the classical EVM, it is instructive and illuminating to consider the closely related problem of n means.[1] Let $y_1, y_2, \ldots, y_n$ be n independent observations drawn from n normal populations, each with the same variance σ^2 but with different means; that is, y_i ($i = 1, 2, \ldots, n$) is assumed to be randomly drawn from a normal population with mean ξ_i and

[1] This problem is discussed in M. G. Kendall and A. Stuart, *The Advanced Theory of Statistics*, Vol. II, New York: Hafner, 1961, p. 61.

variance σ^2. It is to be noted that we have n observations and $n + 1$ unknown parameters, n means, $\xi_1, \xi_2, \ldots, \xi_n$, and σ^2. Just from this count of observations and parameters we may guess that it will be difficult to estimate all $n + 1$ unknown parameters. We wish to analyze the nature of this difficulty because it arises in the functional form of the classical EVM, albeit in a slightly more complicated manner.

The likelihood function for the problem of n means is given by

$$l(\boldsymbol{\xi}, \sigma^2|\mathbf{y}) \propto \frac{1}{\sigma^n} \exp\left[-\frac{1}{2\sigma^2}(\mathbf{y} - \boldsymbol{\xi})'(\mathbf{y} - \boldsymbol{\xi})\right], \tag{5.1}$$

where $\boldsymbol{\xi}' = (\xi_1, \xi_2, \ldots, \xi_n)$, a vector of unknown means, and $\mathbf{y}' = (y_1, y_2, \ldots, y_n)$, the observation vector. On differentiating the logarithm of the likelihood function partially with respect to σ^2 and the ξ_i's and setting the derivatives equal to zero in an effort to obtain maximum likelihood (ML) estimates, we find

$$\boldsymbol{\xi} = \mathbf{y} \tag{5.2a}$$

and

$$\hat{\sigma}^2 = \frac{(\mathbf{y} - \boldsymbol{\xi})'(\mathbf{y} - \boldsymbol{\xi})}{n}. \tag{5.2b}$$

Thus from (5.2*a*), the y_i's are apparently the ML estimates of the corresponding ξ_i's. On inserting the "ML estimate" $\boldsymbol{\xi} = \mathbf{y}$ for $\boldsymbol{\xi}$ in (5.2*b*), it appears that the "ML estimate" for σ^2 is $\hat{\sigma}^2 = 0$. As noted by Kendall and Stuart, this is obviously an absurd result.[2]

The basic defect with the above "ML analysis" is that the likelihood function in (5.1) does not possess a finite maximum in the admissible region of the parameter space $0 < \sigma^2 < \infty$ and $-\infty < \xi_i < \infty$, $i = 1, 2, \ldots, n$. This is easily established by substituting from (5.2*a*) in (5.1) to obtain $l(\sigma^2|\mathbf{y}, \boldsymbol{\xi} = \mathbf{y}) \propto 1/\sigma^n$, which clearly does not possess a maximum for $0 < \sigma^2 < \infty$. Alternatively, if we substitute from (5.2*b*) in (5.1), we obtain $l(\boldsymbol{\xi}|\mathbf{y}, \sigma^2 = \hat{\sigma}^2) \propto [(\mathbf{y} - \boldsymbol{\xi})'(\mathbf{y} - \boldsymbol{\xi})]^{n/2}$, which again has no finite maximum value in the admissible parameter space. Thus, although (5.2*a*) is the ML estimate for $\boldsymbol{\xi}$ for given finite σ^2 [3] and (5.2*b*) is the ML estimate for σ^2 for

[2] They state, *op. cit.*, p. 61, that this is an example in which "... the ML method may become ineffective."

[3] C. M. Stein, "Inadmissibility of the Usual Estimator for the Mean of a Multivariate Normal Distribution," in J. Neyman, Ed., *Proc. Third Berkeley Symp. Math. Statist. Probab.*, Vol. 1, Berkeley: University of California Press, 197–206 (1956), shows that the ML estimator in this case is inadmissible relative to a quadratic loss function for $n \geq 3$. See also W. James and C. M. Stein, "Estimation with Quadratic Loss," in J. Neyman, Ed., *Proc. Fourth Berkeley Symp. Math. Statist. Probab.*, Vol. 1, Berkeley: University of California Press, 361–379 (1961), and C. M. Stein, "Confidence Sets for the Mean of a Multivariate Normal Distribution," *J. Roy. Statist. Soc.*, Series B, **24**, 265–285 (1962).

given $\boldsymbol{\xi}$, (5.2*a*) and (5.2*b*) do not jointly yield ML estimates for $\boldsymbol{\xi}$ and σ^2. Further, it is not hard to show that the likelihood function in (5.1) does not approach a limit as $\sigma^2 \to 0$ and $\boldsymbol{\xi} \to \mathbf{y}$; therefore 0 and $\mathbf{y}$ are not ML estimates.[4]

It is interesting to explore the results of a Bayesian analysis of the n means problem. Let us for present purposes employ the following prior pdf:

$$(5.3) \qquad p(\boldsymbol{\xi}, \sigma) \propto \frac{1}{\sigma}, \qquad \begin{array}{ll} -\infty < \xi_i < \infty, & i = 1, 2, \ldots, n, \\ 0 < \sigma < \infty. & \end{array}$$

On combining (5.3) with (5.1), the posterior pdf is

$$(5.4) \qquad p(\boldsymbol{\xi}, \sigma|\mathbf{y}) \propto \frac{1}{\sigma^{n+1}} \exp\left[-\frac{1}{2\sigma^2}(\boldsymbol{\xi} - \mathbf{y})'(\boldsymbol{\xi} - \mathbf{y})\right].$$

From (5.4) we note that the conditional posterior pdf for $\boldsymbol{\xi}$, given σ is a *proper* multivariate normal pdf with mean vector $\mathbf{y}$ and covariance matrix $\sigma^2 I_n$. Further, for given $\boldsymbol{\xi}$, the conditional posterior pdf for σ is a *proper* inverted gamma pdf. Thus, as with the ML approach, *conditional* inferences can be readily made. However, joint inferences about $\boldsymbol{\xi}$ and σ cannot be made, since (5.4) is an improper pdf; for example, if (5.4) is integrated with respect to the elements of $\boldsymbol{\xi}$, the result is $p(\sigma|\mathbf{y}) \propto 1/\sigma$, $0 < \sigma < \infty$, which is improper and precisely in the same form as our diffuse prior pdf in (5.3). Thus the sample information in this problem gives us no information about σ. Also, on integrating (5.4) with respect to σ, the result is $p(\boldsymbol{\xi}|\mathbf{y}) \propto [(\boldsymbol{\xi} - \mathbf{y})'(\boldsymbol{\xi} - \mathbf{y})]^{-n/2}$ which is an improper pdf. Therefore there is not enough sample information to make joint inferences about σ and the elements of $\boldsymbol{\xi}$. On the other hand, with prior information about one or some of the parameters, for example, given σ, inferences can easily be made about the elements of $\boldsymbol{\xi}$. Thus exact prior information, say $\sigma = \sigma_0$ with σ_0 known, enables us to make inferences about the elements of $\boldsymbol{\xi}$.

It is extremely important to appreciate that less precise information about σ, less precise than $\sigma = \sigma_0$, also permits us to make inferences about the elements of $\boldsymbol{\xi}$; for example, if we use the following prior pdf,

$$(5.5) \quad p(\boldsymbol{\xi}, \sigma) \propto \frac{1}{\sigma^{\nu_0+1}} \exp\left(-\frac{\nu_0 s_0^2}{2\sigma^2}\right), \qquad \begin{array}{ll} -\infty < \xi_i < \infty, & i = 1, 2, \ldots, n, \\ 0 < \sigma < \infty, & \end{array}$$

[4] Observe that $\lim l$ as $\boldsymbol{\xi} \to \mathbf{y}$ and then $\sigma^2 \to 0$ is infinite. On the other hand, $\lim l$ as $\sigma^2 \to 0$ and then $\mathbf{y} \to \boldsymbol{\xi}$ approaches a finite limiting value of zero; note $\lim_{\sigma \to 0} (\sigma^n)^{-1} \exp[-(2\sigma^2)^{-1}(\mathbf{y} - \boldsymbol{\xi})'(\mathbf{y} - \boldsymbol{\xi})] = 0$. Since the limits are different, depending on how we approach the point $\boldsymbol{\xi} = \mathbf{y}$ and $\sigma^2 = 0$, the function does not exist at this point. Also note that $\hat{\sigma}^2 = 0$ is not in the admissible parameter space $0 < \sigma^2 < \infty$.

where ν_0 and s_0 are prior parameters, $\nu_0, s_0 > 0$, the posterior pdf is

$$(5.6) \qquad p(\boldsymbol{\xi}, \sigma|\mathbf{y}) \propto \frac{1}{\sigma^{n+\nu_0+1}} \exp\left\{-\frac{1}{2\sigma^2}[\nu_0 s_0^2 + (\boldsymbol{\xi} - \mathbf{y})'(\boldsymbol{\xi} - \mathbf{y})]\right\},$$

which is a proper pdf. The marginal posterior pdf for $\boldsymbol{\xi}$ is

$$(5.7) \qquad p(\boldsymbol{\xi}|\mathbf{y}) \propto [\nu_0 s_0^2 + (\boldsymbol{\xi} - \mathbf{y})'(\boldsymbol{\xi} - \mathbf{y})]^{-(n+\nu_0)/2},$$

which is in the form of a multivariate Student-t pdf with mean vector $\mathbf{y}$. Thus, rather than assuming that $\sigma = \sigma_0$, a known value, we are able to incorporate less restrictive prior information which permits us to make inferences about the elements of $\boldsymbol{\xi}$. However, on integrating (5.6) with respect to the elements of $\boldsymbol{\xi}$, we obtain the prior pdf for σ as the result. Thus the sample information does not add to our knowledge of σ in the present problem.[5]

Let us take up next the problem of n means in which we have m observations for each mean; that is, our model for the observations is

$$(5.8) \qquad \mathbf{y}_i = \boldsymbol{\iota}\xi_i + \mathbf{u}_i, \qquad i = 1, 2, \ldots, n,$$

where $\mathbf{y}_i' = (y_{i1}, y_{i2}, \ldots, y_{im})$, $\boldsymbol{\iota}$ is an $m \times 1$ vector of ones, that is, $\boldsymbol{\iota}' = (1, 1, \ldots, 1)$, ξ_i is the ith unknown mean, and $\mathbf{u}_i$ is an $m \times 1$ error vector. We assume that $E\mathbf{u}_i = \mathbf{0}$, $E\mathbf{u}_i\mathbf{u}_i' = \sigma^2 I_m$ for $i = 1, 2, \ldots, n$, and $E\mathbf{u}_i\mathbf{u}_j' = 0$, a null matrix, for $i \neq j$. Further, we assume that the elements of the $\mathbf{u}_i$ vectors are jointly normally distributed. To simplify the notation, we can write

$$(5.9a) \qquad \begin{pmatrix} \mathbf{y}_1 \\ \mathbf{y}_2 \\ \vdots \\ \mathbf{y}_n \end{pmatrix} = \begin{pmatrix} \boldsymbol{\iota} & & & \\ & \boldsymbol{\iota} & & \\ & & \ddots & \\ & & & \boldsymbol{\iota} \end{pmatrix} \begin{pmatrix} \xi_1 \\ \xi_2 \\ \vdots \\ \xi_n \end{pmatrix} + \begin{pmatrix} \mathbf{u}_1 \\ \mathbf{u}_2 \\ \vdots \\ \mathbf{u}_n \end{pmatrix}$$

or

$$(5.9b) \qquad \mathbf{y} = W\boldsymbol{\xi} + \mathbf{u},$$

where $\mathbf{y}$ denotes the vector on the lhs of (5.9a), W, the block diagonal matrix on the rhs of (5.9a), $\boldsymbol{\xi}' = (\xi_1, \xi_2, \ldots, \xi_n)$, and $\mathbf{u}' = (\mathbf{u}_1', \mathbf{u}_2', \ldots, \mathbf{u}_n')$.

Note that in the present problem we have nm observations and $n + 1$ unknown parameters, so that for n and m each larger than 1 we have more observations than unknown parameters in contrast to the situation in which $m = 1$, analyzed above. Still, as shown below, there is a fundamental complication with the ML approach.

[5] Also, as n increases, (5.7) does not become concentrated about the mean vector $\mathbf{y}$.

The likelihood function for the system in (5.9) is

$$
\begin{aligned}
l(\boldsymbol{\xi}, \sigma|\mathbf{y}) &\propto \frac{1}{\sigma^{nm}} \exp\left[-\frac{1}{2\sigma^2}(\mathbf{y} - W\boldsymbol{\xi})'(\mathbf{y} - W\boldsymbol{\xi})\right] \\
&\propto \frac{1}{\sigma^{nm}} \exp\left\{-\frac{1}{2\sigma^2}[(\mathbf{y} - W\hat{\boldsymbol{\xi}})'(\mathbf{y} - W\hat{\boldsymbol{\xi}}) + (\boldsymbol{\xi} - \hat{\boldsymbol{\xi}})'W'W(\boldsymbol{\xi} - \hat{\boldsymbol{\xi}})]\right\},
\end{aligned} \tag{5.10}
$$

where $\hat{\boldsymbol{\xi}} = (W'W)^{-1}W'\mathbf{y}$. Since $W'W$ is positive definite, it is clear that $\boldsymbol{\xi} = \hat{\boldsymbol{\xi}}$ is the ML estimate. As is easily established, $\hat{\xi}_i = (\boldsymbol{\iota}'\boldsymbol{\iota})^{-1}\boldsymbol{\iota}'\mathbf{y}_i = \bar{y}_i$, the ith sample mean; that is

$$\bar{y}_i = \frac{1}{m}\sum_{j=1}^{m} y_{ij}.$$

Further, the ML estimate for σ^2 is

$$\hat{\sigma}^2 = \frac{1}{nm}(\mathbf{y} - W\hat{\boldsymbol{\xi}})'(\mathbf{y} - W\hat{\boldsymbol{\xi}}), \tag{5.11}$$

which is obtained by differentiating the logarithm of the likelihood function with respect to σ^2 and setting the derivative equal to zero.[6] As can easily be established, $\boldsymbol{\xi} = \hat{\boldsymbol{\xi}}$ and $\sigma^2 = \hat{\sigma}^2$ are indeed values for $\boldsymbol{\xi}$ and σ^2 associated with a finite maximum of the likelihood function. Yet, as Neyman and Scott and Kendall and Stuart point out,[7] there is still a problem. The ML estimator for σ^2 has a bias that does not disappear as $n \to \infty$ *with m fixed*; that is,

$$
\begin{aligned}
E\hat{\sigma}^2 &= \frac{nm - n}{nm}\sigma^2 \\
&= \left(1 - \frac{1}{m}\right)\sigma^2.
\end{aligned} \tag{5.12}
$$

Thus, as $n \to \infty$ with m fixed, the bias of the ML estimator does not disappear. If, for example, $m = 2$, $E\hat{\sigma}^2 = \frac{1}{2}\sigma^2$ for all n. Heuristically, Kendall and Stuart interpret this situation as the persistence of the small sample bias of the ML estimator. Note that the number of unknown parameters increases as n increases. Indeed, the ratio of the number of parameters to the number of observations, $(n + 1)/nm$ approaches $1/m$ which can be appreciable, $\frac{1}{2}$ for the case $m = 2$. Thus we do not get out of the small sample situation as n increases in the present problem.

[6] From (5.10), $\log l = \text{const} - nm \log \sigma - (2\sigma^2)^{-1}(\mathbf{y} - W\boldsymbol{\xi})'(\mathbf{y} - W\boldsymbol{\xi})$. Then $(d \log l)/d\sigma = -nm/\sigma + (\sigma^3)^{-1}(\mathbf{y} - W\boldsymbol{\xi})'(\mathbf{y} - W\boldsymbol{\xi}) = 0$ yields $\sigma^2 = (nm)^{-1}(\mathbf{y} - W\boldsymbol{\xi})'(\mathbf{y} - W\boldsymbol{\xi})$, which, with $\boldsymbol{\xi} = \hat{\boldsymbol{\xi}}$, yields (5.11).

[7] J. Neyman and E. Scott, "Consistent Estimates Based on Partially Consistent Observations," *Econometrica*, **16**, 1–16 (1948).

An *ad hoc* method for correcting this defect of the ML method in this "incidental parameter" problem,[8] as suggested by Kendall and Stuart, would be to make a "degrees-of-freedom" correction to the estimator for σ^2 in (5.11). We have nm observations and estimate $n\xi_i$'s. Therefore the degrees of freedom left to estimate σ^2 is $nm - n = n(m - 1)$. If we define

$$\tilde{\sigma}^2 = \frac{1}{n(m-1)}(\mathbf{y} - W\hat{\boldsymbol{\xi}})'(\mathbf{y} - W\hat{\boldsymbol{\xi}}), \tag{5.13}$$

then $E\tilde{\sigma}^2 = \sigma^2$ for all n. Below we shall see that a similar "incidental parameter" problem is present in the classical EVM.

For the Bayesian analysis of the model in (5.9) let us employ the prior assumptions given in (5.3). Then the posterior pdf is

$$\begin{aligned} p(\boldsymbol{\xi}, \sigma|\mathbf{y}) &\propto \frac{1}{\sigma^{nm+1}} \exp\left[-\frac{1}{2\sigma^2}(\mathbf{y} - W\boldsymbol{\xi})'(\mathbf{y} - W\boldsymbol{\xi})\right] \\ &\propto \frac{1}{\sigma^{nm+1}} \exp\left\{-\frac{1}{2\sigma^2}[(\mathbf{y} - W\hat{\boldsymbol{\xi}})'(\mathbf{y} - W\hat{\boldsymbol{\xi}}) + (\boldsymbol{\xi} - \hat{\boldsymbol{\xi}})'W'W(\boldsymbol{\xi} - \hat{\boldsymbol{\xi}})]\right\}. \end{aligned} \tag{5.14}$$

The marginal posterior pdf for $\boldsymbol{\xi}$ is

$$p(\boldsymbol{\xi}|\mathbf{y}) \propto [(\mathbf{y} - W\hat{\boldsymbol{\xi}})'(\mathbf{y} - W\hat{\boldsymbol{\xi}}) + (\boldsymbol{\xi} - \hat{\boldsymbol{\xi}})'W'W(\boldsymbol{\xi} - \hat{\boldsymbol{\xi}})]^{-nm/2}, \tag{5.15}$$

which is in the form of a proper multivariate Student t pdf with mean vector $E\boldsymbol{\xi} = \hat{\boldsymbol{\xi}}$ and covariance matrix $(W'W)^{-1}\nu's^2/(\nu' - 2)$.[9] It should be noted that with m fixed as $n \to \infty$ this covariance matrix does *not* approach a null matrix; that is, as $n \to \infty$ with m fixed, the marginal posterior pdf for $\boldsymbol{\xi}$ does not become concentrated about $\hat{\boldsymbol{\xi}}$. This Bayesian result is an analogue of the sampling theory result that the ML estimators for the elements of $\boldsymbol{\xi}$ have variances that do not tend to zero as $n \to \infty$ *with m fixed.*

We can obtain the marginal posterior pdf for σ by integrating (5.14) with respect to the n elements of $\boldsymbol{\xi}$, an operation that yields

$$p(\sigma|\mathbf{y}) \propto \frac{1}{\sigma^{\nu'+1}} \exp\left(-\frac{\nu's^2}{2\sigma^2}\right), \tag{5.16}$$

where $\nu' = n(m - 1)$ and $s^2 = (\mathbf{y} - W\hat{\boldsymbol{\xi}})'(\mathbf{y} - W\hat{\boldsymbol{\xi}})/\nu'$. Note that the integration with respect to the elements of $\boldsymbol{\xi}$ leads automatically to a reduction of the exponent of σ in the denominator of (5.14) from $nm + 1$ to $n(m - 1) + 1$, which is analogous to the degrees-of-freedom correction discussed in connection with (5.11). Also from (5.16) we have $E(\sigma^2|\mathbf{y}) = \nu's^2/(\nu' - 2)$, the posterior mean of σ^2. On viewing this quantity as an estimator, it is seen to be a consistent estimator in contrast to that shown in (5.11).

[8] Neyman and Scott have called the ξ_i's incidental parameters.

[9] Here $\nu' = n(m - 1)$ and $\nu's^2 = (\mathbf{y} - W\hat{\boldsymbol{\xi}})'(\mathbf{y} - W\hat{\boldsymbol{\xi}})$.

Next, we take up the problem of n means with $m = 2$ observations for each unknown mean, one with variance σ_1^2 and the other with variance σ_2^2; that is, our $n \times 1$ observation vectors, $\mathbf{y}_1$ and $\mathbf{y}_2$, are assumed to be generated as follows:

$$\mathbf{y}_1 = \boldsymbol{\xi} + \mathbf{u}_1, \tag{5.17a}$$

$$\mathbf{y}_2 = \boldsymbol{\xi} + \mathbf{u}_2, \tag{5.17b}$$

where $\boldsymbol{\xi}' = (\xi_1, \xi_2, \ldots, \xi_n)$ and $E\mathbf{u}_1 = E\mathbf{u}_2 = \mathbf{0}$, $E\mathbf{u}_1\mathbf{u}_1' = \sigma_1^2 I_n$, $E\mathbf{u}_2\mathbf{u}_2' = \sigma_2^2 I_n$ and $E\mathbf{u}_1\mathbf{u}_2' = 0$, an $n \times n$ null matrix. We also assume that the elements of $\mathbf{u}_1$ and $\mathbf{u}_2$ are normally distributed. Under these assumptions the likelihood function is given by

$$\begin{aligned} l(\boldsymbol{\xi}, \sigma_1, \sigma_2 | \mathbf{y}_1, \mathbf{y}_2) \propto \frac{1}{\sigma_1^n} \frac{1}{\sigma_2^n} \exp \Big[&-\frac{1}{2\sigma_1^2} (\mathbf{y}_1 - \boldsymbol{\xi})'(\mathbf{y}_1 - \boldsymbol{\xi}) \\ &- \frac{1}{2\sigma_2^2} (\mathbf{y}_2 - \boldsymbol{\xi})'(\mathbf{y}_2 - \boldsymbol{\xi}) \Big]. \end{aligned} \tag{5.18}$$

Let us attempt to find ML estimates. By differentiating $\log l$ with respect to σ_1 and σ_2 and setting these derivatives equal to zero we obtain

$$\tilde{\sigma}_1^2 = \frac{(\mathbf{y}_1 - \boldsymbol{\xi})'(\mathbf{y}_1 - \boldsymbol{\xi})}{n} \quad \text{and} \quad \tilde{\sigma}_2^2 = \frac{(\mathbf{y}_2 - \boldsymbol{\xi})'(\mathbf{y}_2 - \boldsymbol{\xi})}{n} \tag{5.19}$$

as the maximizing values of σ_1^2 and σ_2^2 which are seen to depend on the unknown vector $\boldsymbol{\xi}$. On differentiating $\log l$ with respect to the elements of $\boldsymbol{\xi}$, we obtain

$$\begin{aligned} \tilde{\xi}_i &= \frac{y_{1i}/\sigma_1^2 + y_{2i}/\sigma_2^2}{1/\sigma_1^2 + 1/\sigma_2^2}, \qquad i = 1, 2, \ldots, n, \\ &= \frac{(y_{1i} + \lambda y_{2i})}{1 + \lambda}, \end{aligned} \tag{5.20}$$

where $\lambda = \sigma_1^2/\sigma_2^2$, as the maximizing values of the ξ_i's which depend on the variance ratio λ. It is clear that $\tilde{\xi}_i$ is a weighted average of y_{1i} and y_{2i}, with the reciprocals of their respective variances as weights.

Now, if $\boldsymbol{\xi}$ is known, the estimators in (5.19) can be computed. On the other hand, if λ is known, the $\tilde{\xi}_i$'s can be computed from (5.20), and on inserting $\tilde{\boldsymbol{\xi}}$ for $\boldsymbol{\xi}$ in (5.19) the estimator for σ_1^2 is $\tilde{\sigma}_1^2 = (\mathbf{y}_1 - \tilde{\boldsymbol{\xi}})'(\mathbf{y}_1 - \tilde{\boldsymbol{\xi}})/n$. However, it is not difficult to show that[10] $E\tilde{\sigma}_1^2 = \sigma_1^2\lambda/(1 + \lambda)$; that is, $\tilde{\sigma}_1^2$ has a bias that does not disappear as $n \to \infty$. Further, if we write the likelihood function in (5.18) as

[10] We have $\hat{\sigma}_1^2 = (1/n) \sum [(y_{1i} - \xi_i) - (\tilde{\xi}_i - \xi_i)]^2$. With $\tilde{\xi}_i - \xi_i = [(y_{1i} - \xi_i)/\sigma_1^2 + (y_{2i} - \xi_i)/\sigma_2^2]/(1/\sigma_1^2 + 1/\sigma_2^2)$, $E\hat{\sigma}_1^2 = \sigma_1^2 - 2(1/\sigma_1^2 + 1/\sigma_2^2)^{-1} + (1/\sigma_2^2 + 1/\sigma_2^2)^{-1} = \sigma_1^2 - (1/\sigma_1^2 + 1/\sigma_2^2)^{-1} = \sigma_1^2\lambda/(1 + \lambda)$.

(5.21)

$$l(\xi, \sigma_1|\lambda, \mathbf{y}_1, \mathbf{y}_2) \propto \frac{1}{\sigma_1^{2n}} \exp\left\{-\frac{1}{2\sigma_1^2}[(\mathbf{y}_1 - \xi)'(\mathbf{y}_1 - \xi) + \lambda(\mathbf{y}_2 - \xi)'(\mathbf{y}_2 - \xi)]\right\},$$

the ML estimator for σ_1^2, given λ, is

$$\tilde{\sigma}_1^2 = \frac{1}{2n}[(\mathbf{y}_1 - \tilde{\xi})'(\mathbf{y}_1 - \tilde{\xi}) + \lambda(\mathbf{y}_2 - \tilde{\xi})'(\mathbf{y}_2 - \tilde{\xi})], \tag{5.22}$$

with the elements of $\tilde{\xi}$ given by (5.20). It is straightforward to show that the expectation of the estimator in (5.22) is $E\tilde{\sigma}_1^2 = \frac{1}{2}\sigma_1^2$. Again the ML estimator has a bias that does not disappear as $n \to \infty$. This difficulty arises because no allowance has been made for the fact that the n elements of the ξ vector have been estimated; an *ad hoc* adjustment for degrees of freedom could, of course, be made to remove the bias just discussed. This problem arises also in the EVM as will be seen below.

To return to (5.18), we may ask if it is possible to obtain ML estimators for all the parameters of the model, the n elements of ξ and the two variances σ_1^2 and σ_2^2, given our $2n$ observations, $\mathbf{y}_1$ and $\mathbf{y}_2$. Intuitively, this does not appear to be possible, since opposite each unknown mean we have just two observations, each with its own unknown variance. On substituting from (5.20) in (5.19), we obtain

$$\tilde{\sigma}_1^2 = \frac{\tilde{\lambda}^2}{n(1 + \tilde{\lambda})^2}\sum (y_{1i} - y_{2i})^2 \quad \text{and} \quad \tilde{\sigma}_2^2 = \frac{1}{n(1 + \tilde{\lambda})^2}\sum (y_{1i} - y_{2i})^2,$$

and thus for these two conditions to be satisfied we must have

$$\tilde{\lambda} = \frac{\tilde{\sigma}_1^2}{\tilde{\sigma}_2^2} = \tilde{\lambda}^2,$$

which can hold only if $\tilde{\lambda}^2 = 1$. Thus the necessary conditions for a maximum cannot in general be satisfied; a maximum of the likelihood function does not exist for $\sigma_1^2 \neq \sigma_2^2$ with both σ_1^2 and σ_2^2 unknown.[11]

The above difficulty arises because σ_1^2 and σ_2^2 are not identified. This is most easily seen by considering the distribution of $\mathbf{w} = \mathbf{y}_1 - \mathbf{y}_2 = \mathbf{u}_1 - \mathbf{u}_2$. The vector $\mathbf{w}$ has a zero mean and covariance matrix $\omega^2 I_n$, where $\omega^2 = \sigma_1^2 + \sigma_2^2$. There are many values of σ_1^2 and σ_2^2 that sum to a particular ω^2. Since the pdf for $\mathbf{w}$ is completely determined by specifying the quantity ω^2, it is not possible to identify σ_1^2 and σ_2^2 without further prior information.

If we approach the present problem from the Bayesian point of view with the following diffuse prior,

$$p(\xi, \sigma_1, \sigma_2) \propto \frac{1}{\sigma_1\sigma_2} \qquad \begin{array}{ll} -\infty < \xi_i < \infty, & i = 1, 2, \ldots, n, \\ 0 < \sigma_i < \infty, & i = 1, 2 \end{array} \tag{5.23}$$

[11] Note that if we substitute from (5.19) into (5.18) the resulting function has no finite maximum.

it is not difficult to show that on combining this prior with the likelihood function in (5.18) the resulting posterior pdf,

$$(5.24)\qquad p(\boldsymbol{\xi}, \sigma_1, \sigma_2|\mathbf{y}_1, \mathbf{y}_2) \propto \frac{1}{\sigma_1^{n+1}\sigma_2^{n+1}} \exp\left[-\frac{1}{2\sigma_1^2}(\mathbf{y}_1 - \boldsymbol{\xi})'(\mathbf{y}_1 - \boldsymbol{\xi}) - \frac{1}{2\sigma_2^2}(\mathbf{y}_2 - \boldsymbol{\xi})'(\mathbf{y}_2 - \boldsymbol{\xi})\right],$$

is improper. However, the conditional posterior pdf for $\boldsymbol{\xi}$, given σ_1 and σ_2, and the conditional posterior pdf's for σ_1 and σ_2, given $\boldsymbol{\xi}$, are all proper[12]; for example, given σ_1 and σ_2, the posterior pdf for $\boldsymbol{\xi}$, is the following proper normal pdf:

$$(5.25)\qquad p(\boldsymbol{\xi}|\sigma_1, \sigma_2, \mathbf{y}_1, \mathbf{y}_2) \propto \exp\left[-\frac{\sigma_1^2 + \sigma_2^2}{2\sigma_1^2\sigma_2^2}(\boldsymbol{\xi} - \hat{\boldsymbol{\xi}})'(\boldsymbol{\xi} - \hat{\boldsymbol{\xi}})\right],$$

where $\hat{\boldsymbol{\xi}} = (\mathbf{y}_1/\sigma_1^2 + \mathbf{y}_2/\sigma_2^2)/(1/\sigma_1^2 + 1/\sigma_2^2)$ is the posterior conditional mean. This conditional posterior distribution's covariance matrix $\sigma_1^2\sigma_2^2/(\sigma_1^2 + \sigma_2^2)I_n$ has elements that do not tend to zero as n increases.

To illustrate the nature of the identification problem associated with (5.24), we complete the square on $\boldsymbol{\xi}$ and integrate with respect to the elements of this vector, an operation that yields the following result:

$$(5.26)\qquad p(\sigma_1, \sigma_2|\mathbf{y}_1, \mathbf{y}_2) \propto \frac{1}{\sigma_1\sigma_2}\frac{1}{(\sigma_1^2 + \sigma_2^2)^{n/2}} \exp\left[-\frac{1}{2(\sigma_1^2 + \sigma_2^2)}(\mathbf{y}_1 - \mathbf{y}_2)'(\mathbf{y}_1 - \mathbf{y}_2)\right].$$

The factor $1/\sigma_1\sigma_2$ comes from our diffuse prior, whereas the second factor is just the normal pdf for $\mathbf{y}_1 - \mathbf{y}_2$. From the form of this latter pdf, it is impossible to identify σ_1 and σ_2 without adding more prior information than we have added with our diffuse prior pdf. Further, by changing variables in (5.26) from σ_1 and σ_2 to σ_1 and $\lambda = \sigma_1^2/\sigma_2^2$ and on integrating with respect to σ_1 the marginal posterior pdf for λ is just $p(\lambda|\mathbf{y}_1, \mathbf{y}_2) \propto 1/\lambda$, $0 < \lambda < \infty$, which is improper.[13]

Thus, when σ_1 and σ_2 are both unknown, there are difficulties in making inferences about all parameters, whether from the ML or Bayesian approach. If, however, the ratio of unknown variances, say $\lambda = \sigma_1^2/\sigma_2^2$, is known, it can be established that the above difficulties disappear. In fact, for $\lambda = 1$ the problem reduces to the one analyzed above in which we have two observations per unknown mean, with their variances equal. Knowing the value of λ is prior information that resolves the identification problem.

[12] Also, the conditional posterior pdf for $\boldsymbol{\xi}$, given $\lambda = \sigma_1^2/\sigma_2^2$, is a proper normal pdf.
[13] Note that $p(\sigma_1, \sigma_2) \propto 1/\sigma_1\sigma_2$ implies $p(\sigma_1, \lambda) \propto (1/\sigma_1)(1/\lambda)$ and thus the posterior pdf for λ is identical to its prior pdf.

5.2 CLASSICAL EVM: ML ANALYSIS OF THE FUNCTIONAL FORM

In the classical EVM we have n pairs of observations, (y_{1i}, y_{2i}), $i = 1, 2, \ldots, n$, which are assumed to be generated under the following conditions:

$$\begin{aligned} &(5.27) \qquad y_{1i} = \xi_i + u_{1i} \\ &(5.28) \qquad y_{2i} = \eta_i + u_{2i} \end{aligned} \Bigg\} \quad i = 1, 2, \ldots, n,$$

with

$$(5.29) \qquad \eta_i = \beta_0 + \beta\xi_i, \qquad i = 1, 2, \ldots, n,$$

where β_0, β, the η_i's, and the ξ_i's are unknown parameters. In (5.27) to (5.28) we assume that (y_{1i}, y_{2i}) are distributed independently of (y_{1j}, y_{2j}), $i \neq j$ and have a normal distribution with $Ey_{1i} = \xi_i$, $Ey_{2i} = \eta_i$, $\text{Var}\, y_{1i} = \sigma_1^2$, $\text{Var}\, y_{2i} = \sigma_2^2$ and $\text{Cov}(y_{1i}, y_{2i}) = 0$. This last condition implies that the error in y_{1i}, namely, u_{1i}, is uncorrelated with (and here independent of) the error in y_{2i}, u_{2i}.[14]

If we combine (5.28) and (5.29), the model can be written as

$$\begin{aligned} &(5.30) \qquad y_{1i} = \xi_i + u_{1i} \\ &(5.31) \qquad y_{2i} = \beta_0 + \beta\xi_i + u_{2i} \end{aligned} \Bigg\} \quad i = 1, 2, \ldots, n.$$

Written in this form, it is clear that this model is closely associated with both the simple regression model and the problem of n-means; that is, if there were no measurement error in (5.30) (i.e., $y_{1i} = \xi_i$), the model would be in precisely the form of a simple regression model. On the other hand, if $\beta_0 = 0$ and $\beta = 1$, the present model becomes the n-mean problem, with two observations per mean having unequal variances σ_1^2 and σ_2^2. This fact suggests that a problem concerning the existence of a maximum of the likelihood function may arise in connection with the present model just as it did in the n-mean problem to which reference has just been made.

The likelihood function for the parameters of the system (5.30) to (5.31) is given by

$$(5.32) \qquad l(\boldsymbol{\beta}, \boldsymbol{\xi}, \sigma_1, \sigma_2 | \mathbf{y}) \propto \frac{1}{\sigma_1^n} \frac{1}{\sigma_2^n} \exp \left[-\frac{1}{2\sigma_1^2} (\mathbf{y}_1 - \boldsymbol{\xi})'(\mathbf{y}_1 - \boldsymbol{\xi}) - \frac{1}{2\sigma_2^2} (\mathbf{y}_2 - \beta_0 \boldsymbol{\iota} - \beta\boldsymbol{\xi})'(\mathbf{y}_2 - \beta_0 \boldsymbol{\iota} - \beta\boldsymbol{\xi}) \right],$$

[14] As is well known, (5.27) to (5.29) can be viewed as a form of Friedman's consumption function model if for the ith household in a sample of n households we let y_{1i} = log of measured income, y_{2i} = log of measured consumption, ξ_i = log of "permanent" income, η_i = log of "permanent" consumption, u_{1i} = "transitory" income, and u_{2i} = "transitory" consumption; ξ_i, η_i, u_{1i}, and u_{2i} are unobserved quantities. Further, Friedman's theory suggests that $\beta = 1$; that is, the elasticity of permanent consumption with respect to permanent income is one.

where $\boldsymbol{\beta}' = (\beta_0, \beta)$, $\boldsymbol{\xi}' = (\xi_1, \xi_2, \ldots, \xi_n)$, $\mathbf{y}' = (\mathbf{y}_1', \mathbf{y}_2')$, $\mathbf{y}_j' = (y_{j1}, y_{j2}, \ldots, y_{jn})$, $j = 1, 2$, and $\boldsymbol{\iota}$ is a $n \times 1$ column vector, with each element equal to one; that is, $\boldsymbol{\iota}' = (1, 1, \ldots, 1)$. On taking the logarithm of both sides of (5.32) and differentiating with respect to the elements of $\boldsymbol{\xi}$, σ_1, and σ_2 we obtain

$$\frac{\partial \log l}{d\boldsymbol{\xi}} = \frac{1}{\sigma_1^2}(\mathbf{y}_1 - \boldsymbol{\xi}) + \frac{1}{\sigma_2^2}[-\beta^2\boldsymbol{\xi} + \beta(\mathbf{y}_2 - \boldsymbol{\iota}\beta_0)], \tag{5.33}$$

$$\frac{\partial \log l}{\partial \sigma_1} = -\frac{n}{\sigma_1} + \frac{1}{\sigma_1^3}(\mathbf{y}_1 - \boldsymbol{\xi})'(\mathbf{y}_1 - \boldsymbol{\xi}), \tag{5.34}$$

$$\frac{\partial \log l}{\partial \sigma_2} = -\frac{n}{\sigma_2} + \frac{1}{\sigma_2^3}(\mathbf{y}_2 - \beta_0\boldsymbol{\iota} - \beta\boldsymbol{\xi})'(\mathbf{y}_2 - \beta_0\boldsymbol{\iota} - \beta\boldsymbol{\xi}). \tag{5.35}$$

As a necessary condition for a maximum, values of the parameters must exist *in the admissible parameter space*,[15] which sets these derivatives equal to zero as well as the derivatives with respect to β_0 and β. On setting (5.33) equal to zero, we have[16]

$$\begin{aligned} \hat{\boldsymbol{\xi}} &= \frac{\mathbf{y}_1/\hat{\sigma}_1^2 + (\hat{\beta}^2/\hat{\sigma}_2^2)(\mathbf{y}_2 - \hat{\beta}_0\boldsymbol{\iota})/\hat{\beta}}{1/\hat{\sigma}_1^2 + \hat{\beta}^2/\hat{\sigma}_2^2} \\ &= \frac{\mathbf{y}_1 + \hat{\theta}\hat{\mathbf{w}}}{1 + \hat{\theta}}, \end{aligned} \tag{5.36}$$

where $\hat{\theta} = \hat{\sigma}_1^2\hat{\beta}^2/\hat{\sigma}_2^2$ and $\hat{\mathbf{w}} = (\mathbf{y}_2 - \hat{\beta}_0\boldsymbol{\iota})/\hat{\beta}$. Substituting from (5.36) in (5.34) and (5.35) and setting these derivatives equal to zero, we have the following results:

$$\hat{\sigma}_1^2 = \frac{1}{n}\frac{\hat{\theta}^2}{(1 + \hat{\theta})^2}(\mathbf{y}_1 - \hat{\mathbf{w}})'(\mathbf{y}_1 - \hat{\mathbf{w}}), \tag{5.37}$$

$$\hat{\sigma}_2^2 = \frac{\hat{\beta}^2}{n(1 + \hat{\theta})^2}(\mathbf{y}_1 - \hat{\mathbf{w}})'(\mathbf{y}_1 - \hat{\mathbf{w}}). \tag{5.38}$$

These two equations can hold simultaneously if, and only if, $\hat{\beta}^2 = \hat{\sigma}_2^2/\hat{\sigma}_1^2$.[17]

[15] For this problem we define the admissible parameter space as follows: $0 < \sigma_i^2 < \infty$, $i = 1, 2$; $-\infty < \xi_i < \infty$, $i = 1, 2, \ldots, n$; $-\infty < \beta_0, \beta < \infty$; $\sigma_1^2 \neq \sigma_2^2$; and $\beta^2 \neq \sigma_2^2/\sigma_1^2$. The reason for introducing this last condition is explained in the text.

[16] Equation 5.36 can be more fully appreciated by writing (5.30) to (5.31) as $y_{1i} = \xi_i + u_{1i}$ and $(y_{2i} - \beta_0)/\beta = \xi_i + u_{2i}/\beta$. The elements of (5.36) are then weighted averages of y_{1i} and $(y_{2i} - \beta_0)/\beta$. For given β_0 and $\beta(\neq 0)$ this is an n-means problem with two observations per mean and variances σ_1^2 and σ_2^2/β^2.

[17] This result was obtained by D. V. Lindley, "Regression Lines and the Linear Functional Relationship," *J. Roy. Statist. Soc.* (Supplement) **9** (1947), 218–244. His interpretation of it, as well as that of J. Johnston, *Econometric Methods*. New York: McGraw-Hill, 1963, p. 152, is somewhat different from ours.

However, in our definition of the admissible parameter space, we explicitly stated that $\beta^2 \neq \sigma_2^2/\sigma_1^2$ and thus the quantity $\hat{\beta}^2 = \hat{\sigma}_2^2/\hat{\sigma}_1^2$ falls in an inadmissible region of the parameter space. Since this is so, the necessary first-order conditions for a maximum of the likelihood function cannot be simultaneously satisfied; hence a maximum of the likelihood function does not exist in the admissible region of the parameter space.[18]

Since there are basic difficulties with the analysis of the model in (5.30) to (5.31) when all parameters are assumed unknown, analyses have often gone forward under the assumption that $\phi = \sigma_2^2/\sigma_1^2$ is known exactly.[19] Under this assumption a unique maximum of the likelihood function exists and thus ML estimates can be obtained. To do so we write the likelihood function as

$$l(\boldsymbol{\beta}, \boldsymbol{\xi}, \sigma_2|\mathbf{y}, \phi) \propto \frac{1}{\sigma_2^{2n}} \exp\left\{-\frac{1}{2\sigma_2^2}[\phi(\mathbf{y}_1 - \boldsymbol{\xi})'(\mathbf{y}_1 - \boldsymbol{\xi}) + (\mathbf{y}_2 - \beta_0\boldsymbol{\iota} - \beta\boldsymbol{\xi})'(\mathbf{y}_2 - \beta_0\boldsymbol{\iota} - \beta\boldsymbol{\xi})]\right\}. \tag{5.39}$$

On differentiating the logarithm of the likelihood function L with respect to the unknown parameters and setting these derivatives equal to zero, we have

$$\frac{\partial L}{\partial \beta_0} = \frac{1}{\sigma_2^2}(\mathbf{y}_2 - \beta_0\boldsymbol{\iota} - \beta\boldsymbol{\xi})'\boldsymbol{\iota} = 0, \tag{5.40}$$

$$\frac{\partial L}{\partial \beta} = \frac{1}{\sigma_2^2}(\mathbf{y}_2 - \beta_0\boldsymbol{\iota} - \beta\boldsymbol{\xi})'\boldsymbol{\xi} = 0, \tag{5.41}$$

$$\frac{\partial L}{\partial \boldsymbol{\xi}} = \frac{1}{\sigma_2^2}[\phi(\mathbf{y}_1 - \boldsymbol{\xi}) + \beta(\mathbf{y}_2 - \beta_0\boldsymbol{\iota} - \boldsymbol{\beta\xi})] = \mathbf{0}, \tag{5.42}$$

and

$$\frac{\partial L}{\partial \sigma_2} = -\frac{2n}{\sigma_2} + \frac{1}{\sigma_2^3}[\phi(\mathbf{y}_1 - \boldsymbol{\xi})'(\mathbf{y}_1 - \boldsymbol{\xi}) + (\mathbf{y}_2 - \beta_0\boldsymbol{\iota} - \beta\boldsymbol{\xi})'(\mathbf{y}_2 - \beta_0\boldsymbol{\iota} - \beta\boldsymbol{\xi})] = 0. \tag{5.43}$$

From (5.40)

$$\beta_0 = \bar{y}_2 - \beta\bar{\xi}, \tag{5.44}$$

[18] If we combine the following diffuse prior pdf, $p(\beta_0, \beta, \boldsymbol{\xi}, \sigma_1, \sigma_2) \propto 1/\sigma_1\sigma_2$, $0 < \sigma_i < \infty$, $i = 1, 2$, and $-\infty < \beta_0, \beta, \xi_i < \infty$, $i = 1, 2, \ldots, n$, with the likelihood function in (5.32), it is not difficult to show that the joint posterior pdf is improper.

[19] In econometrics this precise knowledge is usually unavailable. Below we present methods for utilizing less precise information about ϕ.

where $\bar{y}_2 = \mathbf{y}_2'\boldsymbol{\iota}/n$ and $\bar{\xi} = \boldsymbol{\xi}'\boldsymbol{\iota}/n$. On substituting this value for β_0 in (5.41),

$$\beta = \frac{(\mathbf{y}_2 - \bar{y}_2\boldsymbol{\iota})'\boldsymbol{\xi}}{(\boldsymbol{\xi} - \bar{\xi}\boldsymbol{\iota})'\boldsymbol{\xi}} = \frac{(\mathbf{y}_2 - \bar{y}_2\boldsymbol{\iota})'(\boldsymbol{\xi} - \bar{\xi}\boldsymbol{\iota})}{(\boldsymbol{\xi} - \bar{\xi}\boldsymbol{\iota})'(\boldsymbol{\xi} - \bar{\xi}\boldsymbol{\iota})}. \tag{5.45}$$

Further from (5.42) we have

$$\boldsymbol{\xi} = \frac{\phi\mathbf{y}_1 + \beta(\mathbf{y}_2 - \beta_0\boldsymbol{\iota})}{\phi + \beta^2} \tag{5.46}$$

and[20]

$$\boldsymbol{\xi} - \bar{\xi}\boldsymbol{\iota} = \frac{\phi(\mathbf{y}_1 - \bar{y}_1\boldsymbol{\iota}) + \beta(\mathbf{y}_2 - \bar{y}_2\boldsymbol{\iota})}{\phi + \beta^2}. \tag{5.47}$$

On substituting from (5.47) in (5.45), the result is

$$\beta = \frac{(\phi + \beta^2)(\phi m_{12} + \beta m_{22})}{\phi^2 m_{11} + 2\phi\beta m_{12} + \beta^2 m_{22}} \tag{5.48}$$

where $m_{ij} = (\mathbf{y}_i - \bar{y}_i\boldsymbol{\iota})'(\mathbf{y}_j - \bar{y}_j\boldsymbol{\iota})/n$ for $i, j = 1, 2$. This last expression yields

$$\beta^2 m_{12} + \beta(\phi m_{11} - m_{22}) - \phi m_{12} = 0 \tag{5.49}$$

as the necessary condition on β. Then the ML estimator for β is a solution of the quadratic equation (5.49), namely,

$$\hat{\beta} = \frac{m_{22} - \phi m_{11} + \sqrt{(m_{22} - \phi m_{11})^2 + 4\phi m_{12}^2}}{2m_{12}}. \tag{5.50}$$

Note that the algebraic sign in front of the square root sign is positive, since this choice leads to a maximum of the likelihood function.[21]

With $\hat{\bar{\xi}} = \bar{y}_1$ from (5.44) and (5.46), the ML estimator for β_0, from (5.44), is

$$\hat{\beta}_0 = \bar{y}_2 - \hat{\beta}\bar{y}_1, \tag{5.51}$$

whereas the ML estimator for $\boldsymbol{\xi}$, from (5.46), is

$$\hat{\boldsymbol{\xi}} = \frac{\phi\mathbf{y}_1 + \hat{\beta}(\mathbf{y}_2 - \hat{\beta}_0\boldsymbol{\iota})}{\phi + \hat{\beta}^2}. \tag{5.52}$$

[20] Equation 5.47 is obtained by multiplying both sides of (5.46) on the left by $\boldsymbol{\iota}'/n$ and on the right by $\boldsymbol{\iota}$ and subtracting the resulting expressions from the quantities on the lhs and rhs of (5.46).

[21] See, for example, A. Madansky, "The Fitting of Straight Lines when Both Variables Are Subject to Error," *J. Am. Statist. Assoc.*, **54**, 173–205 (1959).

Last, from (5.43), we have for the ML estimator for σ_2^2:

$$(5.53) \quad \hat{\sigma}_2^2 = \frac{1}{2n} [\phi(\mathbf{y}_1 - \hat{\boldsymbol{\xi}})'(\mathbf{y}_1 - \hat{\boldsymbol{\xi}}) + (\mathbf{y}_2 - \hat{\beta}_0\boldsymbol{\iota} - \hat{\beta}\hat{\boldsymbol{\xi}})'(\mathbf{y}_2 - \hat{\beta}_0\boldsymbol{\iota} - \hat{\beta}\hat{\boldsymbol{\xi}})].$$

It has been pointed out in the literature[22] that although $\hat{\beta}_0$ and $\hat{\beta}$ are consistent estimators $\hat{\sigma}_2^2$ is not. In fact, plim $\hat{\sigma}_2^2 = \frac{1}{2}\sigma_2^2$. This result is completely analogous to that obtained in Section 5.1 in which we analyzed the problem of n means with two observations and the same variance per mean. In the present problem we know the value of ϕ, the ratio of the variances and in effect have just one unknown variance; see the likelihood function in (5.39) which can be written just in terms of σ_2^2 when ϕ is known.

The inconsistency of the estimator $\hat{\sigma}_2^2$ appears to be due to the fact that the ML method makes no allowance for the fact that in forming the estimator, the n elements of $\boldsymbol{\xi}$ have been estimated. Since the number of elements of $\boldsymbol{\xi}$ grows with the sample size, an appreciable fraction of the sample is employed in estimating the elements of $\boldsymbol{\xi}$ in small as well as large samples. As Kendall and Stuart[23] put it, the finite sample bias of the ML estimator for σ_2^2 does not disappear as n grows, since we never leave the small sample situation. They suggest the following procedure to correct the inconsistency. There are $2n$ observations, and in (5.53) we have inserted estimates for n elements of $\boldsymbol{\xi}$, β_0, and β, that is, for $n + 2$ parameters. Then $2n - (n + 2) = n - 2$ represents the degrees of freedom remaining for the estimation of σ_2^2. A "corrected" consistent estimator is $\tilde{\sigma}_2^2 = 2n\hat{\sigma}_2^2/(n - 2)$.

5.3 ML ANALYSIS OF STRUCTURAL FORM OF EVM

The model in (5.30) to (5.31) in which the vector $\boldsymbol{\xi}$ is assumed to be stochastic is often referred to as the "structural form" of the EVM. Perhaps the most basic general result for this case is due to Kiefer and Wolfowitz who proved that if the parameters are identified then "... under the usual regularity conditions, the ML estimator of a structural parameter is strongly consistent, when the (infinitely many) incidental parameters are independently distributed chance variables with a common unknown distribution function."[24] Among other results, they establish that if the ξ_i's are independently distributed and each has the same non-normal distribution the ML estimators

[22] See, for example, Kendall and Stuart, *op. cit.*, p. 386.

[23] *Op. cit.*, p. 61 and p. 387.

[24] J. Kiefer and J. Wolfowitz, "Consistency of the Maximum Likelihood Estimator in the Presence of Infinitely Many Incidental Parameters," *Ann. Math. Statist.*, **27**, 887–906, p. 887 (1956).

for the structural parameters will be consistent. The condition of non-normality is required to identify the structural parameters when nothing is assumed to be known about the parameters, a result due to Reiersol.[25]

To indicate the nature of the identification problem for the model in (5.30) to (5.31) when the ξ_i's are assumed to be independent of the y's and normally and independently distributed, each with mean μ and variance τ^2, we note that under this assumption, along with the other distributional assumptions introduced in connection with (5.30) to (5.31), the pairs of variables (y_{1i}, y_{2i}), $i = 1, 2, \ldots, n$, will be independently and identically distributed, each pair with a bivariate normal distribution. The following are the moments of (y_{1i}, y_{2i}) for $i = 1, 2, \ldots, n$:

$$Ey_{1i} = \mu, \tag{5.54}$$

$$Ey_{2i} = \beta_0 + \beta\mu, \tag{5.55}$$

$$\text{Var } y_{1i} = \tau^2 + \sigma_1^{\,2}, \tag{5.56}$$

$$\text{Var } y_{2i} = \beta^2\tau^2 + \sigma_2^{\,2}, \tag{5.57}$$

$$\text{Cov}(y_{1i}, y_{2i}) = \beta\tau^2. \tag{5.58}$$

Since these five moments completely determine a bivariate normal distribution and since sample moments are sufficient statistics in this instance, we can equate sample moments to population moments in an effort to obtain estimates. There is, however, a basic difficulty with this approach, namely, that although we have five relations (5.54) to (5.58) there are six unknown parameters, μ, β_0, β, $\sigma_1^{\,2}$, $\sigma_2^{\,2}$, and τ^2. Thus we cannot obtain estimates of all parameters unless prior information is available to reduce the number of unknown parameters.

Let us first consider the case in which we know that $\beta_0 = 0$. When this information is available, we can equate sample moments of the y's to their respective population moments to obtain estimates as follows. From (5.54) $\hat{\mu} = \bar{y}_1$ and from (5.54) to (5.55) $\hat{\beta} = \bar{y}_2/\bar{y}_1$, where $\bar{y}_1$ and $\bar{y}_2$ are sample means. The estimator $\hat{\beta}$ is in the form of the ratio of two correlated normal random variables; hence its mean and higher moments do not exist.[26] Using

[25] O. Reiersol, "Identifiability of a Linear Relation Between Variables Which Are Subject to Error," *Econometrica*, **18**, 375–389 (1950).

[26] For the derivation of the distribution of the ratio of two correlated normal variables see E. C. Fieller, "The Distribution of the Index in a Normal Bivariate Population," *Biometrika*, **24**, 428–440 (1932); R. C. Geary, "The Frequency Distribution of the Quotient of Two Normal Variates," *J. Roy. Statist. Soc.*, **93**, 442–446 (1930); and G. Marsaglia, "Ratios of Normal Variables and Ratios of Sums of Uniform Variables," *J. Am. Statist. Assn.*, **60**, 193–204 (1965).

this estimate for β, we have from (5.56) to (5.58)

$$\hat{\tau}^2 = \frac{m_{12}}{\hat{\beta}},$$

$$\hat{\sigma}_2^2 = m_{22} - \hat{\beta}^2\hat{\tau}^2,$$

and

$$\hat{\sigma}_1^2 = m_{11} - \hat{\tau}^2,$$

where

$$m_{ij} = \sum_{l=1}^{n} \frac{(y_{il} - \bar{y}_i)(y_{jl} - \bar{y}_j)}{n}, \qquad i, j = 1, 2.$$

Although the information $\beta_0 = 0$ enables us to obtain estimates of the remaining parameters, it must be noted that this approach can lead to negative estimates for any or all of the following variances: σ_1^2, σ_2^2 and τ^2; that is, the prior information that these variances are positive has not been introduced explicitly and thus meaningless variance estimates can be obtained.[27] Further, from (5.56) and (5.58) we have $\text{Var}\, y_{1i} = \text{Cov}(y_{1i}, y_{2i})/\beta + \sigma_1^2$ and from (5.57) and (5.58), $\text{Var}\, y_{2i} = \beta\, \text{Cov}(y_{1i}, y_{2i}) + \sigma_2^2$. Since $\sigma_1^2 > 0$ and $\sigma_2^2 > 0$, we have

$$\text{Var}\, y_{1i} - \frac{\text{Cov}(y_{1i}, y_{2i})}{\beta} > 0 \tag{5.59}$$

and

$$\text{Var}\, y_{2i} - \beta\, \text{Cov}(y_{1i}, y_{2i}) > 0. \tag{5.60}$$

From (5.59) to (5.60), with $\text{Cov}(y_{1i}, y_{2i}) > 0$, we have

$$\frac{\text{Cov}(y_{1i}, y_{2i})}{\text{Var}\, y_{1i}} < \beta < \frac{\text{Var}\, y_{2i}}{\text{Cov}(y_{1i}, y_{2i})}. \tag{5.61}$$

If $\text{Cov}(y_{1i}, y_{2i}) < 0$, the inequalities in (5.61) are reversed. Thus the prior information that $\sigma_1^2 > 0$ and $\sigma_2^2 > 0$, combined with the relations (5.56) to (5.58), implies that β falls in one of two *finite* intervals, given by (5.61) or (5.61) with the inequalities reversed.[28] Since our estimator for β is $\bar{y}_2/\bar{y}_1$, it has a range $-\infty$ to ∞ and thus can violate the bounds set by (5.61).

In summary, when it is known that $\beta_0 = 0$, point estimates for the remaining parameters can be obtained by equating sample moments to their respective population moments. However, estimates so obtained may not be

[27] Similar problems occur in "random effects" models. See, for example, G. C. Tiao and W. Y. Tan, "Bayesian Analysis of Random-Effect Models in the Analysis of Variance. I. Posterior Distribution of Variance Components," *Biometrika*, **52**, 37–53 (1965).

[28] This condition has been generally recognized and is stressed in D. V. Lindley and G. M. El-Sayyad, "The Bayesian Estimation of a Linear Functional Relationship," *J. Roy. Statist. Soc.*, Series B, **30**, 190–202 (1968).

consistent with basic prior information; for example, estimates of variances may be negative contrary to the prior information that variances are non-negative. Sampling theory procedures for dealing with this "negative variance" problem for the EVM are not yet available.[29]

If, rather than β_0, $\phi = \sigma_2^2/\sigma_1^2$ is known, the sample moments m_{11}, m_{22}, and m_{12} can be inserted in (5.56) to (5.58) for their population counterparts and estimates for β, τ^2, and σ_1^2, obtained; that is, if we let $\theta_{11} = \text{Var}\, y_{1i}$, $\theta_{22} = \text{Var}\, y_{2i}$ and $\theta_{12} = \text{Cov}(y_{1i}, y_{2i})$, then (5.58) yields $\tau^2 = \theta_{12}/\beta$ and (5.56) yields $\sigma_1^2 = \theta_{11} - \tau^2 = \theta_{11} - \theta_{12}/\beta$. Equation 5.57 is $\theta_{22} = \beta^2\tau^2 + \sigma_1^2\phi$, and on substituting for τ^2 and σ_1^2 the result is

$$\beta^2\theta_{12} + \beta(\phi\theta_{11} - \theta_{22}) - \phi\theta_{12} = 0.$$

On replacing the θ_{ij}'s by their sample counterparts, we have

$$\hat{\beta}^2 m_{12} + \hat{\beta}(\phi m_{11} - m_{22}) - \phi m_{12} = 0,$$

which is in precisely the same form as (5.49). The ML estimate for β is then given by (5.50).[30] Estimates for the remaining parameters are given by $\hat{\tau}^2 = m_{12}/\hat{\beta}$, $\hat{\sigma}_1^2 = m_{11} - \hat{\tau}^2$ and $\hat{\sigma}_2^2 = \hat{\sigma}_1^2\phi$. Also, from (5.54) $\hat{\mu} = \bar{y}_1$ and from (5.55) $\hat{\beta}_0 = \bar{y}_2 - \hat{\beta}\hat{\mu}$.

Above, we have gone forward under the assumption that the ξ_i's are normally and independently distributed, each with mean μ and variance τ^2. With this assumption, additional prior information must be added to identify the parameters. In line with Reiersol's results, however, if the ξ_i's have a non-normal distribution, the parameters will be identified. To illustrate one case of non-normality assume

$$p(\boldsymbol{\xi}) \propto \text{const}, \qquad -\infty < \xi_i < \infty, \qquad i = 1, 2, \ldots, n. \tag{5.62}$$

In (5.62) we assume that the ξ_i's are uniformly and independently distributed. Below we shall see how this assumption about the ξ_i's affects ML estimates.

To formulate the likelihood function in general we consider the joint pdf of $\mathbf{y}_1$, $\mathbf{y}_2$, and $\boldsymbol{\xi}$:

$$p(\mathbf{y}_1, \mathbf{y}_2, \boldsymbol{\xi}|\boldsymbol{\psi}) = p(\mathbf{y}_1, \mathbf{y}_2|\boldsymbol{\xi}, \boldsymbol{\psi}_1)\, g(\boldsymbol{\xi}|\boldsymbol{\psi}_2), \tag{5.63}$$

where $\boldsymbol{\psi}' = (\boldsymbol{\psi}_1', \boldsymbol{\psi}_2')$ denotes the vector of parameters, $\boldsymbol{\psi}_1$, the vector of

[29] A "standard" approach would be to maximize the likelihood function subject to inequality constraints on the parameters. Of course, if the estimates obtained by the procedure described in the text satisfy the inequality constraints, they constitute a solution to the constrained problem. If they violate the constraints, a nonlinear programming problem has to be solved to obtain estimates that satisfy the inequality constraints.

[30] Note that taking a positive sign before the square root in (5.50) gives $\hat{\beta}$ the same algebraic sign as m_{12}. This ensures that the estimate for τ^2, namely, $\hat{\tau}^2 = m_{12}/\hat{\beta}$, is positive.

parameters in the conditional pdf for $\mathbf{y}_1$ and $\mathbf{y}_2$, given $\boldsymbol{\xi}$, and $\boldsymbol{\psi}_2$, the subvector of $\boldsymbol{\psi}$ appearing in the marginal pdf for the elements of $\boldsymbol{\xi}$. To obtain the marginal pdf for $\mathbf{y}_1$ and $\mathbf{y}_2$ (5.63) must be integrated with respect to the elements of $\boldsymbol{\xi}$; that is

$$h(\mathbf{y}_1, \mathbf{y}_2|\boldsymbol{\psi}) = \int p(\mathbf{y}_1, \mathbf{y}_2|\boldsymbol{\xi}, \boldsymbol{\psi}_1)\, g(\boldsymbol{\xi}|\boldsymbol{\psi}_2)\, d\boldsymbol{\xi}. \tag{5.64}$$

Then $h(\mathbf{y}_1, \mathbf{y}_2|\boldsymbol{\psi})$, viewed as a function of the elements of $\boldsymbol{\psi}$, is the likelihood function.

For the EVM, with the vector $\boldsymbol{\xi}$ assumed to have a pdf in the form of (5.62), the likelihood function is obtained by integrating the following expression with respect to the elements of $\boldsymbol{\xi}$:

$$\begin{aligned} p(\mathbf{y}_1, \mathbf{y}_2, \boldsymbol{\xi}|\boldsymbol{\psi}) &\propto \frac{1}{\sigma_1^{\,n}\sigma_2^{\,n}} \exp\left[-\frac{1}{2\sigma_1^{\,2}}(\mathbf{y}_1 - \boldsymbol{\xi})'(\mathbf{y}_1 - \boldsymbol{\xi}) \right. \\ &\qquad \left. - \frac{1}{2\sigma_2^{\,2}}(\mathbf{y}_2 - \beta_0\boldsymbol{\iota} - \beta\boldsymbol{\xi})'(\mathbf{y}_2 - \beta_0\boldsymbol{\iota} - \beta\boldsymbol{\xi})\right] \\ &\propto \frac{1}{\sigma_1^{\,n}\sigma_2^{\,n}} \exp\left\{-\frac{1}{2\sigma_2^{\,2}}[\phi(\mathbf{y}_1 - \boldsymbol{\xi})'(\mathbf{y}_1 - \boldsymbol{\xi}) \right. \\ &\qquad \left. + (\mathbf{y}_2 - \beta_0\boldsymbol{\iota} - \beta\boldsymbol{\xi})'(\mathbf{y}_2 - \beta_0\boldsymbol{\iota} - \beta\boldsymbol{\xi})]\right\}. \end{aligned}$$

To perform the integration with respect to the elements of $\boldsymbol{\xi}$ we can complete the square in the exponent and use properties of the normal distribution to obtain[31]

$$\begin{aligned} h(\mathbf{y}_1, \mathbf{y}_2|\boldsymbol{\psi}) &\propto \frac{\phi^{n/2}}{\sigma_2^{\,n}(\phi + \beta^2)^{n/2}} \\ &\quad \times \exp\left[-\frac{\phi}{2\sigma_2^{\,2}(\phi + \beta^2)}(\mathbf{y}_2 - \beta_0\boldsymbol{\iota} - \beta\mathbf{y}_1)'(\mathbf{y}_2 - \beta_0\boldsymbol{\iota} - \beta\mathbf{y}_1)\right], \end{aligned} \tag{5.65}$$

where $\boldsymbol{\psi}' = (\beta_0, \beta, \sigma_1, \phi)$. On maximizing $\log h$ with respect to σ_1, β_0, and β, we find the maximizing values for β_0 and β to be

$$\hat{\beta}_0 = \bar{y}_2 - \hat{\beta}\bar{y}_1 \quad \text{and} \quad \hat{\beta} = \frac{(\mathbf{y}_1 - \bar{y}_1\boldsymbol{\iota})'(\mathbf{y}_2 - \bar{y}_2\boldsymbol{\iota})}{(\mathbf{y}_1 - \bar{y}_1\boldsymbol{\iota})'(\mathbf{y}_1 - \bar{y}_1\boldsymbol{\iota})}. \tag{5.66}$$

It is seen that $\hat{\beta}_0$ and $\hat{\beta}$ are just the simple least squares estimates obtained from regressing $\mathbf{y}_2$ on $\mathbf{y}_1$. This surprising result[32] is vitally dependent on the

[31] Note that $\phi(\mathbf{y}_1 - \boldsymbol{\xi})'(\mathbf{y}_1 - \boldsymbol{\xi}) + (\mathbf{y}_2 - \beta_0\boldsymbol{\iota} - \beta\boldsymbol{\xi})'(\mathbf{y}_2 - \beta_0\boldsymbol{\iota} - \beta\boldsymbol{\xi}) = (\phi + \beta^2)\boldsymbol{\xi}'\boldsymbol{\xi} - 2\boldsymbol{\xi}'[\phi\mathbf{y}_1 + \beta(\mathbf{y}_2 - \beta_0\boldsymbol{\iota})] + \phi\mathbf{y}_1'\mathbf{y}_1 + (\mathbf{y}_2 - \beta_0\boldsymbol{\iota})'(\mathbf{y}_2 - \beta_0\boldsymbol{\iota}) = (\phi + \beta^2)(\boldsymbol{\xi} - \hat{\boldsymbol{\xi}})'(\boldsymbol{\xi} - \hat{\boldsymbol{\xi}}) + \phi\mathbf{y}_1'\mathbf{y}_1 + (\mathbf{y}_2 - \beta_0\boldsymbol{\iota})'(\mathbf{y}_2 - \beta_0\boldsymbol{\iota}) - (\phi + \beta^2)\hat{\boldsymbol{\xi}}'\hat{\boldsymbol{\xi}} = (\phi + \beta^2)(\boldsymbol{\xi} - \hat{\boldsymbol{\xi}})'(\boldsymbol{\xi} - \hat{\boldsymbol{\xi}}) + \phi[(\mathbf{y}_2 - \beta_0\boldsymbol{\iota} - \beta\mathbf{y}_1)'(\mathbf{y}_2 - \beta_0\boldsymbol{\iota} - \beta\mathbf{y}_1)]/(\phi + \beta^2)$, where $\hat{\boldsymbol{\xi}} = [\phi\mathbf{y}_1 + \beta(\mathbf{y}_2 - \beta_0\boldsymbol{\iota})]/(\phi + \beta^2)$.

[32] This result appeared in a preliminary paper by A. Zellner and U. Sankar, "Errors in the Variables," manuscript, 1967.

assumption about the pdf for $\boldsymbol{\xi}$ in (5.62), an assumption which implies that the elements of $\boldsymbol{\xi}$ have infinite variances. With this assumption the measurement errors' variances for the elements of $\mathbf{y}_1$ are negligible with respect to the variances of the elements of $\boldsymbol{\xi}$ and $\hat{\beta}$ is a consistent estimator.[33] The information about the spread of the ξ_i's contained in (5.62) is reflected by having the ML estimators take the form shown in (5.66).

5.4 BAYESIAN ANALYSIS OF THE FUNCTIONAL FORM OF THE EVM

In the Bayesian analysis of the functional form of the EVM information required to identify the parameters of (5.30) to (5.31) is introduced by means of a prior pdf for the parameters. As will be seen, there is no need to assume, for example, that the value of $\phi = \sigma_2^2/\sigma_1^2$ is known exactly in order to make inferences about parameters of interest. Further, it should be appreciated that since prior information is needed to identify unknown parameters no matter how large the sample size is this prior information will exert an important influence on posterior inferences.[34]

To provide the Bayesian analogue of the ML results in (5.66), let us employ the following prior pdf:

$$p(\boldsymbol{\xi}, \beta_0, \beta, \sigma_1, \sigma_2) \propto \frac{1}{\sigma_1\sigma_2}, \tag{5.67}$$

with $-\infty < \beta_0, \beta, \xi_i < \infty$, $i = 1, 2, \ldots, n$, and $0 < \sigma_i < \infty$, $i = 1, 2$. In (5.67) we are assuming a priori that the ξ_i's, β_0, β, $\log \sigma_1$, and $\log \sigma_2$ are uniformly and independently distributed. The pdf in (5.67) represents our subjective prior beliefs about the unknown parameters, whereas (5.62) is usually not given this interpretation by sampling theorists.

On combining the prior pdf in (5.67) with the likelihood function in (5.32), we have the following posterior pdf:

$$\begin{aligned} p(\boldsymbol{\beta}, \boldsymbol{\xi}, \sigma_1, \sigma_2|\mathbf{y}) &\propto \frac{1}{\sigma_1^{n+1}}\frac{1}{\sigma_2^{n+1}} \exp\left[-\frac{1}{2\sigma_1^2}(\mathbf{y}_1 - \boldsymbol{\xi})'(\mathbf{y}_1 - \boldsymbol{\xi})\right. \\ &\qquad \left. - \frac{1}{2\sigma_2^2}(\mathbf{y}_2 - \beta_0\boldsymbol{\iota} - \beta\boldsymbol{\xi})'(\mathbf{y}_2 - \beta_0\boldsymbol{\iota} - \beta\boldsymbol{\xi})\right] \\ &\propto \frac{1}{\sigma_1^{n+1}\sigma_2^{n+1}} \exp\left\{-\frac{1}{2\sigma_2^2}[\phi(\mathbf{y}_1 - \boldsymbol{\xi})'(\mathbf{y}_1 - \boldsymbol{\xi})\right. \\ &\qquad \left. + (\mathbf{y}_2 - \beta_0\boldsymbol{\iota} - \beta\boldsymbol{\xi})'(\mathbf{y}_2 - \beta_0\boldsymbol{\iota} - \beta\boldsymbol{\xi})]\right\}, \end{aligned} \tag{5.68}$$

[33] Note that, in general, plim $\hat{\beta} = \beta(1 + \sigma_1^2/\tau^2)^{-1}$, where $\hat{\beta}$ is given in (5.66) and τ^2 is the common variance of the elements of $\boldsymbol{\xi}$. As $\tau^2 \to \infty$, plim $\hat{\beta} \to \beta$.

[34] Of course, prior assumptions, say about the ratio of error variances, also affect sampling theory analyses of the EVM, no matter how large the sample size.

where $\boldsymbol{\beta}' = (\beta_0, \beta)$, $\mathbf{y}' = (\mathbf{y}_1', \mathbf{y}_2')$, and $\phi = \sigma_2^2/\sigma_1^2$. Completing the square on $\boldsymbol{\xi}$ in the exponent and integrating with respect to the elements of $\boldsymbol{\xi}$, we have

$$p(\boldsymbol{\beta}, \sigma_1, \sigma_2|\mathbf{y}) \propto \frac{1}{\sigma_1^{n+1}\sigma_2(\beta^2 + \phi)^{n/2}} \times \exp\left[-\frac{1}{2\sigma_1^2(\beta^2 + \phi)}(\mathbf{y}_2 - \beta_0\boldsymbol{\iota} - \beta\mathbf{y}_1)'(\mathbf{y}_2 - \beta_0\boldsymbol{\iota} - \beta\mathbf{y}_1)\right]. \tag{5.69}$$

Changing variables to $\boldsymbol{\beta}$, σ_1, and ϕ, note that $d\sigma_2/\sigma_2 \propto d\phi/\phi$ and thus

$$p(\boldsymbol{\beta}, \sigma_1, \phi|\mathbf{y}) \propto \frac{1}{\phi\sigma_1^{n+1}(\beta^2 + \phi)^{n/2}} \times \exp\left[-\frac{1}{2\sigma_1^2(\beta^2 + \phi)}(\mathbf{y}_2 - \beta_0\boldsymbol{\iota} - \beta\mathbf{y}_1)'(\mathbf{y}_2 - \beta_0\boldsymbol{\iota} - \beta\mathbf{y}_1)\right]. \tag{5.70}$$

Then, on integrating with respect to σ_1,

$$\begin{aligned} p(\boldsymbol{\beta}, \phi|\mathbf{y}) &\propto \frac{1}{\phi}[(\mathbf{y}_2 - \beta_0\boldsymbol{\iota} - \beta\mathbf{y}_1)'(\mathbf{y}_2 - \beta_0\boldsymbol{\iota} - \beta\mathbf{y}_1)]^{-n/2} \\ &\propto \frac{1}{\phi}[\nu s^2 + (\boldsymbol{\beta} - \hat{\boldsymbol{\beta}})'X'X(\boldsymbol{\beta} - \hat{\boldsymbol{\beta}})]^{-n/2}, \end{aligned} \tag{5.71}$$

where $\nu = n - 2$, $X = (\boldsymbol{\iota} \vdots \mathbf{y}_1)$, $\hat{\boldsymbol{\beta}} = (X'X)^{-1}X'\mathbf{y}_2$, and $\nu s^2 = (\mathbf{y}_2 - X\hat{\boldsymbol{\beta}})'(\mathbf{y}_2 - X\hat{\boldsymbol{\beta}})$. From the form of (5.71) we see that $\boldsymbol{\beta}' = (\beta_0, \beta)$ has a posterior pdf in the bivariate Student-t form with mean $\hat{\boldsymbol{\beta}}$, precisely the least squares quantity obtained from regressing $\mathbf{y}_2$ on $\mathbf{y}_1$.[35] As with the ML analysis yielding (5.66), the present result is critically dependent on the assumption about the form of the pdf for $\boldsymbol{\xi}$ in (5.67). In addition, note that the posterior pdf for ϕ in (5.71) is improper and exactly in the form implied by the prior assumptions about σ_1 and σ_2 in (5.67). Thus with the prior assumptions in (5.67), no new information is provided about ϕ from the sample.

Although the assumptions embodied in (5.67) permit us to make posterior inferences about β_0 and β which are appropriate in certain situations, they, of course, are not appropriate for all problems. Since it is often difficult to know what to assume about the n elements of $\boldsymbol{\xi}$, we shall develop a conditional analysis wherein we analyze the EVM, given that

$$\boldsymbol{\xi} = \tilde{\boldsymbol{\xi}} = \frac{\phi\mathbf{y}_1 + \beta(\mathbf{y}_2 - \beta_0\boldsymbol{\iota})}{\phi + \beta^2}. \tag{5.72}$$

The quantity on the rhs of (5.72) is exactly in the form of (5.46), the "ML equation" for $\boldsymbol{\xi}$.[36] The joint pdf for $\mathbf{y}_1$ and $\mathbf{y}_2$, given the parameters, is

[35] This result appeared in A. Zellner and U. Sankar, *loc. cit.*

[36] See also the discussion of (5.36) above in which it is pointed out that the elements of $\tilde{\boldsymbol{\xi}}$ are weighted averages of estimates of ξ_i, $i = 1, 2, \ldots, n$.

$$p(\mathbf{y}_1, \mathbf{y}_2|\boldsymbol{\xi}, \boldsymbol{\beta}, \sigma_2, \phi) \propto \frac{\phi^{n/2}}{\sigma_2^{2n}} \exp\left\{-\frac{1}{2\sigma_2^2}\left[\phi \sum (y_{1i} - \xi_i)^2 + \sum (y_{2i} - \beta_0 - \beta\xi_i)^2\right]\right\}, \tag{5.73}$$

with the summations taken from $i = 1$ to $i = n$. If we use (5.72) to conditionalize (5.73), the result is

$$p(\mathbf{y}_1, \mathbf{y}_2|\boldsymbol{\xi} = \hat{\boldsymbol{\xi}}, \boldsymbol{\beta}, \sigma_2, \phi) \propto \frac{\phi^{n/2}}{\sigma_2^{2n}} \exp\left\{-\frac{1}{2\sigma_2^2}\left[\phi\left(\frac{\beta^2/\phi}{1 + \beta^2/\phi}\right)^2 \sum \left[y_{1i} - \frac{1}{\beta}(y_{2i} - \beta_0)\right]^2 + \left(\frac{1}{1 + \beta^2/\phi}\right)^2 \sum (y_{2i} - \beta_0 - \beta y_{1i})^2\right]\right\}. \tag{5.74}$$

Before proceeding to introduce a prior pdf for the parameters of (5.74), it would be instructive to study the properties of (5.74).[37]

1. Given $\phi = \phi_0$, a given value, (5.74) has a unique mode at the ML estimates for β_0, β, and σ_2. This follows, since $\boldsymbol{\xi} = \hat{\boldsymbol{\xi}}$ is the conditional ML estimate for $\boldsymbol{\xi}$.

2. For finite β^2, as $\phi = \sigma_2^2/\sigma_1^2$ gets large, the second term in the exponential dominates likelihood function.[38] Under these conditions the modal values for β_0 and β will be close to what is obtained from a least squares regression of y_{2i} on y_{1i}. The assumption that ϕ is large implies that the variance of the measurement error in y_{1i} is small relative to the variance of u_{2i} in $y_{2i} = \beta_0 + \beta\xi_i + u_{2i}$.

3. For finite β^2, as $\phi = \sigma_2^2/\sigma_1^2 \to 0$, the first term in the exponential of (5.74) dominates the likelihood function. Note that this first term involves the sum of squares $\sum [y_{1i} - (1/\beta)(y_{2i} - \beta_0)]^2$. Thus when ϕ is very small ML estimates can be obtained approximately by regressing y_{1i} on y_{2i}. The ML estimate of β will be close to the reciprocal of the least squares slope coefficient estimate, whereas the negative of the intercept estimate times the estimate for β yields approximately the ML estimate of β_0 when ϕ is very small.

4. If $\beta = 0$, the first term in the exponential is zero and the second can be used to make inferences about β_0.

5. Although the form of (5.74) is of interest in showing how the two sums of squares associated with the regressions of y_{1i} on y_{2i} and y_{2i} on y_{1i} appear in

[37] Since $\boldsymbol{\xi} = \hat{\boldsymbol{\xi}}$ is the conditional ML estimate for $\boldsymbol{\xi}$, the properties of (5.74) are relevant for likelihood analyses of the EVM.

[38] We can write the exponential of (5.74) as $\exp\{-[2\sigma_2^2(1 + \beta^2/\phi)^2]^{-1}[(\beta^4/\phi) \times \sum [y_{1i} - (1/\beta)(y_{2i} - \beta_0)]^2 + \sum (y_{2i} - \beta_0 - \beta y_{1i})^2]\}$, and as $\phi \to \infty$ the first term becomes relatively less important than the second. On the other hand, as $\phi \to 0$, the first term becomes relatively more important than the second.

the likelihood function conditional on $\boldsymbol{\xi} = \hat{\boldsymbol{\xi}}$, it is the case that (5.74) can be expressed more simply as

(5.75)

$$p(\mathbf{y}_1, \mathbf{y}_2|\boldsymbol{\xi} = \hat{\boldsymbol{\xi}}, \boldsymbol{\beta}, \sigma_2, \phi) \propto \frac{\phi^{n/2}}{\sigma_2^{2n}} \exp\left[-\frac{1}{2\sigma_2^2(1 + \beta^2/\phi)} \sum (y_{2i} - \beta_0 - \beta y_{1i})^2\right].$$

This is the form of the likelihood function, given $\boldsymbol{\xi} = \hat{\boldsymbol{\xi}}$, which we shall combine below with prior pdf's for the parameters. Again, it is important to emphasize that (5.75) has a unique mode at the ML estimates for β, β_0, and σ_2, given ϕ, the location of which is dependent on what is assumed about ϕ.[39]

As regards a prior pdf for the unknown parameters of (5.75), namely, $\boldsymbol{\beta}' = (\beta_0, \beta)$, σ_2, and ϕ, we shall first make the following assumptions:

$$p(\boldsymbol{\beta}, \sigma_2, \phi) \propto \frac{p_1(\beta)\, p_2(\phi)}{\sigma_2}, \tag{5.76}$$

where $p_2(\phi)$ is the prior pdf for ϕ, with still unspecified form. In (5.76) we are assuming that β_0, β, ϕ, and σ_2 are a priori independently distributed. The ranges of the parameters are $0 < \sigma_2,\ \phi < \infty$ and $-\infty < \beta_0 < \infty$. With respect to β, it is important to realize that its range is finite, a priori, as the discussion of (5.61) above indicates. Thus we shall assume that an investigator knows the algebraic sign of β and assigns an a priori range for this parameter, guided by the analysis leading to (5.61). Over this range, say β_L to β_U, the prior pdf for β is $p_1(\beta)$. With respect to σ_2, we are being diffuse on this parameter by taking $\log \sigma_2$ uniformly distributed.[40]

On combining the prior assumptions embodied in (5.76) with the conditional likelihood function in (5.75), the resulting posterior pdf is

(5.77)

$$p(\boldsymbol{\beta}, \sigma_2, \phi|\mathbf{y}_1, \mathbf{y}_2, \boldsymbol{\xi} = \hat{\boldsymbol{\xi}}) \propto \frac{\phi^{n/2} p_1(\beta)\, p_2(\phi)}{\sigma_2^{2n+1}} \times \exp\left[-\frac{1}{2\sigma_2^2(1 + \beta^2/\phi)} \sum (y_{2i} - \beta_0 - \beta y_{1i})^2\right].$$

We can integrate (5.77) analytically with respect to β_0, $-\infty < \beta_0 < \infty$, and σ_2, $0 < \sigma_2 < \infty$, to obtain the marginal posterior pdf for β and ϕ:

(5.78)

$$p(\beta, \phi|\mathbf{y}_1, \mathbf{y}_2, \boldsymbol{\xi} = \hat{\boldsymbol{\xi}}) \propto \phi^{n/2} p_1(\beta)\, p_2(\phi) \frac{(1 + \beta^2/\phi)^n}{\{\sum [y_{2i} - \bar{y}_2 - \beta(y_{1i} - \bar{y}_1)]^2\}^{n - 1/2}},$$

[39] We have pointed out that as $\phi \to \infty$ the modal value for β will approach $\hat{\beta}_{21} = m_{12}/m_{11}$, and as $\phi \to 0$ it will approach $\hat{\beta}_{12} = m_{22}/m_{12}$. Since $\hat{\beta}_{12} - \hat{\beta}_{21} = (m_{11}m_{22} - m_{12}^2)/m_{11}m_{12}$, $\hat{\beta}_{12}$ is larger in absolute value than $\hat{\beta}_{21}$. Thus for $m_{12} > 0$ assuming ϕ to be large will lead to a smaller positive point estimate for β than assuming ϕ to be small.

[40] The analysis can be extended to the case in which we use an informative prior pdf for σ_2 in the inverted gamma form; see Appendix A for properties of this pdf.

with $0 < \phi < \infty$ and $\beta_L \leq \beta \leq \beta_U$, where β_L and β_U are the a priori bounds placed on β.

The denominator of (5.78) has a minimum at $\hat{\beta}_{21} = m_{12}/m_{11}$, the least squares quantity from a regression of y_{2i} on y_{1i}, and thus were it not for the factors involving β in the numerator (5.78) would have a mode at $\hat{\beta}_{21}$, provided $\beta_L \leq \hat{\beta}_{21} \leq \beta_U$. The factor $(1 + \beta^2/\phi)^n$ produces a modal value for β larger in absolute value than $\hat{\beta}_{21}$.[41] The amount by which the modal value for β is increased absolutely will depend on what is assumed about ϕ. If the prior pdf for ϕ favors large values, the modal value for β will be close to $\hat{\beta}_{21}$, abstracting from the information about β included in $p_1(\beta)$.[42]

To use (5.78) in practice prior pdf's for β and ϕ must be assigned. With respect to the prior pdf for β, $p_1(\beta)$, $\beta_L \leq \beta \leq \beta_U$, we can assign a beta pdf of the following form[43]:

$$p(z|a, b) = \frac{1}{B(a, b)} z^{a-1}(1 - z)^{b-1} \Bigg\} \quad \begin{array}{l} a, b > 0, \\ 0 \leq z \leq 1, \end{array} \tag{5.79}$$

where $z = (\beta - \beta_L)/(\beta_U - \beta_L)$, a and b are prior parameters to be assigned by an investigator, and $B(a, b)$ denotes the beta function with arguments a and b. The pdf in (5.79) is a rather rich one that accommodates the prior information that $\beta_L \leq \beta \leq \beta_U$. If, for example, $a = b = 1$, (5.79) gives us a uniform pdf for β. With respect to ϕ, the prior pdf $p_2(\phi)$ might be taken in the following inverted gamma form[44]:

$$p_2(\phi|\nu_0, s_0) \propto \frac{1}{\phi^{\nu_0+1}} \exp\left(-\frac{\nu_0 s_0^2}{2\phi^2}\right), \quad \begin{array}{l} 0 < \phi < \infty, \\ \nu_0, s_0 > 0. \end{array} \tag{5.80}$$

where ν_0 and s_0 are prior parameters.

Although (5.79) and (5.80) are not the only forms of prior pdf's that can be used for the present problem, they appear to be rich enough to be capable of representing available prior information in a wide range of circumstances. Substituting from (5.79) and (5.80) into (5.78), we have a bivariate posterior pdf which can be analyzed using bivariate numerical integration techniques. Marginal posterior pdf's for β and ϕ can be computed along with measures characterizing them. In addition, it is possible to compute joint posterior regions for β and ϕ.

In order to illustrate application of the above techniques, data have be generated from the model

$$y_{1i} = \xi_i + u_{1i}, \qquad y_{2i} = 2.0 + 1.0\xi_i + u_{2i}, \qquad i = 1, 2, \ldots, 20,$$

[41] This can be interpreted heuristically as a correction for the inconsistency of $\hat{\beta}_{21}$.

[42] Note that, given ϕ, the value of β which maximizes $(1 + \beta^2/\phi)/(m_{22} - 2\beta m_{12} + \beta^2 m_{11})$, is the ML estimate. Thus the modal value of (5.78) for given ϕ will be close to the ML estimate if $p_1(\beta) \propto$ constant.

[43] See Appendix A for a review of the properties of the beta pdf.

[44] See Appendix A for properties of the inverted gamma pdf.

under the following conditions. The values of u_{1i} and u_{2i} were drawn independently from normal distributions with zero means and variances 4 and 1, respectively; that is $\phi = \sigma_2^2/\sigma_1^2 = \frac{1}{4}$. The ξ_i's were drawn independently from a normal distribution with mean $\mu = 5$ and variance $\tau^2 = 16$. The 20 pairs of observations are shown in Table 5.1 and plotted in Figure 5.1.

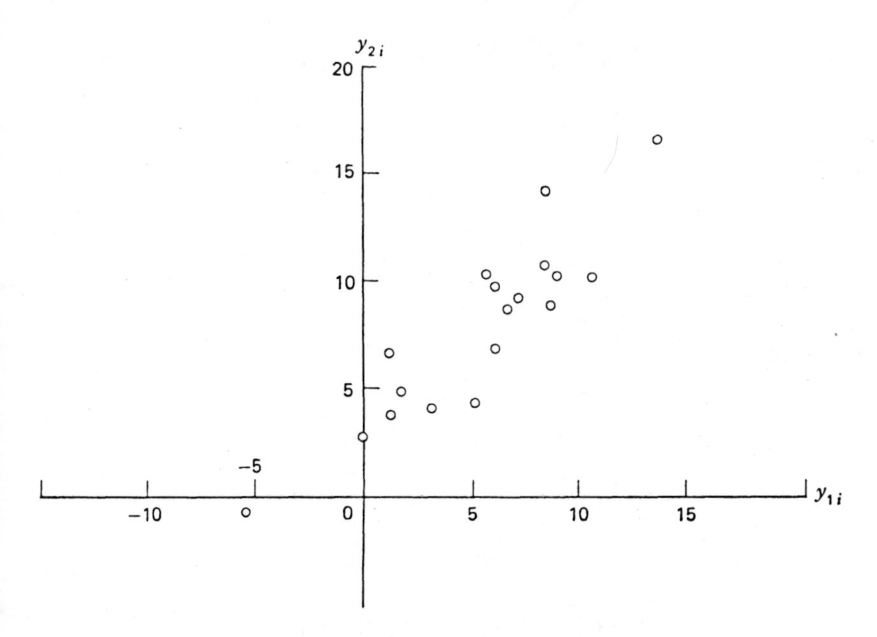

Figure 5.1 Plot of generated data.

Table 5.1 GENERATED OBSERVATIONS

i	y_{1i}	y_{2i}	i	y_{1i}	y_{2i}
1	1.420	3.695	11	1.964	4.987
2	6.268	6.925	12	1.406	6.647
3	8.854	8.923	13	0.002	2.873
4	8.532	14.043	14	3.212	4.015
5	−5.398	−0.836	15	9.042	10.204
6	13.776	16.609	16	1.474	1.953
7	5.278	4.405	17	8.528	10.672
8	6.298	9.823	18	7.348	9.157
9	9.886	12.611	19	6.690	8.552
10	11.362	10.174	20	5.796	10.250

$\bar{y}_1 = 5.587 \qquad \bar{y}_2 = 7.784$

$m_{11} = 19.332 \qquad m_{22} = 17.945 \qquad m_{12} = 16.925$

Using the data in Table 5.1, first conditional posterior pdf's for β, given $\phi = 0.25$ and $\phi = 1.0$, were computed[45] from (5.78) and are shown in Figure 5.2. It is seen that the location and spread of the posterior pdf's are

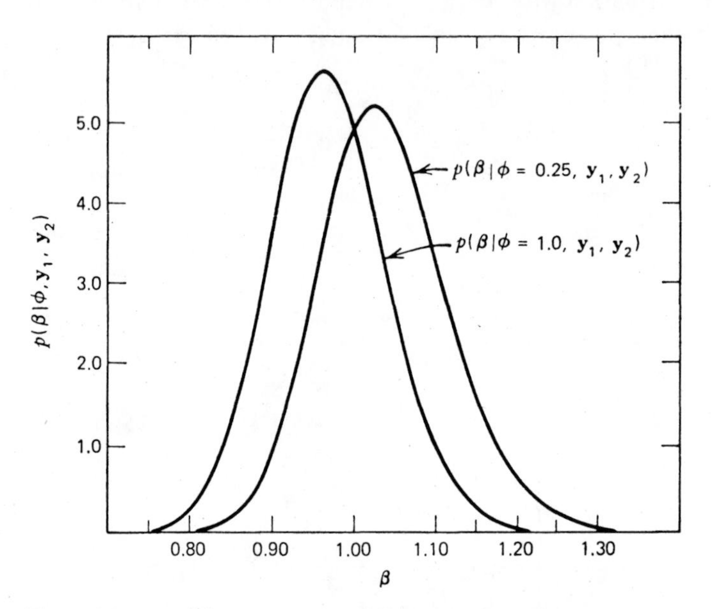

Figure 5.2 Conditional posterior pdf's for β, given ϕ.

not too sensitive to what is assumed about ϕ. In this instance, when ϕ is assumed to be equal to 0.25, the value used to generate the data, the conditional posterior pdf for β, has its mode at $\beta = 1.02$ close to the value used to generate the data, namely, 1.0. With ϕ assumed to be equal to 1.0, the modal value of the conditional posterior pdf for β is located at $\beta = 0.96$. These results contrast with what is obtained from an analysis of the model under the assumption that $\sigma_1^2 = 0$, that is, no measurement error in y_{1i}. In this case, with a diffuse prior pdf on β_0 and β, the posterior pdf for β is centered at $\hat{\beta}_{21} = m_{12}/m_{11} = 0.876$, the least squares quantity.[46]

Although the conditional pdf for β, given ϕ, is of interest, it is often the case that we lack prior information that is precise enough to assign a specific value to ϕ. However, it may be possible to choose a prior pdf to

[45] A uniform prior pdf for β with a rather large range was employed.

[46] A simple regression of y_{2i} on y_{1i} results in

$$\hat{y}_{2i} = \underset{(0.674)}{2.893} + \underset{(0.0948)}{0.876}y_{1i},$$

where the figures in parentheses are conventional standard errors.

represent the information available about ϕ; for example, the inverted gamma pdf, shown in (5.80), could be used for this purpose. To illustrate, let us assign the following values to the parameters of (5.80): $\nu_0 = 8$ and $s_0^2 = \frac{1}{16}$. With these values assigned, the prior pdf for ϕ has its modal value at $\phi = 0.236$, mean equal to 0.246, and variance equal to 0.0230 (standard deviation = 0.152).[47] On inserting this prior pdf for ϕ for $p_2(\phi)$ in (5.78) and using the data shown in Table 5.1, we computed the joint posterior pdf for β and ϕ with a uniform prior for β. Also the marginal posterior pdf's for β and for ϕ were computed. The results of these computations are shown in Figure 5.3. It is seen from the plots in Figure 5.3 that the marginal posterior

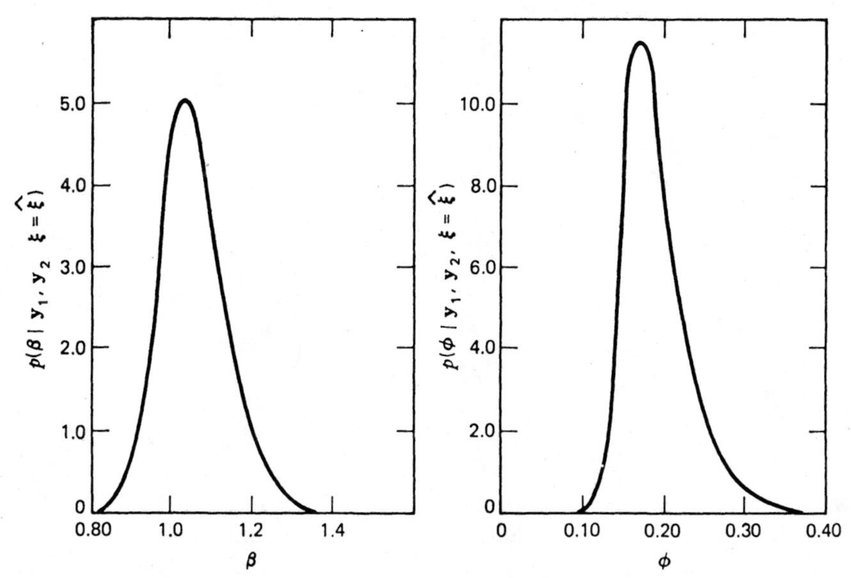

Figure 5.3 Marginal posterior pdf's for β and ϕ based on generated data and utilizing inverted gamma prior pdf for ϕ.

pdf's for β and ϕ are unimodal. As regards the pdf for ϕ, it appears that the information in the particular sample employed has reduced the modal value from 0.236 in the prior pdf to about 0.18 in the posterior pdf. The posterior pdf for β has a modal value at about 1.03, a value close to that used in generating the data.

We have gone ahead conditional on $\boldsymbol{\xi} = \hat{\boldsymbol{\xi}}$, an assumption that obviates the need for a distributional assumption regarding the ξ_i's. On the other hand,

[47] See Appendix A for algebraic expressions for the modal value, mean, variance, etc., of the inverted gamma pdf.

in certain circumstances we may find it appropriate to assume that the n elements of $\boldsymbol{\xi}$ are, a priori, normally and independently distributed with a common mean and variance[48]; that is,

$$p(\boldsymbol{\xi}|\mu, \tau) \propto \frac{1}{\tau^n} \exp\left[-\frac{1}{2\tau^2}(\boldsymbol{\xi} - \mu\boldsymbol{\iota})'(\boldsymbol{\xi} - \mu\boldsymbol{\iota})\right], \tag{5.81}$$

where μ and τ^2 denote the common mean and variance, respectively. If we assume further that μ and τ are to be unknown[49] and distributed a priori as $p(\mu, \tau) \propto$ const, with $-\infty < \mu < \infty$ and $0 < \tau < \infty$, then the marginal prior pdf for $\boldsymbol{\xi}$ is[50]:

$$p(\boldsymbol{\xi}) \propto [(\boldsymbol{\xi} - \bar{\xi}\boldsymbol{\iota})'(\boldsymbol{\xi} - \bar{\xi}\boldsymbol{\iota})]^{-(n-2)/2}, \qquad \begin{array}{c} -\infty < \xi_i < \infty, \\ i = 1, 2, \ldots, n, \end{array} \tag{5.82}$$

where $\bar{\xi} = \boldsymbol{\iota}'\boldsymbol{\xi}/n$. Below we shall combine this prior pdf and prior pdf's for other parameters with the likelihood function.

The likelihood function is

$$\begin{aligned} l(\boldsymbol{\beta}, \phi, \sigma_2, \boldsymbol{\xi}|\mathbf{y}_1, \mathbf{y}_2) &\propto \frac{\phi^{n/2}}{\sigma_2^{2n}} \exp\bigg\{-\frac{1}{2\sigma_2^2}[\phi(\mathbf{y}_1 - \boldsymbol{\xi})'(\mathbf{y}_1 - \boldsymbol{\xi}) \\ &\qquad + (\mathbf{y}_2 - \beta_0\boldsymbol{\iota} - \beta\boldsymbol{\xi})'(\mathbf{y}_2 - \beta_0\boldsymbol{\iota} - \beta\boldsymbol{\xi})]\bigg\} \\ &\propto \frac{\phi^{n/2}}{\sigma_2^{2n}} \exp\bigg\{-\frac{1}{2\sigma_2^2}\bigg[(\phi + \beta^2)(\boldsymbol{\xi} - \hat{\boldsymbol{\xi}})'(\boldsymbol{\xi} - \hat{\boldsymbol{\xi}}) \\ &\qquad + \frac{\phi}{\phi + \beta^2}(\mathbf{y}_2 - \beta_0\boldsymbol{\iota} - \beta\mathbf{y}_1)'(\mathbf{y}_2 - \beta_0\boldsymbol{\iota} - \beta\mathbf{y}_1)\bigg]\bigg\}, \end{aligned} \tag{5.83}$$

where $\boldsymbol{\beta}' = (\beta_0, \beta)$, $\phi = \sigma_2^2/\sigma_1^2$, and $\hat{\boldsymbol{\xi}} = [\phi\mathbf{y}_1 + \beta(\mathbf{y}_2 - \beta_0\boldsymbol{\iota})]/(\phi + \beta^2)$. The second line of (5.83) is obtained from the first by completing the square on $\boldsymbol{\xi}$.[51] Note that $\hat{\boldsymbol{\xi}}$ is the conditional maximum likelihood estimate for $\boldsymbol{\xi}$ used in the conditional Bayesian approach leading to the posterior pdf in (5.77).

As prior pdf for the parameters, we shall employ

$$p(\boldsymbol{\beta}, \phi, \sigma_2, \boldsymbol{\xi}) \propto p_1(\boldsymbol{\beta}, \phi, \sigma_2)\, p_2(\boldsymbol{\xi}), \tag{5.84}$$

[48] If, for example, $\xi_i = \log Y_i^P$, where Y_i^P is the ith individual's permanent income, the prior assumption that the ξ_i's have the normal pdf in (5.81) is, of course, an income distribution assumption.

[49] The parameters μ and τ in (5.81) could be assigned values on a priori grounds. The resulting posterior pdf in this case has not yet been analyzed, nor in the case that an informative prior pdf is employed for μ and τ.

[50] This prior pdf for the elements of $\boldsymbol{\xi}$ is close to one that Stein indicates can be used to generate his estimator for n means when $n \geq 3$. See C. M. Stein, "Confidence Sets for the Mean of a Multivariate Normal Distribution," *op. cit.*, p. 281.

[51] For details see footnote relating to (5.65) above.

where $p_2(\boldsymbol{\xi})$ is given by (5.82) and $p_1(\boldsymbol{\beta}, \phi, \sigma_2)$, not yet specified, is discussed below. Formally, the posterior pdf is

$$p(\boldsymbol{\beta}, \phi, \sigma_2, \boldsymbol{\xi}|\mathbf{y}_1, \mathbf{y}_2) \propto p_1(\boldsymbol{\beta}, \phi, \sigma_2)\, p_2(\boldsymbol{\xi})\, l(\boldsymbol{\beta}, \phi, \sigma_2, \boldsymbol{\xi}|\mathbf{y}_1, \mathbf{y}_2), \tag{5.85}$$

with the likelihood function as shown in (5.83).

On integrating (5.85) with respect to the elements of $\boldsymbol{\xi}$,[52] we obtain

$$\begin{aligned} p(\boldsymbol{\beta}, \phi, \sigma_2|\mathbf{y}_1, \mathbf{y}_2) &\propto p_1(\boldsymbol{\beta}, \phi, \sigma_2)\, \frac{\phi^{n/2}}{\sigma_2^{2n-2}(\phi + \beta^2)} \\ &\quad \times \exp\left[-\frac{1}{2\sigma_2^2}\frac{\phi}{\phi + \beta^2}(\mathbf{y}_2 - \beta_0\boldsymbol{\iota} - \beta\mathbf{y}_1)'(\mathbf{y}_2 - \beta_0\boldsymbol{\iota} - \beta\mathbf{y}_1)\right] \\ &\quad \times e^{-\delta/2}\sum_{\alpha=0}^{\infty}\left(\frac{\delta}{2}\right)^{\alpha}\frac{\Gamma(\alpha + \frac{1}{2})}{\alpha!\,\Gamma[\alpha + (n-1)/2]}, \end{aligned} \tag{5.86}$$

where[53]

$$\delta = \frac{\phi + \beta^2}{\sigma_2^2}\sum_{i=1}^{n}(\hat{\xi}_i - \bar{\hat{\xi}})^2 = \frac{n}{\sigma_2^2(\phi + \beta^2)}(\phi^2 m_{11} + \beta^2 m_{22} + 2\phi\beta m_{12}).$$

As regards the prior pdf $p_1(\boldsymbol{\beta}, \phi, \sigma_2)$ in (5.86), we shall first analyze the case in which β_0, β, ϕ, and σ_2 are assumed to be independently distributed. If our information about β_0 and σ_2 is vague,[54] then the prior pdf can be taken as[55]

$$p_1(\boldsymbol{\beta}, \phi, \sigma_2) \propto \left.\frac{g_1(\beta)\, g_2(\phi)}{\sigma_2}\right\} \quad \begin{array}{c} -\infty < \beta_0 < \infty, \\ 0 < \sigma_2 < \infty, \end{array} \tag{5.87}$$

with $g_1(\beta)$ and $g_2(\phi)$ still unspecified prior pdf's for β and ϕ, respectively.

On substituting from (5.87) into (5.86) and integrating with respect to β_0, the resulting posterior pdf is

$$\begin{aligned} p(\beta, \phi, \sigma_2|\mathbf{y}_1, \mathbf{y}_2) &\propto \frac{g_1(\beta)\, g_2(\phi)\phi^{(n-1)/2}}{\sigma_2^{2n-2}(\phi + \beta^2)^{1/2}} \\ &\quad \times \exp\left\{-\frac{1}{2\sigma_2^2}\frac{\phi}{\phi + \beta^2}\sum_{i=1}^{n}[y_{2i} - \bar{y}_2 - \beta(y_{1i} - \bar{y}_1)]^2\right\} \\ &\quad \times e^{-\delta_1/2\sigma_2^2}\sum_{\alpha=0}^{\infty}\left(\frac{\delta_1}{2\sigma_2^2}\right)^{\alpha}\frac{\Gamma(\alpha + \frac{1}{2})}{\alpha!\,\Gamma(\alpha + (n-1)/2]}, \end{aligned} \tag{5.88}$$

[52] The appendix to this chapter describes this integration in detail.

[53] In the following expressions $\bar{\hat{\xi}} = \sum_{i=1}^{n} \hat{\xi}_i/n$, $m_{11} = \sum_{i=1}^{n} (y_{1i} - \bar{y}_1)^2/n$, $m_{22} = \sum_{i=1}^{n} (y_{2i} - \bar{y}_2)^2/n$ and $m_{12} = \sum_{i=1}^{n} (y_{1i} - \bar{y}_1)(y_{2i} - \bar{y}_2)/n$.

[54] Alternatively, we can pursue the analysis with an informative prior pdf for σ_2 in the inverted gamma form.

[55] Lindley and El-Sayyad, *op. cit.*, also assume that β, ϕ, and σ_2 are independently distributed a priori. In an analysis of the present model they provide valuable discussion and obtain approximations to the posterior pdf.

where $\delta_1/\sigma_2^2 = \delta$, which has been defined above in connection with (5.86). On integrating (5.88) termwise with respect to σ_2, $0 < \sigma_2 < \infty$, the result is

$$p(\beta, \phi|\mathbf{y}_1, \mathbf{y}_2) \propto \frac{g_1(\beta)\, g_2(\phi)\phi^{(n-1)/2}}{(\phi + \beta^2)^{1/2}} \frac{1}{A^{n-3/2}} \sum_{\alpha=0}^{\infty} \left(\frac{\delta_1}{A}\right)^{\alpha} d_\alpha \tag{5.89}$$

with

$$A = \frac{\phi}{\phi + \beta^2} \sum_{i=1}^{n} [y_{2i} - \bar{y}_2 - \beta(y_{1i} - \bar{y}_1)]^2 + \delta_1$$

and

$$d_\alpha = \frac{\Gamma(\alpha + n - \frac{3}{2})\Gamma(\alpha + \frac{1}{2})}{\alpha!\, \Gamma[\alpha + (n-1)/2]}.$$

By straightforward algebra it is the case that

$$A = n(m_{22} + \phi m_{11}) \quad \text{and} \quad \frac{\delta_1}{A} = \frac{\phi^2 m_{11} + \beta^2 m_{22} + 2\phi\beta m_{12}}{(m_{22} + \phi m_{11})(\phi + \beta^2)}.$$

Thus (5.89) can be expressed as[56]

$$\begin{aligned} p(\beta, \phi|\mathbf{y}_1, \mathbf{y}_2) &\propto \frac{g_1(\beta)\, g_2(\phi)}{(\phi + \beta^2)^{1/2}} \frac{\phi^{(n-1)/2}}{(m_{22} + \phi m_{11})^{n-3/2}} \\ &\quad \times \sum_{\alpha=0}^{\infty} \left[\frac{\phi^2 m_{11} + \beta^2 m_{22} + 2\phi\beta m_{12}}{(m_{22} + \phi m_{11})(\phi + \beta^2)}\right]^{\alpha} d_\alpha. \end{aligned} \tag{5.90}$$

Given explicit forms for the prior pdf's $g_1(\beta)$ and $g_2(\phi)$, bivariate numerical integration techniques can be employed to obtain the normalizing constant for (5.90), the marginal posterior pdf's for β and ϕ, and measures pertaining to the joint and marginal posterior pdf's; for example, we might employ a beta prior pdf for β and an inverted gamma or F pdf for ϕ. Since numerical methods are employed in the analysis of (5.90), the choice of prior pdf's for β and for ϕ is not very restricted.

The bivariate posterior pdf for β and ϕ in (5.90) was analyzed by using the generated data presented in Table 5.1. The prior pdf for β, $g_1(\beta)$, was taken to be uniform over a large range, whereas the prior pdf for ϕ, $g_2(\phi)$, was taken in the inverted gamma form [see (5.80)], with prior parameter values $\nu_0 = 8$ and $s_0^2 = \frac{1}{16}$, the same values employed in the calculations underlying Figure 5.3. In Figure 5.4 the marginal posterior pdf for β is presented.[57] It is centered close to 1.0. Further note that it is more spread out than the

[56] Note that $\delta_1/A = (\phi^2 m_{11} + \beta^2 m_{22} + 2\phi\beta m_{12})/(m_{22} + \phi m_{11})(\phi + \beta^2)$, for given ϕ, has a maximum at $\beta = \tilde{\beta}$, where $\tilde{\beta}$ is the ML estimate; that is, for given ϕ $(d/d\beta)(\delta_1/A) = (m_{22} + \phi m_{11})^{-1}[2(\beta m_{22} + \phi m_{12})/(\phi + \beta^2) - 2\beta(\phi^2 m_{11} + \beta^2 m_{22} + 2\phi\beta m_{12})/(\phi + \beta^2)^2]$. On setting this derivative equal to zero, the necessary condition for a maximum is $\beta^2 m_{12} + \beta(\phi m_{11} - m_{22}) - \phi m_{12} = 0$, which is identical to (5.49).

[57] In these calculations the first 501 terms in the series shown in (5.86) were employed.

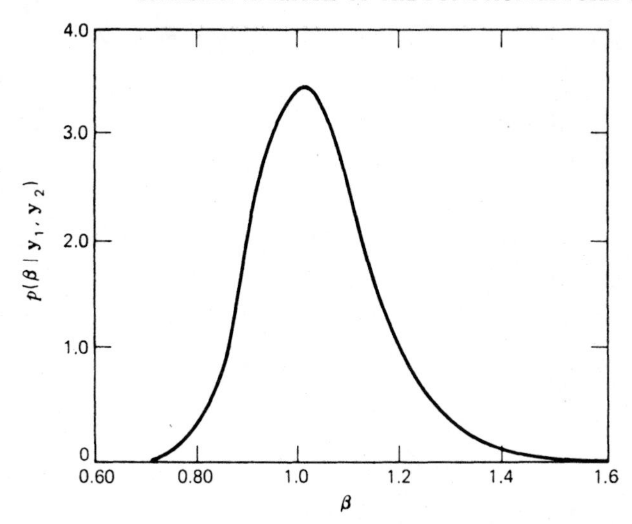

Figure 5.4 Marginal posterior pdf for β computed from (5.86) and generated data. See text for explanation of the prior information utilized.

posterior pdf for β, given $\boldsymbol{\xi} = \hat{\boldsymbol{\xi}}$, which is shown in Figure 5.3. The integration over the elements of $\boldsymbol{\xi}$ appears to be responsible for the greater spread of the posterior pdf in the present case.

As an alternative to the prior assumptions in (5.87), wherein it is postulated that ϕ and σ_2 are a priori independent, it may be appropriate in some situations to assume that σ_1 and σ_2 are independently distributed a priori[58]; for example, if we assume that our prior information about these two independent parameters can be represented by inverted gamma pdf's,

$$(5.91) \quad p(\sigma_i|\nu_i, s_i) \propto \frac{1}{\sigma_i^{\nu_i+1}} \exp\left(-\frac{\nu_i s_i^2}{2\sigma_i^2}\right), \qquad 0 < \sigma_i < \infty, \qquad i = 1, 2,$$

where (ν_i, s_i), $i = 1, 2$, are prior parameters, then on transforming to $\phi = \sigma_2^2/\sigma_1^2$ and σ_2, we have

$$(5.92) \qquad g_3(\phi, \sigma_2) \propto \frac{\phi^{(\nu_1-2)/2}}{\sigma_2^{\nu_1+\nu_2+1}} \exp\left[-\frac{1}{2\sigma_2^2}(\phi\nu_1 s_1^2 + \nu_2 s_2^2)\right]$$

[58] This was assumed in Zellner and Sankar, *op. cit.*, and in R. L. Wright, "A Bayesian Analysis of Linear Functional Relations," manuscript, University of Michigan, 1969. If, for example, different measuring instruments were employed to generate the y_{1i}'s and the y_{2i}'s, the assumption that σ_1 and σ_2 are independent may be appropriate. In the context of Friedman's permanent income hypothesis, if individuals who chose occupations with a high variance of transitory income also tend to have a high variance of transitory consumption, it is not appropriate to assume that σ_1^2 and σ_2^2 are independent a priori. Here it is probably better to assume that $\phi = \sigma_2^2/\sigma_1^2$ and σ_2^2 are independent, as above.

as the joint prior pdf. It is seen that the conditional prior pdf for σ_2, given ϕ, is in the inverted gamma form. Also, the marginal pdf for $(s_1^2/s_2^2)\phi$ is in the form of a Fisher-Snedecor F pdf with ν_1 and ν_2 degrees of freedom.[59] Then, if in place of (5.87) we use the prior pdf

$$p_1(\boldsymbol{\beta}, \phi, \sigma_2) \propto g_1(\beta)\, g_3(\phi, \sigma_2), \tag{5.93}$$

with $g_3(\phi, \sigma_2)$, as shown in (5.92) and $g_1(\beta)$ not yet specified, we can substitute from (5.93) into (5.86) and perform integrations with respect to β_0, $-\infty < \beta_0 < \infty$, and σ_2, $0 < \sigma_2 < \infty$, in much the same way as shown above. The resulting posterior pdf for β and ϕ is

$$p(\beta, \phi | \mathbf{y}_1, \mathbf{y}_2) \propto \frac{g_1(\beta)\phi^{(n+\nu_1-3)/2}}{(\phi + \beta^2)^{1/2}} \frac{1}{B^{n+(\nu_0-3)/2}} \sum_{\alpha=0}^{\infty} \left(\frac{\delta_1}{B}\right)^{\alpha} d_\alpha', \tag{5.94}$$

with $\nu_0 = \nu_1 + \nu_2$,

$$d_\alpha' = \frac{\Gamma[(\alpha + n + (\nu_0 - 3)/2]\Gamma(\alpha + \frac{1}{2})}{\alpha!\, \Gamma[\alpha + (n-1)/2]},$$

$$B = A + \phi\nu_1 s_1^2 + \nu_2 s_2^2 = n\left[\frac{(nm_{22} + \nu_2 s_2^2)}{n} + \frac{\phi(nm_{11} + \nu_1 s_1^2)}{n}\right]$$

and

$$\frac{\delta_1}{B} = \frac{\phi^2 m_{11} + \beta^2 m_{22} + 2\phi\beta m_{12}}{(\phi + \beta^2)[(nm_{22} + \nu_2 s_2^2)/n + \phi(nm_{11} + \nu_1 s_1^2)/n]},$$

where δ_1 and A have been defined in connection with (5.88) and (5.89). Thus we can write the joint posterior pdf for β and ϕ as follows:

$$\begin{aligned} p(\beta, \phi | \mathbf{y}_1, \mathbf{y}_2) \propto & \frac{g_1(\beta)}{(\phi + \beta^2)^{1/2}} \frac{\phi^{(n+\nu_1-3)/2}}{(a_0 + a_1\phi)^{(2n+\nu_0-3)/2}} \\ & \times \sum_{\alpha=0}^{\infty} \left[\frac{\phi^2 m_{11} + \beta^2 m_{22} + 2\phi\beta m_{12}}{(\phi + \beta^2)(a_0 + a_1\phi)}\right]^{\alpha} d_\alpha', \end{aligned} \tag{5.95}$$

where $a_0 = (nm_{22} + \nu_2 s_2^2)/n$ and $a_1 = (nm_{11} + \nu_1 s_1^2)/n$.[60]

As mentioned above, for given ϕ the quantity $(\phi^2 m_{11} + \beta^2 m_{22} + 2\phi\beta m_{12})/(\phi + \beta^2)$ has a maximum at the ML estimate. Thus for given ϕ (5.95) will have a conditional modal value for β close to the ML estimate. If, however, we integrate (5.95) with respect to ϕ, both the prior and sample information regarding ϕ will play a role in determining the location of the marginal posterior pdf for β.

[59] See the section in Appendix A dealing with the Fisher-Snedecor F pdf for a proof.
[60] In (5.95) the factors $\phi^{(n+\nu_1-3)/2}/(a_0 + a_1\phi)^{(2n+\nu_0-3)/2+\alpha}$, $\alpha = 0, 1, 2, \ldots$, are in the form of F pdf's with $\nu_1 + n - 1$ and $\nu_2 + n - 2 + 2\alpha$ degrees of freedom. Note that the prior pdf for ϕ, obtainable from (5.92), is in the F form with ν_1 and ν_2 degrees of freedom.

5.5 BAYESIAN ANALYSIS OF THE STRUCTURAL FORM OF THE EVM

In the structural form of the EVM the vector $\boldsymbol{\xi}$ is regarded as random in formulating the likelihood function; that is, the pdf for $\boldsymbol{\xi}$ is introduced as part of the model and not as subjective prior information. The analysis shown in (5.63) to (5.64) is relevant for obtaining the marginal pdf for the observations $\mathbf{y}_1$ and $\mathbf{y}_2$, which serves as a basis for the likelihood function. If the elements of $\boldsymbol{\xi}$ are assumed to be normally and independently distributed, each with unknown mean μ and unknown variance τ^2, then with a prior pdf for these two parameters and other parameters of the model the formal analysis goes forward in exactly the same way as in the preceding section. Similarly, a conditional analysis based on the assumption that $\boldsymbol{\xi} = \tilde{\boldsymbol{\xi}}$ is possible in the present case, and it too will be identical in form to that presented in the preceding section. To repeat, the main difference in analyzing the functional and structural forms of the model lies in the interpretation given to the pdf for the elements of $\boldsymbol{\xi}$.[61]

5.6 ALTERNATIVE ASSUMPTION ABOUT THE INCIDENTAL PARAMETERS[62]

In this section we analyze the EVM under the assumption that the incidental parameters, the elements of $\boldsymbol{\xi}$, can be represented by a linear combination of observable independent variables; that is $\xi_i = \pi_1 + \pi_2 x_{2i} + \cdots + \pi_k x_{ki}$, $i = 1, 2, \ldots, n$, where the x's are given observed values of independent variables and the $k\pi$'s are unknown parameters.[63] The key importance of this assumption is that the number of parameters in the model is no longer dependent on the sample size n; that is, the n unknown ξ_i's are replaced by k unknown π's and thus complications associated with the incidental

[61] Similar considerations apply in general to comparisons of sampling theory random parameter models and to Bayesian analyses of corresponding fixed parameter models. See, for example, P. A. V. B. Swamy, *Statistical Inference in Random Coefficient Regression Models*, doctoral dissertation, University of Wisconsin, Madison, 1968, p. 75 ff., for further discussion of this point in the context of regression models.

[62] The material in this section is based on A. Zellner, "Estimation of Regression Relationships Containing Unobservable Independent Variables," *Int. Econ. Rev.*, **11**, 441-454 (1970).

[63] This assumption has been utilized in many econometric studies which include J. Crockett, "Technical Note," in I. Friend and R. Jones (Eds.), *Proc. Conf. Consumption and Saving*, Vol. II. Philadelphia: University of Pennsylvania, 1960, 213–222, and M. H. Miller and F. Modigliani, "Some Estimates of the Cost of Capital to the Electric Utility Industry," 1954–1957, *Am. Econ. Rev.*, **56**, 333–391 (1966).

parameter problem, some of which have been analyzed above, are circumvented. Our model then is

$$\mathbf{y}_1 = \boldsymbol{\xi} + \mathbf{u}_1, \tag{5.96}$$

$$\mathbf{y}_2 = \beta\boldsymbol{\xi} + \mathbf{u}_2, \tag{5.97}$$

$$\boldsymbol{\xi} = X\boldsymbol{\pi}_1 \tag{5.98}$$

or

$$\mathbf{y}_1 = X\boldsymbol{\pi}_1 + \mathbf{u}_1, \tag{5.99}$$

$$\mathbf{y}_2 = X\boldsymbol{\pi}_1\beta + \mathbf{u}_2, \tag{5.100}$$

where $\mathbf{y}_1$ and $\mathbf{y}_2$ are $n \times 1$ vectors of observations, X is an $n \times k$ matrix, with rank k, of observations on k independent variables, $\boldsymbol{\pi}_1$ is a $k \times 1$ vector of parameters, β is a scalar parameter,[64] and $\mathbf{u}_1$ and $\mathbf{u}_2$ are each $n \times 1$ vectors of error terms. We assume that $E\mathbf{u}_1 = E\mathbf{u}_2 = \mathbf{0}$, $E\mathbf{u}_1\mathbf{u}_1' = \sigma_1^2 I_n$, $E\mathbf{u}_2\mathbf{u}_2' = \sigma_2^2 I_n$ and $E\mathbf{u}_1\mathbf{u}_2' = 0$, an $n \times n$ matrix with zero elements; that is, we assume that the error terms have zero means and are homoscedastic and uncorrelated. Later on we shall require a normality assumption for the error terms.

Before turning to a Bayesian analysis of the model in (5.99) to (5.100), it would be instructive to consider sampling theory approaches which have been employed quite widely, particularly a so-called "instrumental variable" approach.[65] In this approach it is generally recognized that if the elements of the vector $\boldsymbol{\pi}_1$ in (5.100) had *known* values (5.100) would be in the form of a simple regression model, the analysis of which would be straightforward. Since $\boldsymbol{\pi}_1$'s value is rarely known in practice, (5.99) is employed to obtain an estimate of $\boldsymbol{\pi}_1$, $\hat{\boldsymbol{\pi}}_1 = (X'X)^{-1}X'\mathbf{y}_1$, and then $\mathbf{y}_2$ is regressed on $\hat{\mathbf{y}}_1 = X\hat{\boldsymbol{\pi}}_1$ to obtain the following estimator for β:

$$\hat{\beta}_{(\infty)} = \frac{\mathbf{y}_2'\hat{\mathbf{y}}_1}{\hat{\mathbf{y}}_1'\hat{\mathbf{y}}_1'} = \frac{\mathbf{y}_2'\hat{\mathbf{y}}_1}{\hat{\mathbf{y}}_1'\hat{\mathbf{y}}_1}, \tag{5.101}$$

where $\hat{\mathbf{y}}_2 = X\hat{\boldsymbol{\pi}}_2 = X(X'X)^{-1}X'\mathbf{y}_2$ and the reason for denoting the estimator $\hat{\beta}_{(\infty)}$ will be made clear in what follows. This estimator, often referred to as a "two-stage least squares" estimator, is thus seen to be one formed conditional on $X\boldsymbol{\pi}_1$ assumed equal to $X\hat{\boldsymbol{\pi}}_1$.

[64] In (5.100) we have suppressed the intercept term. Below we shall take up the analysis of the model with (5.100) expanded to include an intercept term and additional observable independent variables with unknown coefficients.

[65] See, for example, A. S. Goldberger, *Econometric Theory*. New York: Wiley, 1964, 284–287; and F. D. Carlson, E. Sobel, and G. S. Watson, "Linear Relationships between Variables Affected by Errors," *Biometrics*, **22**, 252–267 (1966).

If, in (5.99) to (5.100), we define $\boldsymbol{\pi}_2 = \boldsymbol{\pi}_1\beta$, the system becomes

$$\mathbf{y}_1 = X\boldsymbol{\pi}_2/\beta + \mathbf{u}_1 \tag{5.102}$$

$$\mathbf{y}_2 = X\boldsymbol{\pi}_2 + \mathbf{u}_2. \tag{5.103}$$

Since the vector $X\boldsymbol{\pi}_2$ has elements whose values are unknown, (5.103) can be employed to get an estimate, namely, $X\hat{\boldsymbol{\pi}}_2$, where $\hat{\boldsymbol{\pi}}_2 = (X'X)^{-1}X'\mathbf{y}_2$, and this can be inserted in (5.102). Then an estimator for β is obtained by regressing $\mathbf{y}_1$ on $\hat{\mathbf{y}}_2 = X\hat{\boldsymbol{\pi}}_2$ and taking the reciprocal of the slope coefficient estimator; that is,

$$\hat{\beta}_{(0)} = \frac{\hat{\mathbf{y}}_2'\hat{\mathbf{y}}_2}{\hat{\mathbf{y}}_2'\mathbf{y}_1} = \frac{\hat{\mathbf{y}}_2'\hat{\mathbf{y}}_2}{\hat{\mathbf{y}}_2'\hat{\mathbf{y}}_1}, \tag{5.104}$$

where $\hat{\mathbf{y}}_1 = X\hat{\boldsymbol{\pi}}_1$.

Some remarks regarding the estimators $\hat{\beta}_{(\infty)}$ and $\hat{\beta}_{(0)}$ follow:

1. Both $\hat{\beta}_{(\infty)}$ and $\hat{\beta}_{(0)}$ are consistent estimators of β.[66]
2. In small samples the distributions of $\hat{\beta}_{(\infty)}$ and $\hat{\beta}_{(0)}$ will be different[67]; for example, under a normality assumption for the error terms $\hat{\beta}_{(\infty)}$ can have a finite mean, whereas in general the mean and higher moments of $\hat{\beta}_{(0)}$ do not exist.[68]
3. When the mean of $\hat{\beta}_{(\infty)}$ exists, as pointed out in the literature, $\hat{\beta}_{(\infty)}$ is biased toward zero; that is $|E\hat{\beta}_{(\infty)}| < |\beta|$. Heuristically, the bias arises because substitution of $X\hat{\boldsymbol{\pi}}_1$ for $X\boldsymbol{\pi}_1$ in (5.100) introduces measurement error in the independent variable for finite sample size.
4. In connection with (5.99) and (5.100) the estimator $\hat{\boldsymbol{\pi}}_1 = (X'X)^{-1}X'\mathbf{y}_1$ uses just the n observations in $\mathbf{y}_1$. Since $\mathbf{y}_2$ contains sample information relating to $\boldsymbol{\pi}_1$, it is the case that not all the sample information is being employed in the estimation of $\boldsymbol{\pi}_1$ when we use $\hat{\boldsymbol{\pi}}_1$ as an estimator.
5. The fact that $\hat{\beta}_{(\infty)}$ and $\hat{\beta}_{(0)}$ are different estimators means that in any practical case numerical results will depend on which one is used.

As an alternative to the approach leading to $\hat{\beta}_{(\infty)}$ and $\hat{\beta}_{(0)}$ as estimators for

[66] Note $\hat{\beta}_{(\infty)} = \hat{\boldsymbol{\pi}}_2'X'X\hat{\boldsymbol{\pi}}_1/\hat{\boldsymbol{\pi}}_1'X'X\hat{\boldsymbol{\pi}}_1$. Since plim $\hat{\boldsymbol{\pi}}_2 = \boldsymbol{\pi}_1\beta$ and plim $\hat{\boldsymbol{\pi}}_1 = \boldsymbol{\pi}_1$, plim $\hat{\beta}_{(\infty)} = \beta$. Similarly, plim $\hat{\beta}_{(0)} =$ plim $(\hat{\boldsymbol{\pi}}_2'X'X\hat{\boldsymbol{\pi}}_2/\hat{\boldsymbol{\pi}}_2'X'X\hat{\boldsymbol{\pi}}_1) = \beta$. It is assumed that $\lim_{n\to\infty} X'X/n = \bar{M}$, a matrix with finite elements.

[67] See D. H. Richardson, "The Exact Distribution of a Structural Coefficient Estimator," *J. Am. Statistical Assn.*, **63**, 1214–1226 (1968), T. Sawa, "The Exact Sampling Distribution of Ordinary Least Squares and Two-Stage Least Squares Estimators," *J. Am. Statistical Assn.*, **64**, 923–937 (1969); and the references cited in these works for analysis bearing on the distribution of $\hat{\beta}_{(\infty)}$.

[68] From (5.104), for given $\hat{\mathbf{y}}_2$, $\hat{\beta}_{(0)}$ is in the form of the reciprocal of a normal variable, $\hat{\mathbf{y}}_2'\hat{\mathbf{y}}_1$.

β, consider the following least squares approach. Our sum of squares (SS) to be minimized with respect to $\boldsymbol{\pi}_1$ and β is

$$\text{(5.105)} \quad \begin{aligned} \text{SS} &= \frac{1}{\sigma_1^2}(\mathbf{y}_1 - X\boldsymbol{\pi}_1)'(\mathbf{y}_1 - X\boldsymbol{\pi}_1) + \frac{1}{\sigma_2^2}(\mathbf{y}_2 - X\boldsymbol{\pi}_1\beta)'(\mathbf{y}_2 - X\boldsymbol{\pi}_1\beta) \\ &= \frac{1}{\sigma_2^2}[\phi(\mathbf{y}_1 - X\boldsymbol{\pi}_1)'(\mathbf{y}_1 - X\boldsymbol{\pi}_1) + (\mathbf{y}_2 - Z\boldsymbol{\pi}_1)'(\mathbf{y}_2 - Z\boldsymbol{\pi}_1)], \end{aligned}$$

where $\phi = \sigma_2^2/\sigma_1^2$ and $Z = X\beta$.[69] If we complete the square on $\boldsymbol{\pi}_1$ in (5.105), we are led to

(5.106)
$$\text{SS} = \frac{1}{\sigma_2^2}[\phi\mathbf{y}_1'\mathbf{y}_1 + \mathbf{y}_2'\mathbf{y}_2 - \mathbf{v}'M^{-1}\mathbf{v} + (\boldsymbol{\pi}_1 - M^{-1}\mathbf{v})'M(\boldsymbol{\pi}_1 - M^{-1}\mathbf{v})],$$

where $M = Z'Z + \phi X'X$ and $\mathbf{v} = \phi X'\mathbf{y}_1 + Z'\mathbf{y}_2$. From the form of (5.106) and the fact that M is positive definite, the conditional minimizing value for $\boldsymbol{\pi}_1$ is given by

$$\text{(5.107)} \quad \begin{aligned} \tilde{\boldsymbol{\pi}}_1 &= (Z'Z + \phi X'X)^{-1}(\phi X'\mathbf{y}_1 + Z'\mathbf{y}_2) \\ &= \frac{1}{\phi + \beta^2}(\phi\hat{\boldsymbol{\pi}}_1 + \beta\hat{\boldsymbol{\pi}}_2), \end{aligned}$$

where $\hat{\boldsymbol{\pi}}_i = (X'X)^{-1}X'\mathbf{y}_i$, $i = 1, 2$. Since $\hat{\boldsymbol{\pi}}_2$ estimates $\boldsymbol{\pi}_1\beta$, we can write (5.107) as

$$\text{(5.108)} \quad \tilde{\boldsymbol{\pi}}_1 = \frac{1}{1 + \beta^2/\phi}\left(\hat{\boldsymbol{\pi}}_1 + \frac{\beta^2}{\phi}\frac{\hat{\boldsymbol{\pi}}_2}{\beta}\right).$$

Thus $\tilde{\boldsymbol{\pi}}_1$ is a weighted average of two estimators for $\boldsymbol{\pi}_1$, namely, $\hat{\boldsymbol{\pi}}_1$ and $\hat{\boldsymbol{\pi}}_2/\beta$. For given β, if $\phi \to \infty$, $\tilde{\boldsymbol{\pi}}_1 \to \hat{\boldsymbol{\pi}}_1$, the value used above to construct the estimator $\hat{\beta}_{(\infty)}$, whereas, if $\phi \to 0$, $\tilde{\boldsymbol{\pi}}_1 \to \hat{\boldsymbol{\pi}}_2/\beta$, and we use just the information in $\mathbf{y}_2$ to get an estimate of $\boldsymbol{\pi}_1$ as in forming the estimator $\hat{\beta}_{(0)}$ above. In general, $\tilde{\boldsymbol{\pi}}_1$ in (5.108) utilizes the information in both $\mathbf{y}_1$ and $\mathbf{y}_2$, $2n$ observations, and is thus a more precise estimator for $\boldsymbol{\pi}_1$ than those based on just n observations.[70]

On putting $\boldsymbol{\pi}_1 = \tilde{\boldsymbol{\pi}}_1 = M^{-1}\mathbf{v}$ in (5.106), we have

$$\text{(5.109)} \quad \text{SS} = \frac{1}{\sigma_2^2}\left[s_{22} + \phi s_{11} + \frac{\phi(\hat{\mathbf{y}}_2 - \beta\hat{\mathbf{y}}_1)'(\hat{\mathbf{y}}_2 - \beta\hat{\mathbf{y}}_1)}{\phi + \beta^2}\right],$$

[69] Note that if the elements of $\mathbf{u}_1$ and $\mathbf{u}_2$ are normally distributed the likelihood function is proportional to $\phi^{n/2}/\sigma_2^{2n}e^{-\text{SS}/2}$. Thus the values of the parameters minimizing SS will maximize the likelihood function.

[70] Since $E\hat{\boldsymbol{\pi}}_1 = \boldsymbol{\pi}_1$ and $E\hat{\boldsymbol{\pi}}_2 = \boldsymbol{\pi}_1\beta$, $E\tilde{\boldsymbol{\pi}}_1 = \boldsymbol{\pi}_1$, given β and ϕ. Also, $E(\tilde{\boldsymbol{\pi}}_1 - \boldsymbol{\pi}_1)(\tilde{\boldsymbol{\pi}}_1 - \boldsymbol{\pi}_1)' = (1 + \beta^2/\phi)^{-1}\sigma_1^2(X'X)^{-1}$, whereas $E(\hat{\boldsymbol{\pi}}_1 - \boldsymbol{\pi}_1)(\hat{\boldsymbol{\pi}}_1 - \boldsymbol{\pi}_1)' = \sigma_1^2(X'X)^{-1}$. Thus, when β^2/ϕ is large, the elements of $\tilde{\boldsymbol{\pi}}_1$ have much smaller variances than those of $\hat{\boldsymbol{\pi}}_1$ for given β and ϕ.

where $\hat{\mathbf{y}}_i = X\hat{\boldsymbol{\pi}}_i$ and $s_{ii} = (\mathbf{y}_i - X\hat{\boldsymbol{\pi}}_i)'(\mathbf{y}_i - X\hat{\boldsymbol{\pi}}_i)$ for $i = 1, 2$. Differentiating (5.109) with respect to β and setting the derivative equal to zero, we obtain

$$\beta^2\hat{\mathbf{y}}_1'\hat{\mathbf{y}}_2 + \beta(\phi\hat{\mathbf{y}}_1'\hat{\mathbf{y}}_1 - \hat{\mathbf{y}}_2'\hat{\mathbf{y}}_2) - \phi\hat{\mathbf{y}}_1'\hat{\mathbf{y}}_2 = 0 \tag{5.110}$$

as the necessary condition on β for SS to be minimized. It is to be noted that (5.110) is in exactly the same form as the necessary condition arising in the classical EVM, except that here sample moments are in terms of "computed" values, $\hat{\mathbf{y}}_1$ and $\hat{\mathbf{y}}_2$. Solving the quadratic in (5.110), we are led to

$$\hat{\beta}_{(\phi)} = \frac{\hat{\mathbf{y}}_2'\hat{\mathbf{y}}_2 - \phi\hat{\mathbf{y}}_1'\hat{\mathbf{y}}_1 + \sqrt{(\phi\hat{\mathbf{y}}_1'\hat{\mathbf{y}}_1 - \hat{\mathbf{y}}_2'\hat{\mathbf{y}}_2)^2 + 4\phi(\hat{\mathbf{y}}_1'\hat{\mathbf{y}}_2)^2}}{2\hat{\mathbf{y}}_1'\hat{\mathbf{y}}_2} \tag{5.111}$$

as our estimator for β, given ϕ. Since we can estimate ϕ in the present problem, that is,[71]

$$\hat{\phi} = \frac{s_{22}}{s_{11}}, \tag{5.112}$$

an estimator for β, $\hat{\beta}_{(\hat{\phi})}$ can be obtained by setting $\phi = \hat{\phi}$ in (5.111). Also $\hat{\phi}$ and $\hat{\beta}_{(\hat{\phi})}$ can be substituted in (5.108) to give us a computable estimator for $\boldsymbol{\pi}_1$.[72]

From (5.109) we see that if $\phi/\beta^2 \to \infty$ the minimizing value for $\beta \to \hat{\beta}_{(\infty)} = \hat{\mathbf{y}}_2'\hat{\mathbf{y}}_1/\hat{\mathbf{y}}_1'\hat{\mathbf{y}}_1$. On the other hand, if $\phi/\beta^2 \to 0$, the minimizing value for $\beta \to \hat{\beta}_{(0)} = \hat{\mathbf{y}}_2'\hat{\mathbf{y}}_2/\hat{\mathbf{y}}_1'\hat{\mathbf{y}}_2$. For any given sample it is not difficult to show that

$$|\hat{\beta}_{(\infty)}| \leq |\hat{\beta}_{(\hat{\phi})}| \leq |\hat{\beta}_{(0)}|. \tag{5.113}$$

Also, as $n \to \infty$, all three estimators converge to the true value β.[73]

The system in (5.99) and (5.100) can be expanded to include an intercept term and other observable independent variables:

$$\mathbf{y}_1 = X\boldsymbol{\pi}_1 + \mathbf{u}_1, \tag{5.114}$$

$$\mathbf{y}_2 = X\boldsymbol{\pi}_1\beta + W\boldsymbol{\theta} + \mathbf{u}_2, \tag{5.115}$$

where W is an $n \times k_1$ matrix, with rank k_1, of observations on k_1 independent variables[74] and $\boldsymbol{\theta}$ is a $k_1 \times 1$ vector of unknown parameters. As above, we consider minimization of

[71] With the error terms normally distributed, $\hat{\phi}$ is a maximum likelihood estimator for $\phi = \sigma_2^2/\sigma_1^2$.

[72] The results of some sampling experiments relating to these estimators have been reported in P. R. Brown, *Some Aspects of Valuation in the Railroad Industry*, doctoral dissertation, University of Chicago, 1968.

[73] In small samples, if ϕ/β^2 is very large or very small, $\hat{\beta}_{(\infty)}$ and $\hat{\beta}_{(0)}$ have distributions centered at values which are quite far apart.

[74] W could be a submatrix of X; that is $X = (X_1 \vdots W)$, where X_1 is of size $n \times k - k_1$.

(5.116)

$$\text{SS} = \frac{1}{\sigma_2^2}[\phi(\mathbf{y}_1 - X\boldsymbol{\pi}_1)'(\mathbf{y}_1 - X\boldsymbol{\pi}_1) + (\mathbf{y}_2 - Z\boldsymbol{\pi}_1 - W\boldsymbol{\theta})'(\mathbf{y}_2 - Z\boldsymbol{\pi}_1 - W\boldsymbol{\theta})],$$

with respect to $\boldsymbol{\pi}_1$, $\boldsymbol{\theta}$, and β, where $Z = X\beta$. Differentiating with respect to $\boldsymbol{\theta}$, we obtain as the conditional minimizing value of $\boldsymbol{\theta}$

$$\hat{\boldsymbol{\theta}} = (W'W)^{-1}W'(\mathbf{y}_2 - Z\boldsymbol{\pi}_1), \tag{5.117}$$

which when substituted in (5.116) yields

$$\text{SS}_1 = \frac{1}{\sigma_2^2}[\phi(\mathbf{y}_1 - X\boldsymbol{\pi}_1)'(\mathbf{y}_1 - X\boldsymbol{\pi}_1) + (\mathbf{y}_2^* - Z^*\boldsymbol{\pi}_1)'(\mathbf{y}_2^* - Z^*\boldsymbol{\pi}_1)], \tag{5.118}$$

where $\mathbf{y}_2^* - Z^*\boldsymbol{\pi}_1 = [I - W(W'W)^{-1}W'](\mathbf{y}_2 - Z\boldsymbol{\pi}_1)$. Since (5.118) is in precisely the form of (5.105), the same steps can be performed to obtain minimizing values for $\boldsymbol{\pi}_1$ and β. The resulting estimator for β will depend on ϕ, and, as above, we can use an estimate of ϕ to obtain a computable estimate for β. Estimates of β and ϕ, so obtained, can be used to obtain an estimate for $\boldsymbol{\pi}_1$, which can be substituted in (5.117) to obtain an estimate of $\boldsymbol{\theta}$.

In the sampling theory results above we have seen that results depend critically on the size of $\phi = \sigma_2^2/\sigma_1^2$.[75] The least squares approach pursued leads to estimators which depend on ϕ. Fortunately in the present model an estimate of ϕ can be obtained from the data which can be employed to get an approximation to the least squares estimator[76] for β. In the Bayesian approach it will be seen that the quantity $\hat{\beta}_{(\hat{\phi})}$ is close to the modal value of the conditional posterior pdf for β when we use a diffuse prior pdf and assume that $\phi = \hat{\phi}$.

In the Bayesian analysis of the model in (5.99) and (5.100), in addition to other assumptions made about the error terms, we assume that they are normally distributed. Then the likelihood function is

(5.119)

$$l(\beta, \boldsymbol{\pi}_1, \phi, \sigma_2|\mathbf{y}_1, \mathbf{y}_2) \propto \frac{\phi^{n/2}}{\sigma_2^{2n}} \exp\left\{-\frac{1}{2\sigma_2^2}[\phi(\mathbf{y}_1 - X\boldsymbol{\pi}_1)'(\mathbf{y}_1 - X\boldsymbol{\pi}_1) + (\mathbf{y}_2 - Z\boldsymbol{\pi}_1)'(\mathbf{y}_2 - Z\boldsymbol{\pi}_1)]\right\},$$

where $Z = X\beta$. As prior pdf we employ

$$p(\beta, \boldsymbol{\pi}_1, \phi, \sigma_2) \propto \frac{p_1(\phi)\,p_2(\beta)}{\sigma_2} \quad \begin{cases} 0 < \sigma_2, \phi < \infty, \\ -\infty < \pi_{1i} < \infty, \quad i = 1, \ldots, k, \end{cases} \tag{5.120}$$

where $p_1(\phi)$ and $p_2(\beta)$ are still of unspecified form. In (5.120) we are assuming that β, the elements of $\boldsymbol{\pi}_1$, ϕ, and $\log \sigma_2$ are independently distributed with the

[75] More accurately on β^2/ϕ.

[76] This is a case in which an optimal (least squares) estimator depends on a nuisance parameter ϕ.

pdf's for the elements of $\boldsymbol{\pi}_1$ and $\log \sigma_2$ uniform.[77] Then the posterior pdf for the parameters is given by

$$p(\beta, \boldsymbol{\pi}_1, \phi, \sigma_2|\mathbf{y}_1, \mathbf{y}_2) \propto \frac{\phi^{n/2}\, p_1(\phi)\, p_2(\beta)}{\sigma_2^{2n+1}} \times \exp\left\{-\frac{1}{2\sigma_2^2}[\phi(\mathbf{y}_1 - X\boldsymbol{\pi}_1)'(\mathbf{y}_1 - X\boldsymbol{\pi}_1) + (\mathbf{y}_2 - Z\boldsymbol{\pi}_1)'(\mathbf{y}_2 - Z\boldsymbol{\pi}_1)]\right\}. \tag{5.121}$$

On completing the square on $\boldsymbol{\pi}_1$ in the exponent, we have

$$p(\beta, \boldsymbol{\pi}_1, \phi, \sigma_2|\mathbf{y}_1, \mathbf{y}_2) \propto \frac{\phi^{n/2}\, p_1(\phi)\, p_2(\beta)}{\sigma_2^{2n+1}} \times \exp\left\{-\frac{1}{2\sigma_2^2}[\phi\mathbf{y}_1'\mathbf{y}_1 + \mathbf{y}_2'\mathbf{y}_2 - \mathbf{v}'M^{-1}\mathbf{v} + (\boldsymbol{\pi}_1 - M^{-1}\mathbf{v})'M(\boldsymbol{\pi}_1 - M^{-1}\mathbf{v})]\right\}, \tag{5.122}$$

where $\mathbf{v}$ and M have been defined in connection with (5.106). Thus, given β, ϕ, and σ_2, the conditional posterior pdf for $\boldsymbol{\pi}_1$ is normal with

$$E(\boldsymbol{\pi}_1|\beta, \phi, \sigma_2, \mathbf{y}_1, \mathbf{y}_2) = M^{-1}\mathbf{v} = (Z'Z + \phi X'X)^{-1}(\phi X'\mathbf{y}_1 + Z'\mathbf{y}_2) = \frac{\phi\hat{\boldsymbol{\pi}}_1 + \beta\hat{\boldsymbol{\pi}}_2}{\phi + \beta^2} \tag{5.123}$$

and

$$\text{Var}(\boldsymbol{\pi}_1|\beta, \phi, \sigma_2, \mathbf{y}_1, \mathbf{y}_2) = M^{-1}\sigma_2^2 = \frac{(X'X)^{-1}\sigma_2^2}{\beta^2 + \phi}, \tag{5.124}$$

where $\hat{\boldsymbol{\pi}}_i = (X'X)^{-1}X'\mathbf{y}_i$, $i = 1, 2$. The expression in (5.123) is seen to be precisely in the form of (5.107), the conditional minimizing value for $\boldsymbol{\pi}_1$ in the least squares sampling theory approach.

On integrating (5.122) with respect to the elements of $\boldsymbol{\pi}_1$, we obtain[78]

$$p(\beta, \phi, \sigma_2|\mathbf{y}_1, \mathbf{y}_2) \propto \frac{\phi^{n/2} p_1(\phi)\, p_2(\beta)}{\sigma_2^{2n-k+1}(\beta^2 + \phi)^{k/2}} \times \exp\left\{-\frac{1}{2\sigma_2^2}[\phi\mathbf{y}_1'\mathbf{y}_1 + \mathbf{y}_2'\mathbf{y}_2 - \mathbf{v}'M\mathbf{v}]\right\}. \tag{5.125}$$

[77] The analysis can be extended to incorporate an informative prior pdf for σ_2 in the inverted gamma form.

[78] Note $|M/\sigma_2^2|^{-1/2} = [\sigma_2^k/(\beta^2 + \phi)^{k/2}]|X'X|^{-1/2}$.

Then, on integrating (5.125) with respect to σ_2, the result is

$$p(\beta, \phi|\mathbf{y}_1, \mathbf{y}_2) \propto \frac{\phi^{n/2} p_1(\phi) p_2(\beta)}{(\beta^2 + \phi)^{k/2}} [(\phi\mathbf{y}_1'\mathbf{y}_1 + \mathbf{y}_2'\mathbf{y}_2 - \mathbf{v}'M\mathbf{v})^{(2n-k)/2}]^{-1}$$

$$(5.126) \qquad \propto \frac{\phi^{n/2} p_1(\phi) p_2(\beta)}{(\beta^2 + \phi)^{k/2}} \times \left\{\left[s_{22} + \phi s_{11} + \frac{\phi(\hat{\mathbf{y}}_2 - \beta\hat{\mathbf{y}}_1)'(\hat{\mathbf{y}}_2 - \beta\hat{\mathbf{y}}_1)}{\phi + \beta^2}\right]^{(2n-k)/2}\right\}^{-1},$$

where $\hat{\mathbf{y}}_i = X\hat{\boldsymbol{\pi}}_i$ and $s_{ii} = (\mathbf{y}_i - X\hat{\boldsymbol{\pi}}_i)'(\mathbf{y}_i - X\hat{\boldsymbol{\pi}}_i)$, $i = 1, 2$. This then, is the joint posterior pdf for β and ϕ. The factor in brackets in the denominator is identical to the quantity in brackets in (5.109). Thus, given ϕ, $\beta = \hat{\beta}_{(\phi)}$, as shown in (5.111), will minimize the quantity in brackets in (5.126). Aside from prior factors, for given ϕ, the posterior pdf in (5.126) will have a modal value at approximately $\beta = \hat{\beta}_{(\phi)}$. Given explicit forms for the prior pdf's $p_1(\phi)$ and $p_2(\beta)$,[79] bivariate numerical integration techniques can be employed to evaluate the normalizing constant of (5.126) and to compute joint and marginal posterior pdf's.[80]

To illustrate results obtained by using (5.126) with a diffuse prior[81] for ϕ and β, we generated data from the following model:

$$y_{1i} = \pi_{10} + \pi_{11}x_i + u_{1i},$$
$$y_{2i} = (\pi_{10} + \pi_{11}x_i)\beta + u_{2i}, \qquad i = 1, 2, \ldots, 30,$$

with

$$\pi_{10} = 5.0, \qquad \pi_{11} = 0.5, \quad \text{and} \quad \beta = 0.9.$$

Further, the u_{1i}'s and u_{2i}'s were drawn independently from a normal distribution with zero mean and variance equal to 225; that is $\sigma_1^2 = \sigma_2^2 = 225$. The x_i's were drawn independently from a normal distribution with mean and variance equal to 5 and 9, respectively.

Shown in Figure 5.5 are the contours of the joint posterior pdf for β and ϕ. Given that the error term variances are rather large, the joint posterior pdf is quite spread out. The mode is located at $\beta = 0.88$ and $\phi = 0.56$. The

[79] If we are diffuse on ϕ and σ_2 in the following way, $p(\phi, \sigma_2) \propto 1/\phi\sigma_2$, this implies $p(\sigma_1, \sigma_2) \propto 1/\sigma_1\sigma_2$, and thus use of the latter diffuse prior will lead to the same posterior pdf as use of the former, given the same prior assumptions about β and $\boldsymbol{\pi}_1$ in both cases. Of course, if information is available about ϕ and β, it can be introduced in (5.126) by choice of appropriate prior pdf's.

[80] Since the Bayesian analysis of (5.114) and (5.115) proceeds in much the same way as that shown above, we do not present it.

[81] That is, $p_1(\phi)\, p_2(\beta) \propto 1/\phi$, $0 < \phi < \infty$, and $-\infty < \beta < \infty$.

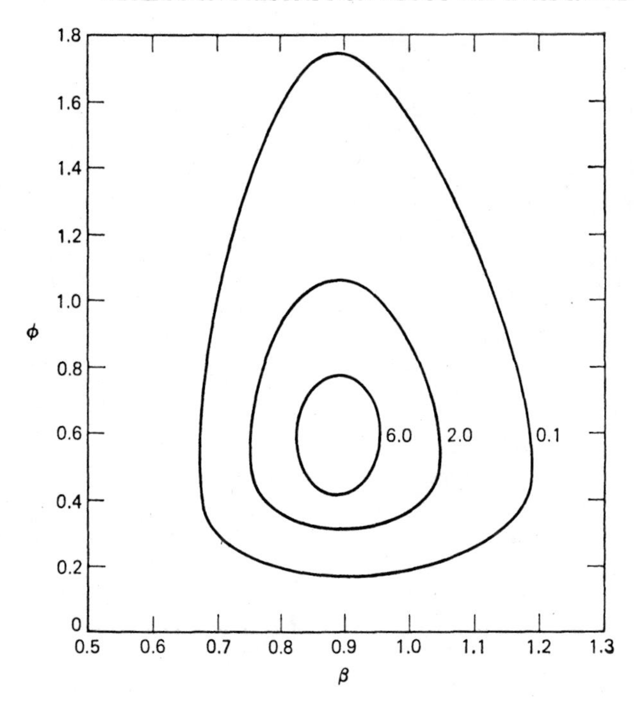

Figure 5.5 Contours of the joint posterior pdf for β and ϕ. The model value is 8.55 at $\beta = 0.88$ and $\phi = 0.56$.

marginal posterior pdf's for ϕ and β are shown in Figure 5.6. The marginal posterior pdf for ϕ has a mean equal to 0.694 and a posterior standard deviation equal to 0.279, thus showing quite a bit of dispersion. On the other hand, the posterior pdf for β is rather sharp, with a mean equal to 0.898 and standard deviation equal to 0.0847. Also, the marginal posterior pdf for β appears to be symmetric.[82]

In concluding this chapter it is worthwhile to point out that just random measurement errors have been considered. In practice it is often the case that both random and systematic measurement errors are present. General methods for treating both kinds of error would be of great value but unfortunately they remain to be developed.[83]

[82] It is interesting to note that from (5.111) $\hat{\beta}_{(\hat{\phi})} = 0.894$ for these generated data, a value close to the posterior mean for β. On the other hand, for these data $\hat{\beta}_{(0)} = 1.033$ and $\hat{\beta}_{(\infty)} = 0.832$, values which depart somewhat from the posterior mean of β.

[83] See Harold Jeffreys, *Theory of Probability* (3rd ed.). Oxford: Clarendon, 1966, pp. 300–307, for an analysis of a special problem that involves both random and systematic measurement errors.

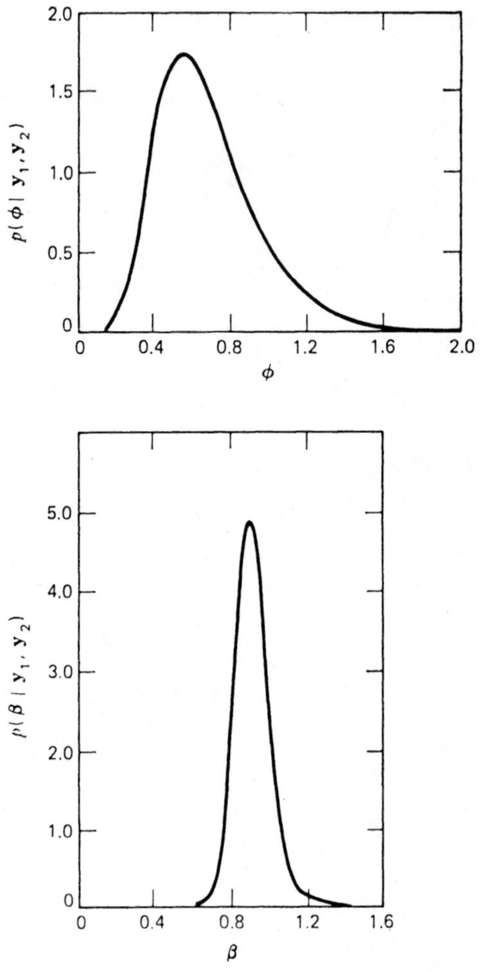

Figure 5.6 Marginal posterior pdf's for ϕ and β.

APPENDIX

In this appendix we take up the problem of integrating (5.85) with respect to the n elements of $\boldsymbol{\xi}$.[84] The factors of (5.85) involving $\boldsymbol{\xi}$ are shown below:

$$(1)\quad [(\boldsymbol{\xi} - \bar{\xi}\boldsymbol{\iota})'(\boldsymbol{\xi} - \bar{\xi}\boldsymbol{\iota})]^{-(n-2)/2} \exp\left[-\frac{1}{2\sigma_2^2}(\phi + \beta^2)(\boldsymbol{\xi} - \hat{\boldsymbol{\xi}})'(\boldsymbol{\xi} - \hat{\boldsymbol{\xi}})\right],$$

[84] The approach employed below was presented in A. Zellner and U. Sankar, "On Errors in the Variables," manuscript, 1967, and is used and studied further in R. L. Wright, "A Bayesian Analysis of Linear Functional Relations," manuscript, University of Michigan, Ann Arbor, 1969.

where $\boldsymbol{\xi}$ has been defined in connection with (5.83). This expression must be integrated with respect to the elements of $\boldsymbol{\xi}$ over $-\infty < \xi_i < \infty$, $i = 1, 2, \ldots, n$.

First, make the change of variables

$$z_i = \frac{(\phi + \beta^2)^{1/2}}{\sigma_2}(\xi_i - \hat{\xi}_i), \qquad i = 1, 2, \ldots, n \tag{2a}$$

or

$$\xi_i = \frac{\sigma_2}{(\phi + \beta^2)^{1/2}}(z_i + q_i), \qquad i = 1, 2, \ldots, n, \tag{2b}$$

where $q_i = (\phi + \beta^2)^{1/2}\hat{\xi}_i/\sigma_2$. Then

$$\prod_{i=1}^{n} d\xi_i = \frac{\sigma_2^{\,n}}{(\phi + \beta^2)^{n/2}} \prod_{i=1}^{n} dz_i$$

and (1) can be written in terms of $\mathbf{z}' = (z_1, z_2, \ldots, z_n)$:

$$\frac{\sigma_2^{\,2}}{(\phi + \beta^2)} \frac{e^{-\frac{1}{2}\mathbf{z}'\mathbf{z}}}{[(\mathbf{z} + \mathbf{q})'M(\mathbf{z} + \mathbf{q})]^{(n-2)/2}}, \tag{3}$$

where $\mathbf{q}' = (q_1, q_2, \ldots, q_n)$ and $M = I_n - \boldsymbol{\iota}\boldsymbol{\iota}'/n$, with I_n an $n \times n$ unit matrix and $\boldsymbol{\iota}$ an $n \times 1$ column vector in which all elements are equal to one. On letting $\mathbf{w} = \mathbf{z} + \mathbf{q}$, (3) becomes

$$\frac{\sigma_2^{\,2}}{(\phi + \beta^2)} \frac{e^{-\frac{1}{2}(\mathbf{w} - \mathbf{q})'(\mathbf{w} - \mathbf{q})}}{[\sum_{i=1}^{n} (w_i - \bar{w})^2]^{(n-2)/2}}. \tag{4}$$

We can view the integration of (4) with respect to the elements of $\mathbf{w}$ as the problem of finding the expectation of $[\sum_{i=1}^{n} (w_i - \bar{w})^2]^{-(n-2)/2}$, with the w_i's having a normal pdf. To get rid of $\bar{w}$ in the denominator, we use Helmert's transformation $\mathbf{c} = B\mathbf{w}$, which is

$$\begin{aligned}
c_1 &= \frac{w_1 - w_2}{\sqrt{2}} \\
c_2 &= \frac{w_1 + w_2 - 2w_3}{\sqrt{6}} \\
c_3 &= \frac{w_1 + w_2 + w_3 - 3w_4}{\sqrt{12}} \\
&\cdots\cdots\cdots\cdots\cdots \\
c_{n-1} &= \frac{w_1 + w_2 + \cdots + w_{n-1} - (n-1)w_n}{\sqrt{n(n-1)}} \\
c_n &= \frac{w_1 + w_2 + \cdots + w_n}{\sqrt{n}}.
\end{aligned} \tag{5}$$

It is known that the matrix B in the Helmert transformation is orthogonal. Thus the Jacobian of the transformation is one. Since the c_i's are linear combinations of the w_i's, they have a normal pdf with $E\mathbf{c} = BE\mathbf{w} = B\mathbf{q}$. Then $\mathbf{c} - E\mathbf{c} = B(\mathbf{w} - \mathbf{q})$ or $\mathbf{w} - \mathbf{q} = B^{-1}(\mathbf{c} - E\mathbf{c})$ and

$$(6) \qquad \begin{aligned}(\mathbf{w} - \mathbf{q})'(\mathbf{w} - \mathbf{q}) &= (\mathbf{c} - E\mathbf{c})'(B^{-1})'B^{-1}(\mathbf{c} - E\mathbf{c}) \\ &= (\mathbf{c} - E\mathbf{c})'(\mathbf{c} - E\mathbf{c}),\end{aligned}$$

since B is orthogonal. Last, from properties of the Helmert transformation in (5) we have

$$(7) \qquad \sum_{i=1}^{n} (w_i - \bar{w})^2 = \sum_{i=1}^{n-1} c_i^2.$$

Using (6) and (7), we can express (4) as

$$\begin{aligned}&\frac{\sigma_2^2}{\phi + \beta^2}\left[\left(\sum_{i=1}^{n-1} c_i^2\right)^{(n-2)/2}\right]^{-1} \exp\left[-\tfrac{1}{2}(\mathbf{c} - E\mathbf{c})'(\mathbf{c} - E\mathbf{c})\right] \\ (8) \qquad &= \frac{\sigma_2^2}{(\phi + \beta^2)}\left[\left(\sum_{i=1}^{n-1} c_i^2\right)^{(n-2)/2}\right]^{-1} \\ &\quad \times \exp\left[-\frac{1}{2}\sum_{i=1}^{n-1}(c_i - \bar{c}_i)^2\right] \exp\left[-\frac{1}{2}(c_n - \bar{c}_n)^2\right],\end{aligned}$$

where $\bar{c}_i = Ec_i$, $i = 1, 2, \ldots, n$.

On integrating (8) with respect to c_n, $-\infty$ to $+\infty$, we get a numerical constant. The integration with respect to the remaining c_i's is viewed as obtaining the expectation of $(\sum_{i=1}^{n-1} c_i^2)^{-(n-2)/2}$, with the c_i's independent and normal, each with its own mean $\bar{c}_i$. Under these conditions $v = \sum_{i=1}^{n-1} c_i^2$ has the following noncentral χ^2 pdf[85]:

$$(9) \qquad \frac{1}{2^{p/2}\sqrt{\pi}} e^{-\frac{1}{2}(\delta + v)} v^{\frac{1}{2}p - 1} \sum_{\alpha=0}^{\infty} \frac{\delta^\alpha v^\alpha}{(2\alpha)!} \frac{\Gamma(\alpha + \frac{1}{2})}{\Gamma(\alpha + \frac{1}{2}p)}, \qquad 0 < v < \infty,$$

with $p = n - 1$. In (9) δ is the noncentrality parameter given by $\delta = \sum_{i=1}^{n-1} \bar{c}_i^2 = \sum_{i=1}^{n} (q_i - \bar{q})^2$, with q_i as defined in connection with (2*b*) above. Then on multiplying (9) by $v^{-[(n-2)/2]}$, the integral of interest is proportional to

$$(10) \qquad \frac{\sigma_2^2}{\phi + \beta^2} e^{-\frac{1}{2}\delta} \int_0^\infty \sum_{\alpha=0}^{\infty} \frac{\delta^\alpha v^{\alpha - \frac{1}{2}}}{(2\alpha)!} e^{-\frac{1}{2}v} \frac{\Gamma(\alpha + \frac{1}{2})}{\Gamma[\alpha + (n - 1)/2]}\, dv.$$

[85] See, for example, T. W. Anderson, *An Introduction to Multivariate Statistical Analysis*. New York: Wiley, 1958, p. 113.

Since term-by-term integration is appropriate in the present instance and we have

$$\int_0^\infty v^{\alpha-1/2}e^{-1/2 v}\,dv = 2^{\alpha+1/2}\Gamma(\alpha+\tfrac{1}{2}), \tag{11}$$

the integral in (10) is proportional to

$$\frac{\sigma_2^{\,2}}{\phi+\beta^2}\,e^{-1/2\delta}\sum_{\alpha=0}^{\infty}\frac{\delta^\alpha 2^\alpha}{(2\alpha)!}\frac{[\Gamma(\alpha+\frac{1}{2})]^2}{\Gamma(\alpha+(n-1)/2]} \tag{12a}$$

or to[86]

$$\frac{\sigma_2^{\,2}}{\phi+\beta^2}\,e^{-1/2\delta}\sum_{\alpha=0}^{\infty}\left(\frac{\delta}{2}\right)^\alpha\frac{\Gamma(\alpha+\frac{1}{2})}{\alpha!\,\Gamma[\alpha+(n-1)/2]}, \tag{12b}$$

where

$$\delta = \sum_{i=1}^{n}(q_i-\bar{q})^2 = \frac{\phi+\beta^2}{\sigma_2^{\,2}}\sum_{i=1}^{n}(\hat{\xi}_i-\bar{\hat{\xi}})^2,$$

with $\bar{\hat{\xi}} = \sum_{i=1}^n \hat{\xi}_i/n$.

QUESTIONS AND PROBLEMS

1. Consider the n-mean problem described in Section 5.1 with the likelihood function given in (5.1). Suppose that in place of (5.3) we employ the following prior pdf for the n elements of $\boldsymbol{\xi}$ and σ:

$$p(\boldsymbol{\xi},\sigma) \propto \frac{1}{\sigma\tau_0^{\,n}}\exp\left[-\frac{1}{2\tau_0^{\,2}}(\boldsymbol{\xi}-\iota\mu_0)'(\boldsymbol{\xi}-\iota\mu_0)\right]\Bigg\} \quad \begin{array}{c} 0<\sigma<\infty, \\ -\infty<\xi_i<\infty, \\ i=1,2,\ldots,n, \end{array}$$

where τ_0 and μ_0 are assigned values by an investigator.

(a) Provide an interpretation of the above prior pdf.

(b) What is the mean of the posterior pdf for $\boldsymbol{\xi}$, given σ?

(c) Comment on properties of the marginal posterior pdf for σ.

2. Consider the analysis of the system in (5.17*a*, *b*), with the likelihood function shown in (5.18), using informative prior pdf's for σ_1 and σ_2 in the form of inverted gamma pdf's. (In this analysis express the posterior pdf in terms of $\boldsymbol{\xi}$, σ_1, and λ, where $\lambda = \sigma_1^{\,2}/\sigma_2^{\,2}$.)

3. For the system in (5.30) and (5.31) to be analyzed appropriately as a simple regression model what assumption about parameter values is sufficient? If this assumption is justified, what is the mean of the posterior pdf for β, using a diffuse prior pdf for β_0, β, and σ_2?

[86] In moving from (12*a*) to (12*b*) the following expressions were used: $\Gamma(\alpha+\frac{1}{2}) = (2\alpha)!\sqrt{\pi}/2^{2\alpha}\Gamma(\alpha+1)$, and $\Gamma(\alpha+1)=\alpha!$.

4. In connection with the inequalities shown in (5.61), under what assumptions about the values of the parameters in (5.56) to (5.58) will the following estimators closely bracket the true value of β in large samples?

$$\hat{\beta}_{12} = \frac{\sum_{i=1}^{n} (y_{1i} - \bar{y}_1)(y_{2i} - \bar{y}_2)}{\sum_{i=1}^{n} (y_{1i} - \bar{y}_1)^2} \quad \text{and} \quad \hat{\beta}_{21} = \frac{\sum_{i=1}^{n} (y_{2i} - \bar{y}_2)^2}{\sum_{i=1}^{n} (y_{1i} - \bar{y}_1)(y_{2i} - \bar{y}_2)}.$$

5. Establish that the root of (5.49), shown in (5.50), is the one that is associated with a maximum of the likelihood function.
6. If, in (5.27), ξ_i represents the logarithm of permanent income for the ith individual, whereas in (5.28) η_i represents the logarithm of permanent consumption, explain in detail the differences in assumptions about these quantities in the functional and structural forms of the EVM.
7. In developing (5.61) we assumed that u_{1i} and u_{2i} in (5.30) and (5.31) were uncorrelated. If $Eu_{1i}u_{2i} = \sigma_{12}$, $i = 1, 2, \ldots, n$, how are the relations in (5.56) to (5.58) and the inequalities in (5.61) altered?
8. Assume that we have measurements of the same quantity, say income per household, from two different sources; for example for the ith region y_{1i} might be a survey estimate and y_{2i}, an estimate derived from a second independent survey. Consider these data in connection with the EVM:

$$\left.\begin{aligned} &\text{(i)} \quad & y_{1i} &= \xi_i + u_{1i} \\ &\text{(ii)} \quad & y_{2i} &= \eta_i + u_{2i} \\ &\text{(iii)} \quad & \eta_i &= \beta_0 + \beta\xi_i \end{aligned}\right\} \qquad i = 1, 2, \ldots, n,$$

   (a) Interpret ξ_i and η_i as well as u_{1i} and u_{2i}.
   (b) If, in (iii), $\beta_0 = 0$ and $\beta = 1$, in what sense are the two sets of measurements consistent?
   (c) Given the assumptions made in connection with (5.30) and (5.31), explain how β_0 and β can be estimated, given that σ_1^2 has a known value or $\sigma_2^2 = 0$.
   (d) Discuss possible sources of prior information about the parameters and indicate how it can be used in estimation.
9. In employing a prior pdf such as that shown in (5.67) in analyzing the EVM, will the influence of the prior pdf be negligible as the sample size increases? Similarly, does the assumption that the ξ_i's are normally and independently distributed affect Bayesian and maximum likelihood estimation results in just small sample situations?
10. Use the information in Table 5.1 to obtain estimates of the bounds on the slope coefficient shown in (5.61).
11. Consider the system (5.27) to (5.30) as representing Friedman's permanent income model with the logarithms of measured income and consumption and of permanent income and consumption given by y_{1i}, y_{2i}, ξ_i, and η_i, respectively. Interpret this model within the context of the functional and structural forms of the EVM.
12. In Problem 11 assume the structural form of the EVM and set forth sufficient conditions for all parameters to be identified. How does the assumption $Eu_{1i}u_{2i} = \sigma_{12} \neq 0$ affect your analysis?

13. Under the identifying assumptions of Problem 12, derive maximum likelihood estimates for the parameters of the permanent income model.

14. Provide a prior pdf for the parameters of the permanent income model, discussed in Problems 11 and 12 above, and indicate how marginal posterior pdf's for the parameters can be computed.

15. Suppose that we have measurable proxies for permanent consumption and permanent income, denoted $\bar{C}_t$ and $\bar{Y}_t$, respectively, and wish to investigate the proportionality hypothesis; that is, the hypothesis that permanent consumption is proportional to permanent income. Is $\bar{C}_t = \beta_0 + \beta \bar{Y}_t + \epsilon_t$, with $\beta_0 \neq 0$ and ϵ_t an error term, an economically reasonable alternative model, particularly over the range of low values for $\bar{Y}_t$? As another alternative to the proportionality hypothesis, consider

$$\log \left(\frac{\bar{C}_t / \bar{Y}_t}{1 - \bar{C}_t / \bar{Y}_t} \right) = \alpha_0 + \alpha_1 \log \bar{Y}_t + v_t,$$

where v_t is an error term. What does this last equation imply about $\bar{C}_t / \bar{Y}_t$, the average propensity to consume, as $\bar{Y}_t \to 0$ with $\alpha_1 < 0$?

16. Shown below and on the next page are per capita data, expressed in U.S. 1955 dollars, for $\bar{Y}_t$ and $\bar{S}_t = \bar{Y}_t - \bar{C}_t$, proxies for permanent income, and permanent savings relating to 26 countries[87]: Use these data, along with diffuse prior assumptions, to analyze the two relations put forward in Problem 15 as alternatives to the proportionality hypothesis within a regression framework. In particular, what is the posterior probability that $\alpha_1 < 0$?

Country	$\bar{Y}_t$	$\bar{S}_t$	$\bar{C}_t = \bar{Y}_t - \bar{S}_t$
United States	1659.1	123.2	1535.9
Canada	1208.1	84.0	1124.1
New Zealand	928.0	81.2	846.8
Australia	905.0	96.6	808.4
Belgium	877.6	95.9	781.7
France	835.7	47.7	788.0
Luxemburg	801.2	107.3	693.9
Sweden	765.0	58.5	706.5
United Kingdom	737.6	30.7	706.9
Denmark	723.2	65.7	657.5
Netherlands	476.4	45.7	430.7
Ireland	416.5	29.5	387.0
Austria	411.9	41.2	370.7

[87] These data appear in H. S. Houthakker, "On Some Determinants of Saving in Developed and Under-Developed Countries," Chapter 10, pp. 212–224, p. 212, of E. A. G. Robinson (Ed.), *Problems in Economic Development*. New York: St. Martin's, 1965. Data for two countries, Panama and Peru, for which $\bar{S}$ is negative, have been omitted. See Houthakker's paper for the method and weights employed in computing the figures presented in the table.

Country	$\bar{Y}_i$	$\bar{S}_i$	$\bar{C}_i = \bar{Y}_i - \bar{S}_i$
Malta	316.8	64.8	252.0
Costa Rica	257.8	13.7	244.1
Jamaica	235.8	8.0	227.8
Spain	234.8	10.7	224.1
Japan	199.2	28.8	170.4
Colombia	198.7	8.6	190.1
Ghana	197.7	9.6	188.1
Mauritius	197.0	18.0	179.0
Honduras	164.0	11.4	152.6
Ecuador	134.5	5.2	129.3
Brazil	127.7	5.6	122.1
Rhodesia	115.5	8.8	106.7
Belgian Congo	58.3	2.4	55.9

17. In the relation $\log (\bar{C}_i/\bar{Y}_i)/(1 - \bar{C}_i/\bar{Y}_i) = \alpha_0 + \alpha_1 \log \bar{Y}_i + v_i$ assume that $\bar{C}_i$ and $\bar{Y}_i$ have common measurement errors, perhaps due to common errors in weighting or in the use of exchange rates in the conversion to U.S. dollars, that is, $\bar{C}_i = c\bar{C}_i^* e^{u_i}$ and $\bar{Y}_i = c\bar{Y}_i^* e^{u_i}$, where $\bar{C}_i^*$ and $\bar{Y}_i^*$ are true values of the variables, c is a constant, and u_i is a random error term with zero mean and variance σ_u^2. How does the presence of such measurement errors affect the results obtained in the calculations in Problem 16?

18. Assuming the measurement error structure for $\bar{C}_i$ and $\bar{Y}_i$ set forth in Problem 17, compute maximum likelihood estimates for the parameter α_1 in the relation,

$$\log \frac{\bar{C}_i/\bar{Y}_i}{1 - \bar{C}_i/\bar{Y}_i} = \alpha_0 + \alpha_1 \log \bar{Y}_i^* + v_i,$$

with $\log \bar{Y}_i = \log c + \log \bar{Y}_i^* + u_i$, using the data in Problem 16 and assuming various values for $\lambda = \sigma_u^2/\sigma_v^2$.

19. Assuming that $\lambda = \sigma_u^2/\sigma_v^2$ has a given value, perform a Bayesian analysis of the model in Problem 18. Determine how properties of the conditional posterior pdf for α_1, given λ, depend on the value assigned to λ.

20. Explain how Problem 18 can be analyzed from the Bayesian point of view with a prior pdf for $\lambda = \sigma_u^2/\sigma_v^2$ and other parameters. Use the data in Problem 16 to compute posterior pdf's.

21. Consider the system, analogous to (5.99) and (5.100),

$$y_{1t} = \mathbf{x}_t'\boldsymbol{\pi}_1 + u_{1t},$$
$$y_{2t} = \mathbf{x}_t'\boldsymbol{\pi}_1\beta + u_{1t},$$

but assume that the $k \times 1$ vector $\mathbf{x}_t$ is stochastic, independent of the u's, and with zero mean and $k \times k$ pds covariance matrix $E\mathbf{x}_t\mathbf{x}_t' = \Phi$. By examining the second moments of y_{1t} and y_{2t}, establish that inequality constraints on β, similar to those shown in (5.61), can be derived. Does the pdf for $\mathbf{x}_t$, $t = 1, 2, \ldots, T$, contain information relating to these bounds, hence to β?

22. Prove the result shown in (5.113).

23. In the forecasting area much attention is given to the comparison of forecasts (F_i) with actual measured outcomes (A_i). Since both F_i and A_i contain errors, consider the following model:

$$\left.\begin{array}{ll} \text{(i)} & F_i = \theta_i + u_i \\ \text{(ii)} & A_i = \eta_i + v_i \\ \text{(iii)} & \eta_i = \beta_0 + \beta\theta_i \end{array}\right\} \qquad i = 1, 2, \ldots, n,$$

with $EF_i = \theta_i$ and $EA_i = \eta_i$. If in (iii) $\beta_0 = 0$ and $\beta = 1$, in what sense can it be said that forecasts are unbiased?

24. In connection with Problem 23, provide examples wherein it would be reasonable to assume that the u_i's are probably independently distributed and examples wherein the u_i's are probably not independently distributed.

25. Consider the analysis of the system in Problem 23 under assumptions of the EVM considered in this chapter.

26. Appraise the following statement: although a regression of A_i on F_i yields inconsistent estimates of β_0 and β in (iii) of Problem 23, such a regression may be valuable in providing systematic corrections to forecasts.

CHAPTER VI

Analysis of Single Equation Nonlinear Models

In some circumstances economic and/or statistical considerations lead us to the problem of analyzing models that are nonlinear in the parameters. In Section 4.1 we encountered a nonlinear relation in the analysis of the regression model with autocorrelated error terms, specifically (4.1*c*) and (4.12*c*). In this chapter we analyze other nonlinear models; for example, the "constant elasticity of substitution" (CES) production function and "generalized production function" (GPF) models. Here nonlinearities develop mainly because of economic considerations; that is, the CES function is a generalization of the Cobb-Douglas (CD) function in the sense that it permits the elasticity of substitution parameter to assume values other than one, its value for the CD function. Similarly, GPF's permit generalization in this respect and also with respect to the behavior of the returns to scale parameter. The economic and statistical importance of taking account of appropriate functional forms of relationships cannot be emphasized too strongly. Use of incorrect functional forms can often lead to serious errors. Thus special attention is given in this chapter to the analysis of several nonlinear forms that appear to be useful in a number of applications.

6.1 THE BOX-COX ANALYSIS OF TRANSFORMATIONS

In the Box-Cox[1] analysis of transformations we encounter relationships nonlinear in one or more parameters; for example, among other transformations Box and Cox consider the following one for the dependent variable y_α in a regression model with $y_\alpha > 0$, $\alpha = 1, 2, \ldots, n$:

$$\text{(6.1)} \qquad \frac{y_\alpha^{\lambda} - 1}{\lambda} = \beta_1 + \beta_2 x_{2\alpha} + \cdots + \beta_k x_{k\alpha} + u_\alpha, \qquad \alpha = 1, 2, \ldots, n,$$

where the β's *and* λ are unknown parameters, the x's are observations on

[1] G. E. P. Box and D. R. Cox, "An Analysis of Transformations," *J. Roy. Statist. Soc., Series B*, **26**, 211–243 (1964).

independent variables, and u_α is the αth error term. They assume that for some *unknown* value of λ the transformed observations $(y_\alpha^\lambda - 1)/\lambda$, $\alpha = 1, 2, \ldots, n$, satisfy the standard assumptions of the normal multiple regression model; that is, they are normally[2] and independently distributed with common (constant) variance σ^2. Thus, by assumption, for some value of λ a transformation on the dependent variable is assumed (a) to induce normality, (b) to stabilize the variance, and (c) to induce simplicity of structure in the sense that $E(y_\alpha^\lambda - 1)/\lambda = \beta_1 + \beta_2 x_{2\alpha} + \cdots + \beta_k x_{k\alpha}$ is a simple function of the β's and x's. Note that the particular power transformation in (6.1) has the following properties. For $\lambda = 1$, $(y_\alpha^\lambda - 1)/\lambda = y_\alpha - 1$ and the model in (6.1) is linear in the y_α. For $\lambda = 0$, $(y_\alpha^\lambda - 1)/\lambda = \log y_\alpha$ and thus the model in (6.1) has $\log y_\alpha$ as the dependent variable. For other values of λ powers of y_α appear as the dependent variable. Since λ is an unknown parameter, it will have to be estimated along with the other unknown parameters, the β's and σ, and thus information in the data is used to determine the appropriate transformation for the dependent variable. Below we show how this estimation problem can be solved by using the maximum likelihood (ML) method[3] and the Bayesian approach.

For convenience, following Box and Cox, we rewrite (6.1) as follows:

$$\mathbf{y}^{(\lambda)} = X\boldsymbol{\beta} + \mathbf{u}, \tag{6.2}$$

where $\mathbf{y}^{(\lambda)}$ is an $n \times 1$ vector with typical element[4] $(y_\alpha^\lambda - 1)/\lambda$, X is an $n \times k$ matrix, with rank k, of given observations on k independent variables, and $\mathbf{u}$ is an $n \times 1$ vector of error terms, assumed to be normally and independently distributed, each with zero mean and common variance σ^2. To write the joint pdf for the y's, given the parameters λ, $\boldsymbol{\beta}$, and σ, and X, we need the Jacobian of the transformation from the n u's to the n y's. Since $\partial u_\alpha/\partial y_\alpha = y_\alpha^{\lambda-1}$ and $J = \prod_{\alpha=1}^n |\partial u_\alpha/\partial y_\alpha|$, we have

$$J = \left(\prod_{\alpha=1}^n y_\alpha\right)^{\lambda-1} = \dot{y}^{n(\lambda-1)} \tag{6.3}$$

where $\dot{y} = (\prod_{\alpha=1}^n y_\alpha)^{1/n}$, the geometric mean of the y's. Note that $J > 0$, since we have assumed $y_\alpha > 0$, $\alpha = 1, 2, \ldots, n$. Then we have for the joint

[2] Note, as pointed out by J. B. Ramsey, that the range of $(y_\alpha^\lambda - 1)/\lambda$ will not be $-\infty$ to ∞ but just a subinterval of this range. If the probability that $(y_\alpha^\lambda - 1)/\lambda$ will fall in the excluded interval is small, the normality assumption will not be vitally affected.

[3] Note from Chapter 2 that *in large samples* maximum likelihood estimates are approximate means of the joint posterior pdf for the parameters, a pdf that will usually be approximately normal.

[4] If a transformation other than the one shown in (6.1) is employed, the typical element of $\mathbf{y}^{(\lambda)}$ will, of course, be different from that employed here. See Box and Cox, *loc. cit.*, for examples of other transformations.

pdf for the elements of **y**

$$p(\mathbf{y}|\lambda, \boldsymbol{\beta}, \sigma^2) \propto \frac{J}{\sigma^n} \exp\left[-\frac{1}{2\sigma^2}(\mathbf{y}^{(\lambda)} - X\boldsymbol{\beta})'(\mathbf{y}^{(\lambda)} - X\boldsymbol{\beta})\right], \tag{6.4}$$

the expression in (6.4) viewed as a function of the parameters is the likelihood function, $l(\lambda, \boldsymbol{\beta}, \sigma^2|\mathbf{y})$. Letting $L = \log l$, we have

$$L = \text{const} + \log J - \frac{n}{2}\log \sigma^2 - \frac{1}{2\sigma^2}(\mathbf{y}^{(\lambda)} - X\boldsymbol{\beta})'(\mathbf{y}^{(\lambda)} - X\boldsymbol{\beta}). \tag{6.5}$$

On differentiating with respect to $\boldsymbol{\beta}$ and σ^2 and setting these derivatives equal to zero, we obtain

$$\hat{\boldsymbol{\beta}}(\lambda) = (X'X)^{-1}X'\mathbf{y}^{(\lambda)} \tag{6.6}$$

and

$$\hat{\sigma}^2(\lambda) = \frac{1}{n}(\mathbf{y}^{(\lambda)} - X\hat{\boldsymbol{\beta}})'(\mathbf{y}^{(\lambda)} - X\hat{\boldsymbol{\beta}}). \tag{6.7}$$

If λ were known, (6.6) and (6.7) could be computed and would be ML estimates. Since λ is assumed unknown, we substitute from (6.6) and (6.7) in (6.5) to obtain the maximized log likelihood function, denoted $L_{\max}(\lambda)$, which is given by

$$\begin{aligned} L_{\max}(\lambda) &= \text{const} + \log J - \frac{n}{2}\log \hat{\sigma}(\lambda) \\ &= \text{const} + (\lambda - 1)\sum_{\alpha=1}^{n} \log y_\alpha - \frac{n}{2}\log \hat{\sigma}^2(\lambda). \end{aligned} \tag{6.8}$$

We now evaluate (and plot) $L_{\max}(\lambda)$ for various values of λ until we find the value, say $\hat{\lambda}$, for which (6.8) attains its maximal value. This is the ML estimate for λ. Then (6.6) and (6.7), evaluated for $\lambda = \hat{\lambda}$ are ML estimates for $\boldsymbol{\beta}$ and σ^2, respectively. Further Box and Cox note that approximate large-sample confidence intervals can be constructed by using the result that in large samples $2[L_{\max}(\hat{\lambda}) - L_{\max}(\lambda)]$ is approximately distributed as χ^2 with one degree of freedom, a fact that follows from general results regarding the large-sample distribution of log likelihood ratios.[5]

In addition to applications reported in the Box-Cox paper, the ML approach described above has been applied in an analysis of the demand for money function.[6] This analysis posits that the demand for money function can be written as

[5] See, for example, M. G. Kendall and A. Stuart, *The Advanced Theory of Statistics*, Vol. II. New York: Hafner, 1961, pp. 230–231.

[6] P. Zarembka, "Functional Form in the Demand for Money," Social Systems Research Institute Workshop Paper, University of Wisconsin, Madison, 1966, *J. Am. Statist. Assocn.*, **63**, 502–511 (1968).

$$(6.9)\qquad \frac{M_\alpha^\lambda - 1}{\lambda} = \beta_1 + \beta_2\left(\frac{Y_\alpha^\lambda - 1}{\lambda}\right) + \beta_3\left(\frac{r_\alpha^\lambda - 1}{\lambda}\right) + u_\alpha, \qquad \alpha = 1, 2, \ldots, n,$$

with the subscript α denoting the value of a variable in the αth year:

M_α = price-deflated currency, demand and time deposits,
Y_α = price-deflated measured income,
r_α = commercial paper interest rate,
u_α = disturbance term.

The data are annual observations for the U.S. economy, 1869 to 1963.

In (6.9) a power transformation involving the parameter λ is applied not only to the dependent variable M_α but also to the variables Y_α and r_α which are given independent variables.[7] If $\lambda = 1$, the relation in (6.9) is linear in the variables. If $\lambda = 0$, it is linear in the logarithms of the variables. As above, the data are employed to estimate λ along with the β's and σ^2, the common variance of the u_α's. The u_α's are assumed to be normally and independently distributed, each with mean zero and variance σ^2. For notational convenience we relabel the variables as follows:

$$(6.10)\qquad y_\alpha^{(\lambda)} = \frac{M_\alpha^\lambda - 1}{\lambda}, \qquad x_{2\alpha}^{(\lambda)} = \frac{Y_\alpha^\lambda - 1}{\lambda}, \qquad x_{3\alpha}^{(\lambda)} = \frac{r_\alpha^\lambda - 1}{\lambda}.$$

Further $\mathbf{y}^{(\lambda)}$, $\mathbf{x}_2^{(\lambda)}$, and $\mathbf{x}_3^{(\lambda)}$ denote $n \times 1$ column vectors with typical elements, as shown in (6.10), and $X^{(\lambda)} = (\boldsymbol{\iota}, \mathbf{x}_2^{(\lambda)}, \mathbf{x}_3^{(\lambda)})$ is an $n \times 3$ matrix with $\boldsymbol{\iota}$ an $n \times 1$ column of ones. With this notation introduced, the likelihood function can be expressed as

$$(6.11)\qquad l(\boldsymbol{\beta}, \lambda, \sigma^2 | \mathbf{y}) \propto \frac{J}{\sigma^n} \exp\left[-\frac{1}{2\sigma^2}(\mathbf{y}^{(\lambda)} - X^{(\lambda)}\boldsymbol{\beta})'(\mathbf{y}^{(\lambda)} - X^{(\lambda)}\boldsymbol{\beta})\right],$$

where $\boldsymbol{\beta}' = (\beta_1, \beta_2, \beta_3)$ and $J = \prod_{\alpha=1}^n y_\alpha^{\lambda-1} = \prod_{\alpha=1}^n M_\alpha^{\lambda-1}$. As above, we maximize $L = \log l$ with respect to $\boldsymbol{\beta}$, λ and σ^2 in a two-step fashion. First, for given λ, the maximizing values for $\boldsymbol{\beta}$ and σ^2 are given by

$$(6.12)\qquad \hat{\boldsymbol{\beta}}(\lambda) = (X^{(\lambda)\prime}X^{(\lambda)})^{-1}X^{(\lambda)\prime}\mathbf{y}^{(\lambda)},$$

$$(6.13)\qquad \hat{\sigma}^2(\lambda) = \frac{1}{n}[\mathbf{y}^{(\lambda)} - X^{(\lambda)}\hat{\boldsymbol{\beta}}(\lambda)]'[\mathbf{y}^{(\lambda)} - X^{(\lambda)}\hat{\boldsymbol{\beta}}(\lambda)].$$

On substituting these values into $\log L$, we obtain

$$(6.14)\qquad \log L_{\max} = -\frac{n}{2}\log \hat{\sigma}^2(\lambda) + (\lambda - 1)\sum_{\alpha=1}^n \log y_\alpha,$$

[7] No attempt has been made to cope with possible "simultaneous equation" problems in this analysis. Also, see the Box-Cox paper for a discussion of transformations for the dependent and independent variables.

which is the maximized log-likelihood function except for a constant. Plots of $\log L_{\max}$ against λ for analyses based on data for the over-all period 1869 to 1963 and for the period 1915 to 1963, using two definitions of money, namely price-deflated currency, demand and time deposits ($C + D + T$) and price-deflated currency and demand deposits ($C + D$), are shown in Figure 6.1. For the period 1869 to 1963 the point estimate for λ is $\hat{\lambda} = 0.19$. An

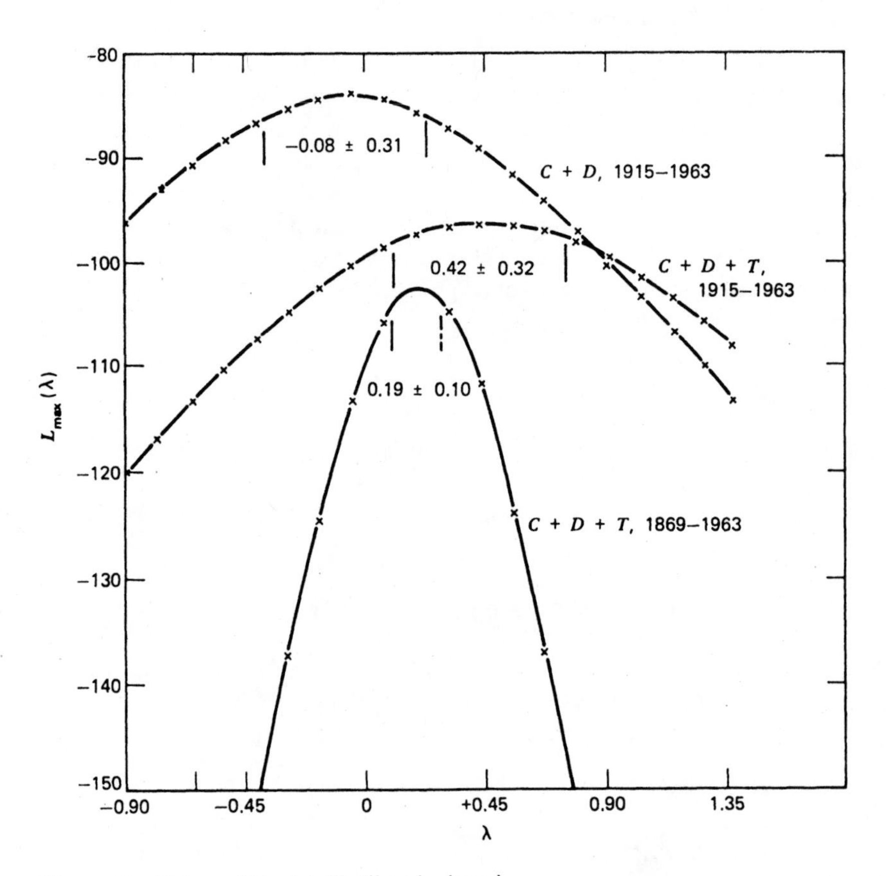

Figure 6.1 Values of the log likelihood, given λ.

approximate 95% confidence interval for λ, which was obtained from $L_{\max}(\hat{\lambda}) - L_{\max}(\lambda) < \frac{1}{2}\chi_1^2(0.05) = 1.92$, is 0.19 ± 0.10. This and the other results in Figure 6.1 indicate that a "log-log" version of the demand for money, formulated in terms of the variables mentioned above, may be approximately in agreement with the information in the data. When $\lambda = \hat{\lambda}$ is substituted in (6.12), this yields[8]

$$\underset{(0.2387)}{\hat{\beta}_1(\hat{\lambda}) = -1.0551}; \qquad \underset{(0.0163)}{\hat{\beta}_2(\hat{\lambda}) = 1.1124}; \qquad \underset{(0.0160)}{\hat{\beta}_3(\hat{\lambda}) = -0.0974}$$

as the ML estimates for β_1, β_2, and β_3; the numbers in parentheses are large-sample standard errors. Similarly, a ML estimate for σ^2 can be obtained by substituting $\hat{\lambda}$ in (6.13). Analyses within the above framework which employ an expected income variable in the money-demand function and which take account of a money-adjustment process are reported in Zarembka's paper. In addition, in handling the autocorrelation problem, if it is assumed that the u_α's are generated by a first-order autoregression process, $u_\alpha = \rho u_{\alpha-1} + \epsilon_\alpha$, in which the ϵ_α's are assumed to be normally and independently distributed, each with zero mean and common variance σ_ϵ^2, then combining this assumption with (6.9) yields

(6.15)
$$y_\alpha^{(\lambda)} - \rho y_{\alpha-1}^{(\lambda)} = \beta_1(1 - \rho) + \beta_2(x_{2\alpha}^{(\lambda)} - \rho x_{2\alpha-1}^{(\lambda)}) + \beta_3(x_{3\alpha}^{(\lambda)} - \rho x_{3\alpha-1}^{(\lambda)}) + \epsilon_\alpha.$$

For given values of λ nonlinear least squares techniques can be employed to obtain conditional estimates of the β's, ρ, and σ_ϵ^2, which can be used, along with the associated values of λ, to evaluate the logarithm of the likelihood function to find the values associated with a maximum as described above.

Having considered the ML analysis of the model in (6.1), which, as already mentioned, can be viewed as an approximate large-sample Bayesian analysis, we now take up the Box-Cox Bayesian analysis of the model. They proceed as follows to set forth a diffuse prior pdf for the parameters of the model, the k elements of $\boldsymbol{\beta}$, σ, and λ. Let

$$p(\boldsymbol{\beta}, \sigma, \lambda) = p_1(\boldsymbol{\beta}, \sigma|\lambda)\, p_2(\lambda) \tag{6.16}$$

be the joint prior pdf, with $p_1(\boldsymbol{\beta}, \sigma|\lambda)$, the conditional prior pdf for $\boldsymbol{\beta}$ and σ, given λ, and $p_2(\lambda)$, the marginal prior pdf for λ. They take $p_2(\lambda)$ uniform; that is $p_2(\lambda) \propto$ const. With respect to $p_1(\boldsymbol{\beta}, \sigma|\lambda)$, they remark that if this conditional pdf were assumed to be independent of λ "nonsensical results would be obtained." This is the case, since the general size and range of the transformed observations $\mathbf{y}^{(\lambda)}$ may depend strongly on λ. Recognizing this, Box and Cox write the diffuse conditional pdf for $\boldsymbol{\beta}$ and $\log \sigma$ as follows:

$$g(\lambda)\, d\boldsymbol{\beta}_\lambda\, d(\log \sigma_\lambda), \tag{6.17}$$

where the subscript λ is introduced to emphasize that this is conditional on

[8] The β's are not invariant with respect to changes in measurement units, whereas λ, t statistics and elasticities, being pure numbers, are.

given λ and where $g(\lambda)$ shows the dependence of $p_1(\boldsymbol{\beta}, \sigma|\lambda)$ on λ. Using an approximate consistency argument,[9] Box and Cox take $g(\lambda) = J^{-k/n}$, where J is the Jacobian shown in (6.3). Thus the final expression for the diffuse prior pdf which Box and Cox employ is

$$(6.18) \qquad p(\boldsymbol{\beta}, \sigma, \lambda) \propto \frac{1}{\sigma J^{k/n}}.$$

On combining (6.18) with the likelihood function in (6.4), we obtain the following posterior pdf for the parameters

$$(6.19) \quad p(\boldsymbol{\beta}, \sigma, \lambda|\mathbf{y}) \propto \frac{J^{(n-k)/n}}{\sigma^{n+1}} \exp\left[-\frac{1}{2\sigma^2}(\mathbf{y}^{(\lambda)} - X\boldsymbol{\beta})'(\mathbf{y}^{(\lambda)} - X\boldsymbol{\beta})\right].$$

Note that we can write

$$(6.20) \quad (\mathbf{y}^{(\lambda)} - X\boldsymbol{\beta})'(\mathbf{y}^{(\lambda)} - X\boldsymbol{\beta}) = \nu s^2(\lambda) + [\boldsymbol{\beta} - \hat{\boldsymbol{\beta}}(\lambda)]'X'X[\boldsymbol{\beta} - \hat{\boldsymbol{\beta}}(\lambda)],$$

where $\nu = n - k$,

$$(6.21) \qquad \hat{\boldsymbol{\beta}}(\lambda) = (X'X)^{-1}X'\mathbf{y}^{(\lambda)}$$

and

$$(6.22) \qquad s^2(\lambda) = \nu^{-1}[\mathbf{y}^{(\lambda)} - X\hat{\boldsymbol{\beta}}(\lambda)]'[\mathbf{y}^{(\lambda)} - X\hat{\boldsymbol{\beta}}(\lambda)].$$

On substituting from (6.20) in (6.19), we have

$$(6.23) \quad p(\boldsymbol{\beta}, \sigma, \lambda|\mathbf{y}) \propto \frac{J^{\nu/n}}{\sigma^{n+1}} \exp\left(-\frac{1}{2\sigma^2}\{\nu s^2(\lambda) + [\boldsymbol{\beta} - \hat{\boldsymbol{\beta}}(\lambda)]'X'X[\boldsymbol{\beta} - \hat{\boldsymbol{\beta}}(\lambda)]\}\right).$$

To obtain the marginal posterior pdf for λ from (6.23) integrate with respect to σ to obtain

$$(6.24) \quad p(\boldsymbol{\beta}, \lambda|\mathbf{y}) \propto J^{\nu/n}\{\nu s^2(\lambda) + [\boldsymbol{\beta} - \hat{\boldsymbol{\beta}}(\lambda)]'X'X[\boldsymbol{\beta} - \hat{\boldsymbol{\beta}}(\lambda)]\}^{-n/2}.$$

[9] Their argument proceeds as follows. Take an arbitrary reference value for λ, say λ_1 and assume provisionally, for fixed λ, that the relation between $y_\alpha^{(\lambda)}$ and $y_\alpha^{(\lambda_1)}$ is effectively linear over the range of the observations; that is, (i) $y_\alpha^{(\lambda)} = \text{const} + l_\lambda y_\alpha^{(\lambda_1)}$. Then choose $g(\lambda)$ in (6.17) so that when the linear relation between $y_\alpha^{(\lambda)}$ and $y_\alpha^{(\lambda_1)}$ holds the conditional prior pdf's for $\boldsymbol{\beta}$ and σ are consistent with one another for different values of λ. From (i), above, we have (ii) $\log \sigma_\lambda{}^2 = \text{const} + \log \sigma_{\lambda_1}{}^2$, and thus, to this order, the prior pdf for $\sigma_\lambda{}^2$ is independent of λ. Further, from (i) we have the β's in $y^{(\lambda)}$ linearly related to those in $y^{(\lambda_1)}$ and $d\beta_\lambda/d\beta_{\lambda_1} = l_\lambda$. Since there are k β's, we take $g(\lambda)$ proportional to $1/l_\lambda{}^k$. Last, to determine l_λ we note that in passing from λ_1 to λ a small element of volume in the n-dimensional sample space is multiplied by $J(\lambda)/J(\lambda_1)$, where J is the Jacobian quantity in (6.3). An average scale change for a single element of $\mathbf{y}$ is the nth root of this ratio. Since λ_1 is just an arbitrary reference value, we have approximately $l_\lambda = [J(\lambda)]^{1/n}$. Thus $g(\lambda) \propto l_\lambda{}^{-k} = [J(\lambda)]^{-k/n}$ is the final expression for $g(\lambda)$ which Box and Cox tentatively employ.

Integrating with respect to $\boldsymbol{\beta}$ yields

$$p(\lambda|\mathbf{y}) \propto J^{\nu/n}[s^2(\lambda)]^{-\nu/2}, \tag{6.25}$$

which is the marginal posterior pdf for λ. This pdf can be analyzed numerically. Note, too, that

$$\begin{aligned} \log p(\lambda|\mathbf{y}) &= \text{const} + \frac{\nu}{n}\left[\log J - \frac{n}{2}\log s^2(\lambda)\right] \\ &= \text{const} + \frac{\nu}{n}\left[(\lambda - 1)\sum_{\alpha=1}^{n} \log y_\alpha - \frac{n}{2}\log s^2(\lambda)\right], \end{aligned} \tag{6.26}$$

and thus, on comparison with (6.8), it is seen that the modal value of the posterior pdf is identical to the ML estimate.

To obtain the marginal pdf for an element of $\boldsymbol{\beta}$, say β_1, (6.24) can be integrated with respect to $\beta_2, \beta_3, \ldots, \beta_k$ by using properties of the multivariate Student t pdf. Then the result is a bivariate posterior pdf for λ and β_1 which can be analyzed numerically. Last, note from (6.24) that the conditional pdf for $\boldsymbol{\beta}$, given λ, is in the form of a multivariate Student t distribution. This fact can be employed to study how sensitive inferences about the β's are to what is assumed about λ.

6.2 CONSTANT ELASTICITY OF SUBSTITUTION (CES) PRODUCTION FUNCTION

In a path-breaking paper[10] Arrow, Chenery, Minhas, and Solow analyzed a class of production functions with a constant elasticity of substitution parameter which we shall denote by $\mathscr{E}$, $0 \leq \mathscr{E} < \infty$. They show that in a two-input function, when $\mathscr{E} = 1$, the CES production function becomes the Cobb-Douglas (CD) function, when $\mathscr{E} = 0$, the Leontief fixed proportion model, and, when $\mathscr{E} = \infty$, a production function with perfect substitutability. Although the parameter $\mathscr{E}$ has most often been estimated by using a necessary condition for profit maximization, here we take up the direct estimation of the nonlinear function with two inputs, as presented by Thornber,[11] and then go on to consider an alternative approach which can accommodate more than two inputs and is closely related to the Box-Cox analysis of transformations considered above.

[10] K. Arrow, H. Chenery, B. Minhas, and R. Solow, "Capital-labor Substitution and Economic Efficiency," *Rev. Econ. Statist.*, **XLIII**, 225–250 (1961).

[11] H. Thornber, "The Elasticity of Substitution: Properties of Alternative Estimators," manuscript, University of Chicago, 1966. See also V. K. Chetty and U. Sankar, "Bayesian Estimation of the CES Production Function," *Rev. Econ. Studies*, **36** (1969).

In the CES function the αth observation on output y_α is related to the capital and labor inputs K_α and L_α as follows:

$$y_\alpha = \gamma[\delta K_\alpha^{-\rho} + (1-\delta)L_\alpha^{-\rho}]^{-v/\rho}e^{u_\alpha}, \qquad \alpha = 1, 2, \ldots, n, \tag{6.27}$$

or

$$\log y_\alpha = \log \gamma + v \log \{[\delta K_\alpha^{-\rho} + (1-\delta)L_\alpha^{-\rho}]^{-1/\rho}\} + u_\alpha, \tag{6.28}$$

where γ, δ, v, and $\rho = -1 + 1/\mathscr{E}$ are parameters that satisfy $0 < \gamma < \infty$, $0 < \delta < 1$, $-\infty < v < \infty$, and $-1 < \rho < \infty$. Further, u_α is the αth disturbance term. We assume that the u_α's are normally and independently distributed, each with mean zero and common variance σ^2. Finally, we assume that either the K_α and L_α are nonstochastic or, if stochastic, are distributed independently of the u_α with a distribution not involving the parameters γ, ρ, δ, v, and σ.

Under the above assumptions the likelihood function is

$$l(\gamma, v, \delta, \rho, \sigma|\mathbf{y}, \mathbf{K}, \mathbf{L}) \propto \frac{1}{\sigma^n} \exp\left\{-\frac{1}{2\sigma^2}[\hat{\mathbf{u}}'\hat{\mathbf{u}} + (\boldsymbol{\theta} - \hat{\boldsymbol{\theta}})'X'X(\boldsymbol{\theta} - \hat{\boldsymbol{\theta}})]\right\}, \tag{6.29}$$

where

$$\boldsymbol{\theta} = \begin{pmatrix} \ln \gamma \\ v \end{pmatrix}, \qquad \hat{\boldsymbol{\theta}} = (X'X)^{-1}X'\mathbf{y},$$

$$X = \begin{bmatrix} 1 & \log \{[\delta K_1^{-\rho} + (1-\delta)L_1^{-\rho}]^{-1/\rho}\} \\ \vdots & \vdots \\ 1 & \log \{[\delta K_n^{-\rho} + (1-\delta)L_n^{-\rho}]^{-1/\rho}\} \end{bmatrix}, \qquad \mathbf{y} = \begin{bmatrix} \log y_1 \\ \vdots \\ \log y_n \end{bmatrix},$$

and

$$\hat{\mathbf{u}}'\hat{\mathbf{u}} = (\mathbf{y} - X\hat{\boldsymbol{\theta}})'(\mathbf{y} - X\hat{\boldsymbol{\theta}}).$$

Note that the $n \times 2$ matrix X, $\hat{\theta}$, and $\hat{\mathbf{u}}'\hat{\mathbf{u}}$ are functions of the parameters δ and ρ.

To generate ML estimates we take the logarithm of (6.29), denoted by $L = \log l$, and maximize with respect to σ and θ, which leads to

$$\tilde{\sigma}^2 = \frac{1}{n}\hat{\mathbf{u}}'\hat{\mathbf{u}} \quad \text{and} \quad \tilde{\theta} = \hat{\theta} \tag{6.30}$$

as the maximizing values. On evaluating L for these values we obtain $L_{\max}$ given by

$$L_{\max} = \text{constant} - \frac{n}{2}\log(\hat{\mathbf{u}}'\hat{\mathbf{u}}), \tag{6.31}$$

where $\hat{\mathbf{u}}'\hat{\mathbf{u}}$, shown above, is a function of ρ and δ. By searching over a grid of values for ρ and δ[12] we can find those values of ρ and δ that minimize $\hat{\mathbf{u}}'\hat{\mathbf{u}}$, if they exist, and they will be ML estimates. Then the quantities in (6.30) can be evaluated for the minimizing values of ρ and δ to provide ML estimates for σ and $\boldsymbol{\theta}$. Thornber points out that the mean and variance of the ML estimator for $\mathscr{E}$, the elasticity of substitution parameter, do not exist in finite samples. However, the mean and variance of its asymptotic normal distribution, given γ, v, δ, and σ, do exist. The mean is $\mathscr{E}$, whereas the variance of this asymptotic conditional distribution is

$$\text{(6.32)} \qquad \begin{aligned} \text{Var}(\hat{\mathscr{E}}) &= \left\{E\left[\frac{\partial \log l(\mathscr{E}|\gamma, v, \delta, \sigma, \mathbf{y})}{\partial\mathscr{E}}\right]^2\right\}^{-1} \\ &= \frac{\mathscr{E}^2(\mathscr{E}-1)^2\sigma^2}{v^2 S(\mathscr{E})}, \end{aligned}$$

with

(6.33)

$$S(\mathscr{E}) = \sum_{\alpha=1}^{n}\left\{\frac{(1/\rho)[\delta K_\alpha^{-\rho} + (1-\delta)L_\alpha^{-\rho}]\log[\delta K_\alpha^{-\rho} + (1-\delta)L_\alpha^{-\rho}] + [\delta K_\alpha^{-\rho}\log K_\alpha + (1-\delta)L_\alpha^{-\rho}\log L_\alpha]}{\delta K_\alpha^{-\rho} + (1-\delta)L_\alpha^{-\rho}}\right\}^2,$$

where $\rho = (1-\mathscr{E})/\mathscr{E}$. This result can be employed to compute an approximate large-sample standard error.

In addition to these large-sample results, Thornber reports the results of some Monte Carlo experiments designed to provide estimates of the risk function associated with alternative estimators for $\mathscr{E}$, including the maximum likelihood estimator, a linearized maximum likelihood estimator, and two estimators generated by the criterion that they minimize expected loss, with the expectation being taken by using the posterior pdf for the parameters. The loss function he employed in this work is

$$\text{(6.34)} \qquad L(\mathscr{E}, \hat{\mathscr{E}}) = \frac{(\mathscr{E} - \hat{\mathscr{E}})^2}{(1+\mathscr{E})^2(1+\hat{\mathscr{E}})^2},$$

a loss function that yields greater relative loss for an underestimate than for an overestimate. Thornber used the following two prior pdf's in his experiments:

First Prior: $\quad p_1(\gamma, v, \sigma, \delta, \mathscr{E}) \propto \dfrac{(1+\mathscr{E})^2 e^{-\mathscr{E}}}{\gamma\sigma}$

Second Prior: $\quad p_2(\gamma, v, \sigma, \delta, \mathscr{E}) \propto \dfrac{\mathscr{E}(1+\mathscr{E})e^{-\mathscr{E}}}{\gamma\sigma},$

[12] Actually reparameterizing in terms of $\lambda = 1/(1+\mathscr{E}) = (1+\rho)/(2+\rho)$ or $\rho = (2\lambda - 1)/(1-\lambda)$ is convenient, since $0 \leq \lambda \leq 1$, and thus the search over δ and λ is confined to the unit square, a point made by Thornber.

with $0 < \sigma, \gamma, v < \infty$, $0 \leq \delta \leq 1$, and $\mathscr{E} \geq 0$. The normalized marginal prior pdf's for $\mathscr{E}$ are

$$f_1(\mathscr{E}) = \tfrac{1}{5}(1 + \mathscr{E})^2 e^{-\mathscr{E}} \quad \text{and} \quad f_2(\mathscr{E}) = \tfrac{1}{11}\mathscr{E}(1 + \mathscr{E})^2 e^{-\mathscr{E}}.$$

The first of these, $f_1(\mathscr{E})$, has its mode at $\mathscr{E} = 1$, whereas the mode associated with $f_2(\mathscr{E})$ is at about $\mathscr{E} = 2.12$.

Shown in Figures 6.2 and 6.3 are the results of Thornber's experiments for

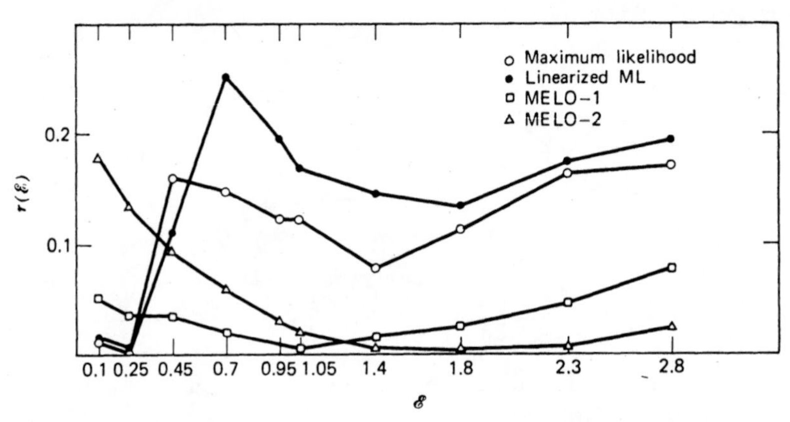

Figure 6.2 Risk functions for $n = 10$. Estimated risk functions under $L(\mathscr{E}, \hat{\mathscr{E}}) = (\mathscr{E} - \hat{\mathscr{E}})^2/[(1 + \mathscr{E})^2 (1 + \hat{\mathscr{E}})^2]$.

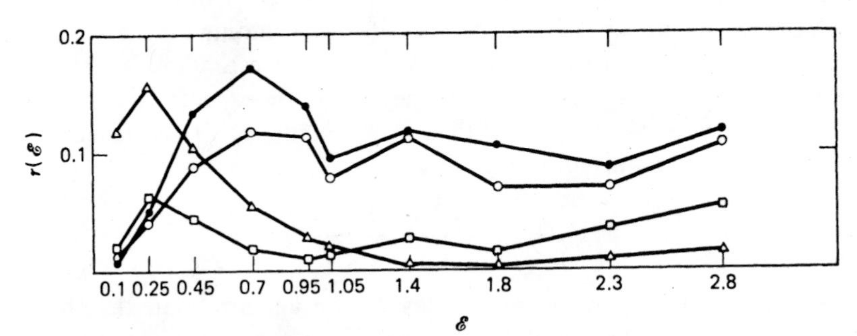

Figure 6.3 Risk functions for $n = 20$.

two sample sizes, 10 and 20. The points labeled MELO-1 and MELO-2 are Bayesian minimum expected loss estimators using the first prior pdf and the second, respectively. It is seen that use of the minimum expected loss procedure to generate estimators that incorporate the prior information has resulted in substantial reduction in risk over almost all of the parameter space. It is just for low values of the parameter $\mathscr{E}$ that the risk associated with the maximum likelihood estimators is lower than that associated with the minimum expected loss estimators.

In appraising these results it should be emphasized that a frequentist criterion is being employed, one that not everyone accepts. For many, the estimate that minimizes expected loss, given the sample information, is optimal in line with the expected utility hypothesis and no frequentist argument is required.

We now turn to an alternative approach to the analysis of the CES production function which illustrates an interesting connection with the Box-Cox analysis of transformations. Let us initially consider the deterministic form of the CES function with two inputs, $x_{1\alpha}$ and $x_{2\alpha}$, and constant returns to scale, namely,

$$V_\alpha = \gamma[\delta_1 x_{1\alpha}{}^g + (1 - \delta_1)x_{2\alpha}{}^g]^{1/g}, \qquad \alpha = 1, 2, \ldots, n, \tag{6.35}$$

where V_α denotes the systematic part of output and $g = -\rho = (\mathscr{E} - 1)/\mathscr{E}$, where $\mathscr{E}$ is the elasticity of substitution parameter. Then, on raising both sides of (6.35) to the power g and rearranging terms, we have

$$V_\alpha^{(g)} = x_{2\alpha}^{(g)} + \gamma^g \delta_1[x_{1\alpha}^{(g)} - x_{2\alpha}^{(g)}] + \gamma^{(g)}(1 + gx_{2\alpha}^{(g)}), \tag{6.36}$$

where $V_\alpha^{(g)} \equiv (V_\alpha{}^g - 1)/g$, $\gamma^{(g)} = (\gamma^g - 1)/g$ and $x_{i\alpha}^{(g)} = (x_{i\alpha}^g - 1)/g$, $i = 1, 2$. Now assume that the observed output y_α satisfies

$$y_\alpha^{(\lambda)} = V_\alpha^{(g)} + u_\alpha, \qquad \alpha = 1, 2, \ldots, n, \tag{6.37}$$

where

$$y_\alpha^{(\lambda)} = \frac{y_\alpha{}^\lambda - 1}{\lambda},$$

or

$$y_\alpha^{(\lambda)} = x_{2\alpha}^{(g)} + \beta_1(x_{1\alpha}^{(g)} - x_{2\alpha}^{(g)}) + \beta_2(1 + gx_{2\alpha}^{(g)}) + u_\alpha, \qquad \alpha = 1, 2, \ldots, n \tag{6.38}$$

where

$$\lambda = \phi g, \qquad \beta_1 = \gamma^g \delta_1, \quad \text{and} \quad \beta_2 = \gamma^{(g)}, \tag{6.39}$$

with ϕ, a free parameter.[13] Note that in (6.38) we do *not* write $y_\alpha^{(g)} = (y_\alpha{}^g - 1)/g$ as the dependent variable, since it does not seem reasonable to assume that a power transformation with parameter $g = (\mathscr{E} - 1)/\mathscr{E}$ will induce normality and stabilize the variance of the disturbance terms. Rather, we introduce a new parameter $\lambda = \phi g$ and use it in the power transformation on the dependent variable. If $g = 0$, (6.38) reduces to the CD form. If $\lambda = 1$ and $g = 1$, we obtain a form linear in the variables. Also, clearly, if $\lambda = 0$ and $g = 1$, we have a semilog relation. It is seen that introducing the new parameter λ widens the range of possible functional forms under consideration.

[13] Equation 6.38 involves power transformations on the dependent and independent variables and is thus an example of Box and Cox's discussion in Section 8 of their paper. What follows below is a presentation of their procedure for analyzing this case.

With $\mathbf{w} = \mathbf{y}^{(\lambda)} - \mathbf{x}_2^{(g)}$, $X = (\mathbf{x}_1^{(g)} - \mathbf{x}_2^{(g)} \vdots \iota + g\mathbf{x}_2^{(g)})$, $\boldsymbol{\beta}' = (\beta_1, \beta_2)$, and when the u_α in (6.38) are assumed to be normally and independently distributed, each with mean zero and common variance σ^2, the likelihood function is

$$
\begin{aligned}
l(\boldsymbol{\beta}, g, \lambda, \sigma|\mathbf{y}) &\propto \frac{J}{\sigma^n} \exp\left[-\frac{1}{2\sigma^2}(\mathbf{w} - X\boldsymbol{\beta})'(\mathbf{w} - X\boldsymbol{\beta})\right] \\
&\propto \frac{J}{\sigma^n} \exp\left\{-\frac{1}{2\sigma^2}[n\hat{\sigma}^2 + (\boldsymbol{\beta} - \hat{\boldsymbol{\beta}})'X'X(\boldsymbol{\beta} - \hat{\boldsymbol{\beta}})]\right\},
\end{aligned} \tag{6.40}
$$

where $J = \prod_{\alpha=1}^n y_\alpha^{\lambda-1}$, the Jacobian of the transformation from the u_α's to the y_α's,

$$\hat{\boldsymbol{\beta}} = (X'X)^{-1}X'\mathbf{w}, \tag{6.41}$$

and

$$\hat{\sigma}^2 = \frac{1}{n}(\mathbf{w} - X\hat{\boldsymbol{\beta}})'(w - X\hat{\boldsymbol{\beta}}). \tag{6.42}$$

It should be emphasized that the quantities $\hat{\boldsymbol{\beta}}$ and $\hat{\sigma}^2$ are functions of λ and g. Also $\hat{\boldsymbol{\beta}}$ and $\hat{\sigma}^2$ are values associated with a maximum of the likelihood function for any given λ and g. On taking the logarithm of (6.40) and substituting (6.41) and (6.42) for $\boldsymbol{\beta}$ and σ^2, respectively, we obtain

$$L_{\max}(g, \lambda) = \text{const} + (\lambda - 1)\sum_{\alpha=1}^{n} \log y_\alpha - \frac{n}{2}\log \hat{\sigma}^2. \tag{6.43}$$

Now we can use the computer to evaluate $L_{\max}(g, \lambda)$ over a grid of values for g and λ to find the pair associated with a maximum of $L_{\max}(g, \lambda)$, if such a pair exists. Then these values say $\hat{g}$ and $\hat{\lambda}$, and the values of $\hat{\boldsymbol{\beta}}$ and $\hat{\sigma}^2$ associated with $\hat{g}$ and $\hat{\lambda}$ are ML estimates. Further from (6.39), given that we have the estimates $\hat{g}$, $\hat{\lambda}$, $\hat{\beta}_1$, and $\hat{\beta}_2$, it is easy to obtain estimates of γ, δ_1, and ϕ.

In situations in which we have more than two inputs and assume returns to scale that may not be constant, a similar approach can be employed to obtain ML estimates. Here in place of (6.34) we have $x_1, \ldots, x_k$ for the k input variables (we suppress the observation subscript α for convenience of notation):

$$V = \gamma[\delta_1 x_1^{g} + \delta_2 x_2^{g} + \cdots + \delta_{k-1}x_{k-1}^{g} + (1 - \delta_1 - \delta_2 - \cdots - \delta_{k-1})x_k^{g}]^{v/g}, \tag{6.44}$$

where again V is the systematic part of output, $g = -\rho = (\mathscr{E} - 1)/\mathscr{E}$, γ and the δ's are parameters, and v is the returns-to-scale parameter. Raising both sides of (6.44) to the power g/v and rearranging terms, we get

$$V^{g/v} = \gamma^{g/v}[\delta_1(x_1^{g} - x_k^{g}) + \delta_2(x_2^{g} - x_k^{g}) + \cdots + \delta_{k-1}(x_{k-1}^{g} - x_k^{g}) + x_k^{g}] \tag{6.45}$$

and, on further rearrangements,

$$(6.46)\quad V^{(g/v)} = vx_k^{(g)} + \beta_1(x_1^{(g)} - x_k^{(g)}) + \beta_2(x_2^{(g)} - x_k^{(g)}) + \cdots + \beta_{k-1}(x_{k-1}^{(g)} - x_k^{(g)}) + \beta_k(1 + gx_k^{(g)}).$$

In (6.46) the β's are defined as follows:

$$(6.47)\quad \beta_i = v\delta_i\gamma^{g/v}, \qquad i = 1, 2, \ldots, k-1, \quad \text{and} \quad \beta_k = \gamma^{(g/v)}.$$

As above, we see no reason why a power transformation involving the parameters g and v should induce normality and stabilize the variance. Rather, we assume that the observed output $\mathbf{y}$ is related to V as follows:

$$(6.48)\quad y^{(\lambda)} = V^{(g/v)} + u,$$

where u is a disturbance term and here $\lambda = \phi g/v$, with ϕ a free parameter. Then, in matrix terms, the model for the observations is

$$(6.49)\quad \mathbf{w} = X\boldsymbol{\beta} + \mathbf{u},$$

where $\mathbf{w} = y^{(\lambda)}$, $\boldsymbol{\beta}' = (v, \beta_1, \beta_2, \ldots, \beta_k)$, $\mathbf{u}' = (u_1, u_2, \ldots, u_n)$, and

$$X = (\mathbf{x}_k^{(g)}, \mathbf{x}_1^{(g)} - \mathbf{x}_k^{(g)}, \mathbf{x}_2^{(g)} - \mathbf{x}_k^{(g)}, \ldots, \mathbf{x}_{k-1}^{(g)} - \mathbf{x}_k^{(g)}, \boldsymbol{\iota} + g\mathbf{x}_k^{(g)}),$$

where $\boldsymbol{\iota}$ is an $n \times 1$ column of ones. Just as above we can formulate the likelihood function, which is

$$(6.50)\quad l(\boldsymbol{\beta}, \lambda, g, v, \sigma|\mathbf{y}) \propto \frac{J}{\sigma^n} \exp\left[-\frac{1}{2\sigma^2}(\mathbf{w} - X\boldsymbol{\beta})'(\mathbf{w} - X\boldsymbol{\beta})\right],$$

where J is the Jacobian factor, and proceed to maximize it in the Box-Cox two-step fashion. For any given values of λ and g the conditional maximizing values for $\boldsymbol{\beta}$ and σ^2 are

$$(6.51)\quad \hat{\boldsymbol{\beta}} = (X'X)^{-1}X'\mathbf{y}^{(\lambda)}$$

and

$$(6.52)\quad \hat{\sigma}^2 = \frac{1}{n}(\mathbf{y}^{(\lambda)} - X\hat{\boldsymbol{\beta}})'(\mathbf{y}^{(\lambda)} - X\hat{\boldsymbol{\beta}}).$$

On substituting these quantities in the logarithm of the likelihood function, we obtain

$$(6.53)\quad L_{\max}(\lambda, g) = \text{const} + (\lambda - 1)\sum_{\alpha=1}^{n} \log y_\alpha - \frac{n}{2}\log \hat{\sigma}^2,$$

which is just a function of the parameters λ and g. By using the computer (6.53) can be evaluated for a range of λ and g values to find the pair of values that yields a maximum and to study the properties of the surface. Given that λ and g are the values associated with the maximum of (6.53), the $\hat{\boldsymbol{\beta}}$ and $\hat{\sigma}^2$

values associated with $\hat{\lambda}$ and $\hat{g}$ are ML estimates for $\boldsymbol{\beta}$ and σ^2. Since the first element of $\boldsymbol{\beta}$ is defined as v, we have the ML estimate of v, $\hat{v}$. Then, on referring back to (6.47), we can determine ML estimates of γ and the δ's. As usual with ML estimation, large-sample standard errors may be obtained from the inverse of the estimated information matrix.[14]

These large-sample ML results will be useful in circumstances in which we have adequate numbers of observations showing enough variation to measure the properties of the highly nonlinear CES function. With small samples of data showing relatively little variation, it will, of course, be difficult to make precise inferences with these large-sample techniques.

6.3 GENERALIZED PRODUCTION FUNCTIONS[15]

Generalized production functions (GPF's) are another broad class of functions which are usually nonlinear in both parameters and variables. These functions have been introduced to permit generalization in two directions. We wish to have production functions with a preassigned elasticity of substitution, say constant, but unknown, or variable, say some function of the capital labor ratio. In addition to this requirement on the elasticity of substitution, we want our production function to have returns-to-scale that vary with the level of output according to a preassigned function. Zellner and Revankar have provided a method, briefly described below, of generating production functions that meet both requirements. Then we take up the problem of estimating the parameters of a particular GPF.

In deterministic terms we consider the following differential equation:

$$\frac{dV}{df} = \frac{V\alpha(V)}{f\alpha_f}, \tag{6.54}$$

with solution

$$V = g(f), \tag{6.55}$$

where $\alpha(V)$ is the returns-to-scale as a function of output $V, f = f(K, L)$ is in the form of a neoclassical production function, and α_f is the returns-to-scale parameter associated with f. The function $\alpha(V)$ is chosen to ensure that $dV/df > 0$ for all f, $0 < f < \infty$. Thus (6.55) is a monotonic transformation of f with the property that the shapes of the isoquants for $g(f)$ will be the

[14] See, for example, M. G. Kendall and A. Stuart, *The Advanced Theory of Statistics*, Vol. 2. New York: Hafner, 1961, pp. 51 ff.

[15] This section draws on the results presented in A. Zellner and N. S. Revankar, "Generalized Production Functions," Social Systems Research Institute Workshop Paper 6607, University of Wisconsin, Madison, 1966, published in the *Rev. Economic Studies*, **36**, 241–250 (1969).

same as those for f. Therefore the elasticity of the substitution parameter, constant or variable, associated with $V = g(f)$ will be the same[16] as that associated with the function f.

As an example,[17] illustrating the analysis of a particular GPF, let us take $\alpha(V)$ in the following form:

$$\alpha(V) = \frac{\alpha}{1 + \theta V}, \tag{6.56}$$

with α and θ parameters. If $\theta = 0$, the returns-to-scale do not depend on V. On the other hand, if $\theta > 0$, the returns-to-scale fall from α $(\alpha > 0)$, as $V \to 0$ and toward zero as $V \to \infty$. Inserting $\alpha(V)$, given in (6.56), into (6.54), the resulting differential equation is

$$\frac{dV}{df} = \frac{V}{f} \frac{\alpha}{\alpha_f(1 + \theta V)}, \tag{6.57}$$

with solution $Ve^{\theta V} = Cf^{\alpha/\alpha_f}$, where C is a constant of integration. If we let $f = AL^{\alpha_f(1-\delta)}K^{\alpha_f \delta}$, then we obtain

$$Ve^{\theta V} = \gamma K^{\alpha(1-\delta)}L^{\alpha\delta} \tag{6.58}$$

as our GPF with $\gamma = CA$. Taking the natural logarithms of both sides of (6.58) and adding a disturbance term, we have[18]

$$\log V_i + \theta V_i = \beta_1 + \beta_2 \log K_i + \beta_3 \log L_i + u_i, \tag{6.59}$$

where the subscript i denotes the ith observation, $i = 1, 2, \ldots, n$, $\beta_1 = \log \gamma$, $\beta_2 = \alpha(1 - \delta)$, and $\beta_3 = \alpha\delta$.

If, in (6.59), we assume that the u_i's are normally and independently distributed, each with mean zero and common variance σ^2, the likelihood function is[19]

$$l(\boldsymbol{\beta}, \theta, \sigma|\text{data}) \propto \frac{J}{\sigma^n} \exp\left[-\frac{1}{2\sigma^2}(\mathbf{z}_\theta - X\boldsymbol{\beta})'(\mathbf{z}_\theta - X\boldsymbol{\beta})\right], \tag{6.60}$$

where $\mathbf{z}_\theta$ is an $n \times 1$ vector, with a typical element $\log V_i + \theta V_i$, $\boldsymbol{\beta}' = (\beta_1, \beta_2, \beta_3)$, X is an $n \times 3$ matrix with a typical row given by $(1, \log K_i,$

[16] See A. Zellner and N. S. Revankar, *loc. cit.*, for an explicit proof.

[17] Other examples are provided in the Zellner-Revankar paper.

[18] Note that β_2 and β_3 are pure numbers, whereas β_1 and θ have values that depend on the units of measurement employed.

[19] We assume that the values of K_i and L_i are fixed or, if random, distributed independently of the disturbance terms with a pdf not involving parameters of (6.59). See Zellner and Revankar, *op. cit.*, p. 246, and A. Zellner, J. Kmenta, and J. Drèze, "Specification and Estimation of Cobb-Douglas Production Function Models," *Econometrica*, **34**, 784–795 (1966), for further discussion of these assumptions.

$\log L_i$), and J denotes the Jacobian of the transformation from the n u_i's to the n V_i's given by (6.59). The explicit expression for J is

$$J = \prod_{i=1}^{n} \frac{1 + \theta V_i}{V_i}. \tag{6.61}$$

We first indicate how the Box-Cox approach can be applied in the present instance to obtain maximum likelihood estimates. We substitute from (6.61) into (6.60) and then take the logarithm of both sides to obtain

$$L = \text{const} - \frac{n}{2}\log \sigma^2 + \sum_{i=1}^{n} \log(1 + \theta V_i) - \frac{1}{2\sigma^2}(\mathbf{z}_\theta - X\boldsymbol{\beta})'(\mathbf{z}_\theta - X\boldsymbol{\beta}), \tag{6.62}$$

where L denotes the log-likelihood, $\log l$. Maximizing with respect to σ^2 leads to

$$\hat{\sigma}^2 = \frac{1}{n}(\mathbf{z}_\theta - X\boldsymbol{\beta})'(\mathbf{z}_\theta - X\boldsymbol{\beta}) \tag{6.63}$$

as the conditional maximizing value for σ^2, given θ and $\boldsymbol{\beta}$. Substituting $\sigma^2 = \hat{\sigma}^2$ in (6.62) yields

$$L_1 = \text{const} - \frac{n}{2}\log(\mathbf{z}_\theta - X\boldsymbol{\beta})'(\mathbf{z}_\theta - X\boldsymbol{\beta}) + \sum_{i=1}^{n} \log(1 + \theta V_i). \tag{6.64}$$

From the form of (6.64) it is clear that, for any *given* θ, L_1 will be maximized if $(\mathbf{z}_\theta - X\boldsymbol{\beta})'(\mathbf{z}_\theta - X\boldsymbol{\beta})$ is minimized with respect to the elements of $\boldsymbol{\beta}$. The minimizing value for $\boldsymbol{\beta}$, given θ, is just

$$\hat{\boldsymbol{\beta}}_\theta = (X'X)^{-1}X'\mathbf{z}_\theta, \tag{6.65}$$

and on substituting this value in (6.64) we have

$$L_2 = \text{const} - \frac{n}{2}\log s_\theta^2 + \sum_{i=1}^{n} \log(1 + \theta V_i), \tag{6.66}$$

where

$$s_\theta^2 = \frac{(\mathbf{z}_\theta - X\hat{\boldsymbol{\beta}}_\theta)'(\mathbf{z}_\theta - X\hat{\boldsymbol{\beta}}_\theta)}{\nu}, \tag{6.67}$$

where $\nu = n - 3$ for the present problem. We can now evaluate the last two terms on the rhs of (6.66) for various values of θ to find the value associated with a maximum of L_2.[20] Let us denote this value by $\tilde{\theta}$. This can be sub-

[20] Note that this can be accomplished by regressing $\mathbf{z}_\theta$ on X for selected values of θ and obtaining s_θ^2. Then the last two terms on the rhs of (6.65) are evaluated. The conditional regressions of $\mathbf{z}_\theta$ on X are often of interest in that they show how sensitive results are to what is assumed about the value of θ.

stituted in (6.66) to obtain the ML estimate for $\boldsymbol{\beta}$, denoted $\tilde{\boldsymbol{\beta}}_{\tilde{\theta}}$. Then, in (6.63), we can take $\boldsymbol{\beta} = \tilde{\boldsymbol{\beta}}_{\tilde{\theta}}$ and $\mathbf{z}_\theta = \mathbf{z}_{\tilde{\theta}}$ to compute the ML estimate for σ^2. Large-sample standard errors associated with ML parameter estimates can be obtained from the inverse of the estimated information matrix.

The parameters associated with (6.59) have been estimated by using the ML approach and 1957 annual cross section observations for the U.S. transportation equipment industry.[21] In this application the ML estimate for θ, based on data for $n = 25$ states, was found to be 0.134, with a large-sample standard error of 0.0638. With $\tilde{\theta} = 0.134$ and $\tilde{\alpha} = 1.49$,[22] the returns-to-scale function in (6.56) can be evaluated, given V. Returns-to-scale were found to vary from a high of 1.45 to a low of 0.76 over the range of V observed in the data.

To pursue a Bayesian analysis of the model in (6.59), we require a prior pdf for the parameters. Given θ, we assume that the prior pdf for β_1, β_2, α, and σ is given by[23]

$$
(6.68a) \qquad p(\beta_1, \beta_2, \alpha, \sigma|\theta) \propto g(\theta)\, p_1(\beta_2|\alpha)\, p_2(\alpha)\, p_3(\sigma) \left\} \begin{array}{l} 0 < \theta, \sigma < \infty, \\ 0 < \beta_2 < \alpha, \\ 0 < \alpha < \infty, \end{array} \right.
$$

with

$$
(6.68b) \qquad g(\theta) \propto J^{-3/n},
$$

$$
(6.68c) \qquad p_1(\beta_2|\alpha) \propto \left(\frac{\beta_2}{\alpha}\right)^{q_1 - 1}\left(1 - \frac{\beta_2}{\alpha}\right)^{q_2 - 1},
$$

$$
(6.68d) \qquad p_2(\alpha) \propto \text{const},
$$

and

$$
(6.68e) \qquad p_3(\sigma) \propto \frac{1}{\sigma}.
$$

In (6.68*b*) we follow Box and Cox's argument, presented in connection with (6.16) and (6.17), to obtain a proportionality factor, $g(\theta)$, in the conditional

[21] See Zellner and Revankar, *loc. cit.*, for a fuller discussion of the data (presented in their paper) and the ML results.

[22] Zellner and Revankar, *loc. cit.*, obtained the following estimates:

$$
\underset{(0.3854)}{\hat{\beta}_1 = 3.0129}, \qquad \underset{(0.1023)}{\hat{\beta}_2 = 0.3330}, \qquad \underset{(0.1564)}{\hat{\beta}_3 = 1.1551},
$$

where figures in parentheses are large-sample standard errors. Since $\alpha = \beta_2 + \beta_3$, the ML estimate of α is given by $\hat{\alpha} = \hat{\beta}_2 + \hat{\beta}_3$.

[23] Since $\beta_2 + \beta_3 = \alpha$, we find it convenient initially to parameterize the prior pdf in terms of β_2 and α, rather than β_2 and β_3. Note that (6.59) can be written as $\log V_i + \theta V_i = \beta_1 + \beta_2 \log K_i/L_i + \alpha \log L_i + u_i$.

prior pdf in (6.68). The prior pdf for β_2, *given* α, shown in (6.68*c*), is a beta pdf with parameters q_1 and q_2, whereas (6.68*d*) and (6.68*e*) represent diffuse prior assumptions about α and σ.[24] Finally, the marginal prior pdf for θ, say $p_4(\theta)$, must be specified. Given that numerical integration techniques are to be employed, $p_4(\theta)$ can be assigned any of a variety of forms to represent the available prior information about θ.

In the present application we assume that our prior information about the parameters is relatively vague. In (6.68*c*) we take $q_1 = q_2 = 1$ and take $p_4(\theta)$, the marginal prior pdf for θ, to be uniform. Thus the prior pdf for the parameters to be employed in the calculations to follow is

$$p(\beta_1, \beta_2, \alpha, \sigma, \theta) \propto \frac{1}{J^{3/n}\sigma}, \tag{6.69}$$

with J given in (6.61). In the present instance we can transform (6.69) to obtain[25]

$$p(\beta_1, \beta_2, \beta_3, \sigma, \theta) \propto \frac{1}{J^{3/n}\sigma}. \tag{6.70}$$

On combining (6.70) with the likelihood function in (6.70), we have the following posterior pdf for the parameters

$$\begin{aligned} p(\boldsymbol{\beta}, \sigma, \theta|\text{data}) &\propto \frac{J^{(n-3)/n}}{\sigma^{n+1}} \exp\left[-\frac{1}{2\sigma^2}(\mathbf{z}_\theta - X\boldsymbol{\beta})'(\mathbf{z}_\theta - X\boldsymbol{\beta})\right] \\ &\propto \frac{J^{\nu/n}}{\sigma^{n+1}} \exp\left\{-\frac{1}{2\sigma^2}[\nu s_\theta^2 + (\boldsymbol{\beta} - \hat{\boldsymbol{\beta}}_\theta)'X'X(\boldsymbol{\beta} - \hat{\boldsymbol{\beta}}_\theta)]\right\}, \end{aligned} \tag{6.71}$$

where $\nu = n - 3$ and $\hat{\boldsymbol{\beta}}_\theta$ and s_θ^2 are shown in (6.65) and (6.66), respectively. As is apparent from the second line of (6.71), the conditional posterior pdf for $\boldsymbol{\beta}$, given θ and σ, is in the multivariate normal form[26] with conditional mean $\hat{\boldsymbol{\beta}}_\theta$ and covariance matrix $(X'X)^{-1}\sigma^2$.

The marginal posterior pdf's for the parameters can be obtained as follows.

[24] Alternatively, the analysis can be performed with an inverted gamma prior pdf for σ and a rather flexible choice of prior pdf's for α, given that numerical integration techniques are employed in analyzing the posterior pdf.

[25] The transformation from the variables of (6.69) to those in (6.70) has a Jacobian equal to 1, since $\alpha = \beta_2 + \beta_3$.

[26] Since the economic theory of the problem tells us $\beta_2, \beta_3 > 0$, this is a truncated normal pdf. For the present analysis we shall not utilize the prior information regarding the non-negativity of β_2 and β_3 but shall let these parameters have ranges from $-\infty$ to $+\infty$. Below it will be seen that for the data utilized the truncation is not important. If it were, a trivariate numerical integration would be needed to obtain marginal posterior pdf's.

If interest centers on σ and θ, (6.71) can be integrated with respect to $\boldsymbol{\beta}$ to obtain the bivariate posterior pdf for θ and σ:

$$p(\sigma, \theta|\text{data}) \propto \frac{J^{\nu/n}}{\sigma^{\nu+1}} \exp\left(-\frac{\nu s_\theta^2}{2\sigma^2}\right). \tag{6.72}$$

The pdf in (6.72) can be analyzed numerically to obtain the marginal posterior pdf's for σ and θ. Alternatively, the marginal posterior pdf for θ can be obtained by integrating (6.72) with respect to σ analytically to yield

$$p(\theta|\text{data}) \propto \frac{J^{\nu/n}}{(s_\theta^2)^{\nu/2}}. \tag{6.73}$$

Univariate numerical integration techniques can be employed to obtain the normalizing constant and to analyze other features[27] of this marginal pdf. Since, as noted above, θ has the dimensions of those of the reciprocal of the output rate [see (6.59)], it must be recognized that both the ML estimate of θ and the pdf in (6.73) will be affected by a change in units of measurement of output. Just as the ratio of the ML estimate to its standard error is free of units of measurement, the mean of θ divided by its standard deviation, the

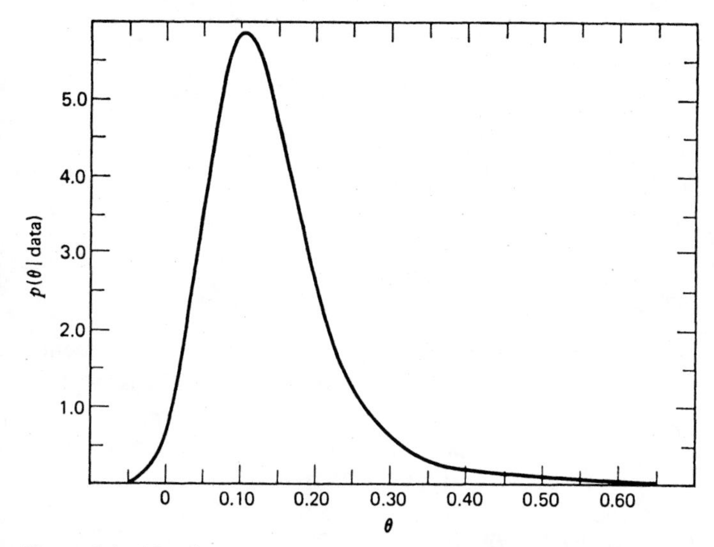

Figure 6.4 Marginal posterior pdf for θ computed from (6.73).

[27] It is interesting to observe that the mode of (6.73) occurs at $\theta = \tilde{\theta}$, the ML estimate; that is $\log p(\theta|\text{data}) = \text{const} + (\nu/n)(\log J - n/2 \log s_\theta^2) = \text{const} + (\nu/n)[\sum_{i=1}^{n} \log (1 + \theta V_i) - n/2 \log s_\theta^2]$. The quantity in square brackets is precisely the same as the last two terms on the rhs of (6.66) and thus the modal value of the posterior pdf is precisely equal to the ML estimate.

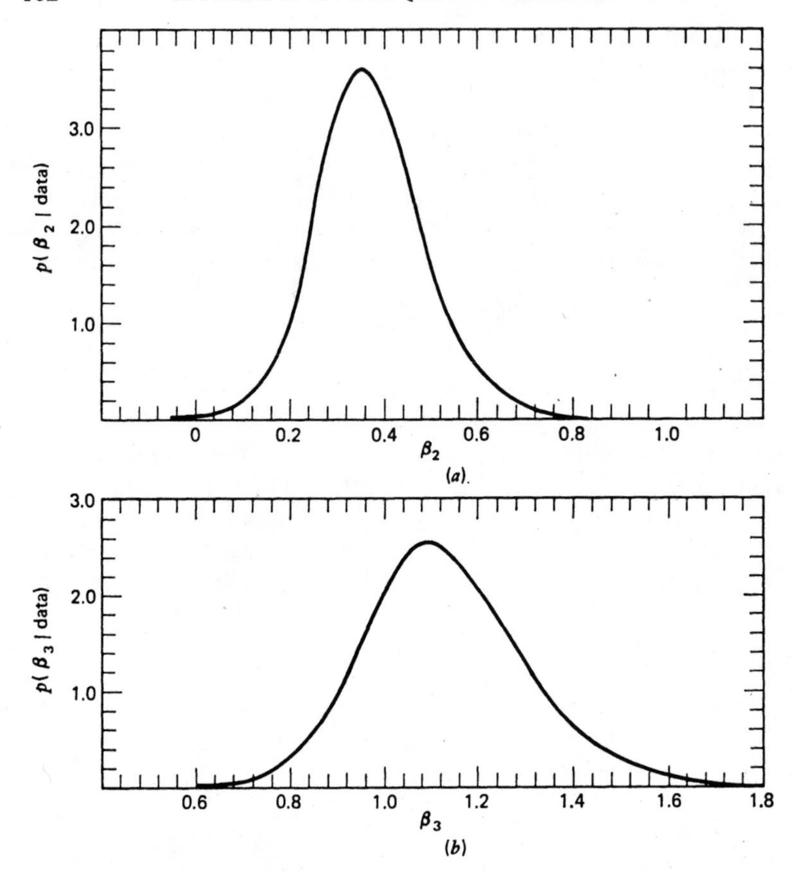

Figure 6.5 Marginal posterior pdf's for β_2 and β_3.

coefficient of variation, is also not dependent on units of measurement. Alternatively, for a given output rate, say V_t, the quantity θV_t is a pure number and its posterior pdf can be obtained from (6.73) by a simple change of variable from θ to $\eta_t = V_t\theta$. The posterior pdf for $\eta_t = \theta V_t$ is of interest because, as can be seen from (6.59), it is precisely the term that reflects a departure from the Cobb-Douglas form of the production function.

To obtain the marginal pdf's for one of the β's, say β_1, integrate (6.71) with respect to σ, β_2, and β_3 analytically. The result is a bivariate posterior pdf for β_1 and θ. Then bivariate numerical integration techniques can be employed to obtain the marginal posterior pdf for β_1. Similar operations yield the marginal posterior pdf's for β_2 and β_3.

The posterior pdf for $\alpha = \beta_2 + \beta_3$ is obtained by integrating (6.71) with respect to β_1 and σ analytically. The result is the posterior pdf for β_2, β_3, and

θ. For given θ this pdf is in the form of a bivariate Student t pdf. Then make a change of variables to $\alpha = \beta_2 + \beta_3$ and β_3 and integrate out β_3, an operation that yields the joint posterior pdf for θ and α. Bivariate numerical integration techniques can be used to calculate the marginal posterior pdf for α.

The operations above have been applied by using the U.S. Census of Manufactures cross-section data for the transportation equipment industry presented in the Zellner-Revankar paper. For each of $n = 25$ states the data are value added, labor input, and capital input, all on a per establishment basis. Shown in Figure 6.4 is the posterior pdf for θ. It is clear that it has most of its density over positive values of θ, which suggests that the returns-to-scale do vary with the level of output. In Figure 6.5 the marginal posterior pdf's for β_2 and β_3 are shown. With the relatively diffuse prior pdf's employed in this analysis, it is seen that the posterior pdf's have modal values that are close to the ML estimates. Although this is the case, it should be noted that the posterior pdf's depart from being normal, which indicates that "large sample" conditions are not yet encountered for $n = 25$.[28]

QUESTIONS AND PROBLEMS

1. Consider a simple regression model, $y_i = \beta_0 + \beta x_i + u_i$, $i = 1, 2, \ldots, n$, with $\eta_i \equiv E(y_i|x_i, \beta_0, \beta) = \beta_0 + \beta x_i$. After providing assumptions sufficient to obtain a posterior pdf for β_0 and β derive the posterior pdf of the elasticity (θ_i) of η_i with respect to x_i, namely,

$$\theta_i \equiv \frac{x_i}{\eta_i}\frac{\partial \eta_i}{\partial x_i} = \frac{x_i\beta}{\beta_0 + x_i\beta} = \frac{1}{1 + \beta_0/\beta x_i},$$

for given $x_i \neq 0$.

2. If β_0 and β have a posterior pdf in the bivariate Student t form, will the posterior mean of θ_i in Problem 1 exist? If the denominator of θ_i has a very small probability of being nonpositive as n grows, justify the approximation $E\theta_i \doteq x_i\hat{\beta}/(\hat{\beta}_0 + x_i\hat{\beta})$ for large n, where $\hat{\beta}_0 = \bar{y} - \hat{\beta}\bar{x}$ and $\hat{\beta} = \sum_{i=1}^n (y_i - \bar{y}) \times (x_i - \bar{x})/\sum_{i=1}^n (x_i - \bar{x})^2$.

3. Let the observation y_i satisfy

$$y_i = f(x_i, \alpha) + \epsilon_i, \qquad i = 1, 2, \ldots, n,$$

where the ϵ_i's are assumed NID$(0, \sigma^2)$ and $f(x_i, \alpha)$ is a continuous, twice differentiable function of an independent variable, x_i, and a scalar parameter,

[28] This suggests that "large-sample" properties of standard errors, based on the inverse of the estimated information matrix, and of sampling theory confidence intervals may not be encountered for the present model and data with $n = 25$.

α. If we expand $f(x_i, \alpha)$ in a Taylor's series about $\hat{\alpha}$, the ML estimate for α, retaining just the linear term, and write

$$y_i \doteq f(x_i, \hat{\alpha}) + (\alpha - \hat{\alpha}) \left.\frac{\partial f}{\partial \alpha}\right|_{\alpha = \hat{\alpha}} + \epsilon_i,$$

explain how linear Bayesian theory can be utilized to analyze this linearized equation. Comment on the form, mean, and variance of the posterior pdf for α, given that a diffuse prior pdf for α and σ is employed.

4. In Problem 3 an approximation to a nonlinear equation was introduced. Comment on the quality of the approximation as n gets large and the likelihood function becomes sharp. In particular, consider the behavior of the second-order term in the Taylor's series expansion as n grows large. When n is small, would it be appropriate to use the approximation for $f(x_i, \alpha)$ along with an informative prior pdf for α?

5. Generalize the considerations in Problems 3 and 4 to the case in which α represents a vector of parameters rather than a scalar parameter.

6. Assume that $\log y_i = \beta_1 + \beta_2 x_{2i} + u_i$, $i = 1, 2, \ldots, n$, where the u_i's are $\text{NID}(0, \sigma^2)$ and the x_{2i}'s are given values of an independent variable. Derive expressions for the mean and median of the pdf for y_i, given x_{2i}, β_1, β_2, and σ, and explain how to obtain posterior pdf's for these two measures of central tendency relating to the pdf for y_i.

7. Assume that we have a discrete random variable, y_i, which assumes the value 1 with probability P_i and the value 0 with probability $1 - P_i$. Then, if, in n independent trials, we observe n_1 1's and $n - n_1$ 0's, the likelihood function (l) is given by

$$l = \prod_{i=1}^{n_1} P_i \prod_{i=n_1+1}^{n} (1 - P_i).$$

Further, suppose that P_i satisfies $P_i = 1 - \alpha_1 e^{-\alpha_2 x_i}$, $i = 1, 2, \ldots, n$, where α_1 and α_2 are parameters and x_i is a non-negative valued given observable "stimulus" variable. (a) Discuss properties of the function, introduced above, giving the locus of the P_i's. (b) Explain how a computer search method can be utilized to obtain ML estimates for α_1 and α_2. (c) Formulate a prior pdf for α_1 and α_2 and indicate how bivariate numerical integration techniques can be used to make posterior inferences.

8. If, in Problem 7, we assume the following alternative, logistic functional form for the P_i's,

$$P_i = (1 + e^{-(\beta_1 + \beta_2 z_i)})^{-1}, \qquad i = 1, 2, \ldots, n,$$

where z_i is a given observable variable that can range from $-\infty$ to $+\infty$, (a) investigate the mathematical properties of the logistic function, particularly the dependence of its shape on the algebraic sign of β_2, (b) present a procedure for computing ML estimates for β_1 and β_2, given sample information, and (c) in terms of a particular suggested application, using the above model, formulate a prior pdf for β_1 and β_2 and indicate how a Bayesian analysis can be performed.

9. Do Problem 8 under the assumption that

$$P_i = \frac{1}{\sqrt{2\pi}} \int_{-\infty}^{\gamma_1 + \gamma_2 z_i} e^{-\frac{1}{2}w^2} \, dw,$$

where γ_1 and γ_2 are parameters with unknown values. (In Part a, investigate the dependence of the shape of the cumulative normal function, shown above, on the algebraic sign of γ_2.)

10. Consider the following "Engel relation":

$$y_i = \alpha x_i^{\beta} + \epsilon_i, \qquad i = 1, 2, \ldots, n,$$

where y_i = expenditure, x_i = income, α and β are parameters with unknown values, ϵ_i is an error term, and the subscript i denotes variables pertaining to the ith household. Assume that the x_i's are given independent variables and the ϵ_i's are NID(0, σ^2) with σ^2 having an unknown value. Provide a convenient algorithm for computing ML estimates of α, β, and σ^2.

11. If, in Problem 10, we have a prior pdf for α, β, and σ, say $p(\alpha, \beta, \sigma) \propto p_1(\alpha, \beta)/\sigma$, where $0 < \sigma < \infty$ and $p_1(\alpha, \beta)$ is a prior pdf for α and β, derive the joint posterior pdf for α and β. Then explain how to compute the posterior mean and variance of αx_i^{β}, given $x_i = x_0$, a known value.

12. If, in Problem 11, we assume that $p_1(\alpha, \beta) \propto$ const, what are the modal values of the joint posterior pdf for α and β? Comment on the assumption $p_1(\alpha, \beta) \propto$ const and provide an alternative that is more in accord with previous experience with Engel curve analysis.

13. Let $y_i = \beta(x_i^* - x_i) + \epsilon_i$ with $x_i^* = \alpha z_i$, $i = 1, 2, \ldots, n$, which implies $y_i = \beta(\alpha z_i - x_i) + \epsilon_i$, where z_i and x_i are given independent variables, α and β are parameters with unknown values, and the ϵ_i's are random error terms assumed to be NID(0, σ^2), with σ^2 the common unknown variance of each ϵ_i. Derive ML estimates of α and β and comment on the sampling properties of the ML estimator for α. In particular, does its mean exist? Evaluate Fisher's information matrix for the parameters α, β, and σ^2 and comment on its properties.

14. Given that $\alpha = \alpha_0$, obtain the posterior pdf for β in Problem 13, using the following prior pdf: $p(\beta, \sigma) \propto p_1(\beta)/\sigma$, with $0 < \sigma < \infty$ and $p_1(\beta)$ is a proper prior pdf for β. As n grows large, what is the mean of the posterior pdf for β, given $\alpha = \alpha_0$ and the sample information?

15. Explain how the model in Problem 13 can be analyzed using the following informative prior pdf: $p(\alpha, \beta, \sigma) = p_1(\alpha)\, p_2(\beta)\, p_3(\sigma)$, with $p_1(\alpha)$ and $p_2(\beta)$ being beta pdf's and $p_3(\sigma)$ an inverted gamma pdf.

CHAPTER VII

Time Series Models: Some Selected Examples

Most, if not all, economic analyses involve time series data.[1] Thus it is important to have good techniques for the analysis of time series models. In this chapter we take up the analysis of selected time series models to demonstrate how the Bayesian approach can be applied in the analysis of time series data. It will be seen that if the likelihood function is given and if we have a prior pdf the general principles of Bayesian analysis, presented in Chapter 2, apply with no special modifications required. This is indeed fortunate, since it means that our general principles are applicable to time series problems as well as to others.

7.1 FIRST ORDER NORMAL AUTOREGRESSIVE PROCESS

The model assumed to generate our observations, $\mathbf{y}' = (y_1, y_2, \ldots, y_T)$ is[2]

$$(7.1) \qquad y_t = \beta_1 + \beta_2 y_{t-1} + u_t, \qquad t = 1, 2, \ldots, T$$

where β_1 and β_2 are unknown parameters and u_t is a disturbance term. We assume that the u_t's are normally and independently distributed, each with zero mean and common variance σ^2. As regards initial conditions, we shall first go ahead conditional on given y_0, the observation for $t = 0$.[3] With these assumptions the likelihood function is

$$(7.2) \qquad l(\beta_1, \beta_2, \sigma|\mathbf{y}, y_0) \propto \frac{1}{\sigma^T} \exp\left[-\frac{1}{2\sigma^2} \sum (y_t - \beta_1 - \beta_2 y_{t-1})^2\right]$$

with the summation extending from $t = 1$ to $t = T$.

[1] The distinction often made between time series and cross-section data does not invalidate the statement made above, since cross-section data are in fact observations on time series variables pertaining to individual units in a cross section. Overlooking the time series nature of cross-section data can at times lead to serious errors in analyzing data.

[2] Here we use the subscript t to denote the tth value of a variable.

[3] See the discussion of initial conditions presented in connection with the problem of autocorrelation in regression analysis, presented in Chapter 4, for other possible assumptions.

As regards a prior pdf for the parameters, we shall assume that our information is diffuse and represent this in the usual way, namely,

$$p(\beta_1, \beta_2, \sigma) \propto \frac{1}{\sigma}, \tag{7.3}$$

with $-\infty < \beta_1 < \infty$, $-\infty < \beta_2 < \infty$, and $0 < \sigma < \infty$. Note that we do not restrict β_2 to be within the interval -1 to $+1$ and thus the analysis applies to both the explosive and nonexplosive cases.[4] In fact, our posterior pdf for β_2 will reflect what the sample information has to say about whether the process is or is not explosive.

On combining (7.2) and (7.3), the posterior pdf for the parameters is

$$p(\boldsymbol{\beta}, \sigma|\mathbf{y}, y_0) \propto \frac{1}{\sigma^{T+1}} \exp\left[-\frac{1}{2\sigma^2} \sum (y_t - \beta_1 - \beta_2 y_{t-1})^2\right], \tag{7.4}$$

where $\boldsymbol{\beta}' = (\beta_1, \beta_2)$, which is in a form exactly similar to that obtained in our analysis of the simple normal regression model in Section 3.1. To obtain the marginal posterior pdf for $\boldsymbol{\beta}$ we integrate (7.4) with respect to σ which yields

$$\begin{aligned} p(\boldsymbol{\beta}|\mathbf{y}, y_0) &\propto [\textstyle\sum (y_t - \beta_1 - \beta_2 y_{t-1})^2]^{-T/2} \\ &\propto [\nu s^2 + (\boldsymbol{\beta} - \hat{\boldsymbol{\beta}})' H (\boldsymbol{\beta} - \hat{\boldsymbol{\beta}})]^{-(\nu+2)/2}, \end{aligned} \tag{7.5}$$

where $\nu = T - 2$,

$$H = \begin{pmatrix} T & \sum y_{t-1} \\ \sum y_{t-1} & \sum y_{t-1}^2 \end{pmatrix}, \tag{7.6}$$

$$\hat{\boldsymbol{\beta}} = \begin{pmatrix} T & \sum y_{t-1} \\ \sum y_{t-1} & \sum y_{t-1}^2 \end{pmatrix}^{-1} \begin{pmatrix} \sum y_t \\ \sum y_{t-1} y_t \end{pmatrix}, \tag{7.7}$$

and $\nu s^2 = \sum (y_t - \hat{\beta}_1 - \hat{\beta}_2 y_{t-1})^2$. It is seen from (7.5) that the joint posterior pdf for β_1 and β_2 is in the bivariate Student t form with mean given by (7.7), the least squares quantity. This fact permits us to make inferences about β_1 and β_2 quite readily. In particular, the marginal posterior pdfs for β_1 and β_2 will each be in the form of univariate Student t pdf. Explicitly, the quantities

$$\frac{\beta_1 - \hat{\beta}_1}{s(h^{11})^{1/2}} \quad \text{and} \quad \frac{\beta_2 - \hat{\beta}_2}{s(h^{22})^{1/2}}$$

will each be distributed as a Student t variable with $\nu = T - 2$ degrees of freedom. Further, by integrating (7.4) with respect to $\boldsymbol{\beta}$, the marginal pdf for σ is given by

$$p(\sigma|\mathbf{y}, y_0) \propto \frac{1}{\sigma^{\nu+1}} \exp\left(-\frac{\nu s^2}{2\sigma^2}\right), \tag{7.8}$$

with $\nu = T - 2$ and $\nu s^2 = \sum (y_t - \hat{\beta}_1 - \hat{\beta}_2 y_{t-1})^2$.

[4] Of course, if we have information that the process is nonexplosive and wish to use it, the prior pdf in (7.3) can be altered to incorporate this information. See below for an example.

As mentioned above, these results for the normal first order autoregressive process parallel those for the simple normal regression model. Also, the predictive pdf for the next future observation y_{T+1} will be in the univariate Student t form with mean $\hat{\beta}_1 + \hat{\beta}_2 y_T$, which depends just on given sample values.

If we have additional information available about the autoregressive process in (7.1) and wish to incorporate it in the analysis, this can be done without much difficulty; for example, we might assume that the process is stationary with $|\beta_2| < 1$. Then the initial observation, y_0, is given by[5]

$$y_0 = \beta_1 \sum_{l=0}^{\infty} \beta_2{}^l + \sum_{l=0}^{\infty} \beta_2{}^l u_{-l}$$

or

$$y_0 = \frac{\beta_1}{1 - \beta_2} + \sum_{l=0}^{\infty} \beta_2{}^l u_{-l}. \tag{7.9}$$

From (7.9) y_0 is normally distributed with mean $\beta_1/(1 - \beta_2)$ and variance $\sigma^2/(1 - \beta_2{}^2)$. Thus the joint pdf for the $T + 1$ observations y_0 and $\mathbf{y}$ is given by

$$p(\mathbf{y}, y_0|\boldsymbol{\beta}, \sigma) \propto \frac{(1 - \beta_2{}^2)^{1/2}}{\sigma^{T+1}} \exp\left\{-\frac{1}{2\sigma^2}\left[(1 - \beta_2{}^2)\left(y_0 - \frac{\beta_1}{1 - \beta_2}\right)^2 + \sum (y_t - \beta_1 - \beta_2 y_{t-1})^2\right]\right\}, \tag{7.10}$$

where $|\beta_2| < 1$, $0 < \sigma < \infty$, and $-\infty < \beta_1 < \infty$. This pdf, viewed as a function of its parameters, is, of course, the likelihood function. As regards prior assumptions, we assume that β_1, β_2, and $\log \sigma$ are independently distributed. With respect to β_1 and $\log \sigma$ we assume that they are uniformly and independently distributed. Our prior pdf for β_2 is designated by $p(\beta_2)$. Thus our joint prior pdf is[6]:

$$p(\beta_1, \beta_2, \sigma) \propto \frac{p(\beta_2)}{\sigma}, \tag{7.11}$$

where $-\infty < \beta_1 < \infty$, $|\beta_2| < 1$, and $0 < \sigma < \infty$.

Under the above prior assumptions the posterior pdf for the parameters is

$$p(\boldsymbol{\beta}, \sigma|\mathbf{y}, y_0) \propto \frac{p(\beta_2)(1 - \beta_2{}^2)^{1/2}}{\sigma^{T+2}} \exp\left\{-\frac{1}{2\sigma^2}\left[(1 - \beta_2{}^2)\left(y_0 - \frac{\beta_1}{1 - \beta_2}\right)^2 + \sum (y_t - \beta_1 - \beta_2 y_{t-1})^2\right]\right\}. \tag{7.12}$$

[5] As usual, with the assumption of stationarity, the series "starts up" in the infinite past. If the series started up at $-T_0$, with T_0 finite, it would not be strictly stationary. However, with modification of (7.9), the model could be analyzed by using the above methods, if T_0 is known.

[6] The analysis can also be carried forward with informative pdf's for σ and β_1.

On completing the square on β_1 in the exponent of (7.12), we have

$$p(\boldsymbol{\beta}, \sigma|\mathbf{y}, y_0) \propto \frac{p(\beta_2)(1 - \beta_2^2)^{1/2}}{\sigma^{T+2}} \exp\left\{-\frac{1}{2\sigma^2}\left[c(\beta_1 - \hat{\beta}_1)^2 + \frac{1}{c}\left(T\textstyle\sum_1 + \frac{1 + \beta_2}{1 - \beta_2}\sum_2\right)\right]\right\}, \tag{7.13}$$

where $c = (1 + \beta_2)/(1 - \beta_2) + T$, $\sum_1 = \sum [y_t - \bar{y} - \beta_2(y_{t-1} - \bar{y}_{-1})]^2$, $\sum_2 = \sum [y_t - \beta_2 y_{t-1} - (1 - \beta_2)y_0]^2$, $\bar{y} = \sum y_t/T$, $\bar{y}_{-1} = \sum y_{t-1}/T$, and $\hat{\beta}_1 = [(1 + \beta_2)y_0 + \sum (y_t - \beta_2 y_{t-1})]/c$. From (7.13) it is seen that the conditional mean of β_1, given β_2 and σ, is $\hat{\beta}_1$. Also, $\hat{\beta}_1$ can be written as follows:

$$\hat{\beta}_1 = \frac{h_1 \sum (y_t - \beta_2 y_{t-1})/T + h_2(1 - \beta_2)y_0}{h_1 + h_2},$$

where $h_1 = T/\sigma^2$ and $h_2 = (1 + \beta_2)/(1 - \beta_2)\sigma^2$. Note that, given β_2, from (7.9) $(1 - \beta_2)y_0$ is an estimate of β_1 and $\sum (y_t - \beta_2 y_{t-1})/T$ is another estimate of β_1. The quantity $\hat{\beta}_1$ is a weighted average of these two estimates with their respective precisions as weights. Also, from the form of (7.13) the conditional posterior pdf for β_1, given β_2 and σ, is normal with mean $\hat{\beta}_1$ and variance σ^2/c. As T grows large $\sigma^2/c \to 0$ and $\hat{\beta}_1 \to \sum (y_t - \beta_2 y_{t-1})/T$; thus the influence of the quantity $(1 - \beta_2)y_0$ on the location of the conditional posterior pdf diminishes as T grows large. To obtain the marginal posterior pdf for β_1, (7.13) can be integrated with respect to σ and the resulting bivariate posterior pdf for β_1 and β_2 can be analyzed by using bivariate numerical integration techniques.

Since interest often centers on β_2, we now discuss its marginal posterior pdf which can be obtained by integrating (7.13) with respect to β_1 and σ. Integration with respect to β_1 yields

$$\begin{aligned} p(\beta_2, \sigma|\mathbf{y}, y_0) &\propto \frac{p(\beta_2)(1 - \beta_2^2)^{1/2}}{\sigma^{T+1}c^{1/2}} \exp\left\{-\frac{1}{2\sigma^2 c}\left[T\textstyle\sum_1 + \left(\frac{1 + \beta_2}{1 - \beta_2}\right)\sum_2\right]\right\} \\ &\propto \frac{p(\beta_2)(1 - \beta_2^2)^{1/2}}{\sigma^{T+2}(h_1 + h_2)^{1/2}} \exp\left[-\frac{1}{2\sigma^2}\left(\frac{h_1\sum_1 + h_2\sum_2}{h_1 + h_2}\right)\right]. \end{aligned} \tag{7.14a}$$

Since $h_1 = T/\sigma^2$, as T gets large this posterior pdf is approximately proportional to $(1/\sigma)^{T+1} \exp[-(2\sigma^2)^{-1}\sum_1]$ in large samples,[7] a form that is free from dependence on information regarding initial conditions except insofar as y_0 appears in $\sum_1$. The pdf in (7.14*a*) can be employed to make joint inferences about β_2 and σ or marginal posterior inferences about σ.

[7] Since the factor $p(\beta_2)(1 - \beta_2^2)^{1/2}$ does not depend on T, it will not be important in large samples.

Next we can integrate (7.14a) with respect to σ to obtain the following marginal posterior pdf for β_2:

$$p(\beta_2|\mathbf{y}, y_0) \propto \frac{p(\beta_2)(1-\beta_2^2)^{1/2}c^{(T-1)/2}}{\{T\sum_1 + (1+\beta_2)/(1-\beta_2)\sum_2\}^{T/2}}, \qquad |\beta_2| < 1$$

$$\text{(7.14}b\text{)} \qquad \propto \frac{p(\beta_2)}{(\sum_1)^{T/2}}\left[\frac{1-\beta_2^2}{1+(1/T)(1+\beta_2)/(1-\beta_2)}\right]^{1/2} \times \left[\frac{1+(1/T)(1+\beta_2)/(1-\beta_2)}{1+(1/T)(1+\beta_2)/(1-\beta_2)(\sum_2/\sum_1)}\right]^{T/2}.$$

This pdf can be analyzed numerically. For large samples it will be approximately proportional to[8] $(\sum_1)^{-T/2} = \{\sum[y_t - \bar{y} - \beta_2(y_{t-1} - \bar{y}_{-1})]^2\}^{-T/2}$ which, as pointed out above, is in the univariate Student t form and does not reflect either the prior pdf $p(\beta_2)$ or the factors in the second line of (7.14b) arising from consideration of the pdf for y_0.

With respect to the prior pdf for β_2, $p(\beta_2)$, often the following beta pdf defined over the range -1 to $+1$ will be flexible enough to represent prior information[9]: $p(\beta_2) \propto (1-\beta_2)^{k_1-1}(1+\beta_2)^{k_2-1}$, where k_1 and k_2 are prior parameters to be assigned by the investigator. If $k_1 = k_2 = \frac{1}{2}$, this prior pdf is identical to what is produced by an approximate application of Jeffreys' invariance theory, namely, $p(\beta_2) \propto (1-\beta_2)^{-1/2}(1+\beta_2)^{-1/2} = (1-\beta_2^2)^{-1/2}$, $|\beta_2| < 1, \ldots$ (see the appendix to this chapter for details). Thus, if we wish to go forward by using an approximate Jeffreys' diffuse prior pdf for this problem, it is $p(\beta_1, \beta_2, \sigma) \propto (1-\beta_2^2)^{-1/2}\sigma^{-1}$.[10] As already explained, as the sample size grows, the influence of the prior pdf on the properties of the posterior pdf diminishes. Also, as seen from the present analysis of the stationary first order process, the influence of the initial conditions, that is, the assumptions about y_0, diminish in importance as the sample size grows.

[8] This approximation can be appreciated most easily by taking the log of both sides of (7.14b) and noting that $-T/2 \log \sum_1$ is the dominant term as T gets large.

[9] This prior pdf is suggested in H. Thornber, "Finite Sample Monte Carlo Studies: An Autoregressive Illustration," *J. Am. Statist. Assoc.* (September 1967), who studied the above system with $\beta_1 = 0$. Note that a change of variable $z = (1+\beta_2)/2$ yields $1 - z = (1-\beta_2)/2$ and thus $p(z) \propto z^{k_2-1}(1-z)^{k_1-1}$ with $0 < z < 1$, since $-1 < \beta_2 < 1$; $p(z)$ is in the usual form of a standard beta pdf and will be proper if $k_1, k_2 > 0$.

[10] See Thornber, *loc. cit.*, and J. B. Copas, "Monte Carlo Results for Estimation in a Stable Markov Time Series," *J. Roy. Statist. Soc.*, Series A, No. 1, 110–116 (1966). These studies show that the sampling properties of Bayesian estimators compare favorably with alternatives. In particular, in Thornber's experiments the estimated average risk of the ML estimator was more than 50% higher than that for the Bayesian estimator. See also G. H. Orcutt and H. S. Winkour, Jr., "First Order Autoregression: Inference, Estimation, and Prediction," *Econometrica*, **37**, 1–14 (1969), for Monte Carlo experiments relating to the finite sample properties of certain sampling theory techniques.

The latter result is not surprising, since the initial condition involves just one observation, y_0, a small fraction of the sample information when T is even moderately large and $|\beta_2| < 1$.

7.2 FIRST ORDER AUTOREGRESSIVE MODEL WITH INCOMPLETE DATA[11]

Suppose that we are interested in making inferences about the parameters in the following autoregressive model for quarterly data,

$$y(t) = \rho y(t-1) + X(t)\boldsymbol{\beta} + u(t), \qquad t = 1, 2, \ldots, 4T, \tag{7.15}$$

where t in parentheses denotes the value of a quantity for the tth quarter, $y(t)$ is a dependent "stock" variable, $u(t)$ is a disturbance term, $X(t) = (x_1(t), x_2(t), \ldots, x_k(t))$, a $1 \times k$ vector of observations on k independent variables, and $\boldsymbol{\beta}' = (\beta_1, \beta_2, \ldots, \beta_k)$ and ρ are unknown coefficients. We assume that the disturbance terms are normally and independently distributed, each with zero mean and common variance σ^2. Although quarterly observations are available for $X(t)$, we assume that only the following $T + 1$ annual observations are available for $y(t)$: $y(0), y(4), y(8), \ldots, y(4T)$; for example, $y(t)$ might be end of quarter stock of capital or money. With just $T + 1$ annual observations on $y(t)$ our problem is to make inferences about the parameters of (7.15), namely, ρ, $\boldsymbol{\beta}$, and σ.

Denoting a quarter for which an observation on the dependent variable is available by t', we find by direct operations that

$$\begin{aligned} y(t') = {} & \rho^4 y(t'-4) + [X(t') + \rho X(t'-1) + \rho^2 X(t'-2) + \rho^3 X(t'-3)]\boldsymbol{\beta} \\ & + u(t') + \rho u(t'-1) + \rho^2 u(t'-2) + \rho^3 u(t'-3), \\ & \qquad t' = 4, 8, \ldots, 4T, \end{aligned} \tag{7.16}$$

which is what the model in (7.15) logically implies for the observations we have on y. If we let

$$w(t') = u(t') + \rho u(t'-1) + \rho^2 u(t'-2) + \rho^3 u(t'-3), \qquad t' = 4, 8, \ldots, 4T, \tag{7.17}$$

it is clear that the $w(t')$ are normally and independently distributed, each with mean zero and common variance $(1 + \rho^2 + \rho^4 + \rho^6)\sigma^2$. Then, if we assume that $y(0)$ is fixed and known, the likelihood function, based on the observations we have on the y's, is

$$\begin{aligned} l(\rho, \boldsymbol{\beta}, \sigma | \mathbf{y}) \propto {} & \frac{1}{[(1 + \rho^2 + \rho^4 + \rho^6)\sigma^2]^{T/2}} \\ & \times \exp\left\{-\frac{1}{2(1 + \rho^2 + \rho^4 + \rho^6)\sigma^2} \sum [w(t')]^2\right\}, \end{aligned} \tag{7.18}$$

[11] The material in this section is drawn from A. Zellner, "On the Analysis of First Order Autoregressive Models with Incomplete Data," *Intern. Econ. Rev.*, **7**, 72–76 (1966).

where $\mathbf{y}$ denotes the $T + 1$ observations, $y(0), y(4), \ldots, y(4T)$, the summation in the exponent is taken over the following values of t', $t' = 4, 8, \ldots, 4T$, and $w(t')$, given by (7.17), represents

$$w(t') = y(t') - \rho^4 y(t' - 4) - [X(t') + \rho X(t' - 1) + \rho^2 X(t' - 2) + \rho^3 X(t' - 3)]\boldsymbol{\beta}. \tag{7.19}$$

First, we shall indicate how ML estimates of the parameters can be obtained. On taking the logarithm of the likelihood function, differentiating with respect to σ^2 and setting the derivative equal to zero, we obtain:

$$\hat{\sigma}^2 = \frac{1}{T(1 + \rho^2 + \rho^4 + \rho^6)} \sum_{t'} [w(t')]^2. \tag{7.20}$$

On substituting $\hat{\sigma}^2$ for σ^2 in the log-likelihood function, the following is the result:

$$L_{\max}(\rho, \boldsymbol{\beta}) = \text{const} - \frac{T}{2} \log \sum_{t'} [w(t')]^2, \tag{7.21}$$

which will be maximized for those values of ρ and $\boldsymbol{\beta}$ that minimize $\sum_{t'} [w(t')]^2$. One method of searching for these values is to evaluate the residual sum of squares, say $s^2(\rho)$, for regressions of $y(t') - \rho^4(y(t' - 4)$ on $[X(t') + \rho X(t' - 1) + \rho^2 X(t' - 2) + \rho^3 X(t' - 3)]$ for various values of ρ. If $\hat{\rho}$ is the value of ρ for which $s^2(\rho)$ is minimal, then $\hat{\rho}$ and the associated $\hat{\boldsymbol{\beta}}$ given by

$$\hat{\boldsymbol{\beta}} = H^{-1} \sum_{t'} [X(t') + \hat{\rho} X(t' - 1) + \hat{\rho}^2 X(t' - 2) + \hat{\rho}^3 X(t' - 3)]' \times [y(t') - \hat{\rho}^4 y(t' - 4)], \tag{7.22}$$

where

$$H = \sum_{t'} [X(t') + \hat{\rho} X(t' - 1) + \hat{\rho}^2 X(t' - 2) + \hat{\rho}^3 X(t' - 3)]' \times [X(t') + \hat{\rho} X(t' - 1) + \hat{\rho}^2 X(t' - 2) + \hat{\rho}^3 X(t' - 3)] \tag{7.23}$$

are ML estimates. On substituting these estimates in (7.20), we obtain a ML estimate for σ^2. Coupled with large-sample standard errors, obtained from the inverse of the estimated information matrix, these results can be employed to make approximate large-sample inferences.

For the finite sample Bayesian analysis of this problem we employ the likelihood function in (7.18) along with the following diffuse prior pdf:

$$p(\rho, \boldsymbol{\beta}, \sigma) \propto \frac{1}{\sigma}, \tag{7.24}$$

with $-\infty < \beta_i < \infty$, $i = 1, 2, \ldots, k$, and $0 < \sigma < \infty$. As regards ρ, we can assume either $-\infty < \rho < \infty$ or $-1 < \rho < 1$, since these assumptions will just affect the range of numerical integrations in what follows. However, it should be recognized that assuming $|\rho| < 1$ implies that the autoregressive process is nonexplosive. With these prior assumptions the posterior pdf for the parameters is

$$p(\rho, \boldsymbol{\beta}, \sigma|\mathbf{y}) \propto \frac{1}{(1 + \rho^2 + \rho^4 + \rho^6)^{T/2}\sigma^{T+1}} \times \exp\left\{-\frac{1}{(1 + \rho^2 + \rho^4 + \rho^6)\sigma^2} \sum [w(t')]^2\right\}, \tag{7.25}$$

where $w(t')$ is given explicitly in (7.19).

To obtain the joint marginal posterior pdf's for ρ and $\boldsymbol{\beta}$ integrate (7.25) with respect to σ to obtain

$$p(\rho, \boldsymbol{\beta}|\mathbf{y}) \propto \{\sum [w(t')]^2\}^{-T/2}. \tag{7.26}$$

It is convenient to write (7.26) as follows:

$$p(\rho, \boldsymbol{\beta}|\mathbf{y}) \propto [(\mathbf{z} - A\boldsymbol{\beta})'(\mathbf{z} - A\boldsymbol{\beta})]^{-T/2}, \tag{7.27}$$

where $\mathbf{z}$ is a $T \times 1$ vector with typical element $y(t') - \rho^4 y(t' - 4)$ and A is a $T \times k$ matrix with typical row $X(t') + \rho X(t' - 1) + \rho^2 X(t - 2) + \rho^3 X(t - 3)$. Letting

$$\hat{\boldsymbol{\beta}}(\rho) = (A'A)^{-1}A'\mathbf{z} \tag{7.28}$$

and

$$s^2(\rho) = (T - k)^{-1}[\mathbf{z} - A\hat{\boldsymbol{\beta}}(\rho)]'[\mathbf{z} - A\hat{\boldsymbol{\beta}}(\rho)], \tag{7.29}$$

we can write (7.27) as

$$p(\rho, \boldsymbol{\beta}|\mathbf{y}) \propto [s^2(\rho)]^{-T/2}\left\{\nu + \frac{[\boldsymbol{\beta} - \hat{\boldsymbol{\beta}}(\rho)]'A'A[\boldsymbol{\beta} - \hat{\boldsymbol{\beta}}(\rho)]}{s^2(\rho)}\right\}^{-(\nu+k)/2}, \tag{7.30}$$

where $\nu = T - k$. From (7.30) we see immediately that the conditional posterior pdf for $\boldsymbol{\beta}$, given ρ, is in the multivariate Student t form with mean given by (7.28).[12] This fact makes it easy to assess how sensitive inferences about the elements of $\boldsymbol{\beta}$ are to what is assumed about ρ.

To derive the marginal posterior pdf for ρ we integrate (7.30) with respect to $\boldsymbol{\beta}$, using properties of the multivariate Student t pdf which yields

$$p(\rho|\mathbf{y}) \propto [s^2(\rho)]^{-\nu/2}|A'A|^{-1/2}, \tag{7.31}$$

which can be analyzed numerically to make inferences about ρ. As regards

[12] If ρ were taken equal to $\hat{\rho}$, the ML estimate, (7.28) would yield the ML estimate for $\boldsymbol{\beta}$ as the mean of the conditional posterior pdf. Since inferences about $\boldsymbol{\beta}$ may be sensitive to what is assumed about ρ, it is better to use the *marginal* posterior pdf for $\boldsymbol{\beta}$ to make inferences. In this connection see, for example, the analysis of the problem of autocorrelation in regression analysis in Chapter 4.

the marginal posterior pdf for an element of $\boldsymbol{\beta}$, say β_1, (7.30) can be integrated with respect to $\beta_2, \beta_3, \ldots, \beta_k$, to yield

$$(7.32)\quad p(\rho, \beta_1|\mathbf{y}) \propto [s^2(\rho)^{-T/2}|V_{22}|^{-1/2} + \{\nu + (V_{11} - V_{12}V_{22}^{-1}V_{21})[\beta_1 - \hat{\beta}_1(\rho)]^2\}^{-(\nu+1)/2},$$

where the V's are submatrices of $A'A/s^2(\rho)$; that is

$$\frac{A'A}{s^2(\rho)} = \begin{bmatrix} V_{11} & V_{12} \\ V_{21} & V_{22} \end{bmatrix},$$

where the partitioning has been done to conform with the partitioning of $\boldsymbol{\beta}$ into β_1 and a vector of remaining elements. Thus V_{11} is a scalar, V_{12}, a $1 \times (k-1)$ row vector, $V_{21} = V_{21}'$, and V_{22} is a $(k-1) \times (k-1)$ matrix. Numerical techniques can be employed to analyze the bivariate pdf in (7.32).

7.3 ANALYSIS OF A SECOND ORDER AUTOREGRESSIVE PROCESS[13]

In this section we show how Bayesian techniques can be employed to make inferences about the dynamic properties of solutions to stochastic difference equations which are often encountered in practice. What is presented is an analysis of a second order linear autoregressive model designed to answer questions of the following kind. On the basis of the data we have, what is the posterior probability that the model's solution will be nonexplosive and oscillatory? Or, what is the posterior probability that the solution will be oscillatory? Clearly, such questions resemble those asked by Samuelson in his well-known paper[14] on the multiplier-accelerator interaction and also those considered by Theil and Boot in their large-sample analysis[15] of Klein's Model I.

Our model for the observations is assumed to be

$$(7.33)\qquad y_t = \alpha_1 y_{t-1} + \alpha_2 y_{t-2} + u_t, \qquad t = 1, 2, \ldots, T,$$

where y_t is the tth observation on a random variable, α_1 and α_2 are unknown coefficients, and u_t is a disturbance term. We assume that the u_t's are normally and independently distributed, each with mean zero and common variance

[13] This material is based on one section of A. Zellner, "Bayesian Inference and Simultaneous Equation Econometric Models," paper presented to the First World Congress of the Econometric Society, Rome, 1965.

[14] P. A. Samuelson, "Interactions between the Multiplier Analysis and the Principle of Acceleration," *Rev. Econ. Statist.*, **21**, 75–78 (1939).

[15] H. Theil and J. C. G. Boot, "The Final Form of Econometric Equation Systems," *Rev. Intern. Statist. Inst.*, **30**, 136–152 (1962), reprinted in A. Zellner (ed.), *Readings in Economic Statistics and Econometrics*. Boston: Little, Brown, 1968, pp. 611–630.

σ^2. Further, assuming that the initial values y_{-1} and y_0 are given, the likelihood function is

$$(7.34) \qquad l(\alpha_1, \alpha_2, \sigma|\mathbf{y}) \propto \frac{1}{\sigma^T} \exp\left[-\frac{1}{2\sigma^2}\sum_{t=1}^{T}(y_t - \alpha_1 y_{t-1} - \alpha_2 y_{t-2})^2\right],$$

where $\mathbf{y}' = (y_{-1}, y_0, y_1, \ldots, y_T)$. As regards prior information about α_1, α_2, and σ, we assume that little is known about these parameters and represent this in the usual way[16]:

$$(7.35) \qquad p(\alpha_1, \alpha_2, \sigma) \propto \frac{1}{\sigma},$$

where $-\infty < \alpha_1, \alpha_2 < \infty$ and $0 < \sigma < \infty$. Then, using Bayes' theorem, the posterior pdf for the parameters is

$$(7.36) \qquad p(\alpha_1, \alpha_2, \sigma|\mathbf{y}) \propto \frac{1}{\sigma^{T+1}} \exp\left[-\frac{1}{2\sigma^2}\sum_{t=1}^{T}(y_t - \alpha_1 y_{t-1} - \alpha_2 y_{t-2})^2\right].$$

On integrating with respect to σ, the marginal posterior pdf for α_1 and α_2 is found to be

$$(7.37) \qquad p(\alpha_1, \alpha_2|\mathbf{y}) \propto [\nu s^2 + (\boldsymbol{\alpha} - \hat{\boldsymbol{\alpha}})'H(\boldsymbol{\alpha} - \hat{\boldsymbol{\alpha}})]^{-T/2},$$

where

$$\boldsymbol{\alpha}' = (\alpha_1, \alpha_2), \qquad \nu = T - 2, \qquad \nu s^2 = \sum_{t=1}^{T}(y_t - \hat{\alpha}_1 y_{t-1} - \hat{\alpha}_2 y_{t-2})^2,$$

$$(7.38) \qquad \hat{\boldsymbol{\alpha}} = \begin{pmatrix}\hat{\alpha}_1\\ \hat{\alpha}_2\end{pmatrix} = \begin{bmatrix}\sum y_{t-1}^2 & \sum y_{t-1}y_{t-2}\\ \sum y_{t-1}y_{t-2} & \sum y_{t-2}^2\end{bmatrix}^{-1}\begin{bmatrix}\sum y_{t-1}y_t\\ \sum y_{t-2}y_t\end{bmatrix},$$

and

$$(7.39) \qquad H = \begin{bmatrix}\sum y_{t-1}^2 & \sum y_{t-1}y_{t-2}\\ \sum y_{t-1}y_{t-2} & \sum y_{t-2}^2\end{bmatrix},$$

with all summations extending from $t = 1$ to $t = T$. It is seen that the posterior pdf for α_1 and α_2 in (7.37) is in the bivariate Student t form with mean vector $\hat{\boldsymbol{\alpha}}$, the least squares quantity shown in (7.38).

Given that we have the observations $\mathbf{y}$, we can use (7.37) to make joint inferences about α_1 and α_2 and thus about properties of solutions; that is, just as Samuelson did in his multiplier-accelerator paper (*cit. supra*), we can determine regions in the (α_1, α_2) plane corresponding to solutions having certain properties. These regions, relating to the present model, are shown in Figure 7.1. Since we have the joint posterior pdf, $p(\alpha_1, \alpha_2|\mathbf{y})$, we can use

[16] The analysis presented below can be extended easily to the case of an informative pdf for α_1 and α_2.

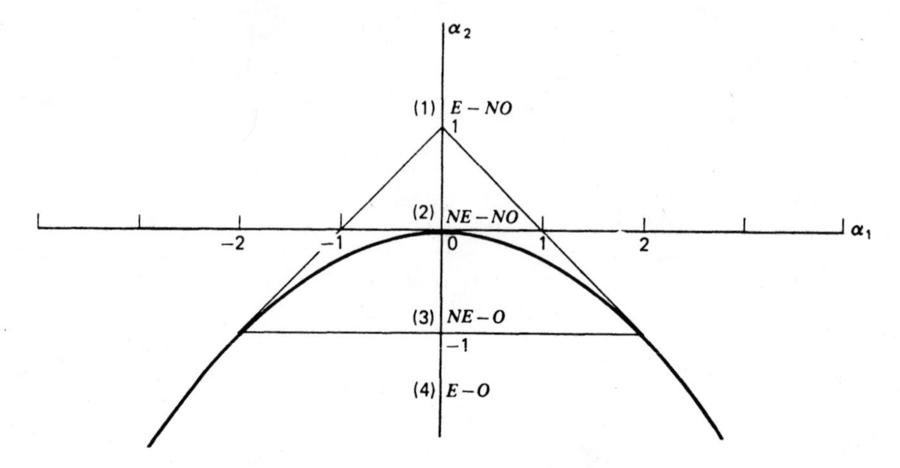

Figure 7.1 Regions in the parameter space for which solution has particular properties. Regions: (1) $E - NO$: Explosive and nonoscillatory; (2) $NE - NO$: nonexplosive and nonoscillatory; (3) $NE - O$: nonexplosive and oscillatory; (4) $E - O$: explosive and oscillatory.

bivariate numerical integration techniques to evaluate its normalizing constant[17] and the volume over each of the regions. Given that the posterior pdf for α_1 and α_2 has been normalized, these volumes are posterior probabilities relating to properties of the solution. If, for example, the volume of the posterior pdf over the "oscillatory-nonexplosive" region were computed to be 0.85, we would say that the probability is 0.85 that the solution will be oscillatory and nonexplosive. Further, by adding the probability that the solution will be oscillatory and nonexplosive and the probability that it will be oscillatory and explosive, we have the probability that the model's solution will be oscillatory. Similarly, by adding the probability that it will be oscillatory and nonexplosive and the probability that it will be nonoscillatory and nonexplosive, we have the probability that the solution will be nonexplosive. Application of this approach, using data generated from known models, is provided below.

We note further that it is possible to derive the posterior pdf's for quantities whose value determines particular properties of the solution; for example, from the characteristic equation for the model $x^2 - \alpha_1 x - \alpha_2 = 0$, we have the following roots:

$$x_1 = \frac{\alpha_1 + \sqrt{\alpha_1^2 + 4\alpha_2}}{2} \quad \text{and} \quad x_2 = \frac{\alpha_1 - \sqrt{\alpha_1^2 + 4\alpha_2}}{2}. \tag{7.40}$$

[17] Alternatively, this normalizing constant is known from properties of the bivariate Student t pdf.

As is well known, the solution will exhibit oscillations if $\alpha_1^2 + 4\alpha_2 < 0$. Thus, on occasion, it will be of interest to have the posterior pdf for the quantity $\alpha_1^2 + 4\alpha_2$. To obtain this pdf we introduce the following transformation:

$$v_1 = \alpha_1 \quad \text{and} \quad v_2 = \alpha_1^2 + 4\alpha_2, \tag{7.41}$$

a transformation from the variables α_1 and α_2 to v_1 and v_2 with a nonzero Jacobian that does not involve any of the variables. Using the posterior pdf in (7.37), we have, for the posterior pdf for v_1 and v_2,

$$\begin{aligned} p(v_1, v_2|\mathbf{y}) \propto \Bigg[\nu s^2 &+ (v_1 - \hat{\alpha}_1)^2 h_{11} + \left(\frac{v_2 - v_1^2}{4} - \hat{\alpha}_2\right)^2 h_{22} \\ &+ 2(v_1 - \hat{\alpha}_1)\left(\frac{v_2 - v_1^2}{4} - \hat{\alpha}_2\right)h_{12}\Bigg]^{-T/2}, \end{aligned} \tag{7.42}$$

where h_{11}, h_{12}, and h_{22} are elements of H, shown in (7.39). Bivariate numerical integration techniques can be employed to integrate (7.42) with respect to v_1 and to normalize the pdf. The result is the normalized marginal posterior pdf for $v_2 = \alpha_1^2 + 4\alpha_2$, denoted $p(v_2|\mathbf{y})$. This distribution can be employed to make inferences about the quantity $\alpha_1^2 + 4\alpha_2$ and thus about whether or not the solution is oscillatory.

To illustrate application of these techniques, we have generated data from the model shown in (7.33) under the conditions given in Table 7.1. In each run the u_t's were independently drawn from a normal population with mean zero and variance one. The sample size for each run was $T = 20$ and the initial values y_{-1} and y_0 were set at zero in all three runs.

Table 7.1

Run	Value of α_1	Value of α_2	Properties of Solution
A	0.500	−0.750	Oscillatory and nonexplosive
B	1.250	−0.375	Nonoscillatory and nonexplosive
C	1.600	−0.550	Nonoscillatory and explosive

The contours of the posterior pdf $p(\alpha_1, \alpha_2|\mathbf{y})$ for runs A, B, and C are shown in Figure 7.2 along with the mean values for α_1 and α_2, namely, $\hat{\alpha}_1$ and $\hat{\alpha}_2$, the least squares quantities shown in (7.38). The computation of volumes above the regions shown in Figure 7.1 produced the results given in Table 7.2. These results square nicely with the known properties of the solutions indicated in Table 7.1. In Run B it should be noted that, from the true values of α_1 and α_2, $\alpha_1^2 + 4\alpha_2 = \frac{1}{8}$, which is a small number. With only 20 observations it is difficult to make precise inferences and this shows up in the results, namely, that the probability that the solution will be nonoscillatory

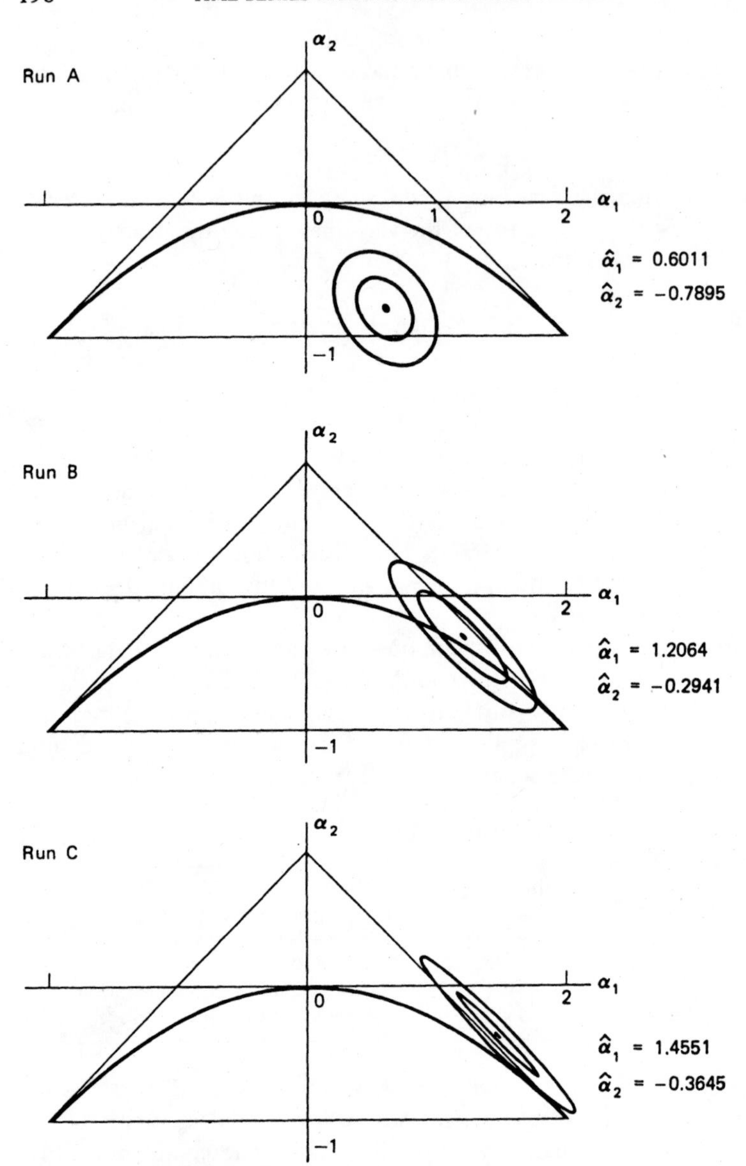

Figure 7.2 Contours of posterior distributions.

Table 7.2

Run	Probability that the solution is: Nonoscillatory and Nonexplosive	Oscillatory and Nonexplosive	Nonoscillatory and Explosive	Oscillatory and Explosive
A	0.000	0.866	0.000	0.134
B	0.593	0.242	0.162	0.003
C	0.025	0.016	0.953	0.006

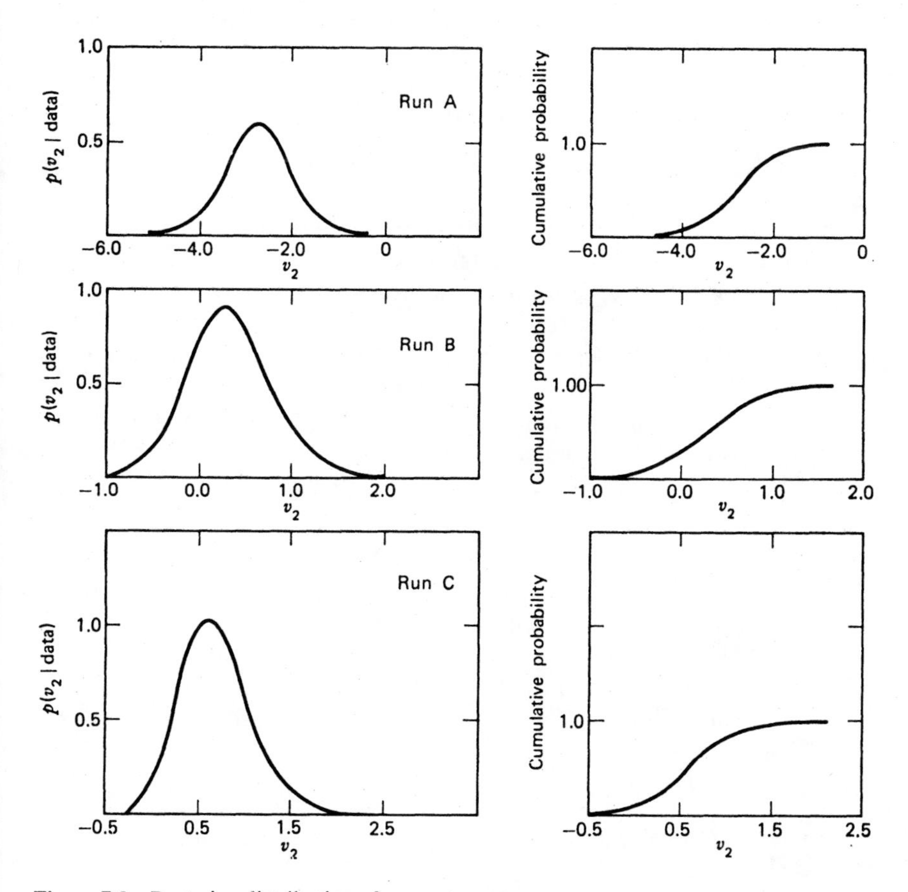

Figure 7.3 Posterior distributions for v_2.

is $0.593 + 0.162 = 0.755$, whereas the posterior probability that it will be oscillatory is 0.245, which is fairly substantial.

Last, the posterior pdf for $v_2 = \alpha_1^2 + 4\alpha_2$ was computed for each of the above runs. The results are shown in Figure 7.3. We see that for Runs A and C there is clear-cut sample evidence that shows whether the solution is or is not oscillatory. As mentioned above, in the case of run B, $v_2 = \alpha_1^2 + 4\alpha_2 = \frac{1}{8}$, a small number, and we see that a non-negligible portion of the pdf is located over negative values. However, this pdf does reflect accurately the information our sample has bearing on v_2.

In closing this section we emphasize the importance of analyzing the dynamic properties of models and hope that generalizations of the methods discussed, appropriate for a range of other models,[18] will be forthcoming.

7.4 "DISTRIBUTED LAG" MODELS

Since the pioneering work of Koyck,[19] distributed lag models have been utilized in a number of econometric studies.[20] These models typically incorporate lagged effects arising from habit persistence, institutional or technological constraints, and/or expectations effects which link anticipations with experience. Further, to conserve on the number of parameters involved in distributed lag models, it is usually assumed that the coefficients of lagged variables are not all independent but functionally related. This functional relation serves to reduce the number of parameters required to represent lagged responses to one or a few parameters. In what follows we take up the analysis of distributed lag models.

The first model we consider is

$$y_t = \alpha \sum_{i=0}^{\infty} \lambda^i x_{t-i} + u_t \tag{7.43}$$

where the subscripts t and $t - i$ denote the values of variables in the tth and $(t - i)$th time periods, respectively, y_t is the observed random "response" variable, u_t is an unobserved random disturbance term, x_{t-i} is the value of a given "stimulus" variable in period $t - i$, and α and λ are unknown

[18] If, for example, (7.33) were elaborated as follows, $y_t = \alpha_1 y_{t-1} + \alpha_2 y_{t-2} + X(t)\boldsymbol{\beta} + u_t$, $t = 1, 2, \ldots, T$, where $X(t)$ is a $1 \times k$ row vector of independent variables and $\boldsymbol{\beta}$ is a $k \times 1$ vector of coefficients, it would be possible to obtain the marginal pdf for α_1 and α_2 and pursue the above analysis.

[19] L. Koyck, *Distributed Lags and Investment Analysis.* Amsterdam: North-Holland, 1954.

[20] See, for example, Z. Griliches, "Distributed Lags: A Survey," *Econometrica*, **35**, 16–49 (1967).

parameters such that $-\infty < \alpha < \infty$ and $0 \leq \lambda < 1$. On subtracting λy_{t-1} from both sides of (7.43), we obtain

$$y_t = \lambda y_{t-1} + \alpha x_t + u_t - \lambda u_{t-1}, \tag{7.44}$$

which is the form of the model usually considered for analysis.

For (7.44) we assume that the u_t's have zero means and a common variance, σ^2, and are normally and independently distributed. Note then that the variance of $u_t - \lambda u_{t-1}$ is $\sigma^2(1 + \lambda^2)$, whereas the first order autocovariance is given by $E(u_t - \lambda u_{t-1})(u_{t-1} - \lambda u_{t-2}) = -\sigma^2\lambda$. All higher order autocovariances are zero, given that the u_t's are independently distributed. Thus for $t = 1, 2, \ldots, T$ the covariance matrix for the first order moving average error term in (7.44) is $E(\mathbf{u} - \lambda\mathbf{u}_{-1})(\mathbf{u} - \lambda\mathbf{u}_{-1})' = \sigma^2 G$, with

$$G = \begin{bmatrix} 1+\lambda^2 & -\lambda & & & \\ -\lambda & 1+\lambda^2 & -\lambda & & \\ \cdot & \cdot & \cdot & & \\ & \cdot & \cdot & \cdot & \\ & & \cdot & \cdot & \cdot \\ & & -\lambda & 1+\lambda^2 & -\lambda \end{bmatrix}; \tag{7.45}$$

all entries not shown are equal to zero. The joint pdf for $\mathbf{y}' = (y_1, y_2, \ldots, y_T)$, given y_0, is[21]

$$p(\mathbf{y}|\lambda, \alpha, \sigma, y_0) \propto \frac{|G|^{-1/2}}{\sigma^T} \exp\left[-\frac{1}{2\sigma^2}(\mathbf{y} - \lambda\mathbf{y}_{-1} - \alpha\mathbf{x})'G^{-1}(\mathbf{y} - \lambda\mathbf{y}_{-1} - \alpha\mathbf{x})\right], \tag{7.46}$$

where $\mathbf{y}_{-1}' = (y_0, y_1, y_2, \ldots, y_{T-1})$ and $\mathbf{x}' = (x_1, x_2, \ldots, x_T)$. Our diffuse prior pdf for the parameters is

$$p(\lambda, \alpha, \sigma) \propto \frac{1}{\sigma}. \tag{7.47}$$

Then the posterior pdf for the parameters is given by

$$p(\lambda, \alpha, \sigma|\mathbf{y}, y_0) \propto \frac{|G|^{-1/2}}{\sigma^{T+1}} \exp\left[-\frac{1}{2\sigma^2}(\mathbf{y} - \lambda\mathbf{y}_{-1} - \alpha\mathbf{x})'G^{-1}(\mathbf{y} - \lambda\mathbf{y}_{-1} - \alpha\mathbf{x})\right]. \tag{7.48}$$

We can simply integrate (7.48) with respect to σ to obtain

$$p(\lambda, \alpha|\mathbf{y}, y_0) \propto \frac{|G|^{-1/2}}{[(\mathbf{y} - \lambda\mathbf{y}_{-1} - \alpha\mathbf{x})'G^{-1}(\mathbf{y} - \lambda\mathbf{y}_{-1} - \alpha\mathbf{x})]^{T/2}}, \tag{7.49}$$

which is the joint posterior pdf for the parameters λ and α. Employing

[21] This analysis is similar to that in Thornber, "Application of Decision Theory to Econometrics," *loc. cit.*, except that our matrix G differs slightly from the one he uses.

bivariate numerical integration techniques, the normalizing constant can be evaluated and the bivariate pdf employed to make joint inferences about λ and α. The marginal posterior pdfs for λ and α can also be obtained numerically.[22]

As an elaboration of (7.43) we may entertain the hypothesis that our data are generated by

$$y_t = \alpha \sum_{i=0}^{\infty} \lambda^i x_{t-i} + \sum_{i=0}^{\infty} \lambda^i u_{t-i} \tag{7.50}$$

with α, λ, y_t, and x_{t-i} as defined above. Here we assume that the response to current and lagged disturbance terms takes the same form as that to current and lagged x's and involves the same parameter λ. Subtraction of λy_{t-1} from both sides of (7.50) yields

$$y_t = \lambda y_{t-1} + \alpha x_t + u_t. \tag{7.51}$$

Now, if we make the standard assumptions about the u_t's, as above, and assume that we have $t = 1, 2, \ldots, T$ observations, then, with y_0 given, the likelihood function is

$$l(\lambda, \alpha, \sigma|\mathbf{y}, y_0) \propto \frac{1}{\sigma^T} \exp\left[-\frac{1}{2\sigma^2}(\mathbf{y} - \lambda\mathbf{y}_{-1} - \alpha\mathbf{x})'(\mathbf{y} - \lambda\mathbf{y}_{-1} - \alpha\mathbf{x})\right]. \tag{7.52}$$

Using the prior assumptions in (7.47) to form the joint posterior pdf and integrating with respect to σ, we have

$$p(\lambda, \alpha|\mathbf{y}, y_0) \propto [(\mathbf{y} - \lambda\mathbf{y}_{-1} - \alpha\mathbf{x})'(\mathbf{y} - \lambda\mathbf{y}_{-1} - \alpha\mathbf{x})]^{-T/2}, \tag{7.53}$$

which would be in the form of a bivariate Student t pdf were it not for the fact that we have assumed $0 \leq \lambda < 1$. With this restriction (7.53) can be analyzed numerically to make joint inferences about λ and α and to obtain the marginal posterior pdf's[23] for α and for λ.

On some occasions we may wish to broaden the model to accommodate the possibility that the u_t's in (7.51) are possibly autocorrelated; for example, suppose that

$$u_t = \rho u_{t-1} + \epsilon_t, \qquad t = 1, 2, \ldots, T, \tag{7.54}$$

where the ϵ_t's are normally and independently distributed, each with mean

[22] As regards the marginal posterior pdf for λ, it can alternatively be obtained from (7.49) by completing the square on α in the denominator and using properties of the univariate Student t pdf to integrate with respect to α analytically. The result is a univariate posterior pdf for λ which can be analyzed numerically.

[23] Alternatively, the marginal pdf for λ can be obtained by completing the square on α in (7.53) and integrating with respect to it by using properties of the univariate Student t pdf.

zero and common variance τ^2, and ρ is a parameter of the first order autoregressive scheme assumed to generate the u_t's. Then, on combining[24] (7.51) and (7.54), the result is

$$y_t = (\lambda + \rho)y_{t-1} - \rho\lambda y_{t-2} + \alpha(x_t - \rho x_{t-1}) + \epsilon_t. \tag{7.55}$$

An alternative way of arriving at (7.55) is to assume[25] that the disturbance $u_t - \lambda u_{t-1}$ in (7.44) satisfies

$$u_t - \lambda u_{t-1} = \rho(u_{t-1} - \lambda u_{t-2}) + \epsilon_t \tag{7.56a}$$

or

$$u_t = (\rho + \lambda)u_{t-1} - \rho\lambda u_{t-2} + \epsilon_t, \tag{7.56b}$$

a fairly general second order process. On combining (7.56) with (7.44), we are led to precisely the equation shown in (7.55). Thus it appears that the assumptions leading to (7.44), coupled with those in (7.56), are equivalent to those underlying (7.51), coupled with those in (7.54).

To analyze (7.55)[26] we adopt the following diffuse prior pdf:

$$p(\lambda, \alpha, \rho, \tau) \propto \frac{1}{\tau}, \tag{7.57}$$

where $-\infty < \rho, \alpha < \infty$, $0 < \lambda < 1$, and $0 < \tau < \infty$ and τ^2 is the common variance of the ϵ_t's. Given these prior assumptions, the assumptions about the ϵ_t's in (7.55), and two initial values, y_0 and y_{-1}, the posterior pdf for the parameters is

$$\begin{aligned} p(\lambda, \alpha, \rho, \tau|\mathbf{y}) \propto \frac{1}{\tau^{T+1}} \exp\Big\{-\frac{1}{2\tau^2} \sum_{t=1}^{T} [y_t - \lambda y_{t-1} - \alpha x_t \\ - \rho(y_{t-1} - \lambda y_{t-2} - \alpha x_{t-1})]^2\Big\}, \end{aligned} \tag{7.58}$$

where $\mathbf{y}' = (y_{-1}, y_0, y_1, \ldots, y_T)$. Integration with respect to τ yields

$$\begin{aligned} p(\lambda, \alpha, \rho|\mathbf{y}) &\propto \Big\{\sum_{t=1}^{T} [y_t - \lambda y_{t-1} - \alpha x_t - \rho(y_{t-1} - \lambda y_{t-2} - \alpha x_{t-1})]^2\Big\}^{-T/2} \\ &\propto \Big\{\sum_{t=1}^{T} [y_t - \rho y_{t-1} - \lambda(y_{t-1} - \rho y_{t-2}) - \alpha(x_t - \rho x_{t-1})]^2\Big\}^{-T/2}. \end{aligned} \tag{7.59}$$

It is seen from the second line of (7.59) that the conditional pdf for λ and α,

[24] We subtract $\rho y_{t-1} = \rho\lambda y_{t-2} + p\alpha x_{t-1} + \rho u_{t-1}$ from (7.51) to obtain (7.55).

[25] The following assumption is utilized in W. A. Fuller and J. E. Martin, "The Effects of Autocorrelated Errors on the Statistical Estimation of Distributed Lag Models," *J. Farm Econ.*, **44**, 71–82 (1961), and A. Zellner and C. J. Park, "Bayesian Analysis of a Class of Distributed Lag Models," *Econometric Ann., Indian Econ. J.*, **13**, 432–444 (1965).

[26] This analysis is presented in A. Zellner and C. J. Park, *loc. cit.*

given ρ, would be in the bivariate Student t form were it not for the fact that we have assumed that $0 < \lambda < 1$. Below we show how sensitive inferences about λ and α are to what is assumed about ρ by analyzing the conditional pdf, $p(\lambda, \alpha|\rho, \mathbf{y})$, numerically.

From the first line of (7.59) we see that we can complete the square on ρ and integrate with respect to it analytically, an operation that yields

$$p(\lambda, \alpha|\mathbf{y}) \propto [\sum (y_{t-1} - \lambda y_{t-2} - \alpha x_{t-1})^2]^{-1/2} \times \left\{\sum (y_t - \lambda y_{t-1} - \alpha x_t)^2 - \frac{[\sum (y_t - \lambda y_{t-1} - \alpha x_t)(y_{t-1} - \lambda y_{t-2} - \alpha x_{t-1})]^2}{\sum (y_{t-1} - \lambda y_{t-2} - \alpha x_{t-1})^2}\right\}^{-(T-1)/2}, \tag{7.60}$$

To determine how sensitive inferences are to incorrect assumptions about the parameter ρ we have computed the conditional posterior pdf's, $p(\alpha|\rho_0, \mathbf{y})$ and $p(\lambda|\rho_0, \mathbf{y})$ [see the second line of (7.59)], for selected values of ρ_0 from the

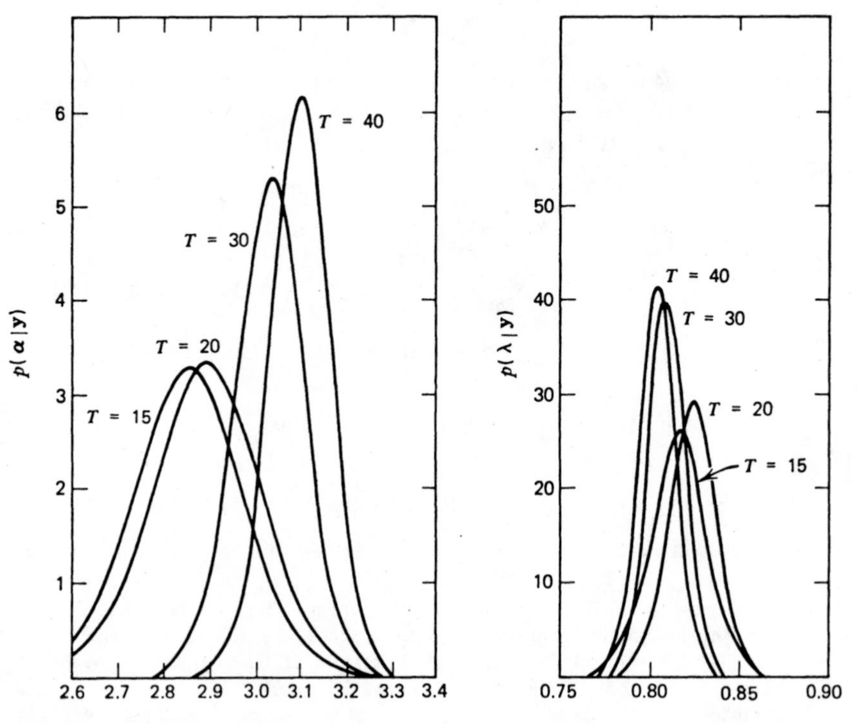

Figure 7.4 Marginal posterior distributions for α and λ computed from data generated with $\rho = -0.5$.

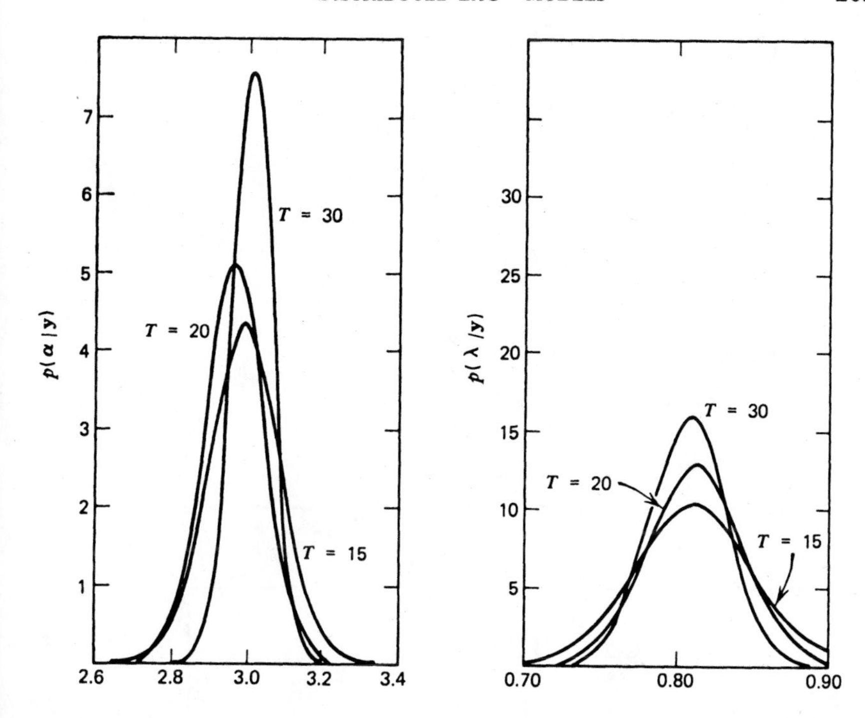

Figure 7.5 Marginal posterior distributions for α and λ computed from data generated with $\rho = 1.25$.

same data ($T = 20$) used to compute the distributions in Figures 7.4 and 7.5. These conditional pdfs are shown in Figure 7.6 for just the data set generated with $\rho = -0.5$. The sensitivity, with respect to location and spread, of these posterior pdfs to what is assumed about ρ is striking.[28] It is vividly seen that an incorrect assumption about ρ can vitally affect an analysis. Thus, when it is suspected that ρ differs from zero, we recommend that the marginal posterior pdf's for α and λ rather than the conditional pdf's be used to make inferences.

Last, we note that since we have the joint posterior pdf for α and λ, $p(\alpha, \lambda|\mathbf{y})$, it is not difficult in general to derive the distribution of a function of the variables α and λ; for example, in some problems interest centers on the "long-run" quantity $\eta = \alpha/(1 - \lambda)$. To derive the posterior pdf for η we change variables in $p(\alpha, \lambda|\mathbf{y})$ from α and λ to η and λ. The Jacobian of this transformation is $1 - \lambda$, which is different from zero for $0 < \lambda < 1$. Thus

[28] The same result applies to the conditional posterior pdf's based on the second data set for which the true $\rho = 1.25$. In fact, greater sensitivity is shown in this case.

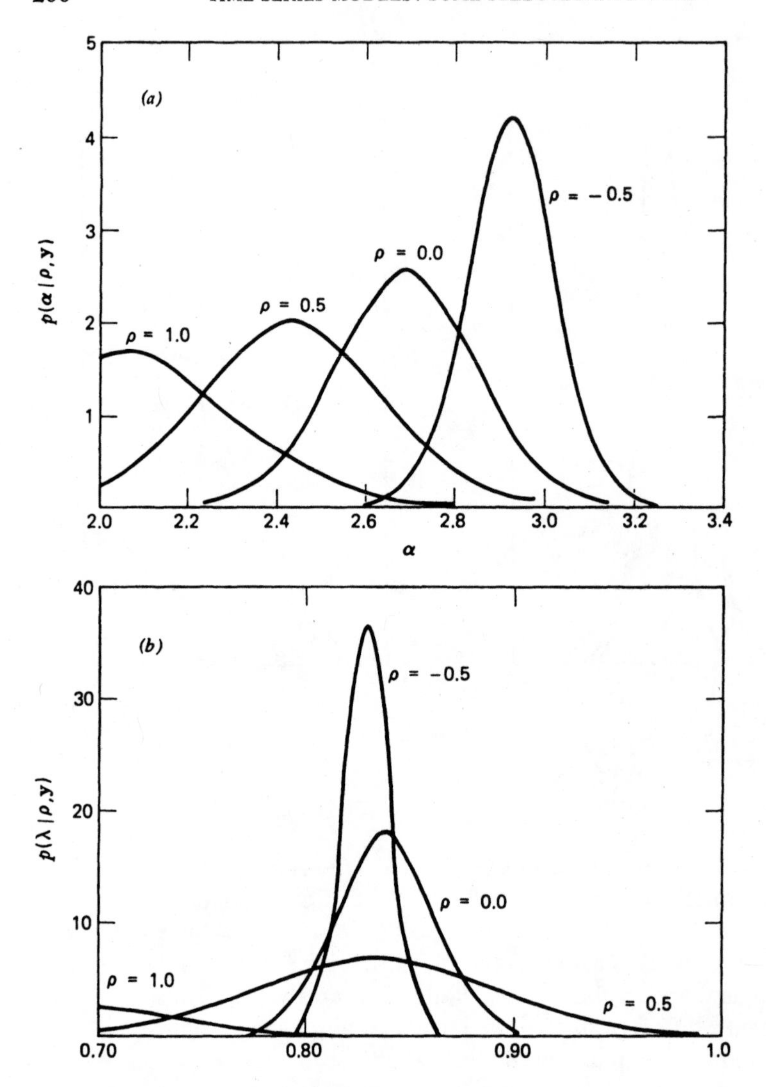

Figure 7.6a Conditional posterior distributions for α computed from data generated with $\rho = -0.5$.

Figure 7.6b Conditional posterior distributions for λ computed from data generated with $\rho = -0.5$.

the marginal posterior pdf for η, say $g(\eta|\mathbf{y})$, is given by performing the following integration with respect to λ numerically:

$$g(\eta|\mathbf{y}) = c \int_0^1 (1 - \lambda)p[\eta(1 - \lambda), \lambda|\mathbf{y}]\, d\lambda, \tag{7.61}$$

where c is a normalizing constant that can be evaluated numerically. Having $g(\eta|\mathbf{y})$, we can use it as a basis for making inferences about the "long-run" parameter $\eta = \alpha/(1 - \lambda)$.

7.5 APPLICATIONS TO CONSUMPTION FUNCTION ESTIMATION[29]

Let

$$C_t = kY_t^* + u_t \tag{7.62}$$

be our consumption function, where, for the tth period, $t = 1, 2, \ldots, T$, C_t is measured real consumption, Y_t^* is "normal" real income, k is a parameter whose value is unknown, and u_t is an error term. Assume that "normal" income satisfies

$$Y_t^* - Y_{t-1}^* = (1 - \lambda)(Y_t - Y_{t-1}^*) \tag{7.63a}$$

or from (7.63*a*), on successive substitution for lagged values of Y_t^*,

$$Y_t^* = (1 - \lambda)(Y_t + \lambda Y_{t-1} + \lambda^2 Y_{t-2} + \cdots + \lambda^n Y_{t-n} + \cdots), \tag{7.63b}$$

where the parameter λ is assumed to have a value such that $0 < \lambda < 1$. On combining (7.62) and (7.63*b*), we obtain[30]

$$C_t = \lambda C_{t-1} + k(1 - \lambda)Y_t + u_t - \lambda u_{t-1}, \tag{7.64}$$

which is the basic equation analyzed under the assumption that Y_t is an exogenous variable; that is, we abstract from "simultaneous equation" complications.

As regards the disturbance term in (7.64), $u_t - \lambda u_{t-1}$, we entertain the following assumptions:

Assumption I. $u_t - \lambda u_{t-1} = \epsilon_{1t}$, $t = 1, 2, \ldots, T$, with the ϵ_{1t}'s normally and independently distributed, each with zero mean and common variance σ_1^2; that is NID$(0, \sigma_1^2)$.

[29] This section draws on results in A. Zellner and M. S. Geisel, "Analysis of Distributed Lag Models with Applications to Consumption Function Estimation," invited paper presented to the Econometric Society meeting in Amsterdam, September 1968, and to appear in *Econometrica*.

[30] Substitute from (7.63*b*) in (7.62). Then subtract λC_{t-1} from both sides to obtain (7.64).

Assumption II. The u_t's in (7.64) are NID$(0, \sigma_2^2)$.

Assumption III. The u_t's in (7.64) satisfy a first order autoregressive scheme, $u_t = \rho u_{t-1} + \epsilon_{3t}$, with the ϵ_{3t}'s NID$(0, \sigma_3^2)$.

Assumption IV. The error term in (7.64) satisfies $u_t - \lambda u_{t-1} = \gamma(u_{t-1} - \lambda u_{t-2}) + \epsilon_{4t}$, with the ϵ_{4t}'s NID$(0, \sigma_4^2)$.

It should be noted that if the parameter ρ in Assumption III were equal to λ III would be indistinguishable from I. Similarly, if $\rho = 0$, III and II would be equivalent assumptions. Also, if $\gamma = 0$ in IV, IV would be equivalent to I.

We now turn to the analysis of (7.64) under Assumptions I to IV by using U.S. quarterly price-deflated, seasonally adjusted data on personal disposable income and consumption expenditures, 1947I–1960IV.[31] Under Assumption I the joint pdf for the observations is[32]

$$(7.65) \quad p(\mathbf{C}|\lambda, k, \sigma_1) \propto \frac{1}{\sigma_1^T} \exp\left\{-\frac{1}{2\sigma_1^2} \sum_{t=1}^{T} [C_t - \lambda C_{t-1} - k(1 - \lambda) Y_t]^2\right\},$$

where $\mathbf{C}' = (C_1, C_2, \ldots, C_T)$. With respect to prior assumptions about the parameters λ, k, and σ_1, we assume that

$$(7.66) \qquad p(\lambda, k, \sigma_1) \propto \frac{1}{\sigma_1} \qquad \begin{array}{l} 0 < \lambda, k < 1, \\ 0 < \sigma_1 < \infty. \end{array}$$

In (7.66) we assume that the parameters are independent, with uniform pdf's[33] on λ, k, and $\log \sigma_1$. Note that the prior information $0 < \lambda < 1$ and $0 < k < 1$ is being employed. On combining (7.65) and (7.66) and integrating with respect to σ_1, the joint posterior pdf for λ and k is

$$(7.67) \quad p(\lambda, k|\mathbf{C}) \propto \left\{\sum_{t=1}^{T} [C_t - \lambda C_{t-1} - k(1 - \lambda) Y_t]^2\right\}^{-T/2}, \quad 0 < \lambda, k < 1.$$

It is interesting to observe that the conditional posterior pdf for λ, given k, and the conditional posterior pdf for k, given λ, are in the form of truncated univariate Student t pdf's. Since these pdf's are truncated, analytical derivation of the marginal posterior pdf's for λ and k is complicated. In view of this fact bivariate numerical integration techniques were employed to obtain marginal pdf's by using the U.S. quarterly data, 1947I–1960IV, referred to above. Some features of the posterior pdf's are described in Table 7.3. From

[31] The data are given in Z. Griliches, G. S. Maddala, R. Lucas, and N. Wallace, "Notes on Estimated Aggregate Quarterly Consumption Functions," *Econometrica*, **30**, 491–500, pp. 499–500 (1962).

[32] Throughout this section we go forward conditional on given initial observations; for example, C_0 in (7.65).

[33] See below for use of nonuniform prior pdf's for λ and k.

the results it is seen that the posterior pdf for k is rather sharp, whereas that for λ has a much greater spread. Also, the results support the belief that λ has a value markedly different from zero.[34]

Table 7.3 POSTERIOR MEASURES ASSOCIATED WITH THE MARGINAL POSTERIOR PDF'S OF (7.67) BASED ON ASSUMPTION I

Posterior Measure	Marginal Posterior Pdf for λ	Marginal Posterior Pdf for k
Mean	0.759	0.959
Modal Value	0.78	0.95
Variance	0.0074	0.00021

Next we take up the analysis of (7.64) under Assumption II about the error terms. Under this assumption the joint pdf for the observations is

$$p(\mathbf{C}|\lambda, k, \sigma_2) \propto \frac{|G|^{-1/2}}{\sigma_2{}^T} \exp\left\{-\frac{1}{2\sigma_2{}^2}[\mathbf{C} - \lambda\mathbf{C}_{-1} - (1-\lambda)k\mathbf{Y}]'G^{-1} \times [\mathbf{C} - \lambda\mathbf{C}_{-1} - (1-\lambda)k\mathbf{Y}]\right\}, \tag{7.68}$$

where $\mathbf{C}_{-1}' = (C_0, C_1, \ldots, C_{T-1})$, $\mathbf{Y}' = (Y_1, Y_2, \ldots, Y_T)$, and G is the band matrix shown in (7.45). As regards prior assumptions about the parameters, we employ (7.66), with σ_1 replaced by σ_2. Using Bayes' theorem to combine (7.66) and (7.68) and integrating with respect to σ_2, we have the following as the joint posterior pdf for λ and k under Assumption II:

$$p(\lambda, k|\mathbf{C}) \propto \frac{|G|^{-1/2}}{\{[\mathbf{C} - \lambda\mathbf{C}_{-1} - (1-\lambda)k\mathbf{Y}]'G^{-1}[\mathbf{C} - \lambda\mathbf{C}_{-1} - (1-\lambda)k\mathbf{Y}]\}^{T/2}}, \tag{7.69}$$

with $0 < \lambda, k < 1$. This posterior pdf was analyzed numerically by using bivariate numerical integration procedures[35] and the quarterly data, 1947I–

[34] If $\lambda = 0$, from (7.63*a*) we have $Y_t^* = Y_t$ and thus (7.62) becomes $C_t = kY_t + u_t$, which is a form of the absolute-income hypothesis. Finding $\lambda \neq 0$ is thus important with respect to assessing the empirical validity of the absolute-income hypothesis.

[35] For each value of λ the method given in R. S. Varga, *Matrix Iterative Analysis*, Englewood Cliffs, N.J.: Prentice-Hall, 1963, p. 195, was used to compute the denominator of (7.69). In A. Zellner and M. S. Geisel, *loc. cit.*, the following approach was also used to analyze the model under Assumption II. With $\eta_t = C_t - u_t$, we have, from (7.64), $\eta_t = \lambda\eta_{t-1} + k(1-\lambda)Y_t = \lambda^t\eta_0 + (1-\lambda)kZ_t(\lambda)$ and then $C_t = \lambda^t\eta_0 + (1-\lambda)kZ_t(\lambda) + u_t$, where $Z_t(\lambda) = Y_t + \lambda Y_{t-1} + \cdots + \lambda^{t-1}Y_1$. With the u_t's NID$(0, \sigma_2{}^2)$, the likelihood function for λ, k, η_0, and σ_2 can be formulated and combined with a prior pdf for the parameters. Note that this approach involves use of a parameter, η_0, associated with initial conditions.

1960IV; the results are given in Table 7.4. In this instance the marginal pdf's for the parameters were found to be quite different from those encountered with Assumption I about the error terms. In fact, the marginal pdf's were found to be bimodal,[36] and use of the prior information $0 < \lambda, k < 1$ produced serious truncation of the posterior pdf's. These results point up the fact that assumptions about error terms can influence the results of an analysis significantly. In the present instance there is evidence that Assumption II, in combination with the other assumptions embedded in our model, is not supported by the data.[37]

Table 7.4 POSTERIOR MEASURES ASSOCIATED WITH THE MARGINAL POSTERIOR PDF'S OF (7.69) BASED ON ASSUMPTION II

Posterior Measure	Marginal Posterior Pdf for λ	Marginal Posterior Pdf for k
Mean	0.508	0.948
Modal Value	0.38 and 0.90	0.94 and 1.0
Variances	0.0643	0.0004

In the analysis of (7.64) under Assumption III it is convenient to note that, with $\eta_t = C_t - u_t$, (7.64) can be expressed as $\eta_t = \lambda\eta_{t-1} + k(1 - \lambda)Y_t$, or $\eta_t - \rho\eta_{t-1} = \lambda(\eta_{t-1} - \rho\eta_{t-2}) + k(1 - \lambda)(Y_t - \rho Y_{t-1})$, which leads to

$$\eta_t(\rho) = \lambda^t\eta_0' + (1 - \lambda)k[Y_t(\rho) + \lambda Y_{t-1}(\rho) + \cdots + \lambda^{t-1}Y_1(\rho)] \tag{7.70a}$$

or

$$C_t(\rho) = \lambda^t\eta_0' + (1 - \lambda)kZ_t(\lambda, \rho) + \epsilon_{3t}, \tag{7.70b}$$

where $\eta_t(\rho) = \eta_t - \rho\eta_{t-1}$, $\eta_0(\rho) = \eta_0'$, $Y_t(\rho) = Y_t - \rho Y_{t-1}$, $C_t(\rho) = C_t - \rho C_{t-1}$, and $Z_t(\rho, \lambda) = Y_t(\rho) + \lambda Y_{t-1}(\rho) + \cdots + \lambda^{t-1}Y_1(\rho)$. Then the joint pdf for the observations under Assumption III is

(7.71)
$$p(\mathbf{C}|\lambda, k, \eta_0', \rho, \sigma_3) \propto \frac{1}{\sigma_3^T}\exp\left\{-\frac{1}{2\sigma_3^2}\sum_{t=1}^{T}[C_t(\rho) - \lambda^t\eta_0' - (1 - \lambda)kZ_t(\lambda, \rho)]^2\right\}.$$

As prior pdf we use

$$p(\lambda, k, \eta_0', \rho, \sigma_3) \propto \frac{1}{\sigma_3}\Bigg\} \quad \begin{array}{l} 0 < \lambda, k < 1, \\ 0 < \sigma_3 < \infty, \; -\infty < \eta_0', \rho < \infty, \end{array} \tag{7.72}$$

[36] Zellner and Geisel, *loc. cit.*, found that the likelihood function was also bimodal with a global maximum at $\lambda = 0.963$ and $k = 1.129$. This large value for k is indeed hard to believe and probably arises because Assumption II is inappropriate.

[37] See Chapter 10 in which posterior probabilities associated with the model under various assumptions about error terms are presented.

which, when combined with (7.71), yields the joint posterior pdf for the parameters. The posterior pdf can be integrated analytically with respect to η_0', ρ, and σ_3 to yield the following marginal bivariate posterior pdf for λ and k:

$$(7.73) \qquad p(\lambda, k|\mathbf{C}) \propto \left.\frac{|R'R|^{-\frac{1}{2}}}{[(\mathbf{v} - R\tilde{\boldsymbol{\beta}})'(\mathbf{v} - R\tilde{\boldsymbol{\beta}})]^{(T-2)/2}}\right\} \qquad \begin{array}{l} 0 \le \lambda < 1, \\ 0 < k < 1, \end{array}$$

where $\mathbf{v}' = (v_1, v_2, \ldots, v_t, \ldots, v_T)$, with $v_t = C_t - (1 - \lambda)k \sum_{i=0}^{t-1} \lambda^i Y_{t-i}$, R is a $T \times 2$ matrix, with tth row $= [\lambda^t : C_{t-1} - (1 - \lambda)k \sum_{i=0}^{t-1} \lambda^i Y_{t-i-1}]$, and $\tilde{\boldsymbol{\beta}} = (R'R)^{-1}R'\mathbf{v}$. The pdf in (7.73) was analyzed numerically; the results[38] are given in Table 7.5.

Table 7.5 POSTERIOR MEASURES ASSOCIATED WITH THE MARGINAL POSTERIOR PDF'S OF (7.73) BASED ON ASSUMPTION III

Posterior Measure	Marginal Posterior Pdf for λ	Marginal Posterior Pdf for k
Mean	0.597	0.878
Modal Value	0.61	0.94
Variance	0.0338	0.0403

Next, we take up the analysis of (7.64) under Assumption IV about the error terms. The joint pdf for the observations is given by[39]

$$(7.74) \qquad p(\mathbf{C}|\lambda, k, \gamma, \sigma_4) \propto \frac{1}{\sigma_4{}^T} \exp\left(-\frac{1}{2\sigma_4{}^2}\boldsymbol{\epsilon}_4'\boldsymbol{\epsilon}_4\right),$$

where $\boldsymbol{\epsilon}_4 = \mathbf{C} - \lambda\mathbf{C}_{-1} - (1 - \lambda)k\mathbf{Y} - \gamma[\mathbf{C}_{-1} - \lambda\mathbf{C}_{-2} - (1 - \lambda)k\mathbf{Y}_{-1}]$, a $T \times 1$ vector. The following prior pdf was employed in the present case:

$$(7.75) \qquad p(\lambda, k, \gamma, \sigma_4) \propto \left.\frac{1}{\sigma_4}\right\} \qquad \begin{array}{l} 0 < \lambda, k < 1, \\ 0 < \sigma_4 < \infty, \\ -\infty < \gamma < \infty. \end{array}$$

On combining (7.74) and (7.75) by means of Bayes' theorem, the posterior pdf for the parameters can be integrated analytically with respect to σ_4 and γ to yield the following bivariate posterior pdf for λ and k:

$$(7.76) \quad p(\lambda, k|\mathbf{C}) \propto \frac{\{\sum_{t=1}^T [C_{t-1} - \lambda C_{t-2} - (1 - \lambda)kY_{t-1}]^2\}^{-\frac{1}{2}}}{\{\sum_{t=1}^T [C_t - \lambda C_{t-1} - (1 - \lambda)kY_t - \tilde{\gamma}(C_{t-1} - \lambda C_{t-2} - (1 - \lambda)kY_t]^2\}^{(T-1)/2}},$$

[38] See the Zellner-Geisel paper for additional discussion and plots of these posterior pdf's.

[39] This analysis is similar to that presented for (7.44) combined with (7.56).

where $0 < \lambda, k < 1$ and where

$$\tilde{\gamma} = \frac{\sum_{t=1}^{T} [C_{t-1} - \lambda C_{t-2} - (1 - \lambda)k Y_{t-1}][C_t - \lambda C_{t-1} - (1 - \lambda)k Y_t]}{\sum_{t=1}^{T} [C_{t-1} - \lambda C_{t-2} - (1 - \lambda)k Y_{t-1}]^2}.$$

The posterior pdf in (7.76) was analyzed numerically by employing U.S. quarterly data. The results are broadly consistent with those above which show λ to be quite different from zero. However, the mean for λ in Table 7.6 is somewhat different from those already reported, thus indicating that results are somewhat sensitive to what is assumed about the properties of error terms.

Table 7.6 POSTERIOR MEASURES ASSOCIATED WITH THE MARGINAL POSTERIOR PDF'S OF (7.76) BASED ON ASSUMPTION IV

Posterior Measure	Marginal Posterior Pdf for λ	Marginal Posterior Pdf for k
Mean	0.769	0.959
Modal Value	0.81	0.95
Variance	0.00937	0.00183

We have worked with relatively diffuse prior pdf's and thus have permitted our posterior pdf's to reflect mainly the information in the data. To illustrate how information in the data affects other than relatively diffuse prior beliefs assume that individuals A and B have differing prior beliefs. Both agree that λ and k are a priori independent but disagree with respect to the value of λ. Assume that their prior pdf's are[40]

$$(7.77) \qquad p_A(\lambda, k, \sigma_1) \propto \frac{\lambda^{34}(1 - \lambda)^{89}k^{89}(1 - k)^9}{\sigma_1} \qquad \begin{matrix} 0 < \lambda, k < 1, \\ 0 < \sigma_1 < \infty, \end{matrix}$$

for A and

$$(7.78) \qquad p_B(\lambda, k, \sigma_1) \propto \frac{\lambda(1 - \lambda)^7 k^{89}(1 - k)^9}{\sigma_1} \qquad \begin{matrix} 0 < \lambda, k < 1, \\ 0 < \sigma_1 < \infty, \end{matrix}$$

for B. Both A and B use the same prior assumptions regarding k and σ_1. For k their prior pdf is a beta pdf with mean 0.9 and variance 0.00089, whereas for σ_1 they employ the same diffuse pdf. As regards λ both use a prior pdf in the beta form but with different prior parameters. A has chosen the parameters of his prior pdf for λ in (7.77) to provide a prior mean equal to 0.7 and prior variance 0.0041; B has assigned his prior parameters values such that the prior mean of λ is 0.2 and its variance is 0.0146.

[40] We shall assume that A and B are working with (7.64) under Assumption I about the error terms.

When we combine the prior pdf's in (7.77) and (7.78) with the likelihood function in (7.65), we can see how the information in our U.S. quarterly data changes the prior beliefs of A and B, as represented by (7.77) and (7.78). In particular, on multiplying (7.65) and (7.77) or (7.78) and integrating analytically with respect to σ_1, we arrive at bivariate posterior pdf's for λ and k for A and B. These posterior pdf's were analyzed numerically with the results shown in Table 7.7.

Table 7.7 MARGINAL POSTERIOR PDF'S FOR λ AND k FOR PRIOR PDF'S FOR A (7.77) AND B (7.78) UNDER ASSUMPTION I[a]

Posterior Measure	Marginal Posterior Pdf for λ		Marginal Posterior Pdf for k	
	A	B	A	B
Mean	0.704	0.581	0.947	0.941
Modal Value	0.71	0.59	0.95	0.94
Variance	0.0025	0.0066	0.00004	0.00003

[a] The prior means and variances are $E\lambda = 0.7$, Var $\lambda = 0.0041$ for A and $E\lambda = 0.2$ and Var $\lambda = 0.0146$ for B. Both A and B have assumed a priori that $Ek = 0.9$ and Var $k = 0.00089$.

The information in Table 7.7 reveals that the sample information has moved both A and B toward a higher value for the parameter k as compared with prior expectations. Also, the sample information has resulted in considerable reduction in the variance of the pdf for k. As regards λ, the views of A and B are brought more in agreement by the sample information. A priori, A gave λ a mean of 0.7, whereas B assigned a value equal to 0.2. Their posterior pdf's for λ have means equal to 0.704 and 0.581, respectively. Thus the common sample information, when combined with the prior pdf's of A and B has diminished the difference of opinion regarding the value of λ.

7.6 SOME GENERALIZATIONS OF THE DISTRIBUTED LAG MODEL

Many generalizations of the simple distributed lag models can be analyzed. Here we take up just two. The first involves the assumption that the data have been generated as follows:

$$y_t = \beta_1 \sum_{i=0}^{\infty} \lambda^i x_{t-i,1} + \beta_2 \sum_{i=0}^{\infty} \lambda^i x_{t-i,2} + \cdots + \beta_k \sum_{i=0}^{\infty} \lambda^i x_{t-i,k} + u_t \tag{7.79a}$$

or

$$y_t = \sum_{i=0}^{\infty} \lambda^i X(t-i)\boldsymbol{\beta} + u_t, \tag{7.79b}$$

where $X(t-i) = (x_{t-i,1}, x_{t-i,2}, \ldots, x_{t-i,k})$, a $1 \times k$ row vector, $\boldsymbol{\beta}' = (\beta_1, \beta_2, \ldots, \beta_k)$, a $1 \times k$ vector of unknown coefficients, and u_t is a disturbance term. In (7.79*a*, *b*) we assume that y_t is influenced by k variables, each with a specific coefficient but with a common form and parameter for the lag structure.

On subtracting λy_{t-1} from both sides of (7.79*b*), we obtain

$$(7.80) \qquad y_t - \lambda y_{t-1} = X(t)\boldsymbol{\beta} + u_t - \lambda u_{t-1}.$$

We now assume that the disturbance term $u_t - \lambda u_{t-1}$ satisfies (7.56), and on combining (7.56) with (7.80) we have

$$(7.81) \qquad y_t - \lambda y_{t-1} - \rho(y_{t-1} - \lambda y_{t-2}) = [X(t) - \rho X(t-1)]\boldsymbol{\beta} + \epsilon_t.$$

Given that we have $t = 1, 2, \ldots, T$ observations, that the ϵ_t's satisfy the standard assumptions, that the initial values y_0 and y_{-1} are taken as given, and that our prior assumptions are

$$(7.82) \qquad p(\lambda, \boldsymbol{\beta}, \rho, \tau) \propto \frac{1}{\tau} \quad \begin{array}{l} 0 < \lambda < 1, \ -\infty < \beta_i, \rho < \infty, \\ i = 1, 2, \ldots, k, \ 0 < \tau < \infty, \end{array}$$

where τ^2 is the variance of the ϵ_t's, the posterior pdf for the parameters is

(7.83)
$$p(\lambda, \boldsymbol{\beta}, \rho, \tau|\mathbf{y}) \propto \frac{1}{\tau^{T+1}} \exp\left\{-\frac{1}{2\tau^2}[\mathbf{w} - (X - \rho X_{-1})\boldsymbol{\beta}]'[\mathbf{w} - (X - \rho X_{-1})\boldsymbol{\beta}]\right\}.$$

Here we have written

$$(7.84) \qquad \mathbf{w} = \mathbf{y} - \lambda\mathbf{y}_{-1} - \rho(\mathbf{y}_{-1} - \lambda\mathbf{y}_{-2}),$$

where $\mathbf{y}' = (y_1, \ldots, y_T)$, $\mathbf{y}_{-1}' = (y_0, y_1, \ldots, y_{T-1})$, $\mathbf{y}_{-2} = (y_{-1}, y_0, y_1, \ldots, y_{T-2})$,

$$X = \begin{pmatrix} x_{11} & x_{12} & \cdots & x_{1k} \\ \vdots & \vdots & \ddots & \vdots \\ x_{T1} & x_{T2} & \cdots & x_{Tk} \end{pmatrix} \quad \text{and} \quad X_{-1} = \begin{pmatrix} x_{01} & x_{02} & \cdots & x_{0k} \\ \vdots & \vdots & \ddots & \vdots \\ x_{T-1,1} & x_{T-1,2} & \cdots & x_{T-1,k} \end{pmatrix}.$$

From the form of (7.83) it is apparent that the elements of $\boldsymbol{\beta}$ and τ can be integrated out analytically. On integration with respect to τ, the following is the result:

$$(7.85) \qquad p(\lambda, \boldsymbol{\beta}, \rho|\mathbf{y}) \propto \{[\mathbf{w} - (X - \rho X_{-1})\boldsymbol{\beta}]'[\mathbf{w} - (X - \rho X_{-1})\boldsymbol{\beta}]\}^{-T/2}.$$

Then, using properties of the multivariate Student t pdf, we can integrate[41] with respect to $\boldsymbol{\beta}$, which gives

[41] This assumes that $(X - \rho X_{-1})'(X - \rho X_{-1})$ is positive definite for any fixed value of ρ. See the appendix to Chapter 4 for the condition that ensures this, a condition on the x's that is not very restrictive.

(7.86) $$p(\lambda, \rho|\mathbf{y}) \propto |H|^{-1/2}\{\mathbf{w}'[I - (X - \rho X_{-1})H^{-1}(X - \rho X_{-1})']\mathbf{w}\}^{-(T-k)/2},$$

where $H = (X - \rho X_{-1})'(X - \rho X_{-1})$. The bivariate posterior pdf in (7.86) can be analyzed numerically to evaluate its normalizing constant, to make joint inferences about ρ and λ, and to obtain marginal posterior pdf's for these parameters. As regards an element of $\boldsymbol{\beta}$, say β_1, its posterior pdf can be obtained from (7.85) by integration with respect to $\beta_2, \beta_3, \ldots, \beta_k$ to yield $p(\lambda, \beta_1, \rho|\mathbf{y})$, a trivariate posterior pdf that will have to be analyzed numerically to obtain the marginal pdf for β_1.

The second generalization of the model to be considered is

(7.87) $$y_t = \alpha \sum_{i=0}^{\infty} \lambda^i x_{t-i} + Z(t)\boldsymbol{\gamma} + u_t;$$

except for the addition of the term $Z(t)\boldsymbol{\gamma}$, this is the model considered previously. In (7.87) $Z(t)$ is a $1 \times k$ vector of observations on m given independent variables and $\boldsymbol{\gamma}$ is an $m \times 1$ vector of coefficients. Subtraction of λy_{t-1} from both sides of (7.87) yields

(7.88*a*) $$y_t - \lambda y_{t-1} = \alpha x_t + [Z(t) - \lambda Z(t-1)]\boldsymbol{\gamma} + u_t - \lambda u_{t-1}$$

or

(7.88*b*) $$\begin{aligned}\mathbf{y} - \lambda \mathbf{y}_{-1} &= (\mathbf{x} \vdots Z - \lambda Z_{-1})\begin{pmatrix}\alpha \\ \cdots \\ \boldsymbol{\gamma}\end{pmatrix} + \mathbf{u} - \lambda \mathbf{u}_{-1} \\ &= W\boldsymbol{\beta} + \mathbf{u} - \lambda \mathbf{u}_{-1},\end{aligned}$$

where $\mathbf{y}' = (y_1, \ldots, y_T)$, $\mathbf{y}_{-1}' = (y_0, y_1, \ldots, y_{T-1})$, $\mathbf{x}' = (x_1, x_2, \ldots, x_T)$, Z is a $T \times m$ matrix of observations on the given independent variables, $W = (\mathbf{x} \vdots Z - \lambda Z_{-1})$, and $\boldsymbol{\beta}' = (\alpha \vdots \boldsymbol{\gamma}')$. If we assume that the u_t's are normally and independently distributed with zero means and common variance σ^2 and go ahead, conditional on given y_0, the likelihood function is given by

(7.89)
$$l(\lambda, \boldsymbol{\beta}, \sigma|\mathbf{y}, y_0) \propto \frac{|G|^{-1/2}}{\sigma^T} \exp\left[-\frac{1}{2\sigma^2}(\mathbf{y} - \lambda \mathbf{y}_{-1} - W\boldsymbol{\beta})'G^{-1}(\mathbf{y} - \lambda \mathbf{y}_{-1} - W\boldsymbol{\beta})\right],$$

where $G\sigma^2$ is the $T \times T$ variance-covariance matrix for $\mathbf{u} - \lambda \mathbf{u}_{-1}$, with G given explicitly in (7.45). With a diffuse prior pdf,

(7.90) $$p(\lambda, \boldsymbol{\beta}, \sigma) \propto \frac{1}{\sigma} \left\} \begin{array}{c} 0 < \lambda < 1 \\ -\infty < \beta_i < \infty \\ 0 < \sigma < \infty \end{array} \right\} \quad i = 1, 2, \ldots, k,$$

the joint posterior pdf is given by

(7.91) $$p(\lambda, \boldsymbol{\beta}, \sigma|\mathbf{y}, y_0) \propto \frac{|G|^{-1/2}}{\sigma^{T+1}} \exp\left[-\frac{1}{2\sigma^2}(\mathbf{w} - W\boldsymbol{\beta})'G^{-1}(\mathbf{w} - W\boldsymbol{\beta})\right],$$

where we have written $\mathbf{w} = \mathbf{y} - \lambda\mathbf{y}_{-1}$. Clearly (7.91) is similar to the expression in (7.48). It can be analyzed as follows: first integrate (7.91) with respect to σ, which yields

$$(7.92) \qquad p(\lambda, \boldsymbol{\beta}|\mathbf{y}, y_0) \propto |G|^{-1/2}[(\mathbf{w} - W\boldsymbol{\beta})'G^{-1}(\mathbf{w} - W\boldsymbol{\beta})]^{-T/2}.$$

Note that, for given λ, the conditional posterior pdf for $\boldsymbol{\beta}$ is in the multivariate Student t form. This fact can be used to determine how sensitive inferences about $\boldsymbol{\beta}$ are to what is assumed about λ.

To derive the marginal posterior pdf for λ we integrate (7.92) with respect to $\boldsymbol{\beta}$ to obtain[42]

$$(7.93) \qquad p(\lambda|\mathbf{y}, y_0) \propto \frac{|G|^{-1/2}|W'G^{-1}W|^{-1/2}}{(\mathbf{w}'G^{-1}\mathbf{w} - \hat{\boldsymbol{\beta}}'W'G^{-1}W\hat{\boldsymbol{\beta}})^{(T-k)/2}},$$

where, it is to be remembered, $\mathbf{w} = \mathbf{y} - \lambda\mathbf{y}_{-1}$ and

$$(7.94) \qquad \hat{\boldsymbol{\beta}} = (W'G^{-1}W)^{-1}W'G^{-1}\mathbf{w},$$

which is a function of λ. The univariate pdf in (7.93) can be analyzed numerically. Finally, if interest centers on an element of $\boldsymbol{\beta}$, say β_1, the other elements of $\boldsymbol{\beta}$ can be integrated out of (7.92) by using properties of the multivariate Student t pdf to yield a bivariate posterior pdf, $p(\lambda, \beta_1|\mathbf{y}, \mathbf{y}_0)$, which can be analyzed numerically.

To close this chapter it is important to emphasize that the general principles of Bayesian analysis apply in the analysis of time series models with no special modification required. Of course, it is imperative that care be taken in formulating time series models to get an adequate representation of economic phenomena; for example, to approximate lag structures adequately, to use appropriate assumptions about initial conditions, and to employ appropriate functional forms. Given that this representation has been made and that the likelihood function can be formulated, as stated above, the analysis will proceed along usual Bayesian lines. Point estimates, intervals, etc., can be derived by using the principles presented in Chapter 2.

APPENDIX DIFFUSE PRIOR PDF FOR STATIONARY AUTOREGRESSIVE PROCESS

In Section 7.1 we considered the first order stationary autoregressive process given by

$$(1) \qquad \begin{aligned} y_t &= \beta_1 + \beta_2 y_{t-1} + u_t \\ &= \frac{\beta_1}{1-\beta_2} + \sum_{l=0}^{\infty} \beta_2{}^l u_{t-l}, \qquad t = 1, 2, \ldots, T, \end{aligned}$$

[42] Note that $(\mathbf{w} - W\boldsymbol{\beta})'G^{-1}(\mathbf{w} - W\boldsymbol{\beta}) = \mathbf{w}'G^{-1}\mathbf{w} + \boldsymbol{\beta}'W'G^{-1}W\boldsymbol{\beta} - 2\boldsymbol{\beta}'W'G^{-1}\mathbf{w} = \mathbf{w}'G^{-1}\mathbf{w} - \hat{\boldsymbol{\beta}}'W'G^{-1}W\hat{\boldsymbol{\beta}} + (\boldsymbol{\beta} - \hat{\boldsymbol{\beta}})'W'G^{-1}W(\boldsymbol{\beta} - \hat{\boldsymbol{\beta}})$, where $\hat{\boldsymbol{\beta}}$ is defined in (7.94).

where the u_t's NID$(0, \sigma^2)$ and $0 \le |\beta_2| < 1$. It is clear from (1) that $\theta = \beta_1/(1 - \beta_2)$ gives the level of the y_t's. In what follows we reparameterize the model in terms of

$$\beta = \beta_2, \qquad \theta = \frac{\beta_1}{1 - \beta_2} \quad \text{and} \quad \sigma; \tag{2}$$

that is

$$y_t = (1 - \beta)\theta + \beta y_{t-1} + u_t. \tag{3}$$

The likelihood function is

$$l(\beta, \theta, \sigma|\mathbf{y}, y_0) \propto \frac{(1 - \beta^2)^{1/2}}{\sigma^{T+1}} \exp\left\{-\frac{1}{2\sigma^2}[(1 - \beta^2)(y_0 - \theta)^2 + \sum \{y_t - \theta - \beta(y_{t-1} - \theta)\}^2]\right\}. \tag{4}$$

As will be recalled from Chapter 2, Jeffreys has suggested that the square root of the determinant of the information matrix be taken as a diffuse prior pdf, although not without great care and thought; that is, Jeffreys' diffuse prior is given by

$$p(\beta, \theta, \sigma) \propto |\text{Inf}|^{1/2}, \tag{5}$$

with

$$\text{Inf} = -E\begin{bmatrix} \dfrac{\partial^2 L}{\partial \sigma^2} & \dfrac{\partial^2 L}{\partial \sigma\, \partial \beta} & \dfrac{\partial^2 L}{\partial \sigma\, \partial \theta} \\ \dfrac{\partial^2 L}{\partial \beta\, \partial \sigma} & \dfrac{\partial^2 L}{\partial \beta^2} & \dfrac{\partial^2 L}{\partial \beta\, \partial \theta} \\ \dfrac{\partial^2 L}{\partial \theta\, \partial \sigma} & \dfrac{\partial^2 L}{\partial \theta\, \partial \beta} & \dfrac{\partial^2 L}{\partial \theta^2} \end{bmatrix}, \tag{6}$$

where E denotes the expectation with respect to the pdf for the observations and $L = \log l$, the log-likelihood. After (5) has been evaluated we can transform to the implied prior pdf for β_1, β_2, and σ, using the relations in (2).

To illustrate the operations that are required to obtain Jeffreys' diffuse prior for this problem we first evaluate the information matrix. From (4),

$$L = \text{const} + \tfrac{1}{2}\log(1 - \beta^2) - (T + 1)\log\sigma - \frac{1}{2\sigma^2}\{(1 - \beta^2)(y_0 - \theta)^2 + \sum[y_t - \theta - \beta(y_{t-1} - \theta)]^2\}. \tag{7}$$

Then, with the quantity in braces in (7) denoted $\{\cdot\}$, we have

$$\frac{\partial L}{\partial \sigma} = -\frac{T + 1}{\sigma} + \frac{1}{\sigma^3}\{\cdot\},$$

$$\frac{\partial^2 L}{\partial \sigma^2} = \frac{T + 1}{\sigma^2} - \frac{3}{\sigma^4}\{\cdot\},$$

$$\frac{\partial^2 L}{\partial\theta\,\partial\sigma} = \frac{1}{\sigma^3}\left\{-2(1-\beta^2)(y_0-\theta) - 2\sum[y_t-\theta-\beta(y_{t-1}-\theta)](1-\beta)\right\},$$

and

$$\frac{\partial^2 L}{\partial\beta\,\partial\sigma} = \frac{1}{\sigma^3}\left\{-2\beta(y_0-\theta)^2 - 2\sum[(y_t-\theta)-\beta(y_{t-1}-\theta)](y_{t-1}-\theta)\right\}.$$

On taking expectations with respect to the y's, we obtain

$$(8)\qquad E\frac{\partial^2 L}{\partial\sigma^2} = -\frac{2(T+1)}{\sigma^2};\qquad E\frac{\partial^2 L}{\partial\theta\,\partial\sigma} = 0;\ \text{and}\ E\frac{\partial^2 L}{\partial\beta\,\partial\sigma} = -\frac{2}{\sigma}\frac{\beta}{1-\beta^2}.$$

In deriving the results in (8), note that $E(y_t-\theta)^2 = \sigma^2/(1-\beta^2)$ and $E(y_t-\theta)(y_{t-1}-\theta) = \beta\sigma^2/(1-\beta^2)$ for all values of t.

Further

$$\frac{\partial L}{\partial\theta} = -\frac{1}{2\sigma^2}\left\{-2(1-\beta^2)(y_0-\theta) - 2\sum[y_t-\theta-\beta(y_{t-1}-\theta)](1-\beta)\right\},$$

$$\frac{\partial^2 L}{\partial\theta^2} = -\frac{1}{\sigma^2}[1-\beta^2+T(1-\beta)^2],$$

and

$$\frac{\partial^2 L}{\partial\beta\,\partial\theta} = -\frac{1}{2\sigma^2}\left\{4\beta(y_0-\theta) + 2\sum[y_t-\theta-\beta(y_{t-1}-\theta) + (1-\beta)(y_{t-1}-\theta)]\right\}.$$

Taking expectations of these last two quantities, we obtain these results:

$$(9)\quad E\frac{\partial^2 L}{\partial\theta^2} = -\frac{1}{\sigma^2}[1-\beta^2+T(1-\beta)^2]\ \text{and}\ E\frac{\partial^2 L}{\partial\beta\,\partial\theta} = 0.$$

Last

$$\frac{\partial L}{\partial\beta} = -\frac{\beta}{1-\beta^2} - \frac{1}{2\sigma^2}\left\{-2\beta(y_0-\theta)^2 - 2\sum[y_t-\theta-\beta(y_{t-1}-\theta)](y_{t-1}-\theta)\right\},$$

and

$$\frac{\partial^2 L}{\partial\beta^2} = -\frac{1+\beta^2}{(1-\beta^2)^2} + \frac{1}{\sigma^2}\left[(y_0-\theta)^2 - \sum(y_{t-1}-\theta)^2\right].$$

The expectation of this last expression is

$$(10)\qquad E\frac{\partial^2 L}{\partial\beta^2} = -\frac{1}{1-\beta^2}\left(T+\frac{2\beta^2}{1-\beta^2}\right).$$

On collecting results from (8) to (10), we evaluate the information matrix in (6) as

$$(11)\quad \text{Inf} = \begin{array}{c} \\ \\ \\ \end{array}\begin{array}{ccc} \sigma & \beta & \theta \end{array}$$

$$(11)\quad \text{Inf} = \left[\begin{array}{c|cc} \dfrac{2(T+1)}{\sigma^2} & \dfrac{2\beta}{\sigma(1-\beta^2)} & 0 \\ \hline \dfrac{2\beta}{\sigma(1-\beta^2)} & \dfrac{T + 2\beta^2/(1-\beta^2)}{1-\beta^2} & 0 \\ 0 & 0 & \dfrac{T(1-\beta)^2 + 1 - \beta^2}{\sigma^2} \end{array}\right] \begin{array}{c} \sigma \\ \beta \\ \theta \end{array}$$

From (11) we see that information about θ is independent of information about either σ or β.

On evaluating the determinant of the matrix in (11) and retaining just the dominant terms, those in T^3, we have the result $2(1 - \beta)^2/(1 - \beta^2)\sigma^4$. On taking the square root of this quantity, we obtain an approximate Jeffreys' diffuse prior, namely,

$$(12)\qquad p(\beta, \theta, \sigma) \propto \frac{1 - \beta}{(1 - \beta^2)^{1/2}} \frac{1}{\sigma^2}, \qquad 0 \leq |\beta| < 1.$$

The appearance of the factor $1/\sigma^2$ in this last expression rather than $1/\sigma$ appears to result from reasons similar to those discussed at the end of the appendix to Chapter 2.[43] If we follow Jeffreys and apply his principle separately for σ and for the other parameters, that is, take the square root of the (1, 1) element of (11) to obtain a diffuse prior for σ and then take the square root of the determinant of the 2×2 information matrix for β and θ, retaining just terms in T^2, the result is

$$(13)\qquad p(\beta, \theta, \sigma) \propto \frac{1 - \beta}{\sigma(1 - \beta^2)^{1/2}},$$

wherein the factor $1/\sigma$ rather than $1/\sigma^2$ appears. To obtain the prior for β_1, β_2, and σ note from (2) that $d\theta = d\beta_1/(1 - \beta_2)$ and thus

$$(14)\qquad p(\beta_1, \beta_2, \sigma) \propto \frac{1}{\sigma(1 - \beta_2{}^2)^{1/2}}.$$

This approximate Jeffreys' prior involves assuming that β_1, $\log \sigma$, and β_2 are independent, with the first two quantities uniformly distributed and the last with a pdf proportional to $(1 - \beta_2{}^2)^{-1/2} = (1 - \beta_2)^{-1/2}(1 + \beta_2)^{-1/2}$, a form of the beta pdf with parameters $(\frac{1}{2}, \frac{1}{2})$. This pdf for β_2 has its greatest density at the end points $\beta_2 = -1$ and $\beta_2 = 1$ and a minimum at $\beta_2 = 0$.

[43] Here, if we had more than one θ, for each additional θ, an additional $1/\sigma$ factor would appear in the prior pdf.

It is interesting to ascertain the form of the "minimal information prior" (see the appendix to Chapter 2) for the present problem. The pdf for an observation is

$$p(y|\beta, \theta, \sigma) = \frac{1}{\sqrt{2\pi}} \frac{(1 - \beta^2)^{1/2}}{\sigma} \exp\left[-\left(\frac{1 - \beta^2}{2\sigma^2}\right)(y - \theta)^2\right].$$

Then the information in the data pdf is

$$I_y = \int_{-\infty}^{\infty} p(y|\beta, \theta, \sigma) \log p(y|\beta, \theta, \sigma)\, dy$$

$$= \log \frac{(1 - \beta^2)^{1/2}}{\sigma} - \tfrac{1}{2} - \tfrac{1}{2} \log 2\pi.$$

If we maximize the prior average information in the data pdf minus the information in a prior pdf, subject to the condition that the prior pdf be proper, the result is

$$p(\beta, \theta, \sigma) \propto \frac{(1 - \beta^2)^{1/2}}{\sigma}, \qquad 0 \le |\beta| < 1. \tag{15}$$

This prior pdf for β, θ, and σ involves similar properties with respect to θ and σ as that in (13).[44] The form for β is somewhat different, however, in that (15) has a unimodal form with a maximum at $\beta = 0$ and falls to zero at $\beta = \pm 1$. The factor $1 - \beta$ in (13) behaves in a similar fashion to $(1 - \beta^2)^{1/2}$; $(1 - \beta)/(1 - \beta^2)^{1/2}$, however, has decidedly different properties.

QUESTIONS AND PROBLEMS

1. Assume that y_t is generated by a first order linear autoregressive scheme; that is, $y_t = \rho y_{t-1} + \epsilon_t$, $t = 1, 2, \ldots, T$, where the ϵ_t's are NID$(0, \sigma^2)$. Specify a prior pdf for ρ and σ in the normal-gamma form; that is, $p(\rho, \sigma) = p_1(\rho|\sigma)p_2(\sigma)$, with $p_1(\rho|\sigma) \propto 1/\sigma \exp[-(c/2\sigma^2)(\rho - \bar{\rho})^2]$, $-\infty < \rho < \infty$, and $p_2(\sigma) \propto (\sigma^{\nu_0+1})^{-1} \exp(-\nu_0 s_0^2/2\sigma^2)$, $0 < \sigma < \infty$, where c, $\bar{\rho}$, ν_0, and s_0^2 are prior parameters whose values are assigned by an investigator to reflect his prior beliefs.
   (a) Given the above prior pdf for ρ and σ, what are the form and properties of the marginal prior pdf for ρ?
   (b) Assuming that y_0 is given, derive the marginal posterior pdf's for ρ and σ and provide a summary of their properties.
2. In Problem 1 formulate and plot a prior pdf for ρ which reflects the prior beliefs that $-1 < \rho < 1$ and that the prior mean and variance of ρ are 0.5 and 0.04, respectively.

[44] Here we are assuming that the ranges for σ and θ in (15) are extremely large.

3. Show how the prior pdf in Problem 2 can be utilized in an analysis of the linear autoregressive scheme described in Problem 1.

4. Assume that the linear autoregressive scheme $y_t = \rho y_{t-1} + \epsilon_t$, $t = 1, 2, \ldots, T$, is known to have "started up" at $t = -T_0$, with T_0 having a known value and $y_{-T_0} = 10$. Derive the pdf of y_0 under these assumptions and with the assumption that the ϵ_t's are NID(0, σ^2) for all t.

5. Consider the process $y_t = \rho y_{t-1} + u_t$, where $u_t = \alpha u_{t-1} + \epsilon_t$, ρ and α are parameters, and u_t and ϵ_t are random error terms with zero means. Further assume that the ϵ_t's are NID(0, σ^2). Note that we can write, $y_t - \alpha y_{t-1} = \rho(y_{t-1} - \alpha y_{t-2}) + \epsilon_t$ or $y_t - \rho y_{t-1} = \alpha(y_{t-1} - \rho y_{t-2}) + \epsilon_t$. Explain the statement, "Without additional prior information about α and ρ, these parameters are not identified." Then provide several examples of prior information about ρ and/or α that is sufficient to identify these parameters.

6. Let $y_t = \sum_{i=0}^{m} w_i x_{t-i} + \epsilon_t$, where w_i is an unknown weight parameter for x_{t-i}, $i = 0, 1, 2, \ldots, m$, where m's value is known, x_{t-i} denotes the value of an independent variable in time period $t - i$, and ϵ_t is a random error term. In the Almon approach it is assumed that w_i can be approximated by a polynomial in i; for example, if we use a second degree polynomial, we have

$$w_i = \gamma_0 + \gamma_1 i + \gamma_2 i^2, \qquad i = 0, 1, 2, \ldots, m,$$

which when substituted in the expression for y_t above yields

$$y_t = \gamma_0 \sum_{i=0}^{m} x_{t-i} + \gamma_1 \sum_{i=0}^{m} i x_{t-i} + \gamma_2 \sum_{i=0}^{m} i^2 x_{t-i} + \epsilon_t.$$

Now assume that the ϵ_t's are NID(0, σ^2) and that we are employing a diffuse prior pdf, $p(\gamma_0, \gamma_1, \gamma_2, \sigma) \propto 1/\sigma$, with the γ's ranging from $-\infty$ to $+\infty$ and $0 < \sigma < \infty$. Derive the joint posterior pdf for the γ's, given sample observations on y_t, $t = 1, 2, \ldots, T$ and $x_{-(m-1)}, x_{-(m-2)}, \ldots, x_0, x_1, \ldots, x_T$. What is the marginal posterior pdf for γ_1?

7. In Problem 6 derive the joint posterior pdf for $w_0 = \gamma_0$, $w_1 = \gamma_0 + \gamma_1 + \gamma_2$ and $w_2 = \gamma_0 + 2\gamma_1 + 4\gamma_2$ and explain how to construct a 95% Bayesian confidence region for w_0 and w_1.

8. Often in applying the Almon technique, see Problem 6, it is assumed that $w_{-1} = \gamma_0 - \gamma_1 + \gamma_2 = 0$ and $w_{m+1} = \gamma_0 + (m + 1)\gamma_1 + (m + 1)^2\gamma_2 = 0$. To investigate these assumptions, use the results of Problem 6 to derive the joint and marginal posterior pdf's for $w_{-1} = \gamma_0 - \gamma_1 + \gamma_2$ and $w_{m+1} = \gamma_0 + (m + 1)\gamma_1 + (m + 1)^2\gamma_2$. If the posterior probability density is very low in the vicinity of the point $w_{-1} = 0$, what, if anything, does this imply about the assumption that $w_{-1} = 0$?

9. Consider the following "multiplier-accelerator" model:

$$\left.\begin{array}{ll} \text{(i)} & C_t = \alpha Y_{t-1} + u_t \\ \text{(ii)} & I_t = \beta(Y_{t-1} - Y_{t-2}) + v_t \\ \text{(iii)} & Y_t = C_t + I_t \end{array}\right\} \qquad t = 1, 2, \ldots, T,$$

where C_t = consumption, I_t = investment, Y_t = income, α and β are scalar parameters, and u_t and v_t are random disturbance terms. By substituting from

(i) and (ii) in (iii) write down the "final equation" for Y_t. After providing assumptions about the error terms' properties and a prior pdf for the parameters, explain how to obtain the joint and marginal posterior pdf's for α and β.

10. Show how the posterior pdf for α and β in Problem 9 can be employed to make probability statements about properties of the solution to the final equation for Y_t obtained in Problem 9.

11. Let net expenditures on housing in the tth period E_t be assumed to satisfy

$$E_t = H_t - H_{t-1} = \beta(H_t^* - H_{t-1}) + \epsilon_t, \qquad t = 1, 2, \ldots, T,$$

where β is an adjustment parameter, H_{t-1} is the given stock of houses at the end of period $t - 1$, ϵ_t is a random disturbance term, and H_t^* is the desired stock of houses for the tth period. Since H_t^* is unobservable, assume that $H_t^* = \alpha_0 + \alpha_1 x_t$, where x_t is an observable independent variable and α_0 and α_1 are parameters with unknown values. On substituting this expression for H_t^* in the equation for E_t, we have

$$\begin{aligned} E_t &= \beta\alpha_0 + \beta\alpha_1 x_t - \beta H_{t-1} + \epsilon_t \\ &= \gamma_0 + \gamma_1 x_t + \gamma_2 H_{t-1} + \epsilon_t, \qquad t = 1, 2, \ldots, T, \end{aligned}$$

where $\gamma_0 = \beta\alpha_0$, $\gamma_1 = \beta\alpha_1$, and $\gamma_2 = -\beta$. Viewing these last three relations as a transformation from γ_0, γ_1, and γ_2 to α_0, α_1, and β, we need the Jacobian of the transformation to be different from zero for the transformation to be regular or "one-to-one." Evaluate the Jacobian and show that the required condition is $\beta^2 \neq 0$. Then, under the assumption that the ϵ_t's are NID$(0, \sigma^2)$ and given the initial observation H_0, what are the ML estimators for γ_0, γ_1, and γ_2? From the ML estimators for the γ's, obtain ML estimators for α_0, α_1, and β and comment on their sampling properties in large samples.

12. If, in Problem 11, we employed the following diffuse prior pdf, $p(\gamma_0, \gamma_1, \gamma_2, \sigma) \propto 1/\sigma$ with $-\infty < \gamma_i < \infty$, $i = 0, 1, 2$, and $0 < \sigma < \infty$, derive and comment on the properties of the implied prior pdf for α_0, α_1, β, and σ, using $\gamma_0 = \beta\alpha_0$, $\gamma_1 = \beta\alpha_1$, and $\gamma_2 = -\beta$. Then obtain the joint posterior pdf for γ_0, γ_1, and γ_2. Show that the marginal posterior pdf for $\beta = -\gamma_2$ is in the form of a univariate Student t pdf. Noting that the marginal posterior pdf's for $\alpha_0 = -\gamma_0/\gamma_2$ and $\alpha_1 = -\gamma_1/\gamma_2$ are in the form of the ratio of correlated Student t variables, show that these quantities will be distributed as the ratio of correlated normal variables in large samples.

13. Write the equation for E_t in Problem 11 as follows:

$$E_t = \beta(\alpha_0 + \alpha_1 x_t - H_{t-1}) + \epsilon_t.$$

Explain how ML estimates of β, α_0, and α_1 can be obtained by minimizing $\sum_{t=1}^{T} \epsilon_t^2$, using a two-parameter search over values of α_0 and α_1. Will this procedure yield the same ML estimates obtained in Problem 11?

14. In Problem 13 assume that the prior pdf for β, α_0, α_1, and σ is given by $p(\beta, \alpha_0, \alpha_1, \sigma) \propto p_1(\beta)/\sigma$, $0 < \sigma < \infty$, and $-\infty < \alpha_i < \infty$, $i = 0, 1$. Derive the conditional posterior pdf for α_0 and α_1, given β and σ. What condition on β is required for this conditional pdf for α_0 and α_1 to be proper?

15. In Problem 11 suppose that we have no data on E_t, net expenditures, but just on gross expenditures, $G_t = E_t + R_t$, where R_t represents replacement expenditures. If it is assumed that $R_t = \delta H_{t-1}$, where δ is a depreciation parameter, we have

$$G_t = \beta\alpha_0 + \beta\alpha_1 x_t + (\delta - \beta)H_{t-1} + \epsilon_t.$$

With δ's value unknown, are the parameters of this equation identified? Alternatively, if δ has a known value, say $\delta = \delta_0$, explain how ML and Bayesian estimates of β, α_0, and α_1 can be obtained.

16. In Problem 15 formulate an informative prior pdf for the parameter δ and show how it can be utilized in deriving marginal posterior pdf's for the parameters in the equation for G_t.

CHAPTER VIII

Multivariate Regression Models

In many circumstances in economics and elsewhere we encounter sets of regression equations. What is more, it is often the case that the disturbance terms in different equations are correlated; for example, if we have a set of regression relations pertaining to firms in the same industry, it is probably the case that the disturbance terms in one firm's regression equations are correlated with those in the equations of other firms.[1] This may be so because firms in the same industry generally experience common random shocks. Or, if we have a set of consumer demand relations, it is the case that disturbance terms in different demand relations are often correlated.[2] It is important that nonindependence of disturbance terms be taken into account in making inferences. If this is not done, inferences may be greatly affected.

In this chapter we first analyze the traditional multivariate regression model. After that, attention is directed to the interpretation and analysis of the "seemingly unrelated" regression model, which is, in certain respects, somewhat more general than the traditional model and has been used in quite a few applications reported in the literature.

8.1 THE TRADITIONAL MULTIVARIATE REGRESSION MODEL[3]

We assume that our observations $Y = (\mathbf{y}_1, \mathbf{y}_2, \ldots, \mathbf{y}_m)$, an $n \times m$ matrix of observations on m variables, have been generated by the following model:

$$Y = XB + U, \tag{8.1}$$

[1] This has actually been observed empirically. See below for an example.

[2] See, for example, A. P. Barten, "Consumer Demand Functions under Conditions of Almost Additive Preferences," *Econometrica*, **32**, 1–38 (1964).

[3] Two papers dealing with the analysis of this model from the Bayesian point of view are S. Geisser, "Bayesian Estimation in Multivariate Analysis," *Ann. Math. Statist.*, **36**, 150–159 (1965), and G. C. Tiao and A. Zellner, "On the Bayesian Estimation of Multivariate Regression," *J. Roy. Statist. Soc.*, Series B, **26** (1964), 277–285.

where X is an $n \times k$ matrix, with rank k of given observations on k independent variables, $B = (\boldsymbol{\beta}_1, \boldsymbol{\beta}_2, \ldots, \boldsymbol{\beta}_m)$ is a $k \times m$ matrix of regression parameters, and $U = (\mathbf{u}_1, \mathbf{u}_2, \ldots, \mathbf{u}_m)$ is an $n \times m$ matrix of unobserved random disturbance terms. We assume that the rows of U are independently distributed,[4] each with an m-dimensional normal distribution with zero vector mean and positive definite $m \times m$ covariance matrix Σ. Under these assumptions the pdf for Y, given X, B, and Σ is

$$p(Y|X, B, \Sigma) \propto |\Sigma|^{-n/2} \exp\left[-\tfrac{1}{2}\operatorname{tr}(Y - XB)'(Y - XB)\Sigma^{-1}\right], \tag{8.2}$$

where "tr" denotes the trace operation. Noting that

$$\begin{aligned}(Y - XB)'(Y - XB) &= (Y - X\hat{B})'(Y - X\hat{B}) + (B - \hat{B})'X'X(B - \hat{B}) \\ &= S + (B - \hat{B})'X'X(B - \hat{B}),\end{aligned} \tag{8.3}$$

where

$$\hat{B} = (X'X)^{-1}X'Y, \tag{8.4}$$

a matrix of least squares quantities, and

$$S = (Y - X\hat{B})'(Y - X\hat{B}), \tag{8.5}$$

a matrix proportional to the sample disturbance covariance matrix, we can write the likelihood function for B and Σ as follows:

$$l(B, \Sigma| Y, X) \propto |\Sigma|^{-n/2} \exp\left[-\tfrac{1}{2}\operatorname{tr} S\Sigma^{-1} - \tfrac{1}{2}\operatorname{tr}(B - \hat{B})'X'X(B - \hat{B})\Sigma^{-1}\right]. \tag{8.6}$$

We assume that little is known, a priori, about the parameters, the elements of B, and the $m(m + 1)/2$ distinct elements of Σ. As our diffuse prior pdf, we assume that the elements of B and those of Σ are independently distributed; that is,

$$p(B, \Sigma) = p(B)\,p(\Sigma). \tag{8.7}$$

In (8.7), using the invariance theory due to Jeffreys,[5] we take

$$p(B) = \text{const} \tag{8.8}$$

and

$$p(\Sigma) \propto |\Sigma|^{-(m+1)/2}. \tag{8.9}$$

[4] This assumption precludes any auto or serial correlation of disturbance terms.

[5] See Harold Jeffreys, *Theory of Probability*, Oxford: Clarendon, 1961, p. 179, and the Appendix at the end of Chapter Two above.

With respect to (8.9), we note that in the special case in which $m = 1$, it reduces to

$$p(\sigma_{11}) \propto \frac{1}{\sigma_{11}}, \tag{8.10}$$

a prior assumption that we have employed many times. Also, it is interesting to observe that if we denote $\sigma^{\mu\mu'}$ as the (μ, μ')th element of the inverse of Σ the Jacobian of the transformation of the $m(m + 1)/2$ variables,

$$(\sigma_{11}, \sigma_{12}, \ldots, \sigma_{mm}) \quad \text{to} \quad (\sigma^{11}, \sigma^{12}, \ldots, \sigma^{mm}),$$

is

$$J = \left| \frac{\partial(\sigma_{11}, \sigma_{12}, \ldots, \sigma_{mm})}{\partial(\sigma^{11}, \sigma^{12}, \ldots, \sigma^{mm})} \right| = |\Sigma|^{m+1}. \tag{8.11}$$

Consequently, the prior pdf in (8.9) implies the following prior pdf on the $m(m + 1)/2$ distinct elements of Σ^{-1}:

$$p(\Sigma^{-1}) \propto |\Sigma^{-1}|^{-(m+1)/2}, \tag{8.12}$$

a diffuse prior pdf used by Savage,[6] who arrived at it through a slightly different argument, and by others. In addition to Jeffreys' invariance theory's approach leading to (8.12), Geisser[7] points out that (8.12) would result from taking an informative prior pdf on Σ^{-1} in the Wishart pdf form and allowing the "degrees of freedom" in the prior pdf to be zero.[8] With respect to the diffuse prior pdf's in (8.8) and (8.9) Geisser[9] also comments,

"These unnormed densities or weight functions presumably may be 'justified' by various rules, e.g., invariance, conjugate families, stable estimation, etc., or heuristic arguments. Although their utilization here does not necessarily preclude other contenders which may also be conceived of as displaying a measure of ignorance, it is our view that no others at present seem to be either more appropriate or as convenient. The fact that their application yields in many instances the same regions as those of classical confidence theory is certainly no detriment to their use, but in fact provides a Bayesian interpretation for these well established procedures."

[6] L. J. Savage, "The Subjective Basis of Statistical Practice," manuscript, University of Michigan, 1961.
[7] S. Geisser, "A Bayes Approach for Combining Correlated Estimates," *J. Am. Statist. Assoc.*, **60**, 602–607, p. 604 (1965).
[8] That is, if we take an informative prior pdf for Σ^{-1}, $p(\Sigma^{-1})\, d\Sigma^{-1} \propto |\Sigma^{-1}|^{-v/2} \times \exp\{-\frac{1}{2}\operatorname{tr}[(m - v + 1)\Lambda\Sigma^{-1}]\}\, d\Sigma^{-1}$, where Λ is positive definite, which is in the Wishart form (see Appendix B), and let the degrees of freedom $m - v + 1 = 0$, then $v = m + 1$ and the Wishart form reduces to (3.12). With zero degrees of freedom we have a "spread out" Wishart pdf which can serve as a diffuse prior pdf in the sense that it is diffuse enough to be substantially modified by a small number of observations.
[9] Geisser, "Bayesian Estimation in Multivariate Analysis," *loc. cit.*

With this said about the diffuse prior pdf's in (8.8), (8.9), and (8.12), we combine them with the likelihood function in (8.6) to obtain the following joint posterior pdf for the parameters:

$$p(B, \Sigma \mid Y, X) \propto |\Sigma|^{-\frac{1}{2}(n+m+1)} \times \exp\{-\tfrac{1}{2}\operatorname{tr}[S + (B - \hat{B})'X'X(B - \hat{B})]\Sigma^{-1}\} \tag{8.13a}$$

or

$$p(B, \Sigma^{-1} \mid Y, X) \propto |\Sigma^{-1}|^{[n-(m+1)]/2} \times \exp\{-\tfrac{1}{2}\operatorname{tr}[S + (B - \hat{B})'X'X(B - \hat{B})]\Sigma^{-1}\}. \tag{8.13b}$$

From (8.13*a*) we can write

$$p(B, \Sigma \mid Y, X) = p(B \mid \Sigma, Y, X)\, p(\Sigma \mid Y, X)$$

with

$$\begin{aligned} p(B \mid \Sigma, Y, X) &\propto |\Sigma|^{-k/2} \exp\{-\tfrac{1}{2}\operatorname{tr}[B - \hat{B})'X'X(B - \hat{B})\Sigma^{-1}]\} \\ &\propto |\Sigma|^{-k/2} \exp[-\tfrac{1}{2}(\boldsymbol{\beta} - \hat{\boldsymbol{\beta}})'\Sigma^{-1} \otimes X'X(\boldsymbol{\beta} - \hat{\boldsymbol{\beta}})], \end{aligned} \tag{8.14}$$

where $\boldsymbol{\beta}' = (\boldsymbol{\beta}_1', \boldsymbol{\beta}_2', \ldots, \boldsymbol{\beta}_m')$, $\hat{\boldsymbol{\beta}}' = (\hat{\boldsymbol{\beta}}_1', \hat{\boldsymbol{\beta}}_2', \ldots, \hat{\boldsymbol{\beta}}_m')$, and $\otimes$ denotes Kronecker or direct matrix multiplication and

$$p(\Sigma \mid Y, X) \propto |\Sigma|^{-\nu/2} \exp(-\tfrac{1}{2}\operatorname{tr}\Sigma^{-1}S), \tag{8.15}$$

with $\nu = n - k + m + 1$.

It is seen from (8.14) that the conditional posterior pdf for B, given Σ, is multivariate normal with mean $\hat{\boldsymbol{\beta}}$ and covariance matrix[10] $\Sigma \otimes (X'X)^{-1}$. If interest centers on a particular equation's coefficient vector, say $\boldsymbol{\beta}_1$, its conditional posterior pdf is

$$p(\boldsymbol{\beta}_1 \mid \Sigma, Y, X) \propto \frac{1}{\sigma_{11}^{k/2}} \exp\left(-\frac{1}{2\sigma_{11}}(\boldsymbol{\beta}_1 - \hat{\boldsymbol{\beta}}_1)'X'X(\boldsymbol{\beta}_1 - \hat{\boldsymbol{\beta}}_1)\right), \tag{8.16}$$

which is, of course, normal with mean vector $\hat{\boldsymbol{\beta}}_1$ and covariance matrix $(X'X)^{-1}\sigma_{11}$.

The marginal posterior pdf for Σ, given in (8.15), is in what Tiao and Zellner call the "inverted" Wishart form. They show (*op. cit.*, p. 280) that the elements of Σ_{11}, the $p \times p$ upper left-hand principal minor matrix[11] of Σ, with $p < m$,

[10] Note that $(\Sigma^{-1} \otimes X'X)^{-1} = \Sigma \otimes (X'X)^{-1}$, which can be verified by direct matrix multiplication, and that $|\Sigma^{-1} \otimes X'X|^{1/2} \propto |\Sigma|^{-k/2}$.

[11] That is, Σ_{11} is defined by

$$\Sigma = \begin{pmatrix} \Sigma_{11} & \Sigma_{12} \\ \Sigma_{21} & \Sigma_{22} \end{pmatrix}.$$

See Appendix B for the analysis leading to the marginal pdf for Σ_{11} shown in (8.17).

also have a posterior pdf in the inverted Wishart form:

$$p(\Sigma_{11}|Y, X) \propto |\Sigma_{11}|^{-[\nu-2(m-p)]/2} \exp(-\tfrac{1}{2}\operatorname{tr}\Sigma_{11}^{-1}S_{11}), \tag{8.17}$$

where S_{11} is the upper left-hand principal minor matrix of S. In particular, if $p = 1$, the posterior pdf for σ_{11} is

$$p(\sigma_{11}|Y, X) \propto \frac{1}{\sigma_{11}^{[\nu-2(m-1)]/2}} \exp\left(-\frac{s_{11}}{2\sigma_{11}}\right), \tag{8.18}$$

which is in the form of an inverted gamma pdf. If we had just one ($m = 1$) regression equation, (8.18) would specialize to

$$p(\sigma_{11}|Y, X, m = 1) \propto \frac{1}{\sigma_{11}^{(n-k+2)/2}} \exp\left(-\frac{s_{11}}{2\sigma_{11}}\right). \tag{8.19}$$

From (8.18) we see that as m increases the posterior pdf for σ_{11} becomes less and less concentrated about s_{11}. This is an intuitively pleasing result because, as m increases, a larger part of the sample information is utilized to estimate $\sigma_{12}, \sigma_{13}, \ldots, \sigma_{1m}$. In fact, the exponent of $\sigma_{11}^{-1/2}$ in (8.18) differs from that in (8.19) by $n - k + 2 - [n - k + m + 1 - 2(m - 1)] = m - 1$. Thus we may say that "one degree of freedom is lost for each of the $m - 1$ elements $\sigma_{12}, \ldots, \sigma_{1m}$."

Further, on specializing (8.17) to the case $p = 2$, we can follow the development in Jeffreys[12] to obtain the posterior pdf for the correlation coefficient $\rho_{12} = \sigma_{12}/(\sigma_{11}\sigma_{22})^{1/2}$, which is

$$p(\rho_{12}|Y, X) \propto \frac{(1 - \rho_{12}^2)^{(n'-3)/2}}{(1 - \rho_{12}r_{12})^{n'-1/2}} S_{n'}(\rho_{12}r_{12}), \tag{8.20}$$

with $n' = n - k - (m - 2)$, $r_{12} = s_{12}/(s_{11}s_{22})^{1/2}$, and

$$S_{n'}(\rho_{12}r_{12}) = 1 + \sum_{l=1}^{\infty} \frac{1}{l!} \frac{1^2 \cdot 3^2 \cdots (2l - 1)^2}{(n' + \frac{1}{2}) \cdots (n' + l - \frac{1}{2})} \left(\frac{1 + \rho_{12}r_{12}}{8}\right)^l. \tag{8.21}$$

The result in (8.20), except for the changes in the "degrees of freedom," is in the same form as that given by Jeffreys for sampling from a bivariate normal population.

To obtain the marginal posterior pdf for a particular equation's coefficient vector, say $\boldsymbol{\beta}_1$, it is pertinent to observe from (8.16) that the conditional posterior pdf for $\boldsymbol{\beta}_1$, given Σ, depends only on σ_{11}. Thus from (8.16) and (8.18) we have for the marginal posterior pdf for $\boldsymbol{\beta}_1$

$$\begin{aligned} p(\boldsymbol{\beta}_1|Y, X) &\propto \int p(\sigma_{11}|Y, X)\, p(\boldsymbol{\beta}_1|\Sigma, Y, X)\, d\sigma_{11} \\ &\propto [s_{11} + (\boldsymbol{\beta}_1 - \hat{\boldsymbol{\beta}}_1)'X'X(\boldsymbol{\beta}_1 - \hat{\boldsymbol{\beta}}_1)]^{-[n-(m-1)]/2}, \end{aligned} \tag{8.22}$$

[12] Jeffreys, *op. cit.*, p. 174. See also Section 3 of Appendix B.

which is in the form of a multivariate Student t pdf. This permits us to make inferences easily about the elements of $\boldsymbol{\beta}_1$. If we had just one regression equation, that is, $m = 1$, (8.22) reduces to the result that we had earlier in Chapter 3. With $m \neq 1$ the only difference is a change in the "degrees of freedom" due to the inclusion of the $m - 1$ parameters $\sigma_{12}, \ldots, \sigma_{1m}$ in the model. A slightly different way of looking at this problem is to consider the case of m regression equations with a *diagonal* covariance matrix containing unknown variances on the main diagonal. Then, with our prior pdf taken as $p(B, \sigma_{11}, \sigma_{12}, \ldots, \sigma_{mm}) \propto \prod_{\mu=1}^{m} \sigma_{\mu\mu}^{-1}$, the marginal posterior pdf for $\boldsymbol{\beta}_1$ would be in the multivariate Student t form, as shown in (8.22), but with exponent $-n/2$. In the case of an $m \times m$ nondiagonal covariance matrix Σ, that is, with correlated observations, there is less information in the data than with uncorrelated observations, and this fact is reflected in (8.22) by a reduction in the degrees of freedom relative to the uncorrelated case.

With respect to the joint posterior pdf for all the regression coefficients B, properties of the Wishart pdf[13] can be utilized to integrate (8.13*b*) with respect to the distinct elements of Σ^{-1}; that is

$$
\begin{aligned}
p(B|Y, X) &\propto \frac{1}{|S + (B - \hat{B})'X'X(B - \hat{B})|^{n/2}} \\
&\quad \times \int \frac{\exp\{-\frac{1}{2}\operatorname{tr}[S + (B-\hat{B})'X'X(B-\hat{B})]\Sigma^{-1}\}|\Sigma^{-1}|^{(n-m-1)/2}}{|S + (B-\hat{B})'X'X(B-\hat{B})|^{-n/2}}\, d\Sigma^{-1}
\end{aligned} \tag{8.23}
$$

which yields

$$
p(B|Y, X) \propto \frac{1}{|S + (B - \hat{B})'X'X(B - \hat{B})|^{n/2}}, \tag{8.24}
$$

since the integral in (8.23) is just equal to the normalizing constant of the Wishart pdf which does not depend on the parameters B. The joint pdf in (8.24) has been called the "generalized multivariate Student t" pdf. Some properties of this pdf follow.

First, we have already shown that the marginal posterior pdf for a vector of regression coefficients, say $\boldsymbol{\beta}_1$, is in the form of a multivariate Student t pdf [see (8.22)]. We now show that if we express the joint pdf of $B = (\boldsymbol{\beta}_1, \boldsymbol{\beta}_2, \ldots, \boldsymbol{\beta}_m)$ as[14]

$$
p(B|Y) = p(\boldsymbol{\beta}_1|Y)\, p(\boldsymbol{\beta}_2|\boldsymbol{\beta}_1, Y) \cdots p(\boldsymbol{\beta}_m|\boldsymbol{\beta}_1, \boldsymbol{\beta}_2, \ldots, \boldsymbol{\beta}_{m-1}, Y) \tag{8.25}
$$

then each of factors on the rhs of (8.25) can be expressed in the form of a multivariate Student t pdf.[15] First we derive an expression for $p(\boldsymbol{\beta}_1, \boldsymbol{\beta}_2, \ldots,$

[13] See Appendix B.

[14] Below it is understood that we are taking X given.

[15] This is proved in Tiao and Zellner, *loc. cit.*

$\boldsymbol{\beta}_{m-1}|Y)\,p(\boldsymbol{\beta}_m|\boldsymbol{\beta}_1, \boldsymbol{\beta}_2, \ldots, \boldsymbol{\beta}_{m-1}, Y)$. Note that the determinant in (8.24) can be expressed as

$$(8.26)\quad |S + Q| = \left|\begin{array}{c:c} \bar{S} + \bar{Q} & \mathbf{s} + \mathbf{q} \\ \hdashline (\mathbf{s} + \mathbf{q})' & s_{mm} + q_{mm} \end{array}\right|$$
$$= |\bar{S} + \bar{Q}|[s_{mm} + q_{mm} - (\mathbf{s} + \mathbf{q})'(\bar{S} + \bar{Q})^{-1}(\mathbf{s} + \mathbf{q})],$$

where $Q = (B - \hat{B})'X'X(B - \hat{B})$, $\bar{S} + \bar{Q}$ is the $(m - 1) \times (m - 1)$ upper lefthand principal minor matrix of $S + Q$, $\mathbf{s}' = (s_{m1}, s_{m2}, \ldots, s_{m(m-1)})$, and $\mathbf{q}' = (q_{m1}, q_{m2}, \ldots, q_{m(m-1)})$. In the second factor in the second line of (8.26) let us write

$$(8.27)\quad q_{\alpha l} = \boldsymbol{\gamma}_\alpha'\boldsymbol{\gamma}_l \quad \text{with} \quad \boldsymbol{\gamma}_l = X(\boldsymbol{\beta}_l - \hat{\boldsymbol{\beta}}_l), \qquad \alpha, l = 1, 2, \ldots, m,$$

and

$$(8.28)\quad \mathbf{q}' = \boldsymbol{\gamma}_m'\gamma, \quad \text{where} \quad \gamma = (\boldsymbol{\gamma}_1, \boldsymbol{\gamma}_2, \ldots, \boldsymbol{\gamma}_{m-1}).$$

Using these definitions and after some algebraic rearrangement, we have that

$$(8.29)\quad |S + Q| = |\bar{S} + \bar{Q}|[c_m + (\boldsymbol{\gamma}_m - \boldsymbol{\mu})'D_m(\boldsymbol{\gamma}_m - \boldsymbol{\mu})],$$

where

$$D_m = I - \gamma(\bar{S} + \bar{Q})^{-1}\gamma',$$
$$\boldsymbol{\mu} = D_m^{-1}\gamma(\bar{S} + \bar{Q})^{-1}\mathbf{s},$$

and

$$c_m = s_{mm} - \mathbf{s}'(\bar{S} + \bar{Q})^{-1}\mathbf{s} - \boldsymbol{\mu}'D_m\boldsymbol{\mu}.$$

We now make use of a theorem due to Tocher[16] which says that if A is an $m \times n$ matrix and B, an $n \times m$ matrix, then

$$(8.30)\quad (I_m - AB)^{-1} = I_m + A(I_n - BA)^{-1}B.$$

Applying (8.30) and noting that $\gamma'\gamma = \bar{Q}$, we obtain

$$(8.31a)\quad D_m^{-1} = I + \gamma\bar{S}^{-1}\gamma',$$

$$(8.31b)\quad \boldsymbol{\mu} = \gamma\bar{S}^{-1}\mathbf{s},$$

and

$$(8.31c)\quad c_m = s_{mm} - \mathbf{s}'\bar{S}^{-1}\mathbf{s}.$$

Thus c_m is the (m, m)th element of S^{-1}. Then the second factor on the rhs of (8.29) is in terms of $\boldsymbol{\beta}_m$:

$$(8.32)\quad [c_m + (\boldsymbol{\gamma}_m - \boldsymbol{\mu})'D_m(\boldsymbol{\gamma}_m - \boldsymbol{\mu})] = [c_m + (\boldsymbol{\beta}_m - \boldsymbol{\eta}_m)'X'D_mX(\boldsymbol{\beta}_m - \boldsymbol{\eta}_m)]$$

with

$$\boldsymbol{\eta}_m = \hat{\boldsymbol{\beta}}_m + d\bar{S}^{-1}\mathbf{s} \quad \text{and} \quad \mathbf{d} = (\boldsymbol{\beta}_1 - \hat{\boldsymbol{\beta}}_1, \ldots, \boldsymbol{\beta}_{m-1} - \hat{\boldsymbol{\beta}}_{m-1}).$$

[16] K. D. Tocher, "Discussion on Mr. Box and Dr. Wilson's Paper," *J. Roy. Statist. Soc.*, Series B, **13**, 39–42 (1951).

Using (8.29) and (8.32), we can write the joint pdf for B, shown in (8.24), as

$$(8.33)\quad p(B|Y) \propto |\bar{S} + \bar{Q}|^{-n/2}\{c_m + (\boldsymbol{\beta}_m - \boldsymbol{\eta}_m)'X'D_mX(\boldsymbol{\beta}_m - \boldsymbol{\eta}_m)\}^{-n/2}.$$

The determinant of the matrix $X'D_mX$ is

$$(8.34)\quad \begin{aligned} |X'D_mX| &= |X'X|\,|I - (X'X)^{-1}X'\gamma(\bar{S} + \bar{Q})^{-1}\gamma'X| \\ &= |X'X|\,|I - \mathbf{d}(\bar{S} + \bar{Q})^{-1}\gamma'X|. \end{aligned}$$

We now make use of a theorem[17] which states that if A and B are two $n \times n$ matrices then $|I - AB| = |I - BA|$. This can be generalized to the case in which A is an $m \times n$ matrix and B is an $n \times m$ matrix. Suppose that $m < n$; then

$$\begin{aligned} |I_m - AB| &= \left|I_n - \begin{bmatrix} A \\ \cdots \\ 0 \end{bmatrix}(B \vdots 0)\right| \\ &= \left|I_n - (B \vdots 0)\begin{bmatrix} A \\ \cdots \\ 0 \end{bmatrix}\right| = |I_n - BA|. \end{aligned}$$

Using this result, we can write the second determinant on the right of (8.34) as

$$|I - \mathbf{d}(\bar{S} + \bar{Q})^{-1}\gamma'X| = |I - (\bar{S} + \bar{Q})^{-1}\gamma'X\mathbf{d}|.$$

Hence

$$(8.35)\quad \begin{aligned} |X'D_mX| &= |X'X|\,|I - (\bar{S} + \bar{Q})^{-1}\gamma'\gamma| \\ &= |X'X|\,|\bar{S}|\,|\bar{S} + \bar{Q}|^{-1}. \end{aligned}$$

Consequently the pdf in (8.33) can be written as

$$(8.36)\quad p(\mathbf{B}|Y) = p(\boldsymbol{\beta}_1, \ldots, \boldsymbol{\beta}_{m-1}|Y)\,p(\boldsymbol{\beta}_n|\boldsymbol{\beta}_1, \ldots, \boldsymbol{\beta}_{m-1}, Y)$$

with

$$(8.37)\quad p(\boldsymbol{\beta}_1, \ldots, \boldsymbol{\beta}_{m-1}|Y) \propto |\bar{S} + \bar{Q}|^{-(n-1)/2}$$

and

$$(8.38)\quad p(\boldsymbol{\beta}_m|\boldsymbol{\beta}_1, \ldots, \boldsymbol{\beta}_{m-1}, Y) \propto |X'D_mX|^{1/2}[c_m + (\boldsymbol{\beta}_m - \boldsymbol{\eta}_m)'X'D_mX(\boldsymbol{\beta}_m - \boldsymbol{\eta}_m)]^{-n/2}.$$

In (8.38) it is seen that the conditional pdf for $\boldsymbol{\beta}_m$ can be expressed in terms of a multivariate Student t pdf, whereas the marginal pdf for $(\boldsymbol{\beta}_1, \ldots, \boldsymbol{\beta}_{m-1})$, shown in (8.37), is of the same form as the original pdf for B [see (8.24)], except, of course, for the changes in the dimensions of the matrix $\bar{S} + \bar{Q}$ and

[17] R. Bellman, *Introduction to Matrix Analysis*. New York: McGraw-Hill, 1962, p. 95.

the value of the exponent of the determinant. Then on repeating the same process $m - 1$ times we can express the joint pdf for B as

$$p(B|Y) \propto [s_{11} + (\boldsymbol{\beta}_1 - \hat{\boldsymbol{\beta}}_1)'X'X(\boldsymbol{\beta}_1 - \hat{\boldsymbol{\beta}}_1)]^{-[n-(m-1)]/2}$$

$$(8.39) \qquad \times \prod_{\alpha=2}^{m} |X'D_\alpha X|^{1/2}[c_\alpha + (\boldsymbol{\beta}_\alpha - \boldsymbol{\eta}_\alpha)'X'D_\alpha X(\boldsymbol{\beta}_\alpha - \boldsymbol{\eta}_\alpha)]^{-(n-m+\alpha)/2},$$

where D_α, η_α, and c_α are defined in exactly the same way as $\alpha = m$, given above.

The factors in (8.39) correspond precisely to those shown in (8.25) and clearly the first factor is the marginal pdf for $\boldsymbol{\beta}_1$ which, of course, is in agreement with (8.22). Thus the generalized multivariate Student t pdf is an interesting example in that, even though conditional pdf's and the marginal pdf's for certain subsets of its variables are in the multivariate Student t form, the joint pdf fails to be of the same form.

As a second property of the generalized multivariate Student t pdf, we shall derive the marginal posterior pdf of B_1 given in

$$(8.40) \qquad \begin{aligned} Y &= (X_1 \vdots X_2)\begin{pmatrix} B_1 \\ \hdashline B_2 \end{pmatrix} + U \\ &= X_1B_1 + X_2B_2 + U; \end{aligned}$$

that is, B_1 is the submatrix of B containing the coefficients of the variables in X_1 for all equations. Following Geisser, *loc. cit.*, we have for the quantity in the determinant in (8.24)

$$(8.41) \qquad \begin{aligned} S + (B - \hat{B})'X'X(B - \hat{B}) &= S + \begin{pmatrix} B_1 - \hat{B}_1 \\ \hdashline B_2 - \hat{B}_2 \end{pmatrix}' \begin{pmatrix} X_1'X_1 & \vdots & X_1'X_2 \\ \hdashline X_2'X_1 & \vdots & X_2'X_2 \end{pmatrix} \begin{pmatrix} B_1 - \hat{B}_1 \\ \hdashline B_2 - \hat{B}_2 \end{pmatrix} \\ &= S + (B_1 - \hat{B}_1)'F(B_1 - \hat{B}_1) \\ &\quad + (B_2 - \Delta)'X_2'X_2(B_2 - \Delta), \end{aligned}$$

where we have completed the square on B_2,

$$F = X_1'X_1 - X_1'X_2(X_2'X_2)^{-1}X_2'X_1,$$

and

$$\Delta = \hat{B}_2 - (X_2'X_2)^{-1}X_2'X_1(B_1 - \hat{B}_1),$$

which do not involve B_2. Thus the joint pdf for B_1 and B_2 is

$$(8.42) \qquad p(B_1, B_2|Y) \propto |S + (B_1 - \hat{B}_1)'F(B_1 - \hat{B}_1) + (B_2 - \Delta)'X_2'X_2(B_2 - \Delta)|^{-n/2},$$

which, when viewed as a function of B_2, given B_1, is in the same form as

(8.24). Using properties of the generalized multivariate Student t pdf,[18] we can perform the integration with respect to the parameters in B_2 readily to yield

$$p(B_1|Y) \propto |S + (B_1 - \hat{B}_1)'F(B_1 - \hat{B}_1)|^{-(n-k_2)/2}, \tag{8.43}$$

where F has been defined above and k_2 denotes the number of rows in B_2. It is seen that the marginal posterior pdf for B_1 is in the generalized multivariate Student t form. Thus the results of Tiao and Zellner can be applied in its analysis to yield marginal pdf's for any column of B_1 and also conditional pdf's.

As a last property of the joint pdf for B, given in (8.24), Geisser points out that the quantity

$$\tilde{U} = \frac{|S|}{|S + (B - \hat{B})'X'X(B - \hat{B})|} \tag{8.44}$$

is distributed like $\tilde{U}_{m,k,n-k}$, that is, as a product of beta variables defined by Anderson.[19] Thus, as Geisser points out, the posterior pdf for $\tilde{U}$, where B is random and the other quantities are fixed, is the same as the sampling distribution of $\tilde{U}$, where B is fixed and S and $\hat{B}$ are the sets of random variables. Then a posterior region for the elements of B is given by

$$P_r[\tilde{U}(\beta) \leq \tilde{U}^{(\alpha)}_{m,k,n-k}] = 1 - \alpha$$

where $\tilde{U}^{(\alpha)}_{m,k,n-k}$ is the αth percentage point.

8.2 PREDICTIVE PDF FOR THE TRADITIONAL MULTIVARIATE REGRESSION MODEL[20]

Assume, as in Section 8.1, that we have sample observations generated by a multivariate normal regression model

$$Y = XB + U \tag{8.45}$$

[18] See A. Ando and G. M. Kaufman, "Bayesian Analysis of the Independent Multinormal Process—Neither Mean nor Precision Known," *J. Am. Statist. Assoc.*, **60**, 347–358, p. 352 (1965); J. M. Dickey, "Matricvariate Generalizations of the Multivariate t Distribution and the Inverted Multivariate t Distribution," *Ann. Math. Statist.*, **38**, 511–518 (1967) and Appendix B to this book.

[19] T. W. Anderson, *An Introduction to Multivariate Statistical Analysis*, New York: Wiley, 1958, p. 194 ff., discusses $\tilde{U}$ which appears as the $n/2$ power of the likelihood ratio for testing a hypothesis on a subset of the elements of B.

[20] Work on various aspects of this problem appears in S. Geisser, *loc. cit.*, A. Ando and G. M. Kaufman, "Bayesian Analysis of Reduced Form Systems," manuscript, MIT, 1964, and A. Zellner and V. K. Chetty, "Prediction and Decision Problems in Regression Models from the Bayesian Point of View," *J. Am. Statist. Assoc.*, **60**, 608–16 (1965).

and wish to derive the predictive pdf for future observations on the dependent variables, say W, a $p \times m$ *matrix*, assumed to be generated by the same model generating Y; that is,

$$W = ZB + V, \tag{8.46}$$

where Z is a $p \times k$ matrix of given values for the independent variables in the next p time periods and V is a $p \times m$ matrix of future normal error terms with independently distributed row vectors having zero means and each having covariance matrix Σ, the same as that assumed for the rows of U. Then the predictive pdf for W is given by

$$p(W|Y, X, Z) = \int \cdots \int p(B, \Sigma^{-1}| Y, X)\, p(W|Z, B, \Sigma^{-1})\, dB\, d\Sigma^{-1}, \tag{8.47}$$

where

$$p(W|Z, B, \Sigma^{-1}) \propto |\Sigma^{-1}|^{p/2} \exp\{-\tfrac{1}{2} \operatorname{tr}[(W-ZB)'(W - ZB)\Sigma^{-1}]\}, \tag{8.48}$$

the joint pdf for W, given Z, B, and Σ^{-1}, and where $p(B, \Sigma^{-1}| Y, X)$ is the joint posterior pdf for B and Σ^{-1}, which we shall take as shown in (8.13). On substituting this expression and (8.48) in (8.47), we have for the integrand

$$|\Sigma^{-1}|^{(n+p-m-1)/2} \exp(-\tfrac{1}{2} \operatorname{tr} A\Sigma^{-1}), \tag{8.49}$$

with $A = (Y - XB)'(Y - XB) + (W - ZB)'(W - ZB)$. From the form of (8.49) properties of the Wishart pdf can be used to integrate with respect to the distinct elements of Σ^{-1}, which yields

$$|A|^{-(n+p)/2} = |(Y - XB)'(Y - XB) + (W - ZB)'(W - ZB)|^{-(n+p)/2}. \tag{8.50}$$

To integrate with respect to the elements of B we complete the square on B, which gives

$$|A|^{-(n+p)/2} = |Y'Y + W'W - \tilde{B}'M\tilde{B} + (B - \tilde{B})'M(B - \tilde{B})|^{-(n+p)/2}, \tag{8.51}$$

where $M = X'X + Z'Z$ and $\tilde{B} = M^{-1}(X'Y + Z'W)$. The pdf in (8.51) is in a form that we have already encountered [see references in connection with (8.42)]. Thus the integration with respect to B can be performed easily and results in the predictive pdf for W,

$$p(W| Y, X, Z) \propto |Y'Y + W'W - \tilde{B}'M\tilde{B}|^{-(n+p-k)/2}. \tag{8.52}$$

To simplify this expression we complete the square on W as follows:

$$\begin{aligned} Y'Y + W'W - \tilde{B}'M\tilde{B} &= Y'Y + W'W - (Y'X + W'Z)M^{-1}(X'Y + Z'W) \\ &= Y'(I - XM^{-1}X')Y + W'(I - ZM^{-1}Z')W \\ &\quad - Y'XM^{-1}Z'W - W'ZM^{-1}X'Y \\ &= Y'(I - XM^{-1}X' - XM^{-1}Z'C^{-1}ZM^{-1}X')Y \\ &\quad + (W - C^{-1}ZM^{-1}X'Y)'C(W - C^{-1}ZM^{-1}X'Y), \end{aligned} \tag{8.53}$$

where $C = I - ZM^{-1}Z'$. Further, we have

$$C^{-1} = (I - ZM^{-1}Z')^{-1} = I + Z(X'X)^{-1}Z',$$

which can be verified by direct multiplication.[21] Also

$$\begin{aligned} C^{-1}ZM^{-1} &= [I + Z(X'X)^{-1}Z']ZM^{-1} \\ &= Z[I + (X'X)^{-1}Z'Z]M^{-1} \\ &= Z(X'X)^{-1}(X'X + Z'Z)M^{-1} \\ &= Z(X'X)^{-1}. \end{aligned} \tag{8.54}$$

Finally,

$$\begin{aligned} XM^{-1}X' + XM^{-1}Z'C^{-1}ZM^{-1}X' &= X[M^{-1} + M^{-1}Z'Z(X'X)^{-1}]X' \\ &= XM^{-1}(X'X + Z'Z)(X'X)^{-1}X' \\ &= X(X'X)^{-1}X', \end{aligned} \tag{8.55}$$

where (8.54) has been used in the first line of (8.55). Substituting from (8.54) and (8.55) in the third line of (8.53), we obtain

$$\begin{aligned} Y'Y + W'W - \tilde{B}'M\tilde{B} = {} & Y'[I - X(X'X)^{-1}X']Y \\ & + (W' - Z\hat{B})'C(W - Z\hat{B}), \end{aligned} \tag{8.56}$$

with $\hat{B} = (X'X)^{-1}X'Y$. Noting that

$$Y'[I - X(X'X)^{-1}X']Y = (Y - X\hat{B})'(Y - X\hat{B}) = S,$$

we can write the predictive pdf in (8.52) as

$$p(W|Y, X, Z) \propto |S + (W - Z\hat{B})'(I - ZM^{-1}Z')(W - Z\hat{B})|^{-(n+p-k)/2}. \tag{8.57}$$

It is thus seen that W, the matrix of future observations, has a pdf in the generalized multivariate Student t form, the same form found for the posterior pdf for B, shown in (8.24). Thus the distributional results established for (8.24) apply as well to (8.57). In particular, the marginal predictive pdf for any column or row vector of W will be in the multivariate Student t form. Further, if we partition W as follows, $W' = (W_1'W_2')$, the marginal predictive pdf for W_1 will be in the generalized multivariate Student t form. Finally, as pointed out by Geisser, just as in (8.44) we have that

$$\tilde{U} = \frac{1}{|S + (W - Z\hat{B})'(I - ZM^{-1}Z')(W - Z\hat{B})|}$$

is distributed as[22] $\tilde{U}_{m,p,n-k}$. In the special case of a single regression, $m = 1$,

[21] $(I - ZM^{-1}Z')[I + Z(X'X)^{-1}Z'] = I - Z[M^{-1} - (X'X)^{-1} + M^{-1}Z'Z(X'X)^{-1}]Z' = I - ZM^{-1}[X'X - M + Z'Z](X'X)^{-1}Z' = I$, since $X'X - M + Z'Z = 0$, given the definition of $M = X'X + Z'Z$.

[22] See T. W. Anderson, *op. cit.*, pp. 194 ff., for a discussion of this distribution. He uses the letter U where we have used $\tilde{U}$.

S is a scalar, and

$$S^{-1}(W - Z\hat{B})'(I - ZM^{-1}Z')(W - Z\hat{B})$$

is a quadratic form distributed as $[p/(n-k)]F_{p,n-k}$. When $p = 1$, that is, W is a $1 \times m$ row vector, the pdf in (8.57) reduces to a multivariate Student t form[23] and the quantity

$$[1 - Z_1(X'X + Z_1'Z_1)^{-1}Z_1'](W_1 - Z_1\hat{B})S^{-1}(W_1 - Z_1\hat{B})'$$

is distributed as $[m/(n-k-m+1)]F_{m,n-k-m+1}$, where Z_1 is the first row of Z and W_1 is the first row of W.

8.3 THE TRADITIONAL MULTIVARIATE MODEL WITH EXACT RESTRICTIONS

In some circumstances we may know that certain coefficients in the B matrix are zero or that elements of B are constrained by exact linear restrictions. In these situations the joint posterior pdf for B, shown in (8.24), can be conditionalized to incorporate this information and the conditionalized posterior pdf can be employed to make inferences about the remaining nonzero coefficients. Several cases are discussed below.

If exact zero restrictions pertain just to the elements of a particular coefficient vector, say $\boldsymbol{\beta}_1$, the posterior distribution for $\boldsymbol{\beta}_1$ in (8.22), which is in the multivariate Student t form, can be analyzed easily to obtain a conditional pdf that incorporates the conditioning information; for example, if we partition $\boldsymbol{\beta}_1$ as $\boldsymbol{\beta}_1' = (\boldsymbol{\beta}_a', \boldsymbol{\beta}_0')$ and it is known that $\boldsymbol{\beta}_0 = \mathbf{0}$, the posterior pdf $p(\boldsymbol{\beta}_a|\mathbf{y}, \boldsymbol{\beta}_0 = \mathbf{0})$ can readily be obtained. Other exact linear restrictions on the elements of $\boldsymbol{\beta}_1$ can also be imposed by using properties of the multivariate Student t pdf.

On the other hand, if restrictions relate to several or all coefficient vectors, the situation is somewhat more complicated. The special case $B_1 = C_1$, where $Y = X_1B_1 + X_2B_2 + U$ and C_1 is known, is easily handled by using the result in (8.42); that is, setting $B_1 = C_1$ yields the conditional posterior pdf for B_2, given $B_1 = C_1$, which is in the generalized Student t form. When zero restrictions pertain to coefficients of *different* subsets of variables in equations of the system, that is,

$$\mathbf{y}_\alpha = X_{\alpha_1}\boldsymbol{\beta}_{\alpha_1} + X_{\alpha_0}\boldsymbol{\beta}_{\alpha_0} + \mathbf{u}_\alpha, \qquad \alpha = 1, 2, \ldots, m, \tag{8.58}$$

[23] This result follows from (8.57) on observing that with W_1 and Z_1, the first *rows* of W and Z, respectively, we can write the determinant in (8.57) as $|1 + a\mathbf{bb}'| = 1 + a\mathbf{b}'\mathbf{b}$, with $a = 1 - Z_1(X'X + Z_1'Z_1)^{-1}Z_1'$ and $\mathbf{b} = A(W_1 - Z_1\hat{B})'$, an $m \times 1$ column vector, where A is a nonsingular matrix such that $ASA' = I_m$ or $S^{-1} = A'A$. Then $p(W_1|X_1 Y, Z_1) \propto |I + a\mathbf{bb}'|^{-(n+1-k)/2} = (1 + a\mathbf{b}'\mathbf{b})^{-(n+1-k)/2} = [1 + a(W_1 - Z_1\hat{B})S^{-1} \times (W_1 - Z_1\hat{B})']^{-(n+1-k)/2}$, which is in the multivariate Student t form.

with $X = (X_{\alpha_1} \vdots X_{\alpha_0})$ and $\boldsymbol{\beta}_\alpha' = (\boldsymbol{\beta}_{\alpha_1}', \boldsymbol{\beta}_{\alpha_0}')$, with $\boldsymbol{\beta}_{\alpha_0} = \mathbf{0}$, the problem is more difficult, since the partitioning of X is not the same for all equations. To analyze this case the joint posterior pdf for B in (8.24) will be expanded and the leading normal term in the expansion will be conditionalized by using the restrictions $\boldsymbol{\beta}_{\alpha_0} = \mathbf{0}$, $\alpha = 1, 2, \ldots, m$.

To expand (8.24) let us write it as

$$p(B|Y) \propto |\bar{S} + (B - \hat{B})'M(B - \hat{B})|^{-n/2}, \tag{8.59}$$

where $\bar{S} = n^{-1}S$ and $M = n^{-1}X'X$. Now let H be a nonsingular matrix such that $H\bar{S}H' = I$ and $H(B - \hat{B})'M(B - \hat{B})H' = D$, where $D = D(\lambda_i)$ is a diagonal matrix with the characteristic roots, the λ_i's, of $(B - \hat{B})'M(B - \hat{B})\bar{S}^{-1}$ on the diagonal. These will be small if n is large and $\lim_{n\to\infty} M = \bar{M}$, a constant. Also, from $H\bar{S}H' = I$, we have $H'H = \bar{S}^{-1}$. Then

$$\begin{aligned}
&|\bar{S} + (B - \hat{B})'M(B - \hat{B})|^{-n/2} \\
&\quad = |HH'|^{n/2}|I + D|^{-n/2} \\
&\quad = |\bar{S}|^{-n/2} \exp\left(-\frac{n}{2}\log|I + D|\right) \\
&\quad = |\bar{S}|^{-n/2} \exp\left[-\frac{n}{2}\sum \log(1 + \lambda_i)\right] \qquad (8.60)\\
&\quad = |\bar{S}|^{-n/2} \exp\left\{-\frac{n}{2}\left[\sum \lambda_i - \tfrac{1}{2}\sum \lambda_i^2 + \tfrac{1}{3}\sum \lambda_i^3 - \cdots\right]\right\} \\
&\quad = |\bar{S}|^{-n/2} \exp\left\{-\frac{n}{2}\left[\text{tr}\, D - \tfrac{1}{2}\text{tr}\, D^2 + \tfrac{1}{3}\text{tr}\, D^3 - \cdots\right]\right\}.
\end{aligned}$$

Note that $\text{tr}\, D = \text{tr}\, H(B - \hat{B})'M(B - \hat{B})H' = \text{tr}\, H'H(B - \hat{B})'M(B - \hat{B}) = \text{tr}\, \bar{S}^{-1}\bar{Q}$, where $\bar{Q} = (B - \hat{B})'M(B - \hat{B})$. Similar operations yield

$$\begin{aligned}
\text{tr}\, D^2 &= \text{tr}\, \bar{S}^{-1}\bar{Q}\bar{S}^{-1}\bar{Q} \\
\text{tr}\, D^3 &= \text{tr}\, \bar{S}^{-1}\bar{Q}\bar{S}^{-1}Q\bar{S}^{-1}\bar{Q} \\
&\cdots\cdots\cdots\cdots
\end{aligned}$$

By use of these results (8.60) becomes

$$\begin{aligned}
p(B|Y) &\propto \exp\left\{-\frac{n}{2}\left[\frac{1}{n}\text{tr}\, \bar{S}^{-1}Q - \frac{1}{2n^2}\text{tr}\, \bar{S}^{-1}Q\bar{S}^{-1}Q\right.\right. \\
&\qquad\qquad \left.\left. + \frac{1}{3n^3}\text{tr}\, \bar{S}^{-1}Q\bar{S}^{-1}Q\bar{S}^{-1}Q - \cdots\right]\right\} \qquad (8.61)\\
&\propto \exp\left(-\tfrac{1}{2}\text{tr}\, \bar{S}^{-1}Q\right) \exp\left(\frac{1}{4n}\text{tr}\, \bar{S}^{-1}Q\bar{S}^{-1}Q\right. \\
&\qquad\qquad \left. - \frac{1}{6n^2}\text{tr}\, \bar{S}^{-1}Q\bar{S}^{-1}Q\bar{S}^{-1}Q - \cdots\right),
\end{aligned}$$

where $Q = (B - \hat{B})'X'X(B - \hat{B})$. On expanding the second factor of the second line of (8.61) as $e^x = 1 + x + x^2/2! + \cdots$, we obtain

$$(8.62)\quad p(B|Y) \mathrel{\dot{\propto}} \exp(-\tfrac{1}{2}\operatorname{tr}\bar{S}^{-1}Q)\left[1 + \frac{1}{4n}\operatorname{tr}\bar{S}^{-1}Q\bar{S}^{-1}Q + \cdots\right],$$

with $\dot{\propto}$ denoting "approximately proportional to." The leading factor of (8.62) is in the multivariate normal form; that is[24]

$$(8.63)\quad \begin{aligned} p(B|Y) &\mathrel{\dot{\propto}} \exp[-\tfrac{1}{2}\operatorname{tr}\bar{S}^{-1}(B - \hat{B})'X'X(B - \hat{B})] \\ &\mathrel{\dot{\propto}} \exp[-\tfrac{1}{2}(\boldsymbol{\beta} - \hat{\boldsymbol{\beta}})'\bar{S}^{-1} \otimes X'X(\boldsymbol{\beta} - \hat{\boldsymbol{\beta}})]. \end{aligned}$$

Then

$$(8.64)\quad p(\boldsymbol{\beta}_{\cdot 1}|Y, \boldsymbol{\beta}_{\cdot 0} = \mathbf{0}) \mathrel{\dot{\propto}} \exp\{-\tfrac{1}{2}\mathbf{d}'[\bar{s}^{\alpha l}(X_{\alpha_1} \vdots X_{\alpha_0})'(X_{l_1} \vdots X_{l_0})]\mathbf{d}\},$$

where $\boldsymbol{\beta}_{\cdot 1}$ denotes the coefficients assumed to be not equal to zero, $\boldsymbol{\beta}_{\cdot 0}$, those equal to zero, $\bar{s}^{\alpha l}$ is the α, lth element of $\bar{S}^{-1}$, $\bar{s}^{\alpha l}(X_{\alpha_1} \vdots X_{\alpha_2})'(X_{l_1} \vdots X_{l_0})$ is a typical element of a partitioned matrix, and

$$(8.65)\quad \mathbf{d}' = [(\boldsymbol{\beta}_{1_1} - \hat{\boldsymbol{\beta}}_{1_1})', -\hat{\boldsymbol{\beta}}_{1_0}{}', \ldots, (\boldsymbol{\beta}_{m_1} - \hat{\boldsymbol{\beta}}_{m_1})', -\hat{\boldsymbol{\beta}}_{m_0}{}']$$

is $(\boldsymbol{\beta} - \hat{\boldsymbol{\beta}})'$ conditionalized to reflect the zero restrictions. With this notation introduced, (8.64) can be expressed as

$$(8.66)\quad \begin{aligned} p(\boldsymbol{\beta}_{\cdot 1}|Y, \boldsymbol{\beta}_{\cdot 0} = \mathbf{0}) \mathrel{\dot{\propto}} \exp[-\tfrac{1}{2}(&\boldsymbol{\beta}_{\cdot 1} - \hat{\boldsymbol{\beta}}_{\cdot 1})'(X_{\alpha_1}{}'X_{l_1}\bar{s}^{\alpha l})(\boldsymbol{\beta}_{\cdot 1} - \hat{\boldsymbol{\beta}}_{\cdot 1}) \\ &- 2\hat{\boldsymbol{\beta}}_{\cdot 0}{}'(X_{\alpha_0}{}'X_{l_1}\bar{s}^{\alpha l})(\boldsymbol{\beta}_{\cdot 1} - \hat{\boldsymbol{\beta}}_{\cdot 1}) \\ &+ \hat{\boldsymbol{\beta}}_{\cdot 0}{}'(X_{\alpha_0}{}'X_{l_0}\bar{s}^{\alpha l})\hat{\boldsymbol{\beta}}_{\cdot 0}]. \end{aligned}$$

Letting $V = (X_{\alpha_1}{}'X_{l_1}\bar{s}^{\alpha l})$ and $R = (X_{\alpha_0}{}'X_{l_1}\bar{s}^{\alpha l})$ and completing the square in the exponent of (8.66), we have

$$(8.67)\quad p(\boldsymbol{\beta}_{\cdot 1}|Y, \boldsymbol{\beta}_{\cdot 0}) \mathrel{\dot{\propto}} \exp[-\tfrac{1}{2}(\boldsymbol{\beta}_{\cdot 1} - \hat{\boldsymbol{\beta}}_{\cdot 1} - V^{-1}R'\hat{\boldsymbol{\beta}}_{\cdot 0})'V(\boldsymbol{\beta}_{\cdot 1} - \hat{\boldsymbol{\beta}}_{\cdot 1} - V^{-1}R'\hat{\boldsymbol{\beta}}_{\cdot 0})].$$

Thus this normal approximation to the conditional posterior pdf has mean $\hat{\boldsymbol{\beta}}_{\cdot 1} + V^{-1}R'\hat{\boldsymbol{\beta}}_{\cdot 0}$ and covariance matrix $V^{-1} = (X_{\alpha_1}{}'X_{l_1}\bar{s}^{\alpha l})^{-1}$. It is seen that the mean is the least squares quantity $\hat{\boldsymbol{\beta}}_{\cdot 1}$, plus another term that includes the vector $\hat{\boldsymbol{\beta}}_{\cdot 0}$, the sample estimate of the zero restrictions.

8.4 TRADITIONAL MODEL WITH AN INFORMATIVE PRIOR PDF

We have gone forward above employing diffuse prior pdf's and considering cases in which we have exact restrictions on some of the elements of the B

[24] It is interesting to note that the leading normal term in this expansion is just the conditional posterior pdf for B given $\Sigma = \bar{S}$.

matrix. In this section we consider the problem of introducing prior information about the elements of B by use of an informative prior pdf. Given that such prior information is reasonably accurate, we, of course, shall improve the precision of our inferences by using it. In addition, we shall learn how the sample information modifies our beliefs by comparing the properties of the prior and posterior pdf's. Our problem is to formulate a prior pdf that will be useful in representing prior information in a broad range of circumstances and that will be relatively convenient mathematically. Clearly, it should be recognized that one class of prior pdf's will not be appropriate for all situations; however, it is believed that the class studied below will be useful in many situations.

As Rothenberg[25] has noted, if we use a "simple" natural conjugate prior distribution[26] for the traditional multivariate regression model, this will involve placing restrictions on the parameters, namely, the variances and covariances of coefficients appearing in equations of the system. This is due to the fact that the matrix $(X'X)^{-1}$ enters the covariance structure in the following way, $\Sigma \otimes (X'X)^{-1}$. Thus, for example, the ratios of variances of corresponding coefficients in the first and second equations will all be equal if we use a simple natural conjugate prior pdf. The way to avoid this problem is to use a general multivariate normal prior pdf for all coefficients of the model.[27] By so doing we will avoid the problem raised by Rothenberg but we will have to pay a price; that is, such a general normal prior pdf does not combine so neatly with the likelihood function as the simple natural conjugate pdf. However, the price is well worth paying, since the restrictions involved in using the natural conjugate prior pdf for the traditional model are not reasonable in most situations.

In view of what has been said above, we introduce the following prior pdf:

$$p(\boldsymbol{\beta}, \Sigma^{-1}) \propto |\Sigma^{-1}|^{-(m+1)/2} \exp\left[-\tfrac{1}{2}(\boldsymbol{\beta} - \bar{\boldsymbol{\beta}})'C^{-1}(\boldsymbol{\beta} - \bar{\boldsymbol{\beta}})\right], \tag{8.68}$$

where $\bar{\boldsymbol{\beta}}$, an $mk \times 1$ vector, is the mean of the prior pdf, assigned by the investigator, and $C = (c_{\alpha l})$ is an $mk \times mk$ matrix, the prior covariance matrix, also assigned by the investigator. As above, we assume that little is known a priori about the elements of Σ and use the diffuse prior pdf introduced and discussed in Section 8.1.

On combining the prior pdf in (8.68) with the likelihood function in (8.6),

[25] T. J. Rothenberg, "A Bayesian Analysis of Simultaneous Equation Systems," Report 6315, Econometric Institute, Netherlands School of Economics, Rotterdam, 1963.

[26] That is, if we took our prior pdf in the same form as the likelihood function, we would have a natural conjugate prior pdf.

[27] From what is presented below it is the case that the prior pdf that we shall use is the natural conjugate prior pdf for the "seemingly unrelated" regression model (see Section 8.5).

we have for the joint posterior pdf

$$p(B, \Sigma^{-1} | Y) \propto |\Sigma^{-1}|^{(n-m-1)/2} \exp\{-\tfrac{1}{2} \operatorname{tr} \Sigma^{-1}[S + (B - \hat{B})'X'X(B - \hat{B})]\}$$
$$\times \exp[-\tfrac{1}{2}(\boldsymbol{\beta} - \bar{\boldsymbol{\beta}})C^{-1}(\boldsymbol{\beta} - \bar{\boldsymbol{\beta}})], \tag{8.69}$$

where $B = (\boldsymbol{\beta}_1, \ldots, \boldsymbol{\beta}_m)$, $\boldsymbol{\beta}' = (\boldsymbol{\beta}_1', \ldots, \boldsymbol{\beta}_m')$, and $\hat{B} = (X'X)^{-1}X'Y$. We can now integrate (8.69) with respect to Σ^{-1} to obtain

$$p(B | Y) \propto |S + (B - \hat{B})'X'X(B - \hat{B})|^{-n/2} \exp[-\tfrac{1}{2}(\boldsymbol{\beta} - \bar{\boldsymbol{\beta}})'C^{-1}(\boldsymbol{\beta} - \bar{\boldsymbol{\beta}})]. \tag{8.70}$$

The posterior pdf for B is seen to be the product of a factor in the generalized multivariate Student t form and a factor in the multivariate normal form. Since (8.70) is rather complicated as it stands, we shall expand the first factor on the right-hand side of (8.70) just as we did in Section 8.3. This yields the following as the leading normal term approximating the posterior pdf:

$$p(B | Y) \mathrel{\dot{\propto}} \exp[-\tfrac{1}{2}(\boldsymbol{\beta} - \hat{\boldsymbol{\beta}})'\bar{S}^{-1} \otimes X'X(\boldsymbol{\beta} - \hat{\boldsymbol{\beta}})\} \exp[-\tfrac{1}{2}(\boldsymbol{\beta} - \bar{\boldsymbol{\beta}})'C^{-1}(\boldsymbol{\beta} - \bar{\boldsymbol{\beta}})]$$
$$\mathrel{\dot{\propto}} \exp[-\tfrac{1}{2}(\boldsymbol{\beta} - \mathbf{b})'F(\boldsymbol{\beta} - \mathbf{b})], \tag{8.71}$$

where

$$\mathbf{b} = (C^{-1} + \bar{S}^{-1} \otimes X'X)^{-1}(C^{-1}\bar{\boldsymbol{\beta}} + \bar{S}^{-1} \otimes X'X\hat{\boldsymbol{\beta}}) \tag{8.72}$$

and

$$F = C^{-1} + \bar{S}^{-1} \otimes X'X. \tag{8.73}$$

The quantity $\mathbf{b}$ in (8.72) is the mean of the leading normal term of the expansion and is seen to be a "matrix weighted average" of the prior mean $\bar{\boldsymbol{\beta}}$ and the least squares quantity $\hat{\boldsymbol{\beta}}$ whose weights are the inverse of the prior covariance matrix C and the sample covariance matrix, $(\bar{S} \otimes X'X^{-1})^{-1}$, respectively. The matrix F in (8.73) is the inverse of the covariance matrix of the leading normal term approximating[28] the posterior pdf for B.

8.5 THE "SEEMINGLY UNRELATED" REGRESSION MODEL

The "seemingly unrelated" regression model[29] is in a certain sense a generalization of the traditional multivariate regression model in that the

[28] Further work to take account of additional terms in the expansion along the lines of Appendix 4.2 will lead to a better approximation of the posterior pdf.

[29] For sampling theory analyses of this model, see A. Zellner, "An Efficient Method of Estimating Seemingly Unrelated Regressions and Tests for Aggregation Bias," *J. Am. Statist. Assoc.*, **57**, 348–368 (1962); A. Zellner, "Estimators for Seemingly Unrelated Regression Equations: Some Exact Finite Sample Results," *ibid.*, **58**, 977–992 (1963); A. Zellner and D. S. Huang, "Further Properties of Efficient Estimators for Seemingly Unrelated Regression Equations," *Intern. Econ. Rev.*, **3**, 300–313 (1962); J. Kmenta and R. F. Gilbert, "Small Sample Properties of Alternative Estimates of Seemingly Unrelated Regressions," *J. Am. Statist. Assoc.*, **63**, 1180–1200 (1968).

matrix X appearing in each equation of the traditional model is permitted to be different in the seemingly unrelated regression model; that is, the model assumed to generate the observations is

$$(8.74)\qquad \begin{pmatrix} \mathbf{y}_1 \\ \mathbf{y}_2 \\ \vdots \\ \mathbf{y}_m \end{pmatrix} = \begin{pmatrix} X_1 & & & \\ & X_2 & & \\ & & \ddots & \\ & & & X_m \end{pmatrix} \begin{pmatrix} \boldsymbol{\beta}_1 \\ \boldsymbol{\beta}_2 \\ \vdots \\ \boldsymbol{\beta}_m \end{pmatrix} + \begin{pmatrix} \mathbf{u}_1 \\ \mathbf{u}_2 \\ \vdots \\ \mathbf{u}_m \end{pmatrix},$$

where $\mathbf{y}_\alpha$, $\alpha = 1, 2, \ldots, m$, is an $n \times 1$ vector of observations on the αth dependent variable, X_α is an $n \times k_\alpha$ matrix, with rank k_α, of observations on k_α independent variables appearing in the αth equation with coefficient vector $\boldsymbol{\beta}_\alpha$, a $k_\alpha \times 1$ column vector, and $\mathbf{u}_\alpha$ is an $n \times 1$ vector of disturbance terms appearing in the αth equation; for example, the subscript α might refer to the αth firm. Thus in (8.74) we have a regression relation for each of m firms with n observations on each of the variables and with each firm having its own independent variables[30] and coefficient vector.

For simplicity we rewrite (8.74) as follows:

$$(8.75)\qquad \mathbf{y} = Z\boldsymbol{\beta} + \mathbf{u},$$

where $\mathbf{y}' = (\mathbf{y}_1', \mathbf{y}_2', \ldots, \mathbf{y}_m')$, $\boldsymbol{\beta}' = (\boldsymbol{\beta}_1', \boldsymbol{\beta}_2', \ldots, \boldsymbol{\beta}_m')$, $\mathbf{u}' = (\mathbf{u}_1', \mathbf{u}_2', \ldots, \mathbf{u}_m')$, and Z denotes the block diagonal matrix on the right-hand side of (8.74). Our distributional assumptions about the nm elements of $\mathbf{u}$ are the same as those employed in the analysis of the traditional model, namely that they are jointly normally distributed with $E\mathbf{u} = \mathbf{0}$ and

$$(8.76)\qquad E\mathbf{u}\mathbf{u}' = \Sigma \otimes I_n,$$

where I_n is an $n \times n$ unit matrix and Σ is a positive definite symmetric $m \times m$ matrix. Then the likelihood function for the parameters $\boldsymbol{\beta}$ and Σ is given by

$$(8.77)\qquad \begin{aligned} l(\boldsymbol{\beta}, \Sigma|\mathbf{y}) &\propto |\Sigma^{-1}|^{n/2} \exp\left[-\tfrac{1}{2}(\mathbf{y} - Z\boldsymbol{\beta})'\Sigma^{-1} \otimes I_n(\mathbf{y} - Z\boldsymbol{\beta})\right] \\ &\propto |\Sigma^{-1}|^{n/2} \exp\left(-\tfrac{1}{2} \operatorname{tr} A\Sigma^{-1}\right), \end{aligned}$$

where in the second line of (8.77) we have written

$$(8.78)\qquad A = \begin{bmatrix} (\mathbf{y}_1 - X_1\boldsymbol{\beta}_1)'(\mathbf{y}_1 - X_1\boldsymbol{\beta}_1) & \cdots & (\mathbf{y}_1 - X_1\boldsymbol{\beta}_1)'(\mathbf{y}_m - X_m\boldsymbol{\beta}_m) \\ \vdots & \ddots & \vdots \\ (\mathbf{y}_m - X_m\boldsymbol{\beta}_m)'(y_1 - X_1\boldsymbol{\beta}_1) & \cdots & (\mathbf{y}_m - X_m\boldsymbol{\beta}_m)'(\mathbf{y}_m - X_m\boldsymbol{\beta}_m) \end{bmatrix},$$

an $m \times m$ symmetric matrix.

[30] That having the X's different in (8.74) leads to results differing from those associated with the traditional model came as a surprise. In fact, the phrase "seemingly unrelated" was chosen to emphasize this point.

We shall use the same diffuse prior assumptions about the parameters employed in the analysis of the traditional multivariate regression model, namely

$$(8.79)\qquad \begin{aligned} p(\boldsymbol{\beta}, \Sigma^{-1}) &= p(\boldsymbol{\beta})\, p(\Sigma^{-1}) \\ &\propto |\Sigma^{-1}|^{-(m+1)/2}. \end{aligned}$$

On combining (8.77) and (8.79), the joint posterior pdf for the parameters is

$$(8.80)\qquad \begin{aligned} p(\boldsymbol{\beta}, \Sigma^{-1}|\mathbf{y}) &\propto |\Sigma^{-1}|^{[n-(m+1)/2]} \exp\left[-\tfrac{1}{2}(\mathbf{y} - Z\boldsymbol{\beta})'\Sigma^{-1} \otimes I_n(\mathbf{y} - Z\boldsymbol{\beta})\right] \\ &\propto |\Sigma^{-1}|^{[n-(m+1)/2]} \exp\left(-\tfrac{1}{2} \operatorname{tr} A\Sigma^{-1}\right), \end{aligned}$$

where A has been defined in (8.78).

From (8.80) it is seen that the conditional posterior pdf for $\boldsymbol{\beta}$, given Σ^{-1}, is in the multivariate normal form with mean

$$(8.81)\qquad E(\boldsymbol{\beta}|\Sigma^{-1}, \mathbf{y}) = [Z'(\Sigma^{-1} \otimes I_n)Z]^{-1}Z'(\Sigma^{-1} \otimes I_n)\mathbf{y}$$

and the conditional covariance matrix

$$(8.82)\qquad \operatorname{Cov}(\boldsymbol{\beta}|\Sigma^{-1}, \mathbf{y}) = [Z'(\Sigma^{-1} \otimes I_n)Z]^{-1}.$$

The conditional mean in (8.81) is precisely the generalized least squares quantity that Zellner obtained in studying the system in (8.74) from the sampling theory point of view; that is, we multiply both sides of (8.75) by a matrix H, $H\mathbf{y} = HZ\boldsymbol{\beta} + H\mathbf{u}$, where H is a nonsingular matrix such that $EH\mathbf{u}\mathbf{u}'H' = H\Sigma \otimes I_nH' = I_{mn}$, which is possible, since $\Sigma \otimes I_n$ is assumed to be a positive definite symmetric matrix. Thus, from $H\Sigma \otimes I_nH' = I_{mn}$, $H'H = \Sigma^{-1} \otimes I_n$. Then the transformed system $Hy = HZ\boldsymbol{\beta} + H\mathbf{u}$ satisfies the conditions of the Gauss-Markoff theorem, which means that the sampling theory estimator

$$(8.83)\qquad \begin{aligned} \hat{\boldsymbol{\beta}} &= (Z'H'HZ)^{-1}Z'H'H\mathbf{y} \\ &= [Z'(\Sigma^{-1} \otimes I_n)Z]^{-1}Z'\Sigma^{-1} \otimes I_n\, \mathbf{y}, \end{aligned}$$

with covariance matrix

$$(8.84)\qquad \operatorname{Cov}(\hat{\boldsymbol{\beta}}) = [Z'(\Sigma^{-1} \otimes I_n)Z]^{-1},$$

is a minimum variance linear unbiased estimator for $\boldsymbol{\beta}$. Also, with a normality assumption, as can be seen from the likelihood function in (8.77), the estimator in (8.83) is a maximum likelihood estimator for $\boldsymbol{\beta}$, given Σ. As can readily be established, if the X's are all the same (or proportional) and/or if Σ is diagonal, the quantities in (8.81) and (8.83) reduce algebraically to vectors of single-equation least squares estimates; that is, with respect to (8.83), under these conditions, $\hat{\boldsymbol{\beta}}_\alpha = (X_\alpha'X_\alpha)^{-1}X_\alpha'\mathbf{y}_\alpha$, $\alpha = 1, 2, \ldots, m$, and similarly for (8.81).

As mentioned above, the sampling theory estimator in (8.83) is identical to the mean of the conditional posterior pdf for $\boldsymbol{\beta}$, given Σ or, equivalently, given Σ^{-1}. Note, however, that the sampling theory estimator in (8.83) depends on the matrix Σ which is usually unknown. Zellner's suggestion that Σ be replaced by a consistent estimate $\hat{\Sigma}$, formed from the residuals of the equations estimated individually by least squares to yield the estimator[31]

$$\mathbf{b} = [Z'(\hat{\Sigma}^{-1} \otimes I_n)Z]^{-1}Z'\hat{\Sigma}^{-1} \otimes I_n\mathbf{y}, \tag{8.85}$$

is equivalent to what is obtained from a Bayesian analysis if we go ahead under the assumption that $\Sigma = \hat{\Sigma}$. With this assumption the conditional mean in (8.81) is precisely the quantity $\mathbf{b}$ in (8.85). In *large samples* $\hat{\Sigma}$ will not differ markedly from Σ, and thus going ahead with the assumption $\Sigma = \hat{\Sigma}$ will produce satisfactory results. In small samples,[32] however, it is better to obtain the marginal posterior pdf for $\boldsymbol{\beta}$ and base inferences on it rather than rely on conditional results.

To obtain the marginal posterior pdf for $\boldsymbol{\beta}$ we can use properties of the Wishart pdf to integrate (8.80) with respect to the distinct elements of Σ^{-1}. This yields

$$p(\boldsymbol{\beta}|\mathbf{y}) \propto |A|^{-n/2}$$

$$\propto \begin{vmatrix} (\mathbf{y}_1 - X_1\boldsymbol{\beta}_1)'(\mathbf{y}_1 - X_1\boldsymbol{\beta}_1) & \cdots & (\mathbf{y}_1 - X_1\boldsymbol{\beta}_1)'(\mathbf{y}_m - X_m\boldsymbol{\beta}_m) \\ \vdots & \ddots & \vdots \\ (\mathbf{y}_m - X_m\boldsymbol{\beta}_m)'(\mathbf{y}_1 - X_1\boldsymbol{\beta}_1) & \cdots & (\mathbf{y}_m - X_m\boldsymbol{\beta}_m)'(\mathbf{y}_m - X_m\boldsymbol{\beta}_m) \end{vmatrix}^{-n/2} \tag{8.86}$$

as the joint posterior pdf for the elements of $\boldsymbol{\beta}$. Although this distribution resembles a generalized multivariate Student t pdf, it does not appear to be possible to bring it into this form because of the fact that not all X's are the same. Further work to provide techniques for analyzing (8.86) is required before it can be used in practical work.

Another way of looking at the seemingly unrelated regression model is to consider it as a "restricted" traditional multivariate regression; that is, we can write

$$(\mathbf{y}_1, \ldots, \mathbf{y}_m) = (X_1X_2\cdots X_m)\begin{pmatrix} \boldsymbol{\beta}_1 & 0 & \cdots & 0 \\ 0 & \boldsymbol{\beta}_2 & \cdots & 0 \\ \vdots & \vdots & \ddots & \vdots \\ 0 & 0 & \cdots & \boldsymbol{\beta}_m \end{pmatrix} + (\mathbf{u}_1, \ldots, \mathbf{u}_m), \tag{8.87}$$

[31] This estimator has the same *large sample* properties as that of the estimator $\hat{\boldsymbol{\beta}}$ in (8.83).

[32] See A. Zellner, "Estimators for Seemingly Unrelated Regression Equations: Some Exact Finite Sample Results," *loc. cit.*, for an analysis of the finite sample properties of the estimator $\mathbf{b}$ in (8.85) for a two-equation model. Kmenta and Gilbert, *loc. cit.*, provide interesting Monte Carlo results, and N. C. Kakwani, "The Unbiasedness of Zellner's Seemingly Unrelated Regression Equations Estimators," *J. Am. Statist. Assoc.* **62,** 141–142 (1967), shows that $\mathbf{b}$ is unbiased.

which shows the (zero) restrictions explicitly. In cases in which the matrix $(X_1 X_2 \cdots X_m)$ has full rank the approach taken in Section 8.3 to incorporate zero restrictions in the traditional model can be applied. However, if we work with just the leading normal term in the expansion, we have seen that this is equivalent to using the conditional posterior pdf $p(\boldsymbol{\beta}|\Sigma = \hat{\Sigma}, y)$. Thus, to this degree of accuracy, using the results in (8.81) and (8.82) with $\Sigma = \hat{\Sigma}$ will be satisfactory in large samples.

To illustrate an application of these large-sample results we consider annual investment data pertaining to 10 large U.S. corporations from 1935 to 1954. For each of the 10 firms we posit a regression relation that "explains" its deflated annual gross investment in terms of two explanatory variables, the

Table 8.1 POSTERIOR MEANS AND STANDARD DEVIATIONS FOR TWO ANALYSES OF TEN FIRMS' INVESTMENT RELATIONS[a]

	Analysis Based on Equations (8.81) and (8.82)			Analysis Based on Individual Firm's Data Alone		
		Slope Coefficients[b]			Slope Coefficients[b]	
Corporation	Intercept	(1)	(2)	Intercept	(1)	(2)
General Electric	−11.2	0.0332	0.124	−9.96	0.0266	0.152
	(21.1)	(0.00928)	(0.0214)	(31.4)	(0.0156)	(0.0257)
Westinghouse	4.11	0.0525	0.0412	−0.509	0.0529	0.0924
	(5.09)	(0.00794)	(0.0347)	(8.02)	(0.0157)	(0.0561)
U.S. Steel	−18.6	0.170	0.320	−49.2	0.175	0.390
	(78.4)	(0.0377)	(0.101)	(148)	(0.0742)	(0.142)
Diamond Match	2.20	−0.0181	0.365)	0.162	0.00457	0.437
	(1.12)	(0.0151)	(0.0578)	(2.07)	(0.0272)	(0.0760)
Atlantic Refining	26.5	0.131	0.0102	22.7	0.162	0.00310
	(6.48)	(0.0473)	(0.0187)	(6.87)	(0.0570)	(0.0220)
Union Oil	−9.67	0.112	0.128	−4.50	0.0875	0.124
	(9.01)	(0.0456)	(0.0155)	(11.3)	(0.0656)	(0.0171)
Goodyear	−2.58	0.0760	0.0641	−7.72	0.0754	0.0821
	(7.59)	(0.0202)	(0.0229)	(9.36)	(0.0340)	(0.0280)
General Motors	−133.0	0.113	0.386	−150	0.119	0.371
	(73.2)	(0.0167)	(0.0312)	(106)	(0.0258)	(0.0371)
Chrysler	2.45	0.0672	0.306	−6.19	0.0780	0.316
	(11.5)	(0.0166)	(0.0271)	(13.5)	(0.0200)	(0.0288)
IBM	−5.56	0.131	0.0571	−8.69	0.132	0.0854
	(3.56)	(0.0167)	(0.0575)	(4.54)	(0.0312)	(0.100)

[a] The data underlying these computations were taken from J. C. G. Boot and G. M. de Wit, "Investment Demand: An Empirical Contribution to the Aggregation Problem," *Intern. Econ. Rev.*, **1**, 3–30 (1960).

[b] The slope coefficient (1) is the coefficient of the value of outstanding shares at the beginning of the year while (2) is that of the firm's beginning of year real capital stock.

value of its outstanding shares at the beginning of the year, and its beginning-of-year real capital stock. In addition, we assume a nonzero intercept term in each relation. We further assume that the model generating the data is the normal seemingly unrelated regression model, that is, (8.74) with $m = 10$. Shown in Table 8.1 are the approximate means of the posterior pdf for the coefficients based on the conditional posterior pdf $p(B|\Sigma = \hat{\Sigma}, Y)$. Shown below each entry in parentheses are the conditional standard deviations of the coefficients computed as square roots of diagonal elements of the matrix in (8.84). Further, the results of an equation-by-equation analysis of the 10 regression equations is presented, an analysis that utilized diffuse prior pdf's, as in Chapter 4, on the multiple regression model. Entries in the table are elements of $\hat{\boldsymbol{\beta}}_\alpha = (X_\alpha' X_\alpha)^{-1} X_\alpha' \mathbf{y}_\alpha$ for each firm's data; the figures in parentheses are square roots of diagonal elements of $(X_\alpha' X_\alpha)^{-1} \bar{s}_{\alpha\alpha}$, $\alpha = 1, 2, \ldots, 10$. On comparing the precision with which coefficients have been determined, it is seen that the analysis using *all* the observations yields sharper posterior pdf's for individual coefficients than did that employing just the individual firm's data in making inferences about its coefficients.[33] To the extent that the analyses using just part of the data fail to incorporate information in all the data, information is being lost. To indicate the extent to which the

Table 8.2 SAMPLE DISTURBANCE CORRELATION MATRIX

	General Electric	Westing-house	U.S. Steel	Diamond Match	Atlantic Refining	Union Oil	Goodyear	General Motors	Chrysler	IBM
General Electric	1.00	0.74	0.45	0.60	−0.02	0.02	0.43	0.29	−0.07	0.48
Westing-house	—	1.00	0.64	0.62	0.00	0.14	0.54	0.16	0.12	0.55
U.S. Steel	—	—	1.00	0.75	0.24	−0.30	0.30	−0.28	0.36	0.39
Diamond Match	—	—	—	1.00	0.13	−0.23	0.28	−0.27	0.12	0.41
Atlantic Refining	—	—	—	—	1.00	0.15	−0.15	−0.32	0.06	0.22
Union Oil	—	—	—	—	—	1.00	0.20	0.53	−0.13	0.14
Goodyear	—	—	—	—	—	—	1.00	0.21	0.07	−0.18
General Motors	—	—	—	—	—	—	—	1.00	−0.28	0.12
Chrysler	—	—	—	—	—	—	—	—	1.00	0.21
IBM	—	—	—	—	—	—	—	—	—	1.00

[33] This has to be qualified in that the analysis using all firms' data is based just on a large-sample result.

observations are correlated the sample disturbance correlation matrix is presented in Table. 8.2.

It is seen that some of these correlations are substantial and that there is some reason to doubt that the observations are independent.

QUESTIONS AND PROBLEMS

1. Given the diffuse prior pdf for the distinct elements σ_{11}, σ_{22}, σ_{12} of a 2×2 pds covariance matrix, that is, $p(\Sigma) \propto |\Sigma|^{-1}$, with $0 < \sigma_{11}, \sigma_{22}$ and $-\infty < \sigma_{12} < \infty$, derive the implied prior pdf for the correlation coefficient $\rho = \sigma_{12}/\sqrt{\sigma_{11}\sigma_{22}}$ and describe its properties.
2. Given that $\mathbf{y}_1, \mathbf{y}_2, \ldots, \mathbf{y}_n$ are n vectors of observations, each 2×1 and drawn independently from a bivariate normal population with mean vector $\boldsymbol{\mu}' = (\mu_1, \mu_2)$ and 2×2 pds covariance matrix Σ, derive the bivariate marginal posterior pdf for μ_1 and μ_2 by employing a diffuse prior pdf for $\boldsymbol{\mu}$ and the distinct elements of Σ.
3. With the information and assumptions of Problem 2, derive the posterior pdf for the difference in means, namely $\mu_1 - \mu_2$. Also obtain the marginal posterior pdf's for σ_{11} and for ρ, the correlation coefficient.
4. Suppose in Problem 2 that the $\mathbf{y}_i$'s are each $m \times 1$ vectors, $\boldsymbol{\mu}' = (\mu_1, \mu_2, \ldots, \mu_m)$ and Σ is an $m \times m$ pds matrix. With diffuse prior assumptions about $\boldsymbol{\mu}$ and the distinct elements of Σ, show that the marginal posterior pdf for $\boldsymbol{\mu}$ is in the multivariate Student t form and obtain the marginal posterior pdf for μ_1.
5. For the conditions of Problem 4 explain how to construct a 90% Bayesian confidence interval for $\mathbf{c}'\boldsymbol{\mu}$, where $\mathbf{c}' = (c_1, c_2, \ldots, c_m)$, a $1 \times m$ vector whose elements have known values.
6. In Problem 4 obtain the marginal posterior pdf for the distinct elements of Σ and comment on its properties. From the posterior pdf for Σ derive the posterior pdf for $\Omega = A'\Sigma A$, where A is an $m \times m$ nonsingular matrix whose elements have known values.
7. Given the posterior pdf for Σ, derived in Problem 6, derive the marginal posterior pdf of the distinct elements of the $m_1 \times m_1$ matrix Σ_{11}, a submatrix of Σ shown below:
$$\Sigma = \begin{pmatrix} \Sigma_{11} & \Sigma_{12} \\ \Sigma_{21} & \Sigma_{22} \end{pmatrix}.$$
8. What are the posterior pdf's for Σ^{-1} and $\Sigma_{11.2}^{-1} = (\Sigma_{11} - \Sigma_{12}\Sigma_{22}^{-1}\Sigma_{21})^{-1} = \Sigma^{11}$ implied by the posterior pdf for Σ obtained in Problem 6?
9. Consider the standard multivariate regression model $Y = XB + U$ which was analyzed in Section 8.3. From the marginal posterior pdf for the elements of B shown in (8.24) derive the posterior pdf for the elements of $\Theta = CB$, where C is a $k \times k$ nonsingular matrix whose elements have known values.

10. Let $Y_1 = X_1B + U_1$ and $Y_2 = X_2B + U_2$, where the matrices Y_i, X_i, and U_i are of sizes $n_i \times m$, $n_i \times k$, and $n_i \times m$, respectively, where $i = 1, 2$, and B is a $k \times m$ matrix of regression coefficients. Assume that the $1 \times m$ rows of U_1 and U_2 are all independently and normally distributed with zero mean vectors and that the rows of U_1 have a common pds $m \times m$ covariance matrix Σ_1 and those of U_2 have common pds $m \times m$ covariance matrix Σ_2. With X_1 and X_2 given, each of rank k, and using the following diffuse prior pdf,
$$p(B, \Sigma_1, \Sigma_2) \propto |\Sigma_1|^{-(m+1)/2}|\Sigma_2|^{-(m+1)/2},$$
where the elements of B range from $-\infty$ to $+\infty$ and $|\Sigma_i| > 0$, $i = 1, 2$, show that the marginal posterior pdf for B is in the form of a product of two factors, each in the generalized Student t form.
11. Provide the leading normal term in an asymptotic expansion of the posterior pdf for B obtained in Problem 10.
12. Consider the following regression system:
$$\left.\begin{aligned} y_{1i} &= \beta_{11}x_{1i} + \beta_{12}x_{2i} + u_{1i} \\ y_{2i} &= \beta_{21}x_{1i} + \beta_{22}x_{2i} + u_{2i} \end{aligned}\right\} \qquad i = 1, 2, \ldots, n,$$
where the β's are regression coefficients, x_{1i} and x_{2i} are given values of two independent variables, y_{1i} and y_{2i} are dependent variables, and u_{1i} and u_{2i} are disturbance terms. Assume that the $n \times 2$ matrix $X = (\mathbf{x}_1\mathbf{x}_2)$ has rank 2 and that the pairs of random variables (u_{1i}, u_{2i}), $i = 1, 2, \ldots, n$, are normally and independently distributed, each with zero mean vector and common 2×2 pds covariance matrix Σ. If the elements of Σ have known values, derive the conditional posterior pdf for the β's by using a diffuse prior pdf.
13. Suppose, in Problem 12, we assume that $\beta_{12} = \beta_{21} = 0$. Obtain the conditional posterior pdf for β_{11} and β_{22}, given Σ and $\beta_{12} = \beta_{21} = 0$; that is, $p(\beta_{11}, \beta_{22}|\Sigma, \beta_{12} = \beta_{21} = 0, \mathbf{y}_1, \mathbf{y}_2)$. What are the mean and the covariance matrix of this conditional posterior pdf? How is the posterior variance of β_{11} affected by the assumption that $\mathbf{x}_1'\mathbf{x}_2 = \sum_{i=1}^{n} x_{1i}x_{2i} = 0$?
14. Given the results in Problem 13, compare the posterior variance of β_{11}, given Σ and $\beta_{12} = \beta_{21} = 0$, with the posterior variance of β_{11}, given σ_{11} obtained from an analysis of $y_{1i} = \beta_{11}x_{1i} + u_{1i}$ with a diffuse prior pdf for β_{11}.
15. Show that the expression in (8.81) reduces to $E(\boldsymbol{\beta}|\Sigma^{-1}, \mathbf{y}) = (Z'Z)^{-1}Z'\mathbf{y}$ if (a) Σ^{-1} is a diagonal matrix and/or (b) $X_1 = X_2 = \cdots = X_m$. Interpret the quantity $(Z'Z)^{-1}Z'\mathbf{y}$. Also consider the form of (8.82) under assumptions (a) and (b).

CHAPTER IX

Simultaneous Equation Econometric Models

Since most economic phenomena involve interactions among several or many variables, it is important to construct and analyze models that incorporate such interactions or feedback effects and are capable of "explaining" the variation of a set of variables. This is one of the main objectives we have in mind when we construct a simultaneous equation model to represent, say, a particular market or national economy. In the former instance the model will provide an explanation of the variation of price and quantity of the commodity or service traded in the market. With respect to models of national economies,[1] one objective of model construction is the explanation of variation in such variables as national income, consumption expenditures, investment, and the price level.

Here, the term "simultaneous equation econometric model" is taken to mean a stochastic model that permits an investigator to make probabilistic statements about a set of random variables, the so-called endogenous variables.[2] The models we shall consider are linear in the parameters and a generalization of the multivariate regression model in the sense that in multivariate regression models we have just one "dependent" variable per equation whereas in simultaneous equation models we may have more than one dependent variable appearing in each equation; that is, in the simultaneous equation model it is assumed that the data are generated by a model in the following form[3]:

$$Y\Gamma = XB + U, \tag{9.1}$$

[1] See M. Nerlove, "A Tabular Survey of Macro-Econometric Models," *Intern. Econ. Rev.*, 7, 127–173 (1966), for a review of some salient features of a number of models which have appeared in the literature.

[2] To make such probability statements information about a model's initial conditions, the form of the pdf for its stochastic variables, exogenous variables, and parameter values is required. Providing information about parameter values is an objective of statistical estimation techniques.

[3] Insofar as possible, the notation in this chapter parallels that used in discussing multivariate regression models in Chapter 8.

where $Y = (\mathbf{y}_1, \mathbf{y}_2, \ldots, \mathbf{y}_m)$, an $n \times m$ matrix of observations on m "dependent" or endogenous variables whose variation is to be explained by the model, Γ is an $m \times m$ nonsingular matrix of coefficients for the endogenous variables, $X = (\mathbf{x}_1, \mathbf{x}_2, \ldots, \mathbf{x}_k)$ is an $n \times k$ matrix of observations on k "predetermined" variables, B is a $k \times m$ matrix of coefficients for the "predetermined" variables, and $U = (\mathbf{u}_1, \mathbf{u}_2, \ldots, \mathbf{u}_m)$ is an $n \times m$ matrix of random disturbance terms. The variables in X, the predetermined variables, may include lagged values of the endogenous and/or exogenous variables. In the latter category we include both nonstochastic and stochastic variables whose variation is determined outside the model. Stochastic exogenous variables, by definition, are assumed to be distributed independently of the elements of U and to have distributions not involving any of the parameters of the model; that is, Γ, B, and the elements of the covariance matrix of the elements of U. Further, in this chapter we assume that the elements of U have zero means and that the rows of U are normally and independently distributed, each with $m \times m$ pds covariance matrix Σ. This independence assumption rules out any form of auto or serial correlation.

Note that if the elements of the matrix Γ in (9.1) had elements with known values the model could be analyzed by using results for multivariate regression models[4] presented in Chapter 8. One special case that deserves mention is $\Gamma = I_m$, where I_m is an $m \times m$ unit matrix. Here the model is in the form of a regression system, except that X usually includes lagged values of the y's. More accurately, then, the system with $\Gamma = I_m$ is usually a set of autoregressive equations. That $\Gamma = I_m$ does not mean that there are no feedback effects in the system; it means that any feedback effects in the system are lagged.

Other special cases, which we distinguish below, are that in which Γ is triangular and Σ, the disturbance covariance matrix, is diagonal, the "fully recursive" case and that in which Γ is triangular and Σ is not diagonal, called the "triangular" case. Others which can be distinguished are combinations of the Γ block diagonal and the disturbance covariance matrix full or block diagonal. Finally, when Γ and the disturbance covariance matrix have no special form, we refer to the system simply as an "interdependent" model.

We discuss first the analysis of fully recursive and triangular models. After that we treat the concept of identification in Bayesian terms. Then several simple models are considered, followed by the presentation of methods for "limited information," single-equation, and full-system Bayesian analysis. Last the results of some Monte Carlo experiments which compare the sampling properties of Bayesian and well known sampling theory estimators are reported.

[4] This assumes that we have gone ahead depending on given initial conditions if the model had autoregressive features.

9.1 FULLY RECURSIVE MODELS

We assume that our observations $Y = (\mathbf{y}_1, \ldots, \mathbf{y}_m)$ have been generated by the following model:

$$
\begin{aligned}
\mathbf{y}_1 &= X_1\boldsymbol{\beta}_1 + \mathbf{u}_1 \\
\mathbf{y}_2 &= \mathbf{y}_1\gamma_{21} + X_2\boldsymbol{\beta}_2 + \mathbf{u}_2 \\
\mathbf{y}_3 &= \mathbf{y}_1\gamma_{31} + \mathbf{y}_2\gamma_{32} + X_3\boldsymbol{\beta}_3 + \mathbf{u}_3 \\
\vdots \quad &\vdots \quad \vdots \qquad\quad \vdots \qquad\quad \vdots \qquad \vdots \\
\mathbf{y}_m &= \mathbf{y}_1\gamma_{m1} + \mathbf{y}_2\gamma_{m2} + \cdots + \mathbf{y}_{m-1}\gamma_{m,m-1} + X_m\boldsymbol{\beta}_m + \mathbf{u}_m,
\end{aligned}
\tag{9.2}
$$

where $\mathbf{y}_\alpha$ is an $n \times 1$ vector of observations on the αth endogenous variable, X_α is an $n \times k_\alpha$ matrix with rank k_α of observations on k_α predetermined variables appearing in the αth equation with coefficient vector $\boldsymbol{\beta}_\alpha$, a $k_\alpha \times 1$ column vector, $\mathbf{u}_\alpha$ is an $n \times 1$ vector of disturbance terms, and the $\gamma_{\alpha l}$ are scalar coefficients.[5] Further, we assume that the elements of the $\mathbf{u}_\alpha$, $\alpha = 1, 2, \ldots, m$, are normally distributed with zero means and covariance matrix

$$
E\mathbf{u}\mathbf{u}' = D(\sigma_\alpha^2) \otimes I_n, \tag{9.3}
$$

where $\mathbf{u}' = (\mathbf{u}_1', \ldots, \mathbf{u}_m')$ and $D(\sigma_\alpha^2)$ is a diagonal matrix with $\sigma_1^2, \sigma_2^2, \ldots, \sigma_m^2$ on the main diagonal. Thus (9.3), along with the assumption of normality, implies that all elements of $\mathbf{u}$ are independently distributed and that the disturbance terms in different equations have different variances.

With the above distributional assumptions and noting that the Jacobian of the transformation from the u's to the y's is 1, we obtain the likelihood function

$$
l(\boldsymbol{\delta}, \boldsymbol{\sigma} \mid Y, Y_0) \propto \prod_{\alpha=1}^{m} \frac{1}{\sigma_\alpha^n} \exp\left[-\frac{1}{2\sigma_\alpha^2}(\mathbf{y}_\alpha - Z_\alpha\boldsymbol{\delta}_\alpha)'(\mathbf{y}_\alpha - Z_\alpha\boldsymbol{\delta}_\alpha)\right], \tag{9.4}
$$

where Y_0 denotes given initial conditions, $\boldsymbol{\sigma}' = (\sigma_1, \sigma_2, \ldots, \sigma_m)$, $Z_\alpha = (\mathbf{y}_1, \mathbf{y}_2, \ldots, \mathbf{y}_{\alpha-1} \vdots X_\alpha)$, $\boldsymbol{\delta}_\alpha' = (\boldsymbol{\gamma}_\alpha' \vdots \boldsymbol{\beta}_\alpha')$ with $\boldsymbol{\gamma}_\alpha' = (\gamma_{\alpha1}, \gamma_{\alpha2}, \ldots, \gamma_{\alpha,\alpha-1})$, and $\boldsymbol{\delta}' = (\boldsymbol{\delta}_1', \boldsymbol{\delta}_2', \ldots, \boldsymbol{\delta}_m')$.

As regards prior assumptions we employ the following diffuse prior assumptions, $p(\boldsymbol{\delta}, \boldsymbol{\sigma}) = p(\boldsymbol{\delta})p(\boldsymbol{\sigma})$, with

$$
p(\boldsymbol{\delta}) \propto \text{constant} \quad \text{and} \quad p(\boldsymbol{\sigma}) \propto \prod_{\alpha=1}^{m} \frac{1}{\sigma_\alpha} \Bigg\} \qquad \begin{aligned} -\infty\boldsymbol{\iota} &< \boldsymbol{\delta} < \infty\boldsymbol{\iota}, \\ 0\boldsymbol{\iota} &< \boldsymbol{\sigma} < \infty\boldsymbol{\iota}, \end{aligned} \tag{9.5}
$$

[5] In (9.2) we have introduced all $\gamma_{\alpha l}$ which could be non-zero, subject to our normalization, and yet yield a triangular structure. If some of these coefficients are known to be zero, a priori, they can be set equal to zero without affecting the analysis in any fundamental way.

where $\boldsymbol{\iota}$ denotes a column vector of ones. Under these assumptions the joint posterior pdf for $\boldsymbol{\delta}$ and $\boldsymbol{\sigma}$ is

$$(9.6)\quad p(\boldsymbol{\delta}, \boldsymbol{\sigma}|Y, Y_0) \propto \prod_{\alpha=1}^{m} \frac{1}{\sigma_\alpha^{n+1}} \exp\left[-\frac{1}{2\sigma_\alpha^2}(\mathbf{y}_\alpha - Z_\alpha\boldsymbol{\delta}_\alpha)'(\mathbf{y}_\alpha - Z_\alpha\boldsymbol{\delta}_\alpha)\right].$$

From the form of (9.6) it is seen immediately that each factor contains a particular equation's parameters, σ_α and $\boldsymbol{\delta}_\alpha$, and thus each equation's parameters are, a posteriori, distributed independently of those of other equations. The posterior pdf for the αth equation's parameters is

$$(9.7)\quad \begin{aligned} p(\boldsymbol{\delta}_\alpha, \sigma_\alpha|Y, Y_0) &\propto \frac{1}{\sigma_\alpha^{n+1}} \exp\left[-\frac{1}{2\sigma_\alpha^2}(\mathbf{y}_\alpha - Z_\alpha\boldsymbol{\delta}_\alpha)'(\mathbf{y}_\alpha - Z_\alpha\boldsymbol{\delta}_\alpha)\right] \\ &\propto \frac{1}{\sigma_\alpha^{n+1}} \exp\left\{-\frac{1}{2\sigma_\alpha^2}[\hat{\mathbf{u}}_\alpha'\hat{\mathbf{u}}_\alpha + (\boldsymbol{\delta}_\alpha - \hat{\boldsymbol{\delta}}_\alpha)'Z_\alpha'Z_\alpha(\boldsymbol{\delta}_\alpha - \hat{\boldsymbol{\delta}}_\alpha)]\right\}, \end{aligned}$$

where

$$(9.8)\quad \hat{\boldsymbol{\delta}}_\alpha = (Z_\alpha'Z_\alpha)^{-1}Z_\alpha'\mathbf{y}_\alpha$$

and $\hat{\mathbf{u}}_\alpha = \mathbf{y}_\alpha - Z_\alpha\hat{\boldsymbol{\delta}}_\alpha$. Thus the conditional posterior pdf for $\boldsymbol{\delta}_\alpha$, given σ_α, is in the multivariate normal form with the mean $\hat{\boldsymbol{\delta}}_\alpha$, given in (9.8). Further, on integrating (9.7) with respect to $\boldsymbol{\delta}_\alpha$, we find that the marginal posterior pdf for σ_α is

$$(9.9)\quad p(\sigma_\alpha|Y, Y_0) \propto \frac{1}{\sigma_\alpha^{n-q_\alpha+1}} \exp\left(-\frac{\hat{\mathbf{u}}_\alpha'\hat{\mathbf{u}}_\alpha}{2\sigma_\alpha^2}\right),$$

where q_α is the number of elements in $\boldsymbol{\delta}_\alpha$, which is in the inverted gamma form. Last, on integrating (9.7) with respect to σ_α, we have for the marginal posterior pdf for $\boldsymbol{\delta}_\alpha$

$$(9.10)\quad p(\boldsymbol{\delta}_\alpha|Y, Y_0) \propto [\hat{\mathbf{u}}_\alpha'\hat{\mathbf{u}}_\alpha + (\boldsymbol{\delta}_\alpha - \hat{\boldsymbol{\delta}}_\alpha)'Z_\alpha'Z_\alpha(\boldsymbol{\delta}_\alpha - \hat{\boldsymbol{\delta}}_\alpha)]^{-n/2},$$

which is in the multivariate Student t form with mean $\hat{\boldsymbol{\delta}}_\alpha$ shown in (9.8). It should be noted that $\hat{\boldsymbol{\delta}}_\alpha$ is also the maximum likelihood estimate for $\boldsymbol{\delta}_\alpha$.

It is seen that the results for the fully recursive model, given initial conditions, are completely analogous to Bayesian results for the multiple regression model. Thus many of the results obtained for the multiple regression model can be carried over to apply to the equations of fully recursive models; for example, if, instead of the diffuse prior pdf for the elements of $\boldsymbol{\delta}_\alpha$, we had employed an informative prior pdf in the multivariate normal form, that is,

$$(9.11)\quad p(\boldsymbol{\delta}_\alpha) \propto \exp\left[-\frac{1}{2\tau_\alpha^2}(\boldsymbol{\delta}_\alpha - \bar{\boldsymbol{\delta}}_\alpha)'R_\alpha(\boldsymbol{\delta}_\alpha - \bar{\boldsymbol{\delta}}_\alpha)\right],$$

where $\bar{\boldsymbol{\delta}}_\alpha$ is the prior mean vector and $\tau_\alpha^2 R_\alpha^{-1}$ is the prior covariance matrix, the joint posterior pdf for $\boldsymbol{\delta}_\alpha$, assuming that our prior pdf for σ_α is $p(\sigma_\alpha) \propto 1/\sigma_\alpha$, is

$$(9.12)\quad \begin{aligned} p(\boldsymbol{\delta}_\alpha | Y, Y_0, PI) &\propto [\hat{\mathbf{u}}_\alpha'\hat{\mathbf{u}}_\alpha + (\boldsymbol{\delta}_\alpha - \hat{\boldsymbol{\delta}}_\alpha)'Z_\alpha'Z_\alpha(\boldsymbol{\delta}_\alpha - \hat{\boldsymbol{\delta}}_\alpha)]^{-n/2} \\ &\quad \times \exp\left[-\frac{1}{2\tau_\alpha^2}(\boldsymbol{\delta}_\alpha - \bar{\boldsymbol{\delta}}_\alpha)'R_\alpha(\boldsymbol{\delta}_\alpha - \bar{\boldsymbol{\delta}}_\alpha)\right], \end{aligned}$$

where PI denotes prior information. It is seen that the posterior pdf in (9.12) is in what Tiao and Zellner[6] have called the "normal-t" form and the methods they have presented (see Appendix 4.1) can be used in its analysis. Also, if an informative prior pdf for σ_α in the inverted gamma form is introduced with the normal prior for $\boldsymbol{\delta}_\alpha$ in (9.11), the joint posterior pdf for $\boldsymbol{\delta}_\alpha$ can easily be shown to be in the normal-t form. Among other results that can be carried over from regression analysis to apply to fully recursive models is the analysis of autocorrelation and the Box-Cox analysis of transformations. Then, too, it is not difficult to obtain the predictive pdf for the y's in period $n + 1$. Going further into the future, however, is somewhat complicated in situations in which the system is autoregressive.

The simplicity of the fully recursive model, both with respect to its triangular form and analysis, is striking. However, it may be that the assumption that the disturbance terms are contemporaneously uncorrelated is not satisfied in all circumstances. We now turn to the analysis of triangular systems without the assumption that the disturbance covariance matrix is diagonal.

9.2 GENERAL TRIANGULAR SYSTEMS

In this section we analyze the system in (9.2) with the assumption in (9.3) replaced by

$$(9.13)\quad E\mathbf{u}\mathbf{u}' = \Sigma \otimes I_n,$$

where Σ is an $m \times m$ positive definite symmetric matrix. The assumption in (9.13) permits nonzero contemporaneous disturbance covariances and differing variances for disturbance terms of different equations but rules out any kind of auto- or serial correlation. Except for (9.13), all other distributional assumptions about the system are the same as those in Section 9.1.

[6] G. C. Tiao and A. Zellner, "Bayes' Theorem and the Use of Prior Knowledge in Regression Analysis," *Biometrika*, **51**, 219–230 (1964).

Note that we can write the system in (9.2) as

$$\begin{pmatrix} \mathbf{y}_1 \\ \mathbf{y}_2 \\ \vdots \\ \mathbf{y}_m \end{pmatrix} = \begin{pmatrix} Z_1 & & & \\ & Z_2 & & \\ & & \ddots & \\ & & & Z_m \end{pmatrix} \begin{pmatrix} \boldsymbol{\delta}_1 \\ \boldsymbol{\delta}_2 \\ \vdots \\ \boldsymbol{\delta}_m \end{pmatrix} + \begin{pmatrix} \mathbf{u}_1 \\ \mathbf{u}_2 \\ \vdots \\ \mathbf{u}_m \end{pmatrix}, \tag{9.14}$$

where the Z's and $\boldsymbol{\delta}$'s have been defined in connection with (9.4). Equation 9.14 is in the form of the "seemingly unrelated" regression model. Thus, given initial conditions, and employing the following diffuse prior pdf

$$p(\boldsymbol{\delta}, \Sigma^{-1}) \propto |\Sigma^{-1}|^{-(m+1)/2}, \qquad -\infty\boldsymbol{\iota} < \boldsymbol{\delta} < \infty\boldsymbol{\iota}, \tag{9.15}$$

the analysis of the system in (9.14) goes through in exactly the same fashion as that for the seemingly unrelated regression model. In particular, the conditional posterior pdf for $\boldsymbol{\delta}$, given Σ, will be multivariate normal with mean

$$E(\boldsymbol{\delta}|\, Y, Y_0, \Sigma) = [Z'(\Sigma^{-1} \otimes I_n)Z]^{-1}Z'(\Sigma^{-1} \otimes I_n)\mathbf{y} \tag{9.16}$$

and conditional covariance matrix

$$V(\boldsymbol{\delta}|\, Y, Y_0, \Sigma) = [Z'(\Sigma^{-1} \otimes I_n)Z]^{-1}, \tag{9.17}$$

where Z denotes the block diagonal matrix on the rhs of (9.14) and $\mathbf{y}$, the vector on the lhs of (9.14). A *large-sample* analysis based on these conditional results can be performed by using a consistent sample estimate of Σ.[7] As regards finite sample results, the joint posterior pdf for the elements of $\boldsymbol{\delta}$ is

$$P(\boldsymbol{\delta}|\, Y, Y_0) \propto |A|^{-n/2}, \tag{9.18}$$

with

$$A = \begin{bmatrix} (\mathbf{y}_1 - Z_1\boldsymbol{\delta}_1)'(\mathbf{y}_1 - Z_1\boldsymbol{\delta}_1) & \cdots & (\mathbf{y}_1 - Z_1\boldsymbol{\delta}_1)'(\mathbf{y}_m - Z_m\boldsymbol{\delta}_m) \\ \vdots & \ddots & \vdots \\ (\mathbf{y}_m - Z_m\boldsymbol{\delta}_m)'(\mathbf{y}_1 - Z_1\boldsymbol{\delta}_1) & \cdots & (\mathbf{y}_m - Z_m\boldsymbol{\delta}_m)'(\mathbf{y}_m - Z_m\boldsymbol{\delta}_m) \end{bmatrix}. \tag{9.19}$$

The posterior pdf in (9.18) is in exactly the same form as that for the coefficients of the seemingly unrelated regression model. Further work is required to produce useful techniques for obtaining marginal pdf's associated with the pdf in (9.18).

9.3 THE CONCEPT OF IDENTIFICATION IN BAYESIAN ANALYSIS

At this point we turn to a consideration of the concept of identification. At the outset, just as in sampling theory, it is important to emphasize that the identification problem is not peculiar to simultaneous equation models

[7] In addition, the results presented in Sections 9.5 and 9.6 can be specialized to apply to triangular systems.

but arises in connection with all statistical models; that is, in sampling theory terms, if the observations $\mathbf{y}$ are assumed to be generated by a pdf $p(\mathbf{y}|\boldsymbol{\theta})$, where $\boldsymbol{\theta}$ is a parameter vector, we may ask if there is another vector of parameters, say $\boldsymbol{\phi}$, such that $p(\mathbf{y}|\boldsymbol{\theta}) = p(\mathbf{y}|\boldsymbol{\phi})$. If this is the case, it is clear that there will be a problem in deciding from the information in any sample whether the model generating the observations is $p(\mathbf{y}|\boldsymbol{\theta})$ or $p(\mathbf{y}|\boldsymbol{\phi})$; that is there will be a problem in identifying the model generating the observations. In this situation the model is said to be not identified. Equivalently, there will be a problem in determining whether the parameters are $\boldsymbol{\theta}$ or $\boldsymbol{\phi}$; that is, the parameters are not identified. Usually in the sampling theory framework exact restrictions are imposed on the parameters of a model to achieve identification; for example, certain elements of $\boldsymbol{\theta}$ may be assumed to be zero. If we denote the restricted parameter vector by $\boldsymbol{\theta}_r$ and $p(\mathbf{y}|\boldsymbol{\theta}_r) \neq p(\mathbf{y}|\boldsymbol{\phi})$ for all $\boldsymbol{\phi}$ not identical to $\boldsymbol{\theta}_r$, the model $p(\mathbf{y}|\boldsymbol{\theta}_r)$ is identified or, equivalently, the parameter vector $\boldsymbol{\theta}_r$ is identified.

Since prior information in the Bayesian framework need not necessarily take the form of exact restrictions but can be introduced flexibly by use of prior pdf's, there is a need to broaden the concept of identification to allow for the more general kind of prior information used in Bayesian analysis. It is this problem that we now consider.

Suppose that we have two models, say M_1 with its parameter vector $\boldsymbol{\theta}_1$ and prior pdf $p(\boldsymbol{\theta}_1|M_1)$ and M_2 with its parameter vector $\boldsymbol{\theta}_2$ and prior pdf $p(\boldsymbol{\theta}_2|M_2)$. If both models, along with their associated prior information, lead to exactly the same marginal pdf for a set of observations,[8] we shall say, by definition, that M_1 and its associated prior information and M_2 and its associated prior information are observationally equivalent and we are unable to use data to discriminate between them. Alternatively phrased, M_1 with its prior information is not identified in relation to M_2 with its prior information. Explicitly, we have

$$p(\mathbf{y}, \boldsymbol{\theta}_1|M_1) = p(\mathbf{y}|\boldsymbol{\theta}_1, M_1)p(\boldsymbol{\theta}_1|M_1), \tag{9.20}$$

where $\mathbf{y}$ is a vector of observations, $p(\mathbf{y}, \boldsymbol{\theta}_1|M_1)$ is the joint pdf for $\mathbf{y}$ and $\boldsymbol{\theta}_1$, given model M_1, $p(\mathbf{y}|\boldsymbol{\theta}_1, M_1)$ is the pdf for $\mathbf{y}$, given $\boldsymbol{\theta}_1$ and M_1, and $p(\boldsymbol{\theta}_1|M_1)$ is the prior pdf for the parameter vector $\boldsymbol{\theta}_1$ associated with M_1. Then the marginal pdf for $\mathbf{y}$ is

$$p(\mathbf{y}|M_1) = \int p(\mathbf{y}|\boldsymbol{\theta}_1, M_1)p(\boldsymbol{\theta}_1|M_1)\, d\boldsymbol{\theta}_1. \tag{9.21}$$

Similarly, for model M_2 with its associated parameter vector $\boldsymbol{\theta}_2$

$$p(\mathbf{y}|M_2) = \int p(\mathbf{y}|\boldsymbol{\theta}_2, M_2)\, p(\boldsymbol{\theta}_2|M_2)\, d\boldsymbol{\theta}_2. \tag{9.22}$$

[8] See Chapter 2 for a definition of the marginal pdf for the observations.

Then M_1 and M_2 and their respective prior information are defined as observationally equivalent if, and only if,

$$p(\mathbf{y}|M_1) = p(\mathbf{y}|M_2). \tag{9.23}$$

If the condition in (9.23) holds, we cannot decide whether M_1 and the information in $p(\boldsymbol{\theta}_1|M_1)$ or M_2 and the information in $p(\boldsymbol{\theta}_2|M_2)$ "explain" the data. Both have exactly the same implications for the distribution of the observations.

In the special case of simultaneous equation models let M_1 be[9]

$$Y\Gamma_1 = XB_1 + U_1, \tag{9.24}$$

with disturbance covariance matrix Σ_1, and let M_2 be

$$Y\Gamma_1 A = XB_1 A + U_1 A, \tag{9.25}$$

where A is *any* nonsingular matrix, or

$$Y\Gamma_2 = XB_2 + U_2, \tag{9.26}$$

with disturbance covariance matrix Σ_2 and where $\Gamma_2 = \Gamma_1 A$, $B_2 = B_1 A$, $U_2 = U_1 A$, and $\Sigma_2 = A'\Sigma_1 A$. Under these conditions which involve no restrictions on coefficients or covariance matrix elements, we have, as is well known,

$$p(Y|\Theta_1, M_1) \equiv p(Y|\Theta_2, M_2), \tag{9.27}$$

where $\Theta_1 = (\Gamma_1, B_1, \Sigma_1)$ and $\Theta_2 = (\Gamma_2, B_2, \Sigma_2)$.

Given the identity in (9.27), let us see what happens when we use "non-informative" prior pdf's for the parameters and assume that our non-informative or diffuse prior pdf's are chosen to be invariant[10] to the class of

[9] The notation and distributional assumptions introduced in Section 9.1 are employed below.

[10] See H. Jeffreys, *Theory of Probability* (3rd ed.). Oxford: Clarendon, p. 179 ff., and J. Hartigan, "Invariant Prior Distributions," *Ann. Math. Statist.*, **35** (1964), 836–845, and Appendix 2.1, for a discussion of invariance theory and procedures for obtaining invariant noninformative prior pdf's. As an example of invariance, suppose that our prior pdf for Σ_1 is $p(\Sigma_1) \propto |\Sigma_1|^{-(m+1)/2}$, a pdf used above. Now consider $\Sigma_2 = A'\Sigma_1 A$, with A nonsingular, as a transformation of the distinct elements of Σ_1 to those of Σ_2. The Jacobian of this transformation is $|A|^{-(m+1)}$. Thus

$$p(\Sigma_2) \propto |A|^{-(m+1)}|(A')^{-1}\Sigma_2 A^{-1}|^{-(m+1)/2} = |\Sigma_2|^{-(m+1)/2}.$$

It is seen that our noninformative prior pdf for Σ_2 is in exactly the same form as that for Σ_1, a result that follows from the fact that the particular prior pdf employed, namely, $p(\Sigma_1) \propto |\Sigma_1|^{-(m+1)/2}$, is an invariant one provided by Jeffreys' invariance theory. If improper prior pdf's are employed, the integrals in (9.28) and (9.29) will not in general converge indicating that not enough prior information has been introduced to obtain proper marginal pdf's for the observations. Below, we assume a large finite range for the parameters which makes Jeffreys' prior pdf's proper.

transformations being considered above, namely, $\Gamma_2 = \Gamma_1 A$, $B_2 = B_1 A$ and $\Sigma_2 = A'\Sigma_1 A$. In this case our prior pdf for Θ_1 will be in exactly the same form as our prior pdf for Θ_2. Then

$$p(Y|M_1) = \int p(Y|\Theta_1, M_1)\, p(\Theta_1|M_1)\, d\Theta_1 \tag{9.28}$$

and

$$p(Y|M_2) = \int p(Y|\Theta_2, M_2)\, p(\Theta_2|M_2)\, d\Theta_2 \tag{9.29}$$

will be identical in view of the relation in (9.27) and our assumptions about the way in which the noninformative or diffuse prior pdf's were obtained.

Of course, with informative prior pdf's in general, the pdf's in (9.28) and (9.29) will not be identical and we would state that M_1 and its associated prior information are not observationally equivalent to M_2 and its associated prior information. Thus, in connection with the general linear simultaneous equation model, $Y\Gamma = XB + U$, it is the case that prior information *must* be introduced to identify it. On the question of how strongly the prior information identifies M_1 in relation to M_2, we suggest use of the information theory quantity J, the "divergence,"[11] as a possible measure; that is,

$$J = \int [p(Y|M_1) - p(Y|M_2)] \log \left[\frac{p(Y|M_1)}{p(Y|M_2)}\right] dY. \tag{9.30}$$

This approach to the identification problem can be extended easily to apply to posterior pdf's for parameters; for example, if the assumptions that produce equality of (9.28) and (9.29) hold, we shall see that the posterior pdf's for Θ_1 and Θ_2 are indistinguishable. For posterior pdf's we have

$$p(\Theta_1|Y, M_1) = \frac{p(\Theta_1|M_1)\, p(Y|\Theta_1, M_1)}{p(Y|M_1)} \tag{9.31}$$

and

$$p(\Theta_2|Y, M_2) = \frac{p(\Theta_2|M_2)\, p(Y|\Theta_2, M_2)}{p(Y|M_2)}. \tag{9.32}$$

From (9.27), the equality of (9.28) and (9.29), and the fact that $p(\Theta_1|M_1)\, d\Theta_1 = p(\Theta_2|M_2)\, d\Theta_2$, given Jeffreys' invariant prior pdf's, (9.31) and (9.32) are exactly the same with respect to form and distributional parameters. Thus there is no basis for knowing whether the posterior pdf relates to Θ_1 or Θ_2, and this means that there is an identification problem.

We have discussed the identification problem in general terms applicable to a wide range of statistical models, including simultaneous equation models as a special case. Since the identification problem for simultaneous equation models is often considered in terms of the relations between reduced form

[11] See Jeffreys, *ibid.*, pp. 179 ff., and S. Kullback, *Information Theory and Statistics*. New York: Wiley, 1959, pp. 6 ff. for a discussion of the properties of this measure.

parameters and structural parameters, we consider this approach in Bayesian terms by making use of the valuable work of Drèze.[12] The reduced-form system associated with the simultaneous equation model $Y\Gamma = XB + U$ is given by

$$Y = X\Pi + V, \tag{9.33}$$

where $\Pi = B\Gamma^{-1}$ and $V = U\Gamma^{-1}$. Let us denote the reduced-form disturbance matrix by $\Omega = (\Gamma^{-1})'\Sigma\Gamma^{-1}$ and assume initially that Ω is known. Then, viewing Π and Γ as our unknown parameters, we have for the joint posterior pdf

$$p(\Gamma, \Pi | Y) \propto p(\Gamma, \Pi) l(\Pi | Y), \tag{9.34}$$

where $p(\Gamma, \Pi)$ is our prior pdf and $l(\Pi | Y)$ is the likelihood function. From (9.34) it is seen immediately that[13]

$$p(\Gamma | \Pi, Y) \propto p(\Gamma | \Pi); \tag{9.35}$$

that is, that the conditional *posterior* pdf for Γ, given Π, is proportional to the *prior* conditional pdf for Γ, given Π. This result, due to Drèze, shows that since the conditional prior pdf $p(\Gamma | \Pi)$ is unaffected by the observations, the posterior conditional pdf $p(\Gamma | \Pi, Y)$ can be affected only by prior information. It is the prior pdf for Π, $p(\Pi)$ that gets modified by the sample information[14] and, of course, the sample information adds to our knowledge of Π.

In the traditional approach exact restrictions on Γ, B, and Σ are introduced to achieve identification. By limiting ourselves to the case in which exact restrictions are placed on Γ and B we can distinguish three others, namely, (a) too few restrictions to get a unique set of values for the elements of Γ and B, given the elements of Π and the restrictions, the case of "under-identification," (b) the case of just an adequate number of restrictions to achieve unique values for the elements of Γ and B, given Π and the restrictions, the case of "just-identification," and (c) too many restrictions to solve for the elements of Γ and B uniquely, given Π and the restrictions, the case of "overidentification." In Bayesian terms, following Drèze's exposition, with nonstochastic a priori restrictions achieving just-identification, case (b), and the prior on Π, $p(\Pi) \propto$ const, $p(\Gamma | \Pi)$ has all its mass concentrated at a

[12] J. Drèze, "The Bayesian Approach to Simultaneous Equations Estimation," Research Memorandum No. 67, Technological Institute, Northwestern University, 1962.

[13] Note that $p(\Gamma, \Pi | Y) = p(\Gamma | \Pi, Y) p(\Pi | Y)$ and $p(\Gamma, \Pi) = p(\Gamma | \Pi) p(\Pi)$.

[14] If the reduced-form disturbance covariance matrix is unknown and our prior pdf $p(\Gamma, \Pi, \Omega) = p(\Gamma, \Pi)\, p(\Omega)$, the result mentioned above follows immediately; $p(\Omega | \Gamma, \Pi) \neq p(\Omega)$ can be shown to lead to this result if $p(\Omega | \Gamma, \Pi) = p(\Omega | \Pi)$, that is, when a priori Ω is independent of Γ for given Π. Although this will not be true in general, it is true when we have a diffuse prior pdf for Ω.

single point.[15] If one (or more generally r) of these restrictions is dropped, we have (a), underidentification, in which case $p(\Gamma|\Pi)$ would not have its mass concentrated at a single point but rather uniformly spread out over a line, or more generally an r-dimensional hyperplane, and there would be one, or r dimensions, in which the posterior pdf $p(\Gamma, \Pi|Y)$ or $p(\Gamma, B|Y)$ would become uniform. Finally, if we add one (or r) additional a priori restrictions, the prior pdf $p(\Pi)$ could not be uniform because not all values of Π are free, given the prior exact restrictions. Examples illustrating aspects of these cases are provided below.

9.4 ANALYSIS OF PARTICULAR SIMULTANEOUS EQUATION MODELS

The first model to be analyzed is the simple Haavelmo consumption model,[16]

$$\left.\begin{aligned} c_t &= \beta + \alpha y_t + u_t \quad (9.36) \\ y_t &= c_t + z_t \quad (9.37) \end{aligned}\right\} \quad t = 1, 2, \ldots, T,$$

where c_t and y_t are endogenous variables—per capita price-deflated consumption and disposable income, respectively—z_t is an exogenous variable, "autonomous expenditures," and u_t is a disturbance term. We assume that the u_t's have zero means and common variance σ^2 and are normally and independently distributed. With these assumptions, the likelihood function for the model is[17]

$$l(\alpha, \beta, \sigma|\text{data}) \propto \frac{|1 - \alpha|^T}{\sigma^T} \exp\left[-\frac{1}{2\sigma^2}\sum (c_t - \beta - \alpha y_t)^2\right] \tag{9.38}$$

with the summation extending from $t = 1$ to $t = T$. As prior assumptions, we shall go forward with the following prior pdf:

$$p(\alpha, \beta, \sigma) \propto \frac{1}{\sigma} \tag{9.39}$$

[15] Note that linear restrictions on Γ and B imply a set of bilinear restrictions on Γ and Π, since $\Pi = B\Gamma^{-1}$. Thus we can carry forward the discussion in terms of Γ and Π rather than in terms of Γ and B.

[16] T. Haavelmo, "Methods of Measuring the Marginal Propensity to Consume," *Journal of the American Statistical Association*, **42**, 105–122 (1947), reprinted as Chapter 4 in Wm. C. Hood and T. C. Koopmans, *Studies in Econometric Method*, New York: Wiley, 1953. This model has been analyzed from the Bayesian point of view by T. J. Rothenberg, "A Bayesian Analysis of Simultaneous Equation Systems," *cit. supra*, and by V. K. Chetty, "Bayesian Analysis of Some Simultaneous Equation Models and Specification Errors," unpublished doctoral dissertation, University of Wisconsin, Madison, 1966, and "Bayesian Analysis of Haavelmo's Models," *Econometrica*, **36**, 582–602 (1968).

[17] From (9.41) and (9.42) we have $c_t = \beta + \alpha(c_t + z_t) + u_t$. Thus the Jacobian of the transformation from each u_t to each c_t is $|1 - \alpha|$. Since there are T such transformations, the factor $|1 - \alpha|^T$ appears in (9.38).

where $-\infty < \beta < \infty, 0 < \alpha < 1$, and $0 < \sigma < \infty$. Note that we are asserting a priori that α, β, and $\log \sigma$ are uniformly and independently distributed and are putting in the information that α, the marginal propensity to consume, is restricted to the interval 0 to 1. This is an example of how inequality constraints can be introduced into an analysis.[18]

Combining the likelihood function in (9.38) with the prior pdf in (9.39), we obtain the posterior pdf for the parameters:

$$(9.40)\qquad p(\alpha, \beta, \sigma|\text{data}) \propto \frac{(1-\alpha)^T}{\sigma^{T+1}} \exp\left[-\frac{1}{2\sigma^2}\sum (c_t - \beta - \alpha y_t)^2\right].$$

If interest centers on making inferences about the marginal propensity to consume, α, we can integrate (9.40) with respect to β and σ to obtain the marginal posterior pdf for α, which is[19]

$$(9.41)\qquad \begin{aligned} p(\alpha|\text{data}) &\propto \frac{(1-\alpha)^T}{\{\sum [c_t - \bar{c} - \alpha(y_t - \bar{y})]^2\}^{(T-1)/2}} \\ &\propto \frac{(1-\alpha)^T}{\{\nu s^2 + (\alpha - \hat{\alpha})^2 \sum (y_t - \bar{y})^2\}^{(T-1)/2}}, \qquad 0 < \alpha < 1, \end{aligned}$$

where $\bar{c} = T^{-1}\sum c_t$, $\bar{y} = T^{-1}\sum y_t$, $\nu = T - 2$, $\nu s^2 = \sum [c_t - \bar{c} - \hat{\alpha}(y_t - \bar{y})]^2$, and

$$(9.42)\qquad \hat{\alpha} = \frac{\sum (c_t - \bar{c})(y_t - \bar{y})}{\sum (y_t - \bar{y})^2},$$

the least squares quantity obtained from a regression of c_t on y_t.

On viewing (9.41), we see that it is the product of one factor in a truncated univariate Student t form centered at the least squares quantity $\hat{\alpha}$, given in (9.42), if $0 < \hat{\alpha} < 1$, and a second Jacobian factor $(1 - \alpha)^T$. The factor $(1 - \alpha)^T$ pushes the center of the posterior pdf toward zero and thus may be interpreted roughly in sampling theory terms as compensating for Haavelmo's finding that plim $\hat{\alpha} > \alpha$, an interpretation put forward by Rothenberg.

Using Haavelmo's annual data, 1929–1941 ($T = 13$), Chetty computed the posterior pdf in (9.41) with results shown in Figure 9.1.[20] The mean of the

[18] If we wished, we could introduce other than flat prior pdf's for α and $\log \sigma$ without greatly modifying the following analysis; for example, we might use a beta pdf for α and an inverted gamma pdf for σ. See Chetty, *loc. cit.*, for the results of computations that employ a beta pdf for the prior pdf for α, the marginal propensity to consume.

[19] The integration with respect to β is done by completing the square on β in the exponent and using properties of the univariate normal pdf. Then the integration with respect to σ can be performed by using properties of the inverted gamma pdf.

[20] The marginal posterior pdf for β, obtained by a bivariate numerical integration, is also shown in Figure 9.1.

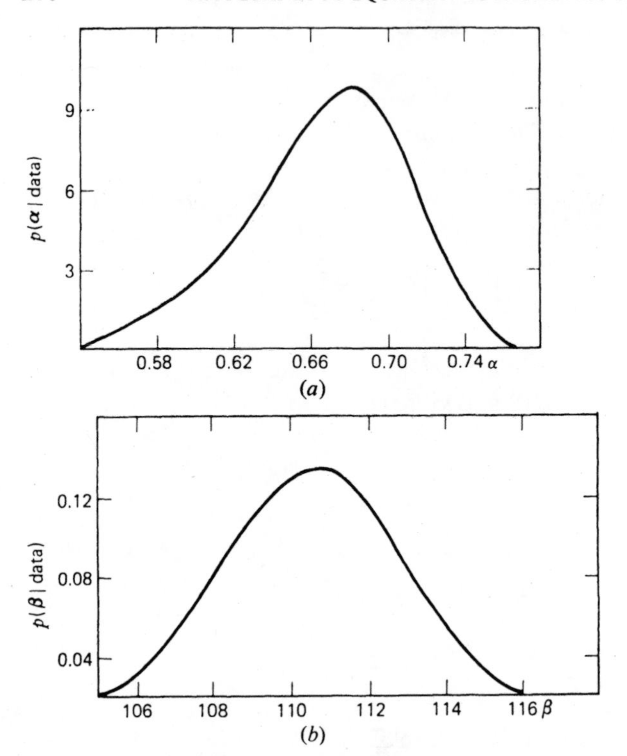

Figure 9.1 Marginal posterior distributions of α and β. (*a*) $E(\alpha) = 0.660$; $E(\alpha - E(\alpha))^2 = 0.004$; $E(\alpha - E(\alpha))^3 = 0.001$; $E(\alpha - E(\alpha))^4 = 0.0002$; (*b*) $E(\beta) = 111.589$; $E(\beta - E(\beta))^2 = 129.642$; $E(\beta - E(\beta))^3 = 71.3034$; $E(\beta - E(\beta))^4 = 49283.172$.

posterior pdf for α was calculated to be 0.660 which is somewhat below $\hat{\alpha}$, the least squares quantity in (9.42), computed by Haavelmo to be 0.732. In addition, since we have the complete posterior pdf for α, posterior probability statements can be made readily. Further, the posterior pdf for the Keynesian "multiplier," $1/(1 - \alpha)$, can be derived from $p(\alpha|\text{data})$ shown in (9.41).[21] This posterior pdf for the multiplier will incorporate the prior information that the marginal propensity to consume is restricted to the range zero to one as well as the sample information, and, of course, if additional prior information about α is introduced, it too will be reflected in the posterior pdf for the multiplier.

[21] Using the above assumptions and assuming that the u_t's are generated by a first order autoregressive model, $u_t = \rho u_{t-1} + \epsilon_t$, Chetty computed conditional posterior pdf's, $p(\alpha|\rho, \text{data})$, and marginal posterior pdf's, $p(\alpha|\text{data})$ and $p(\rho|\text{data})$, using Haavelmo's data. These results indicated that inferences about α were somewhat sensitive to departures of ρ from zero. In fact, the posterior pdf for ρ was highly concentrated about 0.898, its mean.

Just as in current discussions revolving about the Friedman-Meiselman testing of the "income-expenditure" and "quantity-theory" models,[22] Haavelmo in his 1947 paper was concerned about the assumption that z_t in (9.37) is an exogenous variable. To investigate this assumption he formulated two broader models in which the assumption was relaxed. Here we consider an analysis of his third model,[23] whose equations are

$$\left.\begin{aligned} &(9.43) \qquad c_t = \beta + \alpha y_t + u_{1t} \\ &(9.44) \qquad r_t = \nu + \mu(c_t + x_t) + u_{2t} \\ &(9.45) \qquad y_t = c_t + x_t - r_t \end{aligned}\right\} \qquad t = 1, 2, \ldots, T.$$

In this model (9.43) is a consumption function exactly the same as that in his first model. Equation 9.44 is a "gross business saving" equation which relates gross business saving, r_t, to the total "gross disposable income," c_t and x_t, of the private sector in which x_t is gross investment.[24] Finally, Haavelmo considers (9.45) as an accounting identity, assumes that x_t is exogenous, and introduces disturbance terms u_{1t} and u_{2t}. In this system c_t, y_t, and r_t are endogenous variables. Note that by definition we have $z_t = x_t - r_t$, where z_t is the variable appearing in (9.37). Thus, if in (9.44) $\mu = 0$ and u_{1t} and u_{2t} are independently distributed, then z_t would be exogenous, as assumed in the first model. Studying how departures from these assumptions affect inferences about parameters of the model is an excellent way of assessing the implications of the assumption that z_t is exogenous.

If we substitute from (9.45) into (9.43), we obtain

$$\left.\begin{aligned} &(9.46) \qquad c_t = \beta + \alpha(c_t + x_t - r_t) + u_{1t} \\ &(9.47) \qquad r_t = \nu + \mu(c_t + x_t) + u_{2t} \end{aligned}\right\} \qquad t = 1, 2, \ldots, T,$$

as a two equation model in two endogenous variables, c_t and r_t, and one exogenous variable, x_t. We assume that u_{1t} and u_{2t} have a bivariate normal pdf with zero mean vector and covariance matrix Σ, a 2×2 positive definite symmetric matrix. Further the disturbance terms are assumed temporally

[22] M. Friedman and D. Meiselman, "The Relative Stability of Velocity and the Investment Multiplier," in *The Commission on Money and Credit Volume, Stabilization Policies*, Englewood Cliffs, N.J.: Prentice-Hall, 1963.

[23] A brief analysis of his second model appears in A. Zellner, "Bayesian Inference and Simultaneous Equation Econometric Models," *cit. supra*. The analysis of the third model was initiated in the author's lecture notes in 1963 and studied further in V. K. Chetty's work, *cit. supra*.

[24] All variables are price-deflated and on a per capita basis. See Haavelmo's paper for further details of the definitions of the variables and their relation to the national income accounts.

independently distributed. Under these assumptions the likelihood function for the model is given by

$$l(\boldsymbol{\theta}, \Sigma|\text{data}) \propto J|\Sigma^{-1}|^{T/2} \exp(-\tfrac{1}{2} \operatorname{tr} U'U\Sigma^{-1}), \tag{9.48}$$

where J is the Jacobian of the transformation from the u's to the c's and r's:

$$J = |1 - \alpha(1 - \mu)|^T, \tag{9.49}$$

$U = (\mathbf{u}_1\mathbf{u}_2)$ is a $T \times 2$ matrix related to the structural parameters and observations by (9.46) and (9.47), and $\boldsymbol{\theta}' = (\alpha, \beta, \mu, \nu)$ is a vector of structural coefficients.

As prior assumptions we assume not only that $0 < \alpha < 1$ but also that $0 < \mu < 1$, since μ is the marginal propensity to save on the part of the private business sector. For present purposes we also assume that

$$p(\boldsymbol{\theta}, \Sigma^{-1}) \propto |\Sigma^{-1}|^{-(m+1)/2}; \tag{9.50}$$

that is, the elements of $\boldsymbol{\theta}$ are uniformly and independently distributed[25] and we have diffuse prior information about the distinct elements of Σ or, equivalently, about those of Σ^{-1}. With these prior assumptions the joint posterior pdf for the parameters is

$$p(\boldsymbol{\theta}, \Sigma^{-1}|\text{data}) \propto J|\Sigma^{-1}|^{(T-m-1)/2} \exp(-\tfrac{1}{2} \operatorname{tr} U'U\Sigma^{-1}). \tag{9.51}$$

On integrating with respect to Σ^{-1}, we obtain

$$p(\boldsymbol{\theta}|\text{data}) \propto \frac{[1 - \alpha(1 - \mu)]^T}{|U'U|^{T/2}}. \tag{9.52}$$

Then a further integration with respect to the intercept parameters, β and ν, yields[26]

$$p(\alpha, \mu|\text{data}) \propto \frac{[1 - \alpha(1 - \mu)]^T}{|\tilde{U}'\tilde{U}|^{(T-1)/2}}, \qquad 0 < \alpha, \mu < 1, \tag{9.53}$$

[25] The analysis can be carried through with nonuniform prior pdf's for α and μ without much difficulty.

[26] β and ν appear just in the denominator of (9.52). Note that $|U'U| = \mathbf{u}_1'\mathbf{u}_1\mathbf{u}_2'\mathbf{u}_2 - (\mathbf{u}_1'\mathbf{u}_2)^2$. From (9.46), with $w_{1t} \equiv c_t - \alpha(c_t + x_t - r_t)$, we have $u_{1t} = w_{1t} - \bar{w}_1 - (\beta - \bar{w}_1)$, where $\bar{w}_1 = \sum_{t=1}^T w_{1t}/T$. Similarly, from (9.47) with $w_{2t} \equiv r_t - \mu(c_t + x_t)$, we have $u_{2t} = w_{2t} - \bar{w}_2 - (\nu - \bar{w}_2)$, where $\bar{w}_2 = \sum_{t=1}^T w_{2t}/T$. Then straightforward algebra yields

(i) $\mathbf{u}_1'\mathbf{u}_1\mathbf{u}_2'\mathbf{u}_2 - (\mathbf{u}_1'\mathbf{u}_2)^2 = m_{11}m_{22} - m_{12}^2 + Tm_{22}(\beta - \bar{w}_1)^2 + Tm_{11}(\nu - \bar{w}_2)^2 - 2Tm_{12}(\beta - \bar{w}_1)(\nu - \bar{w}_2),$

where $\tilde{U} = (\tilde{\mathbf{u}}_1 \tilde{\mathbf{u}}_2)$ is a $T \times 2$ matrix in which $\tilde{\mathbf{u}}_1$ and $\tilde{\mathbf{u}}_2$ have typical elements given by

$$\tilde{u}_{1t} = c_t - \bar{c} - \alpha[c_t - \bar{c} + x_t - \bar{x} - (r_t - \bar{r})] \tag{9.54}$$

and

$$\tilde{u}_{2t} = r_t - \bar{r} - \mu(c_t - \bar{c} + x_t - \bar{x}),$$

respectively. In the definition of $\tilde{u}_{1t}$ and $\tilde{u}_{2t}$, $\bar{c}$, $\bar{x}$, and $\bar{r}$ are sample means.

Using Haavelmo's data and bivariate numerical integration techniques, we calculated marginal posterior pdf's for α and μ from (9.53). Both marginal pdf's are unimodal. That for α had a mean of 0.705 and a variance of 0.00137, whereas that for μ had a mean of 0.158 and a variance of 0.00050. Shown in Figure 9.2 are the contours of the joint posterior pdf. The results strongly suggest that the parameter μ has a non-zero value.

Also, it should be noted that (9.53) can be employed to study how sensitive inferences about α are to assumptions about μ; that is, if it is assumed that $\mu = \mu_0$, where μ_0 is a given value, numerical methods can be employed to analyze (9.53) under this assumption. Further, if it is assumed that the disturbance terms in (9.46) and (9.47) are uncorrelated ($\sigma_{12} = 0$) and if diffuse prior assumptions are introduced for σ_{11} and σ_{22}, posterior pdf's for α and μ can readily be obtained under the prior assumptions that we have made about other parameters, assumptions that incorporate prior information regarding the ranges of α and μ.

To illustrate some aspects of systems that are "overidentified" in the traditional sense we consider the following simple model:

$$\left.\begin{aligned} y_{1t} &= \gamma y_{2t} + u_{1t} && (9.55)\\ y_{2t} &= \beta_1 x_{1t} + \beta_2 x_{2t} + u_{2t} && (9.56) \end{aligned}\right\} \quad t = 1, 2, \ldots, T,$$

where y_1 and y_2 are endogenous variables, x_1 and x_2 are exogenous variables, γ, β_1, and β_2 are scalar parameters, and u_{1t} and u_{2t} are disturbance terms with covariance matrix Σ, a 2×2 full matrix. The reduced form system is

$$\left.\begin{aligned} y_{1t} &= \pi_{11} x_{1t} + \pi_{12} x_{2t} + v_{1t} && (9.57)\\ y_{2t} &= \pi_{21} x_{1t} + \pi_{22} x_{2t} + v_{2t} && (9.58) \end{aligned}\right\} \quad t = 1, 2, \ldots, T,$$

---

where $m_{ij} = \sum_{t=1}^{T} (w_{it} - \bar{w}_i)(w_{jt} - \bar{w}_j)$, $i, j = 1, 2$, and $\bar{w}_i = \sum w_{it}/T$, $i = 1, 2$. On substituting from (i) in (9.57),

$$p(\boldsymbol{\theta}|\text{data}) \propto \frac{[1 - \alpha(1 - \mu)]^T}{(m_{11} m_{22} - m_{12}{}^2)^{T/2}} \times \left\{\left[1 + T(\beta - \bar{w}_1 \vdots \nu - \bar{w}_2)\begin{pmatrix} m^{11} & m^{12} \\ m^{21} & m^{22} \end{pmatrix}(\beta - \bar{w}_1 \vdots \nu - \bar{w}_2)'\right]^{T/2}\right\}^{-1}.$$

The integration with respect to β and ν may now be performed by using properties of the bivariate Student t pdf to yield (9.53). The posterior pdf given in Chetty, *op. cit.*, p. 593, for this problem is incorrect.

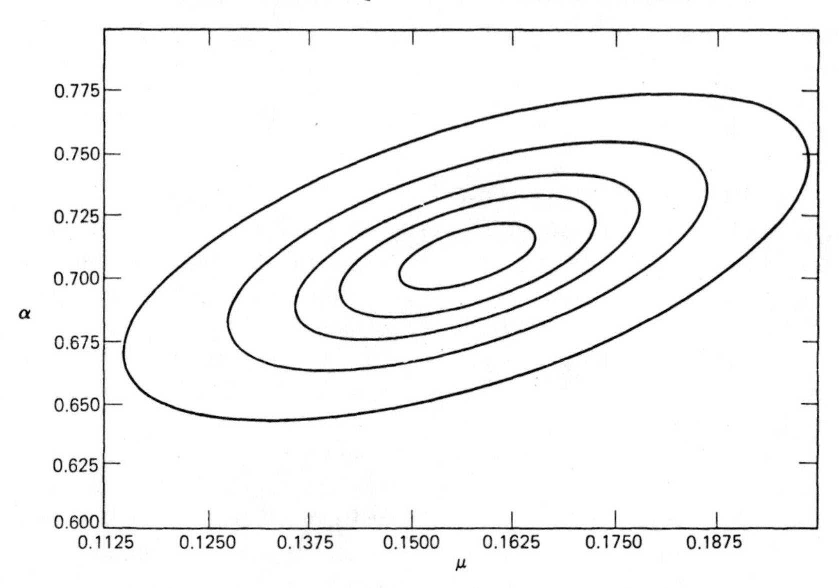

Figure 9.2 Contours of joint posterior pdf for α and μ computed from (9.53).

with

$$\begin{aligned} \pi_{11} &= \gamma\beta_1, & \pi_{12} &= \gamma\beta_2, \\ \pi_{21} &= \beta_1, & \pi_{22} &= \beta_2. \end{aligned} \tag{9.59}$$

It is apparent that the relations in (9.64) imply that $\gamma = \pi_{11}/\pi_{21} = \pi_{12}/\pi_{22}$ or

$$\pi_{11}\pi_{22} - \pi_{12}\pi_{21} = 0, \tag{9.60}$$

which is a condition on the π's. Thus not all four π's are capable of independent variation, given that (9.60) is to be employed as prior information in our analysis. Care must be exercised in the choice of a prior pdf for the π's in this situation. In larger systems restrictions on the structural coefficients usually imply quite a few restrictions on reduced-form cc ficients and should be taken into account in analyzing systems. Of course our prior pdf is in terms of the structural parameters Γ, B, and Σ, an analysis can go forward, as above, without bringing in reduced-form parameters.

In the next section we present Bayesian analogues to single-equation sampling theory estimation approaches. In this single-equation approach we take account only of the a priori identifying information pertaining to the parameters of the equation under consideration. Such analysis will be useful, for example, when we are uncertain about the formulation of some or all of the other structural equations of a model or when we wish to build into our analysis just that part of the identifying prior information relating to the parameters of one equation.

9.5 "LIMITED INFORMATION" BAYESIAN ANALYSIS

Consider a particular structural equation of an m-equation model, say the first:

$$\mathbf{y}_1 - Y_1\boldsymbol{\gamma}_1 = X_1\boldsymbol{\beta}_1 + \mathbf{u}_1, \tag{9.61}$$

where $\mathbf{y}_1$ is an $n \times 1$ vector of observations on the endogenous variable whose coefficient is set equal to one by our normalization, Y_1 is an $n \times m_1$ matrix of observations on m_1 other endogenous variables appearing in the first equation with coefficients not assumed equal to zero, X_1 is an $n \times k_1$ matrix of observations on k_1 predetermined variables appearing in the first equation with coefficients not assumed equal to zero, $\boldsymbol{\gamma}_1$ and $\boldsymbol{\beta}_1$ are $m_1 \times 1$ and $k_1 \times 1$ coefficient vectors, respectively, and $\mathbf{u}_1$ is an $n \times 1$ vector of serially uncorrelated normal disturbance terms, each with mean zero and variance σ_{11}. We assume that the parameters of (9.61) are identified by virtue of the restrictions imposed.[27]

The reduced-form equations for $(\mathbf{y}_1 \vdots Y_1)$ are given by

$$(\mathbf{y}_1 \vdots Y_1) = (X_1 \vdots X_0)\begin{bmatrix} \boldsymbol{\pi}_{11} & \Pi_{01} \\ \boldsymbol{\pi}_{10} & \Pi_{00} \end{bmatrix} + (\mathbf{v}_1 \vdots V_1) \tag{9.62}$$

where $X = (X_1 \vdots X_0)$ is an $n \times k$ matrix of observations with rank k on the k predetermined variables of the system, with X_1, an $n \times k_1$ matrix of observations on the k_1 predetermined variables appearing in (9.66), and X_0, an $n \times k_0$ matrix of observations on k_0 predetermined variables excluded from (9.61) on a priori grounds. The partitioned reduced-form coefficient matrix in (9.62) has elements $\boldsymbol{\pi}_{11}$, a $k_1 \times 1$ vector, $\boldsymbol{\pi}_{10}$, a $k_0 \times 1$ vector, Π_{01}, a $k_1 \times m_1$ matrix, and Π_{00}, a $k_0 \times m_1$ matrix. The $n \times (m_1 + 1)$ matrix $(\mathbf{v}_1 \vdots V_1)$ contains reduced-form disturbance terms, where $\mathbf{v}_1$ is an $n \times 1$ vector and V_1 is an $n \times m_1$ matrix.

On postmultiplying both sides of (9.62) by $(1 \vdots -\boldsymbol{\gamma}_1')'$ and equating the resulting coefficients with those appearing in (9.61), we obtain

$$\boldsymbol{\pi}_{11} - \Pi_{01}\boldsymbol{\gamma}_1 = \boldsymbol{\beta}_1 \tag{9.63}$$

and

$$\boldsymbol{\pi}_{10} - \Pi_{00}\boldsymbol{\gamma}_1 = \boldsymbol{\beta}_0 = \mathbf{0}, \tag{9.64}$$

[27] That is, we can write (9.61) as $\mathbf{y}_1 - Y_1\boldsymbol{\gamma}_1 - Y_0\boldsymbol{\gamma}_0 = X_1\boldsymbol{\beta}_1 + X_0\boldsymbol{\beta}_0 + \mathbf{u}_1$ with $\boldsymbol{\gamma}_0 = \mathbf{0}$ and $\boldsymbol{\beta}_0 = \mathbf{0}$, where $Y = (\mathbf{y}_1 \vdots Y_1 \vdots Y_0)$, an $n \times m$ matrix of observations on all endogenous variables, and $X = (X_1 \vdots X_0)$, an $n \times k$ matrix of observations on all predetermined variables. The identifying restrictions are $\boldsymbol{\gamma}_0 = \mathbf{0}$ and $\boldsymbol{\beta}_0 = \mathbf{0}$.

where $\boldsymbol{\beta}_0$ is assumed a priori to be a zero vector. Using (9.63) and (9.64), we can express the reduced-form equations, given in (9.62), as

$$\begin{aligned}\mathbf{y}_1 &= (X_1\Pi_{01} + X_0\Pi_{00})\boldsymbol{\gamma}_1 + X_1\boldsymbol{\beta}_1 + \mathbf{v}_1 \\ &= X\Pi_0\boldsymbol{\gamma}_1 + X_1\boldsymbol{\beta}_1 + \mathbf{v}_1\end{aligned} \tag{9.65a}$$

and

$$Y_1 = X\Pi_0 + V_1, \tag{9.65b}$$

where $\Pi_0' = (\Pi_{01}' \vdots \Pi_{00}')$. We have brought the reduced-form system (9.62) into the form of (9.65) to illustrate that for given Π_0, say $\Pi_0 = \hat{\Pi}_0 = (X'X)^{-1}X'Y_1$, (9.65*a*) is in the form of a multiple regression model. With a diffuse prior pdf for the elements of $\boldsymbol{\gamma}_1$, $\boldsymbol{\beta}_1$, and ω_{11} the common variance of the elements of $\mathbf{v}_1$, assumed to be normally and independently distributed, each with zero mean, the conditional posterior pdf for $\boldsymbol{\gamma}_1$ and $\boldsymbol{\beta}_1$ is in the multivariate Student *t* form with mean vector given by

$$\tilde{\boldsymbol{\delta}}_1 = \begin{pmatrix} \tilde{\boldsymbol{\gamma}}_1 \\ \text{--} \\ \tilde{\boldsymbol{\beta}}_1 \end{pmatrix} = \begin{pmatrix} \hat{\Pi}_0'X'X\hat{\Pi}_0 & \hat{\Pi}_0'X'X_1 \\ X_1'X\hat{\Pi}_0 & X_1'X_1 \end{pmatrix}^{-1} \begin{pmatrix} \hat{\Pi}_0'X'\mathbf{y}_1 \\ X_1'\mathbf{y}_1 \end{pmatrix}, \tag{9.66}$$

which is just the two-stage least squares (2SLS) estimate. It must be appreciated, however, that $\tilde{\boldsymbol{\delta}}_1$ is a conditional posterior mean, given $\Pi_0 = \hat{\Pi}_0$, and may be a poor approximation to the unconditional posterior mean of $\boldsymbol{\delta}_1$ in small samples. Also, it is interesting to observe that since $\mathbf{y}_1 - \mathbf{v}_1 = X\boldsymbol{\pi}_1$, where $\boldsymbol{\pi}_1' = (\boldsymbol{\pi}_{11}' \vdots \boldsymbol{\pi}_{10}')$, we can write (9.65*a*) as

$$X\boldsymbol{\pi}_1 = (X\Pi_0 \vdots X_1)\begin{pmatrix} \boldsymbol{\gamma}_1 \\ \boldsymbol{\beta}_1 \end{pmatrix}. \tag{9.67}$$

Then on premultiplying both sides of (9.67) by $(X\Pi_0 \vdots X_1)'$ and solving for the coefficient vector, we have

$$\boldsymbol{\delta}_1 = \begin{pmatrix} \boldsymbol{\gamma}_1 \\ \text{--} \\ \boldsymbol{\beta}_1 \end{pmatrix} = \begin{pmatrix} \Pi_0'X'X\Pi_0 & \Pi_0'X'X_1 \\ X_1'X\Pi_0 & X_1'X_1 \end{pmatrix}^{-1} \begin{pmatrix} \Pi_0'X'X\boldsymbol{\pi}_1 \\ X_1'X\boldsymbol{\pi}_1 \end{pmatrix}, \tag{9.68}$$

an algebraic expression for the elements of $\boldsymbol{\delta}_1$ in terms of reduced-form parameters. Further, if we expand the rhs of (9.68) about $\hat{\Pi}_0 = (X'X)^{-1}X'Y_1$ and $\hat{\boldsymbol{\pi}}_1 = (X'X)^{-1}X'\mathbf{y}_1$, the mean of the posterior pdf for the reduced-form parameters if we use a diffuse prior pdf in connection with the system in (9.62),[28] we obtain the following result for the posterior mean of $\boldsymbol{\delta}_1$, which is assumed to exist:

$$E(\boldsymbol{\delta}_1 \mid \mathbf{y}_1, Y_1) = \begin{pmatrix} \hat{\Pi}_0'X'X\hat{\Pi}_0 & \hat{\Pi}_0'X'X_1 \\ X_1'X\hat{\Pi}_0 & X_1'X_1 \end{pmatrix}^{-1} \begin{pmatrix} \hat{\Pi}_0'X'X\hat{\boldsymbol{\pi}}_1 \\ X_1'X\hat{\boldsymbol{\pi}}_1 \end{pmatrix} + \text{remainder}. \tag{9.69}$$

[28] That is, (9.62) is viewed as a normal multivariate regression system, studied in Chapter 8, and a diffuse prior pdf for the elements of $(\boldsymbol{\pi}_1, \Pi_0)$, and the disturbance covariance matrix Ω in the form of (8.7–8.9) is used.

The zeroth-order term in this expansion is precisely the 2SLS estimate. It is not known how well the 2SLS estimate approximates the posterior mean of $\boldsymbol{\delta}_1' = (\boldsymbol{\gamma}_1'\boldsymbol{\beta}_1')$ when our prior information about $\boldsymbol{\delta}_1$ is diffuse; however, it may be conjectured that neglect of the remainder in (9.69) can yield a poor approximation when the sample size is small.

To provide an explicit posterior pdf for the parameters of (9.61) we employ an approach similar in certain respects to that put forward by Drèze.[29] Combining (9.61) and that part of (9.62) relating to Y_1, we have

$$(9.70)\qquad (\mathbf{y}_1 - Y_1\boldsymbol{\gamma}_1 \vdots Y_1) = (X_1 \vdots X_0)\begin{pmatrix} \boldsymbol{\beta}_1 & \Pi_{01} \\ \boldsymbol{\beta}_0 & \Pi_{00} \end{pmatrix} + (\mathbf{u}_1 \vdots V_1),$$

where we have already incorporated the prior information that certain endogenous variables do not appear in (9.61) and will later introduce $\boldsymbol{\beta}_0 = \mathbf{0}$. Note that the disturbance matrix in (9.70) is given by

$$(9.71)\qquad (\mathbf{u}_1 \vdots V_1) = (\mathbf{v}_1 \vdots V_1)\begin{pmatrix} 1 & \mathbf{0}' \\ -\boldsymbol{\gamma}_1 & I \end{pmatrix},$$

since $\mathbf{u}_1 = \mathbf{v}_1 - V_1\boldsymbol{\gamma}_1$.[30] Under the assumption that the rows of $(\mathbf{v}_1 \vdots V_1)$ are normally and independently distributed, each with zero mean vector and $(m_1 + 1) \times (m_1 + 1)$ pds covariance matrix Ω_1, the covariance matrix for each row of $(\mathbf{u}_1 \vdots V_1)$ is $\Omega_* = \Lambda'\Omega_1\Lambda$, where Λ is the triangular matrix on the rhs of (9.71). Then the likelihood function for the system in (9.70) is[31]

$$(9.72)\qquad l(\boldsymbol{\gamma}_1, \boldsymbol{\beta}, \Pi_0, \Omega_* | \mathbf{y}_1, Y_1) \propto |\Omega_*|^{-n/2} \exp\left[-\tfrac{1}{2}\operatorname{tr}(W - X\Pi_*)'(W - X\Pi_*)\Omega_*^{-1}\right],$$

where $W = (\mathbf{y}_1 - Y_1\boldsymbol{\gamma}_1 \vdots Y_1)$, $\Pi_* = (\boldsymbol{\beta} \vdots \Pi_0)$, and $\boldsymbol{\beta}' = (\boldsymbol{\beta}_1' \vdots \boldsymbol{\beta}_0')$, with the prior information $\boldsymbol{\beta}_0 = \mathbf{0}$ to be incorporated in our prior pdf. Note that for given $\boldsymbol{\gamma}_1$ the likelihood function in (9.72) is in a form encountered earlier in the analysis of the multivariate regression model. The prior pdf we shall employ is given by

$$(9.73)\qquad p(\boldsymbol{\gamma}_1, \boldsymbol{\beta}_1, \Pi_0, \Omega_* | \boldsymbol{\beta}_0 = \mathbf{0}) \propto p_1 |\Omega_*|^{-(m'+1)/2},$$

[29] J. Drèze, "Limited Information Estimation from a Bayesian Viewpoint," Discussion Paper 6816, University of Louvain, 1968. See also A. Zellner, "Bayesian and Non-Bayesian Analyses of Simultaneous Equation Models," paper presented to the Second World Congress of the Econometric Society, Cambridge, 1970.

[30] Note from (9.33) that $U = V\Gamma$. With $U = (\mathbf{u}_1, \mathbf{u}_2, \ldots, \mathbf{u}_m)$, $V = (\mathbf{v}_1 \vdots V_1 \vdots V_0)$, and the first column of Γ given by $(1 \vdots -\boldsymbol{\gamma}_1' \vdots \mathbf{0}')'$ $\mathbf{u}_1 = \mathbf{v}_1 - V_1\boldsymbol{\gamma}_1$, as stated above.

[31] The Jacobian of the transformation from $(\mathbf{u}_1 \vdots V_1)$ in (9.70) to $(\mathbf{y}_1 \vdots Y_1)$ is equal to one.

where $p_1 \equiv p_1(\boldsymbol{\gamma}_1, \boldsymbol{\beta}_1 | \boldsymbol{\beta}_0 = \mathbf{0})$ and $m' = m_1 + 1$. In (9.73) we are assuming that $(\boldsymbol{\gamma}_1, \boldsymbol{\beta}_1)$, Π_0, and Ω_* are independently distributed. The elements of Π_0 are assumed to be uniformly and independently distributed, and for Ω_* we use a diffuse prior pdf in a form introduced in Chapter 8. Then the posterior pdf is

$$(9.74) \quad p(\boldsymbol{\gamma}_1, \boldsymbol{\beta}, \Pi_0, \Omega_* | D, P) \propto p_1 |\Omega_*|^{-(n+m'+1)/2} \times \exp\left[-\tfrac{1}{2}\operatorname{tr}(W - X\Pi_*)'(W - X\Pi_*)\Omega_*^{-1}\right],$$

where D denotes given data and P prior assumptions.

In the analysis of (9.74) we first integrate with respect to the elements of Ω_* and then with respect to those of Π_0, a submatrix of Π_*. The resulting pdf will then be conditionalized to reflect the identifying prior information $\boldsymbol{\beta}_0 = \mathbf{0}$. On integrating (9.74) with respect to the elements of Ω_*, we obtain

$$(9.75) \quad \begin{aligned} p(\boldsymbol{\gamma}_1, \boldsymbol{\beta}, \Pi_0 | D, P) &\propto p_1 |(W - X\Pi_*)'(W - X\Pi_*)|^{-n/2} \\ &\propto p_1 |A + (\Pi_* - \hat{\Pi}_*)'X'X(\Pi_* - \hat{\Pi}_*)|^{-n/2}, \end{aligned}$$

where $\hat{\Pi}_* = (X'X)^{-1}X'W$ and $A = (W - X\hat{\Pi}_*)'(W - X\hat{\Pi}_*)$. Since (9.75) has the form of a generalized Student t pdf (see Appendix B), we can use this fact to integrate it with respect to the elements of Π_0, a submatrix of Π_*, to obtain[32]

$$(9.76) \quad \begin{aligned} p(\boldsymbol{\gamma}_1, \boldsymbol{\beta} | D, P) &\propto p_1 a_{11}{}^{(n-m_1-k)/2}[a_{11} + (\boldsymbol{\beta} - \hat{\boldsymbol{\beta}})'X'X(\boldsymbol{\beta} - \hat{\boldsymbol{\beta}})]^{-(n-m_1)/2} \\ &\propto p_1 a_{11}^{-k/2}\left[1 + (\boldsymbol{\beta} - \hat{\boldsymbol{\beta}})'\frac{X'X}{a_{11}}(\boldsymbol{\beta} - \hat{\boldsymbol{\beta}})\right]^{-(n-m_1)/2}. \end{aligned}$$

In (9.76) $\hat{\boldsymbol{\beta}} = (X'X)^{-1}X'(\mathbf{y}_1 - Y_1\boldsymbol{\gamma}_1)$ and a_{11}, the (1, 1) element of A, is given by

$$(9.77) \quad \begin{aligned} a_{11} &= (\mathbf{y}_1 - Y_1\boldsymbol{\gamma}_1 - X\hat{\boldsymbol{\beta}})'(\mathbf{y}_1 - Y_1\boldsymbol{\gamma}_1 - X\hat{\boldsymbol{\beta}}) \\ &= (\hat{\mathbf{v}}_1 - \hat{V}_1\boldsymbol{\gamma}_1)'(\hat{\mathbf{v}}_1 - \hat{V}_1\boldsymbol{\gamma}_1), \end{aligned}$$

where $(\hat{\mathbf{v}}_1 \vdots \hat{V}_1) = (\mathbf{y}_1 \vdots Y_1) - X(\hat{\boldsymbol{\pi}}_1 \vdots \hat{\Pi}_0)$.

With respect to the quadratic form in $\boldsymbol{\beta}$ in (9.76) we have[33]

$$(\boldsymbol{\beta} - \hat{\boldsymbol{\beta}})'X'X(\boldsymbol{\beta} - \hat{\boldsymbol{\beta}}) = (\boldsymbol{\delta}_1 - \tilde{\boldsymbol{\delta}}_1)'\hat{Z}_1{}'\hat{Z}_1(\boldsymbol{\delta}_1 - \tilde{\boldsymbol{\delta}}_1),$$

[32] Note that

$$\boldsymbol{\beta} - \hat{\boldsymbol{\beta}} = \begin{pmatrix} \boldsymbol{\beta}_1 \\ \boldsymbol{\beta}_0 \end{pmatrix} - \begin{pmatrix} \hat{\boldsymbol{\pi}}_{11} - \hat{\Pi}_{10}\boldsymbol{\gamma}_1 \\ \hat{\boldsymbol{\pi}}_{10} - \hat{\Pi}_{00}\boldsymbol{\gamma}_1 \end{pmatrix} = \begin{pmatrix} \boldsymbol{\beta}_1 - (\hat{\boldsymbol{\pi}}_{11} - \hat{\Pi}_{10}\boldsymbol{\gamma}_1) \\ \hat{\Pi}_{00}\boldsymbol{\gamma}_1 - \hat{\boldsymbol{\pi}}_{10} \end{pmatrix},$$

where $\boldsymbol{\beta}_0 = \mathbf{0}$, reflects for given $\boldsymbol{\beta}_1$ and $\boldsymbol{\gamma}_1$ the extent to which the restrictions in (9.63–64) are not met. Also, it can be shown that with $\boldsymbol{\beta}_0 = \mathbf{0}$ the values for $\boldsymbol{\gamma}_1$ and $\boldsymbol{\beta}_1$ which minimize $(\boldsymbol{\beta} - \hat{\boldsymbol{\beta}})'X'X(\boldsymbol{\beta} - \hat{\boldsymbol{\beta}})/a_{11}$ are the limited-information maximum likelihood estimates.

[33] This algebraic result is derived as follows:

$$\begin{aligned} X(\boldsymbol{\beta} - \hat{\boldsymbol{\beta}}) &= X_1\boldsymbol{\beta}_1 + X_0\boldsymbol{\beta}_0 - X(X'X)^{-1}X'(\mathbf{y}_1 - Y_1\boldsymbol{\gamma}_1) \\ &= X_1\boldsymbol{\beta}_1 - (\hat{\mathbf{y}}_1 - \hat{Y}_1\boldsymbol{\gamma}_1) \\ &= -(\hat{\mathbf{y}}_1 - \hat{Z}_1\boldsymbol{\delta}_1), \end{aligned}$$

where $\tilde{\boldsymbol{\delta}}_1$ is the 2SLS quantity shown in (9.66). Thus (9.76) can be expressed as

$$(9.78) \qquad p(\boldsymbol{\delta}_1|D,P) \propto p_1 a_{11}^{-k/2}[1 + (\boldsymbol{\delta}_1 - \tilde{\boldsymbol{\delta}}_1)'H(\boldsymbol{\delta}_1 - \tilde{\boldsymbol{\delta}}_1)]^{-(n-m_1)/2},$$

where $H = \hat{Z}_1'\hat{Z}_1/a_{11}$.

If in (9.78) we take $p_1 \equiv p_1(\boldsymbol{\gamma}_1, \boldsymbol{\beta}_1|\boldsymbol{\beta}_0 = \mathbf{0}) \propto$ const, the posterior pdf for $\boldsymbol{\delta}_1$ would be in the multivariate Student t form centered at the quantity $\tilde{\boldsymbol{\delta}}_1$, were it not for the fact that a_{11} depends on $\boldsymbol{\gamma}_1$.[34] However, since the conditional posterior pdf for $\boldsymbol{\beta}_1$, given $\boldsymbol{\gamma}_1$, is in the multivariate Student t form, (9.78) can be integrated with respect to the k_1 elements of $\boldsymbol{\beta}_1$ to yield the following marginal posterior pdf for $\boldsymbol{\gamma}_1$

$$(9.79) \quad p(\boldsymbol{\gamma}_1|D,P) \propto a_{11}^{-(k-k_1)/2}[1 + (\boldsymbol{\gamma}_1 - \tilde{\boldsymbol{\gamma}}_1)'H_1(\boldsymbol{\gamma}_1 - \tilde{\boldsymbol{\gamma}}_1)]^{-(\nu+m_1)/2},$$

where $\nu = n - 2m_1 - k_1$ and $H_1 = [\hat{Y}_1'\hat{Y}_1 - \hat{Y}_1'X_1(X_1'X_1)^{-1}X_1'\hat{Y}_1]/a_{11}$. When, as is often the case in applications, $\boldsymbol{\gamma}_1$ has just a small number of elements, say one or two, numerical integration techniques can be employed to evaluate the normalizing constant and to analyze other features of (9.79). With regard to the posterior pdf for an element of $\boldsymbol{\beta}_1$, say β_{11}, (9.78) can be integrated analytically with respect to the elements of $\boldsymbol{\beta}_1$, other than β_{11}, to provide the joint posterior pdf for $\boldsymbol{\gamma}_1$ and β_{11} which can be analyzed

---

where $\hat{\mathbf{y}}_1 = X(X'X)^{-1}X'\mathbf{y}_1$, $\hat{Y}_1 = X(X'X)^{-1}X'Y_1 = X\hat{\Pi}_0$, $\hat{Z}_1 = (\hat{Y}_1 \vdots X_1)$, and $\boldsymbol{\beta}_0 = \mathbf{0}$ have been used. Then $(\boldsymbol{\beta} - \hat{\boldsymbol{\beta}})'X'X(\boldsymbol{\beta} - \hat{\boldsymbol{\beta}}) = (\hat{\mathbf{y}}_1 - \hat{Z}_1\boldsymbol{\delta}_1)'(\mathbf{y}_1 - \hat{Z}_1\boldsymbol{\delta}_1)$ and

$$\begin{aligned}(\hat{\mathbf{y}}_1 - \hat{Z}_1\boldsymbol{\delta}_1)'(\hat{\mathbf{y}}_1 - \hat{Z}_1\boldsymbol{\delta}_1) &= [\hat{\mathbf{y}}_1 - \hat{Z}_1\tilde{\boldsymbol{\delta}}_1 - \hat{Z}_1(\boldsymbol{\delta}_1 - \tilde{\boldsymbol{\delta}}_1)]'[\hat{\mathbf{y}}_1 - \hat{Z}_1\tilde{\boldsymbol{\delta}}_1 - \hat{Z}_1(\boldsymbol{\delta}_1 - \tilde{\boldsymbol{\delta}}_1)] \\ &= (\hat{\mathbf{y}}_1 - \hat{Z}_1\tilde{\boldsymbol{\delta}}_1)'(\hat{\mathbf{y}}_1 - \hat{Z}_1\tilde{\boldsymbol{\delta}}_1) + (\boldsymbol{\delta}_1 - \tilde{\boldsymbol{\delta}}_1)'\hat{Z}_1'\hat{Z}_1(\boldsymbol{\delta}_1 - \tilde{\boldsymbol{\delta}}_1) \\ &= (\boldsymbol{\delta}_1 - \tilde{\boldsymbol{\delta}}_1)'\hat{Z}_1'\hat{Z}_1(\boldsymbol{\delta}_1 - \tilde{\boldsymbol{\delta}}_1),\end{aligned}$$

since $\hat{Z}_1'(\hat{\mathbf{y}}_1 - \hat{Z}_1\tilde{\boldsymbol{\delta}}_1) = 0$ from (9.66) and $\hat{\mathbf{y}}_1 - \hat{Z}_1\tilde{\boldsymbol{\delta}}_1 = \hat{\mathbf{y}}_1 - \hat{Y}_1\tilde{\boldsymbol{\gamma}}_1 - X_1\tilde{\boldsymbol{\beta}}_1 = \mathbf{y}_1 - Y_1\tilde{\boldsymbol{\gamma}}_1 - X_1\tilde{\boldsymbol{\beta}}_1 - (\hat{\mathbf{v}}_1 - \hat{V}_1\tilde{\boldsymbol{\gamma}}_1) = \hat{\mathbf{u}}_1 - (\hat{\mathbf{v}}_1 - \hat{V}_1\tilde{\boldsymbol{\gamma}}_1) = 0$, since $\hat{\mathbf{u}}_1 = \mathbf{y}_1 - Y_1\tilde{\boldsymbol{\gamma}}_1 - X_1\tilde{\boldsymbol{\beta}}_1$, the 2SLS structural residual vector, equals $\hat{\mathbf{v}}_1 - \hat{V}_1\tilde{\boldsymbol{\gamma}}_1$.

[34] It is not obvious that (9.79) is a proper pdf. If we write it as

$$\frac{a_{11}^{(n-m_1-k)/2}}{[a_{11} + (\boldsymbol{\gamma}_1 - \tilde{\boldsymbol{\gamma}}_1)'H_2(\boldsymbol{\gamma}_1 - \tilde{\boldsymbol{\gamma}}_1)]^{(\nu+m_1)/2}},$$

where $H_2 = a_{11}H_1$, and note that a_{11} is a quadratic form in the elements of $\boldsymbol{\gamma}_1$, then for (9.78) to integrate to a constant we need the existence of the $(n - m_1 - k)$th moment of the elements of $\boldsymbol{\gamma}_1$. The $(n - m_1 - k)$th moment will exist if $\nu > n - m_1 - k$ or $k - k_1 > m_1$. Since $k - k_1$ is the number of predetermined variables excluded from the first structural equation by the identifying restriction $\boldsymbol{\beta}_0 = \mathbf{0}$ and $m_1 + 1$ is the number of endogenous variables assumed to appear in the first structural equation, the condition $k - k_1 > m_1$ is formally the same as the usual "order condition" for identifiability. Further, the condition that H_2 be pds can be shown to be formally equivalent to the usual "rank condition" for identifiability.

conveniently by using numerical integration techniques when the dimensionality of $\boldsymbol{\gamma}_1$ is small.[35]

If the prior pdf p_1 in (9.78) implies a multivariate normal prior pdf for the elements of $\boldsymbol{\delta}_1$, the posterior pdf is in the normal-t form encountered in Chapter 4, except for the dependence of a_{11} on $\boldsymbol{\gamma}_1$. The asymptotic expansion technique, explained in Appendix 4.2 along with expansions of negative powers of a_{11}, appears to be an approach to an approximation of the posterior pdf which is useful; however, the details of this approach have not yet been fully worked out.

9.6 FULL SYSTEM ANALYSIS

In contrast to the approach presented in Section 9.5 which led to the posterior pdf for the parameters of a single equation, using just the identifying prior information for those parameters, we now concentrate attention on the problem of deriving the joint posterior pdf for the parameters appearing in all structural equations, a joint posterior pdf that will incorporate the prior identifying information for all parameters of an m equation system, $Y\Gamma = XB + U$. Having the joint posterior pdf for the unrestricted elements of Γ and B, the marginal pdf for the parameters of a single equation can be obtained. Since this full-system marginal posterior pdf incorporates more prior and sample information than the corresponding single-equation marginal posterior pdf for an equation's parameters, it will in general exhibit less dispersion than the corresponding single equation posterior pdf.[36]

If our identifying information is in the form of exact zero restrictions on elements of Γ and B in $Y\Gamma = XB + U$ and our prior pdf for the remaining parameters is not dogmatic or degenerate, then in large-sized samples the likelihood function will approximate the posterior pdf, as explained in general in Chapter 2. Since the likelihood function assumes a normal form in large samples, the large-sample posterior pdf for the unrestricted parameters then takes the multivariate normal form which has a mean vector with elements that are the "full information" maximum likelihood (FIML) estimates and

[35] Another approach to the analysis of (9.78) is the development of an asymptotic expansion along the lines explained in Appendix 4.2. In this approach the fact that a_{11} depends on $\boldsymbol{\gamma}_1$ must be taken into account. It is useful to write $\hat{\mathbf{v}}_1 - \hat{V}_1\boldsymbol{\gamma}_1 = \hat{\mathbf{v}}_1 - \hat{V}_1\tilde{\boldsymbol{\gamma}}_1 - \hat{V}_1(\boldsymbol{\gamma}_1 - \tilde{\boldsymbol{\gamma}}_1) = \hat{\mathbf{u}}_1 - \hat{V}_1(\boldsymbol{\gamma}_1 - \tilde{\boldsymbol{\gamma}}_1)$ and to write $a_{11} = \hat{\mathbf{u}}_1'\hat{\mathbf{u}}_1[1 + \Delta]$, where $\Delta = [-2\hat{\mathbf{u}}_1'\hat{V}_1(\boldsymbol{\gamma}_1 - \tilde{\boldsymbol{\gamma}}_1) + (\boldsymbol{\gamma}_1 - \tilde{\boldsymbol{\gamma}}_1)'\hat{V}_1'\hat{V}_1(\boldsymbol{\gamma}_1 - \tilde{\boldsymbol{\gamma}}_1)]/\hat{\mathbf{u}}_1'\hat{\mathbf{u}}_1$. Then $a_{11}^{-(k-k_1)/2}$ and a_{11}^{-1} can be expanded in a series involving powers of Δ. The leading normal term in the asymptotic series has mean $\tilde{\boldsymbol{\delta}}_1$ and covariance matrix $(\hat{Z}_1'\hat{Z}_1)^{-1}\hat{\sigma}_{11}$, where $\hat{\sigma}_{11} = \hat{\mathbf{u}}_1'\hat{\mathbf{u}}_1/(n - m_1)$.

[36] This will be true for small and large sample sizes, since the identifying prior information exerts an influence on the posterior pdf for the elements of Γ and B in both small- and large-sample situations.

a covariance matrix equal to the inverse of Fisher's information matrix, evaluated with FIML estimates.[37] Given that the computation of FIML estimates has been discussed in the literature, we shall not pursue this topic here. However, it must be emphasized that this approximation to the posterior pdf is appropriate just for large-sample sizes and not much is known about how large a sample must be in relation to a given model for the normal approximation to be reasonably good.

Another interesting large-sample approximation to the mean of the full-system posterior pdf can be obtained by writing the relation in (9.67) for each equation of an m-equation system; that is

$$\begin{pmatrix} \bar{\mathbf{y}}_1 \\ \bar{\mathbf{y}}_2 \\ \vdots \\ \bar{\mathbf{y}}_m \end{pmatrix} = \begin{pmatrix} \bar{Z}_1 & & & \\ & \bar{Z}_2 & & \\ & & \ddots & \\ & & & \bar{Z}_m \end{pmatrix} \begin{pmatrix} \boldsymbol{\delta}_1 \\ \boldsymbol{\delta}_2 \\ \vdots \\ \boldsymbol{\delta}_m \end{pmatrix} \tag{9.80a}$$

or

$$\bar{\mathbf{y}} = \bar{Z}\boldsymbol{\delta}, \tag{9.80b}$$

where in (9.80a), with $\alpha = 1, 2, \ldots, m$, $\bar{\mathbf{y}}_\alpha = X\boldsymbol{\pi}_\alpha$, the systematic part of the reduced form equation for $\mathbf{y}_\alpha$, $\bar{Z}_\alpha = (\bar{Y}_\alpha \vdots X_\alpha)$, with $\bar{Y}_\alpha$ the systematic part of the reduced form equations for Y_α, a matrix of observations for the endogenous variables assumed to appear in the αth equation with nonzero coefficients, and X_α, the observation matrix for the predetermined variables assumed to appear in the αth equation, with nonzero coefficients, and $\boldsymbol{\delta}_\alpha' = (\boldsymbol{\gamma}_\alpha', \boldsymbol{\beta}_\alpha')$, the coefficient vector for the αth equation. In (9.80b) $\bar{\mathbf{y}}' = (\bar{\mathbf{y}}_1', \bar{\mathbf{y}}_2', \ldots, \bar{\mathbf{y}}_m')$, $\bar{Z}$ represents the block diagonal matrix on the rhs of (9.80a), and $\boldsymbol{\delta}' = (\boldsymbol{\delta}_1', \boldsymbol{\delta}_2', \ldots, \boldsymbol{\delta}_m')$. Then, given that H is a square nonsingular matrix, we have $H\bar{\mathbf{y}} = H\bar{Z}\boldsymbol{\delta}$ or

$$\begin{aligned} \boldsymbol{\delta} &= (\bar{Z}'H'H\bar{Z})^{-1}\bar{Z}'H'H\bar{\mathbf{y}} \\ &= [\bar{Z}'(\Sigma^{-1} \otimes I_n)\bar{Z}]^{-1}\bar{Z}'(\Sigma^{-1} \otimes I_n)\bar{\mathbf{y}}, \end{aligned} \tag{9.81}$$

if H is taken so that $H'H = \Sigma^{-1} \otimes I_n$. Since (9.81) is an algebraic relation connecting elements of $\boldsymbol{\delta}$ with elements of Π, the reduced-form coefficient matrix, and of Σ, the structural-disturbance covariance matrix, we can approximate the posterior mean of $\boldsymbol{\delta}$, $E\boldsymbol{\delta}$, assumed to exist, by expanding the rhs of (9.81) around consistent sample estimates of Σ, $\bar{Z}$, and $\bar{\mathbf{y}}$, say $\hat{\Sigma}$, $\hat{Z}$, and

[37] See, for example, T. J. Rothenberg and C. T. Leenders, "Efficient Estimation of Simultaneous Equation Systems," *Econometrica*, **32**, 57–76 (1964), for an explicit evaluation of the information matrix.

$\hat{\mathbf{y}}$, respectively.[38] Then the zeroth-order term in the expansion yields the following approximation:

$$E\boldsymbol{\delta} \doteq [\hat{Z}'(\hat{\Sigma}^{-1} \otimes I_n)\hat{Z}]^{-1}\hat{Z}'(\hat{\Sigma}^{-1} \otimes I_n)\mathbf{y}. \tag{9.82}$$

It will be noted that the rhs of (9.82) is in the form of the three-stage least squares estimate (3SLS).[39] Below we show that the leading normal term in an asymptotic expansion approximating the posterior pdf is centered at (9.82) with covariance matrix $[\hat{Z}'(\hat{\Sigma}^{-1} \otimes I_n)\hat{Z}]^{-1}$. It must be remembered, however, that these are large-sample approximations and that little is known about how good these approximations are for a given model when the sample size is not large.

We now turn to an explicit derivation of the posterior pdf for the parameters of the model $Y\Gamma = XB + U$. Under the assumption that the rows of U are normally and independently distributed, each with zero mean vector and $m \times m$ pds covariance matrix Σ, the likelihood function is[40]

$$l(\Gamma, B, \Sigma | D) \propto |\Sigma|^{-n/2}\, |\Gamma|^n \exp\left[-\tfrac{1}{2} \operatorname{tr}(Y\Gamma - XB)'(Y\Gamma - XB)\Sigma^{-1}\right] \tag{9.83}$$

where D denotes given data. Our prior pdf for the parameters is

$$p(\Gamma, B, \Sigma) \propto p_1(\Gamma, B)|\Sigma|^{-(m+1)/2}. \tag{9.84}$$

In (9.84) we assume that the elements of Σ are a priori independent of those of Γ and B and use a diffuse prior pdf in a form employed and explained in Chapter 8. The prior pdf for Γ and B, $p_1(\Gamma, B)$, is assumed to incorporate identifying prior information which, as indicated above, must be provided to estimate the model's parameters. Then the posterior pdf is given by

$$\begin{aligned} p(\Gamma, B, \Sigma | D, P) &\propto p_1(\Gamma, B)|\Sigma|^{-(n+m+1)/2}|\Gamma|^n \\ &\quad \times \exp\left[-\tfrac{1}{2} \operatorname{tr}(Y\Gamma - XB)'(Y\Gamma - XB)\Sigma^{-1}\right] \\ &\propto p_1(\Gamma, B)|\Sigma|^{-(n+m+1)/2}|\Gamma'\hat{\Omega}\Gamma|^{n/2} \\ &\quad \times \exp\{-\tfrac{1}{2} \operatorname{tr}[n\Gamma'\hat{\Omega}\Gamma + (B - \hat{B})'X'X(B - \hat{B})]\Sigma^{-1}\}, \end{aligned} \tag{9.85}$$

where P denotes prior assumptions, $\hat{B} = (X'X)^{-1}X'Y\Gamma = \hat{\Pi}\Gamma$, and $n\hat{\Omega} = (Y - X\hat{\Pi})'(Y - X\hat{\Pi}) = \hat{V}'\hat{V}$.

[38] For example, we could take $\hat{\Sigma} = \hat{U}'\hat{U}/T$, where $\hat{U}$ is a matrix of 2SLS structural residuals and use elements of $\hat{\Pi} = (X'X)^{-1}X'Y$ to obtain $\hat{Z}$ and $\hat{\mathbf{y}}$. Alternatively, FIML estimates of Σ, $\bar{Z}$, and $\bar{y}$ could be employed.

[39] See A. Zellner and H. Theil, "Three-Stage Least Squares: Simultaneous Estimation of Simultaneous Equations," *Econometrica*, **30**, 54–78 (1962).

[40] Note that $|\Gamma|^n$ appears in (9.83), since the Jacobian of the transformation from each row of U to the corresponding row of Y is $|\Gamma|$. Since there are n rows in U and in Y, $|\Gamma|^n$ appears in (9.83). Also, it is understood that the Jacobian factor is taken to be positive.

On integrating (9.85) with respect to the elements of Σ, we find that the marginal posterior pdf for Γ and B is

$$(9.86) \qquad p(\Gamma, B|D, P) \propto \frac{p_1(\Gamma, B)|\Gamma'\hat{\Omega}\Gamma|^{n/2}}{|n\Gamma'\hat{\Omega}\Gamma + (B - \hat{B})'X'X(B - \hat{B})|^{n/2}}.$$

From the form of (9.86) we see that, if the conditional prior pdf for B, given Γ, $p_2(B|\Gamma) \propto$ const, the conditional posterior pdf for the elements of B, given Γ, is in the generalized Student t form with mean $\hat{B} = (X'X)^{-1}X'Y\Gamma$, which is to be expected, since $Y\Gamma = XB + U$ for given Γ is, as pointed out above, in the form of a multivariate regression system. Further analysis of (9.86) for a particular choice of $p_1(\Gamma, B)$ appears possible but has not yet been carried through.[41]

If we use the relation $\Sigma = \Gamma'\Omega\Gamma$, we can rewrite the second line of (9.85) in terms of Γ, B, and Ω as follows:

$$(9.87) \qquad \begin{aligned} p(\Gamma, B, \Omega|D, P) \propto\, & p_1(\Gamma, B)|\Omega|^{-(n+m+1)/2} \exp\left(-\frac{n}{2}\operatorname{tr}\hat{\Omega}\Omega^{-1}\right) \\ & \times \exp\left[-\tfrac{1}{2}\operatorname{tr}(\hat{Y}\Gamma - XB)'(\hat{Y}\Gamma - XB)(\Gamma'\Omega\Gamma)^{-1}\right], \end{aligned}$$

since $(B - \hat{B})'X'X(B - \hat{B}) = (XB - X\hat{\Pi}\Gamma)'(XB - X\hat{\Pi}\Gamma) = (\hat{Y}\Gamma - XB)' \times (\hat{Y}\Gamma - XB)$, with $\hat{Y} = X\hat{\Pi}$. Further, we assume that our prior pdf for Γ and B, $p_1(\Gamma, B)$, incorporates the prior identifying information that certain elements of (Γ, B) are equal to zero. If we denote the nonzero elements of (Γ, B), other than the m elements of Γ set equal to one in our normalization, by $\boldsymbol{\delta}' = (\boldsymbol{\delta}_1', \boldsymbol{\delta}_2', \ldots, \boldsymbol{\delta}_m')$, with $\boldsymbol{\delta}_\alpha' = (\boldsymbol{\gamma}_\alpha', \boldsymbol{\beta}_\alpha')$, $\alpha = 1, 2, \ldots, m$, and let $p(\boldsymbol{\delta})$ be the prior pdf, the posterior pdf can be written as

$$(9.88) \qquad \begin{aligned} p(\boldsymbol{\delta}, \Omega|D, P) \propto\, & p(\boldsymbol{\delta})|\Omega|^{-(n+m+1)/2} \exp\left(-\frac{n}{2}\operatorname{tr}\hat{\Omega}\Omega^{-1}\right) \\ & \times \exp\left[-\tfrac{1}{2}\operatorname{tr}(\hat{Y}\Gamma_r - XB_r)'(\hat{Y}\Gamma_r - XB_r)(\Gamma_r'\Omega\Gamma_r)^{-1}\right], \end{aligned}$$

where (Γ_r, B_r) represents (Γ, B) with identifying and normalizing conditions imposed.

We shall now develop an asymptotic expansion of (9.88). In the last factor on the rhs of (9.88) write

$$(9.89) \qquad \begin{aligned} \Gamma_r'\Omega\Gamma_r &= (\hat{\Gamma}_r + \Delta\Gamma_r)'(\hat{\Omega} + \Delta\Omega)(\hat{\Gamma}_r + \Delta\Gamma_r) \\ &= \hat{\Sigma} + C, \end{aligned}$$

where $\hat{\Gamma}_r$ and $\hat{\Omega}$ are consistent estimates of Γ_r and Ω, respectively, $\Delta\Gamma_r = \Gamma_r - \hat{\Gamma}_r$ and $\Delta\Omega = \Omega - \hat{\Omega}$, both having elements of $0(n^{-1/2})$, $\hat{\Sigma} = \hat{\Gamma}_r'\hat{\Omega}\hat{\Gamma}_r$,

[41] If $p_1(\Gamma, B)$ implies zero identifying restrictions, a normalization rule and a diffuse prior pdf on the remaining parameters, the mode of (9.86) is located close to the FIML estimate.

and $C = \Gamma_r'\Omega\Gamma_r - \hat{\Gamma}_r'\hat{\Omega}\hat{\Gamma}_r$. Then[42]

$$(\Gamma_r'\Omega\Gamma_r)^{-1} = \hat{\Sigma}^{-1} + R, \tag{9.90}$$

where $R = \hat{\Sigma}^{-1}C\hat{\Sigma}^{-1} + \hat{\Sigma}^{-1}C\hat{\Sigma}^{-1}C\hat{\Sigma}^{-1} - \cdots$. On substituting from (9.90) in the second exponential factor in (9.88), we have

$$\begin{aligned} &\exp[-\tfrac{1}{2}\operatorname{tr}(\hat{Y}\Gamma_r - XB_r)'(\hat{Y}\Gamma_r - XB_r)(\Gamma_r'\Omega\Gamma_r)^{-1}] \\ &= \exp[-\tfrac{1}{2}\operatorname{tr}(\hat{Y}\Gamma_r - XB_r)'(\hat{Y}\Gamma_r - XB_r)\hat{\Sigma}^{-1} - \operatorname{tr} K] \\ &= \exp[-\tfrac{1}{2}(\hat{\mathbf{y}} - \hat{Z}\boldsymbol{\delta})'(\hat{\Sigma}^{-1} \otimes I_n)(\hat{\mathbf{y}} - \hat{Z}\boldsymbol{\delta})]\exp(-\operatorname{tr} K), \end{aligned} \tag{9.91}$$

where $K = (\hat{Y}\Gamma_r - XB_r)'(\hat{Y}\Gamma_r - XB_r)R/2$.[43] Further, if we complete the square on $\boldsymbol{\delta}$ in the last line of (9.91) and expand $\exp(-\operatorname{tr} K)$ as $e^x = 1 + x + x^2/2! + x^3/3! + \cdots$,[44] we obtain (9.91) proportional to

$$\begin{aligned} &\exp[-\tfrac{1}{2}(\boldsymbol{\delta} - \tilde{\boldsymbol{\delta}})'\hat{Z}'(\hat{\Sigma}^{-1} \otimes I_n)\hat{Z}(\boldsymbol{\delta} - \tilde{\boldsymbol{\delta}})] \\ &\qquad \times [1 - \operatorname{tr} K + \tfrac{1}{2}(\operatorname{tr} K)^2 - \tfrac{1}{6}(\operatorname{tr} K)^3 + \cdots], \end{aligned} \tag{9.92}$$

where $\tilde{\boldsymbol{\delta}} = [\hat{Z}'(\hat{\Sigma}^{-1} \otimes I_n)\hat{Z}]^{-1}\hat{Z}'(\hat{\Sigma}^{-1} \otimes I_n)\hat{\mathbf{y}}$. Using (9.92), we find that the posterior pdf in (9.88) takes the following form:

$$\begin{aligned} p(\boldsymbol{\delta}, \Omega | D, P) \propto p(\boldsymbol{\delta})&|\Omega|^{-(n+m+1)/2}\exp\left(-\frac{n}{2}\operatorname{tr}\hat{\Omega}\Omega^{-1}\right) \\ &\times \exp[-\tfrac{1}{2}(\boldsymbol{\delta} - \tilde{\boldsymbol{\delta}})'M(\boldsymbol{\delta} - \tilde{\boldsymbol{\delta}})] \\ &\times [1 - \operatorname{tr} K + \tfrac{1}{2}(\operatorname{tr} K)^2 - \tfrac{1}{6}(\operatorname{tr} K)^3 + \cdots], \end{aligned} \tag{9.93}$$

where $M = \hat{Z}'(\hat{\Sigma}^{-1} \otimes I_n)\hat{Z}$.

[42] Equation 9.90 is obtained by writing the rhs of (9.89) as

$$(\hat{\Sigma} + C)^{-1} = \hat{\Sigma}^{-1}(I_m + C\hat{\Sigma}^{-1})^{-1}$$

and expanding $(I_m + C\hat{\Sigma}^{-1})^{-1}$ in a power series. Note that the elements of C are $0(n^{-1/2})$.

[43] In going from the second to the third line of (9.91)

$$\operatorname{tr}(\hat{Y}\Gamma_r - XB_r)'(\hat{Y}\Gamma_r - XB_r)\hat{\Sigma}^{-1} = (\hat{\mathbf{y}} - \hat{Z}\boldsymbol{\delta})'(\hat{\Sigma}^{-1} \otimes I_n)(\hat{\mathbf{y}} - \hat{Z}\boldsymbol{\delta})$$

has been used where $\hat{\mathbf{y}}$ and $\hat{Z}$ have been defined in connection with (9.81).

[44] In the expression for K the elements of $(\hat{Y}\Gamma_r - XB_r)'(\hat{Y}\Gamma_r - XB_r) = (\hat{\Pi}\Gamma_r - B_r)'X'X \times (\hat{\Pi}\Gamma_r - B_r)$ are $0(1)$, since from $\Pi\Gamma_r = B_r$ we have $\hat{\Pi}\Gamma_r - B_r = -\Delta\Pi\Gamma_r$, where $\Delta\Pi = \Pi - \hat{\Pi}$. Thus $\hat{\Pi}\Gamma_r - B_r$ has elements of $0(n^{-1/2})$ because those of $\Delta\Pi$ are of $0(n^{-1/2})$ and $(\hat{\Pi}\Gamma_r - B_r)'X'X(\hat{\Pi}\Gamma_r - B_r)$ has elements of $0(1)$, given that those of $X'X$ are $0(n)$.

If only the leading term of (9.93) is retained, we have[45]

$$
\begin{aligned}
p(\boldsymbol{\delta}, \Omega \mid D, P) &\propto |\Omega|^{-(n+m+1)/2} \exp\left(-\frac{n}{2} \operatorname{tr} \hat{\Omega}\Omega^{-1}\right) \\
&\quad \times \exp\left[-\tfrac{1}{2}(\boldsymbol{\delta} - \bar{\boldsymbol{\delta}})' M (\boldsymbol{\delta} - \bar{\boldsymbol{\delta}})\right].
\end{aligned} \tag{9.94}
$$

We see that in (9.94) Ω and $\boldsymbol{\delta}$ are independently distributed, with the elements of Ω having an inverted Wishart pdf and those of $\boldsymbol{\delta}$ having a multivariate normal pdf, with mean vector $\bar{\boldsymbol{\delta}} = [\hat{Z}'(\hat{\Sigma}^{-1} \otimes I_n)\hat{Z}]^{-1}\hat{Z}'(\hat{\Sigma}^{-1} \otimes I_n)\hat{\mathbf{y}}$ and covariance matrix $M^{-1} = [\hat{Z}'(\hat{\Sigma}^{-1} \otimes I_n)\hat{Z}]^{-1}$, quantities analogous to the large-sample sampling theory 3SLS estimate and its large-sample covariance matrix estimate.

If, for example, (9.94) is used as a prior pdf in the analysis of another set of data, (Y_a, X_a), which is assumed to satisfy $Y_a\Gamma_r = X_aB_r + U_a$, with the n_a rows of U_a independently and normally distributed with zero mean vector $m \times m$ pds covariance matrix Σ, then, to the order of the approximation involved in (9.94), the posterior pdf is given by

$$
\begin{aligned}
p(\boldsymbol{\delta}, \Omega \mid D, D_a, P) &\propto |\Omega|^{-(n'+m+1)/2} \exp\left(-\frac{n'}{2} \operatorname{tr} \tilde{\Omega}\Omega^{-1}\right) \\
&\quad \times \exp\left[-\tfrac{1}{2}(\boldsymbol{\delta} - \check{\boldsymbol{\delta}})' \tilde{M} (\boldsymbol{\delta} - \check{\boldsymbol{\delta}})\right],
\end{aligned} \tag{9.95}
$$

where D_a denotes the new data, $n' = n + n_a$, $\tilde{\Omega} = (n\hat{\Omega} + n_a\hat{\Omega}_a)/(n + n_a)$, $\hat{\Omega}_a = (Y_a - X_a\hat{\Pi}_a)'(Y_a - X_a\hat{\Pi}_a)$, with $\hat{\Pi}_a = (X_a'X_a)^{-1}X_a'Y_a$,

$$
\check{\boldsymbol{\delta}} = \tilde{M}^{-1}(M\bar{\boldsymbol{\delta}} + M_a\bar{\boldsymbol{\delta}}_a), \tag{9.96}
$$

with $\bar{\boldsymbol{\delta}}_a = M_a^{-1}\hat{Z}_a'(\hat{\Sigma}_a^{-1} \otimes I_{n_a})\hat{\mathbf{y}}_a$, $M_a = \hat{Z}_a'(\hat{\Sigma}_a^{-1} \otimes I_{n_a})\hat{Z}_a$, and

$$
\tilde{M}^{-1} = (M + M_a)^{-1}. \tag{9.97}
$$

The sample quantities $\hat{Z}_a$, $\hat{\mathbf{y}}_a$, and $\hat{\Sigma}_a$ are based on the new data and are defined precisely in the same way as $\hat{Z}$, $\hat{\mathbf{y}}$, and $\hat{\Sigma}$.

From (9.95) and (9.96) we see that to this order of approximation the posterior mean of $\boldsymbol{\delta}$, $\check{\boldsymbol{\delta}}$ in (9.96), is a matrix weighted average of the two sample quantities $\bar{\boldsymbol{\delta}}$ and $\bar{\boldsymbol{\delta}}_a$ with their respective precision matrices, M and M_a, as weights. Also, $\tilde{M}^{-1}$ in (9.97) is the posterior covariance matrix for $\boldsymbol{\delta}$ if we rely on the leading term in the expansion of the posterior pdf.

Since dependence on the leading term of the expansion (9.93) may not be an accurate enough approximation in many circumstances, we now indicate how further terms in the expansion can be taken into account. Let us assume that the prior pdf for $\boldsymbol{\delta}$ in (9.93), $p(\boldsymbol{\delta})$, is given by

$$
p(\boldsymbol{\delta}) \propto \exp\left[-\tfrac{1}{2}(\boldsymbol{\delta} - \bar{\boldsymbol{\delta}})' A (\boldsymbol{\delta} - \bar{\boldsymbol{\delta}})\right], \tag{9.98}
$$

[45] To this order of approximation the prior factor $p(\boldsymbol{\delta})$ must be omitted; that is, if we expand $p(\boldsymbol{\delta})$ about $\bar{\boldsymbol{\delta}}$, $p(\boldsymbol{\delta}) = p(\bar{\boldsymbol{\delta}}) + R'$, where R' is the remainder, the elements of R' are of the same order as terms omitted in (9.94).

where $\bar{\boldsymbol{\delta}}$ is the prior mean vector and A^{-1} is the prior covariance matrix. On inserting this in (9.93) and completing the square on $\boldsymbol{\delta}$ in the exponent, we have

$$
\begin{aligned}
p(\boldsymbol{\delta}, \Omega | D, P) \propto |\Omega|^{-(n+m+1)/2} &\exp\left(-\frac{n}{2}\operatorname{tr}\hat{\Omega}\Omega^{-1}\right) \\
&\times \exp\left[-\tfrac{1}{2}(\boldsymbol{\delta} - \bar{\bar{\boldsymbol{\delta}}})'(M + A)(\boldsymbol{\delta} - \bar{\bar{\boldsymbol{\delta}}})\right] \\
&\times \left[1 - \operatorname{tr} K + \tfrac{1}{2}(\operatorname{tr} K)^2 - \tfrac{1}{6}(\operatorname{tr} K)^3 + \cdots\right],
\end{aligned}
\tag{9.99}
$$

with $\bar{\bar{\boldsymbol{\delta}}} = (M + A)^{-1}(M\hat{\boldsymbol{\delta}} + A\bar{\boldsymbol{\delta}})$. Then, to evaluate the normalizing constant of (9.98), we can integrate with respect to $\boldsymbol{\delta}$ and Ω, taking acccount of the terms involving powers of tr K. These integrations can be viewed as evaluating the expectation of tr K and its powers with respect to the inverted Wishart-multivariate normal pdf for Ω and $\boldsymbol{\delta}$ shown as the leading factor in (9.98). Given that these integrations have been performed,[46] we have the normalization constant for (9.98) and can use this normalized pdf to provide a better approximation to the posterior pdf than that provided by relying on just the leading term.

9.7 RESULTS OF SOME MONTE CARLO EXPERIMENTS

In this section we analyze a simple two-equation simultaneous-equation model from the Bayesian and sampling-theory approaches. The model has the property that almost all sampling-theory approaches, including limited-information maximum likelihood, two-stage least squares, full-information maximum likelihood, three-stage least squares, indirect least squares, etc., yield the same estimators for the coefficients of the model. Thus the experiments described compare the Bayesian approach with almost *all* well-known sampling-theory estimation techniques.

In the experiments samples of data were generated from a two-equation model under known conditions and Bayesian and sampling-theory techniques were applied in the analysis of each generated sample. We then compared the relative performance of the two approaches in a number of trials. The rationale of this procedure is that we envisage investigators using the model on a number of occasions to analyze different sets of data. What we want to measure are various properties of the performance of the Bayesian and sampling-theory approaches in repeated trials.[47] It should be recognized that this criterion of "performance in repeated trials" is one that is commonly

[46] Terms of a given order in n, say $0(n^{-\alpha})$, $\alpha > 0$, are retained, and those of higher order of smallness in n are neglected.

[47] See H. V. Roberts, "Statistical Dogma: One Response to a Challenge," *The American Statistician*, **20**, 25–27 (1966) for further comment on the relevance of Monte Carlo experimental results for appraising alternative approaches.

employed to rationalize sampling-theory approaches. We do not regard it as the ultimate or even the most appropriate for judging alternative approaches. However, it does appear to have some relevance in the sense mentioned above and has certainly been given extremely heavy weight in the sampling-theory literature. With these remarks made, let us turn to the specific details and results of the experiments.

9.7.1 The Model and Its Specifications

The simple model we analyze is

$$\left.\begin{aligned} y_{1t} &= \gamma y_{2t} + u_{1t} \qquad (9.100)\\ y_{2t} &= \beta x_t + u_{2t} \qquad (9.101)\end{aligned}\right\} \quad t = 1, 2, \ldots, T,$$

where y_{1t} and y_{2t} are observations on two endogenous variables, x_t is an observation on an exogenous variable, u_{1t} and u_{2t} are disturbance terms, and γ and β are scalar parameters. The reduced-form equations for this model are

$$\left.\begin{aligned} y_{1t} &= \pi_1 x_t + v_{1t} \qquad (9.102)\\ y_{2t} &= \pi_2 x_t + v_{2t} \qquad (9.103)\end{aligned}\right\} \quad t = 1, 2, \ldots, T,$$

where $\pi_1 = \beta\gamma$, $\pi_2 = \beta$, and v_{1t} and v_{2t} are reduced form disturbance terms.

This is the model we have used to generate data for our Monte Carlo experiments under the conditions summarized in Table 9.1. In all three runs the parameters γ and β were assigned values of 2.0 and 0.5, respectively. In Run I the x_t were obtained by independent drawings from a normal distribution with mean zero and variance one, and in Runs II and III they were similarly obtained from normal distributions with zero means and variances equal to two and nine, respectively. In all runs u_{1t} and u_{2t} had a bivariate

Table 9.1 CONDITIONS UNDER WHICH DATA WERE GENERATED

Run I	Run II	Run III
$\gamma = 2.0$	$\gamma = 2.0$	$\gamma = 2.0$
$\beta = 0.5$	$\beta = 0.5$	$\beta = 0.5$
x_t:NID(0, 1)	x_t:NID(0, 2)	x_t:NID(0, 9)
(u_{1t}, u_{2t}):	(u_{1t}, u_{2t}):	(u_{1t}, u_{2t}):
NID$(0, 0; \sigma_{11}, \sigma_{22}, \sigma_{12})$	$(0, 0; \sigma_{11}, \sigma_{12}, \sigma_{22})$	NID$(0, 0; \sigma_{11}, \sigma_{22}, \sigma_{12})$
$\sigma_{11} = 1.0$	$\sigma_{11} = 1.0$	$\sigma_{11} = 1.0$
$\sigma_{22} = 4.0$	$\sigma_{22} = 4.0$	$\sigma_{22} = 4.0$
$\sigma_{12} = -1.0$	$\sigma_{12} = 1.0$	$\sigma_{12} = 1.0$
$T = 20$; $N = 50$	$T = 20$; $N = 50$	$T = 20$; $N = 50$
$T = 40$; $N = 50$	$T = 40$; $N = 50$	$T = 40$; $N = 50$
$T = 60$; $N = 50$	$T = 60$; $N = 50$	$T = 60$; $N = 50$
$T = 100$; $N = 50$	$T = 100$; $N = 50$	$T = 100$; $N = 50$

normal distribution with zero means and variances equal to 1.0 and 4.0, respectively. In Run I the covariance of u_{1t} and u_{2t}, σ_{12}, was equal to -1.0; thus the correlation between u_{1t} and u_{2t} was $-\frac{1}{2}$. In Runs II and III the covariance σ_{12} was set equal to 1.0; thus the correlation between u_{1t} and u_{2t} on these runs was $\frac{1}{2}$. For all runs we first generated $N = 50$ samples, each of size $T = 20$, then $N = 50$ new samples, each of size $T = 40$, then $N = 50$ new samples, each of size $T = 60$, and finally $N = 50$ new samples, each of size $T = 100$.

9.7.2 Sampling-Theory Analysis of the Model

For most principles of sampling-theory estimation, including maximum-likelihood, two-stage least squares and three-stage least squares, the following is the sampling-theory estimator for γ:

$$\hat{\gamma} = \frac{\sum_{t=1}^{T} \hat{y}_{2t} y_{1t}}{\sum_{t=1}^{T} \hat{y}_{2t}^2} = \frac{\hat{\pi}_1}{\hat{\pi}_2}, \tag{9.104}$$

where $\hat{y}_{2t} = \hat{\pi}_2 x_t$, $\hat{\pi}_2 = \sum_{t=1}^{T} x_t y_{2t} / \sum_{t=1}^{T} x_t^2$, and $\hat{\pi}_1 = \sum_{t=1}^{T} x_t y_{1t} / \sum_{t=1}^{T} x_t^2$. The quantity in (9.104) was computed for each of our samples. Also, in an effort to check the validity of sampling-theory confidence intervals, which are employed frequently in practice, we computed two sets of 95% confidence intervals constructed in the following ways: in the first instance we assumed that $(\hat{\gamma} - \gamma)/s_{\hat{\gamma}}$ is normally distributed where $s_{\hat{\gamma}}^2 = (\sum_{t=1}^{T} \hat{y}_{2t}^2)^{-1} s^2$ with $s^2 = \sum_{t=1}^{T} (y_{1t} - \hat{\gamma} y_{2t})^2/(T-1)$. On this assumption the probability statement we wished to check in our experiments is

$$\Pr\{\hat{\gamma} - 1.96 s_{\hat{\gamma}} \le \gamma \le \hat{\gamma} + 1.96 s_{\hat{\gamma}}\} = 0.95. \tag{9.105}$$

The interval $\hat{\gamma} \pm 1.96 s_{\hat{\gamma}}$ was computed for each sample. Also, we computed an interval based on the assumption that $(\hat{\gamma} - \gamma)/s_{\hat{\gamma}}$ has a univariate Student t distribution with $\nu = T - 1$ degrees of freedom. Here our interval is $\hat{\gamma} \pm t_\nu s_{\hat{\gamma}}$ where t_ν was selected from the t-tables so that

$$\Pr\{\hat{\gamma} - t_\nu s_{\hat{\gamma}} \le \gamma \le \hat{\gamma} + t_\nu s_{\hat{\gamma}}\} = 0.95 \tag{9.106}$$

would be a valid probability statement if, in fact, $(\hat{\gamma} - \gamma)/s_{\hat{\gamma}}$ had a t distribution with $\nu = T - 1$ degrees of freedom, a result that, to the best of our knowledge, has not been established analytically. Since *approximate* intervals of this kind are widely used in econometrics, we think that it would be of interest to study their properties.

9.7.3 Bayesian Analysis of the Model

Viewing (9.102) and (9.103) as a simple bivariate regression model and using the results on the multivariate regression model in Chapter 8, based on diffuse prior distributions for π_1, π_2 and the distinct elements of the reduced form disturbance covariance matrix, we obtain the following posterior pdf for π_1 and π_2:

(9.107)

$$p(\pi_1, \pi_2|\mathbf{y}_1, \mathbf{y}_2) \propto \left[1 + (\pi_1 - \hat{\pi}_1 \vdots \pi_2 - \hat{\pi}_2)\begin{pmatrix} s^{11} & s^{12} \\ s^{21} & s^{22} \end{pmatrix}(\pi_1 - \hat{\pi}_1 \vdots \pi_2 - \hat{\pi}_2)'\right]^{-T/2},$$

a pdf in the bivariate Student t form, where $\hat{\pi}_1$ and $\hat{\pi}_2$ are the least squares quantities and $s^{\alpha l} = \hat{w}^{\alpha l} \sum_{t=1}^{T} x_t^2$, with $\hat{w}^{\alpha l}$ the α, lth element of $[\sum_{t=1}^{T} (y_{\alpha t} - \hat{\pi}_\alpha x_t)(y_{lt} - \hat{\pi}_l x_t)]$ α, $l = 1, 2$. Now introduce the following transformation:

$$\gamma = \frac{\pi_1}{\pi_2}, \qquad \beta = \pi_2,$$

which has Jacobian $|\beta|$. Then the posterior distribution of γ and β is

$$p(\gamma, \beta|\mathbf{y}_1, \mathbf{y}_2) \propto |\beta|[1 + (\beta\gamma - \hat{\pi}_1)^2 s^{11} + (\beta - \hat{\pi}_2)^2 s^{22} + 2(\beta\gamma - \hat{\pi}_1)(\beta - \hat{\pi}_2)s^{12}]^{-T/2}. \tag{9.108}$$

By standard methods we integrated β out of (9.108) to obtain[48]

$$p(\gamma|\mathbf{y}_1, \mathbf{y}_2) \propto \frac{b_0^{-\nu/2}}{b_1^{1/2}}\left\{b_2[1 - 2F(d)] - 2\left(\frac{b_0}{\nu b_1}\right)^{1/2} F_1(d)\right\}, \tag{9.109}$$

where $\nu = T - 1$, $F(d) = \int_{-\infty}^{d} p(t)\,dt$, $F_1(d) = \int_{-\infty}^{d} tp(t)\,dt$, $p(t)$ is the Student-t pdf with ν degrees of freedom, $d = -(\nu b_1/b_0)^{1/2} b_2$,

$$b_1 = \gamma^2 s^{11} + 2\gamma s^{12} + s^{22},$$

$$b_2 = \frac{\gamma\hat{\pi}_1 s^{11} + (\hat{\pi}_1 + \gamma\hat{\pi}_2)s^{12} + \hat{\pi}_2 s^{22}}{b_1},$$

and

$$b_0 = 1 + \hat{\pi}_1^2 s^{11} + 2\hat{\pi}_1\hat{\pi}_2 s^{12} + \hat{\pi}_2^2 s^{22} - \frac{[\gamma\hat{\pi}_1 s^{11} + (\hat{\pi}_1 + \gamma\hat{\pi}_2)s^{12} + \hat{\pi}_2 s^{22}]^2}{b_1}.$$

Using standard numerical techniques and a high-speed electronic computer, we computed the normalizing constant for the posterior pdf shown in (9.109) and the complete posterior pdf for each sample we generated. We also had the computer plot each posterior pdf.[49]

In addition, we have computed 95% Bayesian confidence intervals for γ from each sample of data. One interval was computed such that the area

[48] For further work on the distribution of the ratio of correlated Student-t variables, such as $\gamma = \pi_1/\pi_2$, see S. James Press, "The t-Ratio Distribution," *J. Am. Statist. Assoc.*, **64**, 242–252 (1969).

[49] For 50 samples, each of size 20, these computations and plots, as well as the sampling-theory computations, listings of data generated, and tabulations of results, took about a minute and a half on an IBM 7094 computer.

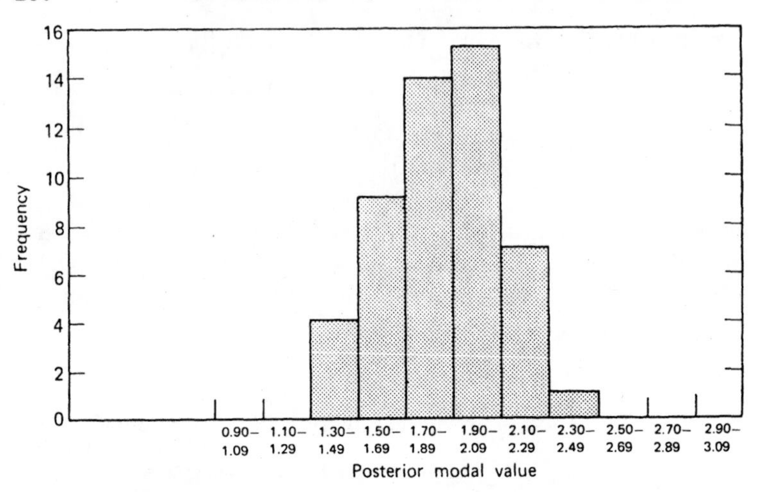

Figure 9.3 Frequency distribution of Bayesian modal values of posterior distributions for γ (Run I: $T = 20$).

under each tail of the posterior pdf was 0.025. We refer to this interval as the "exact central" interval. Another interval, the shortest 95% interval, was also computed for each sample and is referred to as the "exact shortest interval." The sampling performance of these Bayesian intervals is compared with approximate sampling-theory intervals, shown above, in what follows.

9.7.4 Experimental Results: Point Estimates

In Table 9.2 we present the distribution of Bayesian posterior modal values and sampling-theory estimates for Run I.[50] For the 50 samples each of size $T = 20$, the distributions of estimates of γ, true value equal to 2.0, are presented in columns (2) and (3) of the table. It is seen that the distribution of Bayesian posterior modal values has a well-defined mode at the interval 1.900 to 2.099, which embraces the true value 2.0, whereas the distribution of sampling-theory estimates in column (3) is almost rectangular over the interval 1.300 to 2.499 (see Figures 9.3 and 9.4). Note also that two of the sampling-theory estimates are negative for this run with $T = 20$. As T is increased to 40, 60, and 100, we note from columns (4) to (9) of Table 9.2, that the distributions of Bayesian posterior modal values and sampling-theory estimates become more similar and more closely concentrated about the true value. We note, however, that even in samples of sizes 40, 60, and 100 "outliers" appear in the sampling-theory approach, a fact discussed below.

[50] See Table 9.1 for a description of conditions underlying Run I. Under our assumptions, the posterior mean of $\gamma = \pi_1/\pi_2$ does not exist.

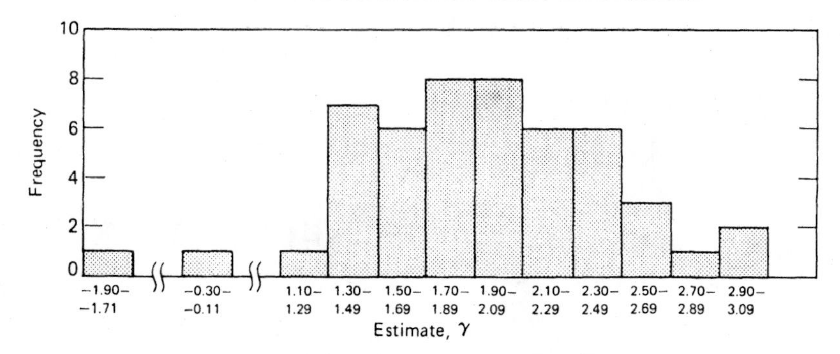

Figure 9.4 Frequency distribution of sampling theory estimates of γ (Run I: $T = 20$).

In Table 9.3 we present results obtained for Run II in which the variance of x_t was increased to two, compared with its value of one on Run I. Also the correlation between u_{1t} and u_{2t} was changed to $\frac{1}{2}$, compared with its value of $-\frac{1}{2}$ on Run I. With these changes it is seen from Table 9.3 that the distributions of the estimates are more sharply concentrated about the true value of γ than in Run I, for both approaches. As in Run I, however, we see that for the smaller sample sizes, $T = 20$ and $T = 40$, the Bayesian modal values are more highly concentrated about the true value of γ. As in Run I, also, the sampling-theory approach has produced a number of outlying estimates uncharacteristic of the Bayesian approach.

As regards "outliers" in the sampling-theory approach, we note that they have been encountered in other Monte Carlo experiments.[51] Although other workers point to possible rounding errors and near singularity of moment matrices as possible explanations for their outliers, such explanations are not relevant for explaining the current "outliers." It just appears that the distribution of the estimator $\hat{\gamma}$, shown in (9.104), is such that under the conditions underlying Runs I and II extreme values can be encountered with a non-negligible probability.[52] That the Bayesian approach is not characterized by extreme values for posterior modal values, at least in this set of experiments, is indeed an important result of the experiments.

Shown in Figure 9.5 are the general shapes of the exact posterior distributions encountered in Run I. Generally speaking, for smaller sample sizes these distributions often departed from the normal, usually being more

[51] See, for example, J. G. Cragg, *loc. cit.*, and R. Summers, "A Capital Intensive Approach to the Small Sample Properties of Various Simultaneous Equation Estimators," *Econometrica*, **33**, 1–41 (1965).

[52] For some recent analysis bearing on the distribution of the ratio of correlated normal random variables, which is what the estimator $\hat{\gamma}$ is, see G. Marsaglia, "Ratios of Normal Variables and Ratios of Sums of Uniform Variables," *J. Am. Statist. Assoc.*, **60**, 193–204 (1965).

Table 9.3 FREQUENCY DISTRIBUTIONS OF BAYESIAN ESTIMATES[a] AND SAMPLING-THEORY ESTIMATES OF γ (TRUE VALUE = 2.0): RUN II

	Sample Size $T = 20$		Sample Size $T = 40$		Sample Size $T = 60$		Sample Size $T = 100$	
Interval (1)	Bayesian Estimates[a] (2)	Sampling-Theory Estimates (3)	Bayesian Estimates[a] (4)	Sampling-Theory Estimates (5)	Bayesian Estimates[a] (6)	Sampling-Theory Estimates (7)	Bayesian Estimates[a] (8)	Sampling-Theory Estimates (9)
Smaller than 0	0	2	0	1	0	0	0	0
⋮								
0.100–0.299	0	1	0	0	0	0	0	0
0.300–0.499	0	1	0	1	0	0	0	0
0.500–0.699	0	0	0	0	0	0	0	0
0.700–0.899	0	0	0	1	0	0	0	0
0.900–1.099	0	1	0	1	0	0	0	0
1.100–1.299	0	2	0	0	0	1	0	0
1.300–1.499	0	1	1	4	3	1	0	0
1.500–1.699	0	7	1	2	2	4	0	1
1.700–1.899	6	5	10	11	12	6	7	13
1.900–2.099	15	11	19	13	23	28	30	27
2.100–2.299	17	7	17	13	9	9	12	9
2.300–2.499	9	7	2	3	1	1	1	0
2.500–2.699	1	3	0	0	0	0	0	0
2.700–2.899	1	0	0	0	0	0	0	0
2.900–3.099	1	1	0	0	0	0	0	0
3.100–3.299	0	1	0	0	0	0	0	0
3.300–3.499	0	0	0	0	0	0	0	0
Total:	50	50	50	50	50	50	50	50

[a] Posterior modal values.

Table 9.2 FREQUENCY DISTRIBUTIONS OF BAYESIAN ESTIMATES[a] AND SAMPLING-THEORY ESTIMATES OF γ (TRUE VALUE = 2.0): RUN I

	Sample Size $T = 20$		Sample Size $T = 40$		Sample Size $T = 60$		Sample Size $T = 100$	
Interval (1)	Bayesian Estimates[a] (2)	Sampling-Theory Estimates (3)	Bayesian Estimates[a] (4)	Sampling-Theory Estimates (5)	Bayesian Estimates[a] (6)	Sampling-Theory Estimates (7)	Bayesian Estimates[a] (8)	Sampling-Theory Estimates (9)
Smaller than 0	0	2	0	2	0	2	0	0
⋮								
0.300–0.499	0	0	0	1	0	0	0	0
0.500–0.699	0	0	0	0	0	0	0	0
0.700–0.899	0	0	0	0	0	0	0	0
0.900–1.099	0	0	0	0	0	0	0	0
1.100–1.299	0	1	1	0	0	0	0	0
1.300–1.499	4	7	0	0	1	0	0	0
1.500–1.699	9	6	5	2	4	4	3	2
1.700–1.899	14	8	17	11	14	9	14	12
1.900–2.099	15	8	15	7	24	17	23	12
2.100–2.299	7	6	11	7	5	9	10	15
2.300–2.499	1	6	1	6	1	1	0	6
2.500–2.699	0	3	0	3	1	3	0	2
2.700–2.899	0	1	0	5	0	1	0	0
2.900–3.099	0	2	0	1	0	2	0	0
3.100–3.299	0	0	0	2	0	0	0	0
3.300–3.499	0	0	0	0	0	1	0	0
3.500–3.699	0	0	0	0	0	0	0	0
3.700–3.899	0	0	0	0	0	0	0	0
3.900–4.099	0	0	0	2	0	0	0	0
Larger than 4.099	0	0	0	1	0	1	0	1
Total:	50	50	50	50	50	50	50	50

[a] Posterior modal values.

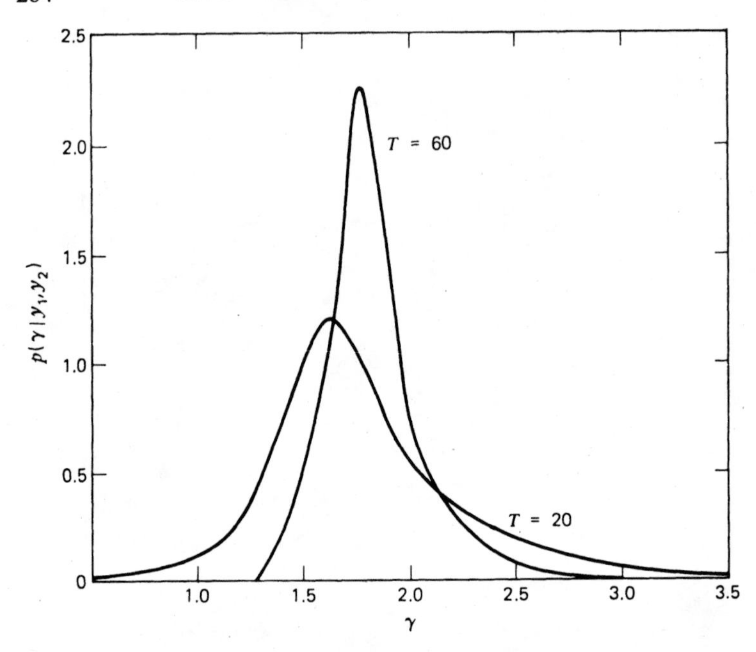

Figure 9.5 Typical posterior probability density functions for γ (Run I: $T = 20$ and $T = 60$).

peaked, with fatter tails, and skewed. In large samples many more cases were encountered in which the distributions were close to being normal and closer to the approximating normal distribution discussed above.

In Run III the only change in the experimental set-up relative to Run II was to raise the variance of x_t to nine, a change that improves the precision with which π_2 can be estimated. As in Run II, the correlation between u_{1t} and u_{2t}, the structural disturbance terms, was set equal to $\frac{1}{2}$. Raising the variance of x_t to nine produced conditions in which it was possible to make much more precise inferences about γ in both the Bayesian and sampling-theory approaches. Shown in Table 9.4 are the distributions of Bayesian posterior modal values and sampling-theory estimates for Run III. Here the results are quite different from those shown in Table 9.2 for Run I. In Run III the distributions of estimates for each sample size are more similar. This is probably because the conditions for Run III may be such that large-sample results take hold even for smaller T. In these experiments, also, the exact posterior distributions were quite close to being normal in many instances and were closer to the approximating normal distribution discussed above.

One point brought out by the results of Runs I, II, and III is that what we

Table 9.4 FREQUENCY DISTRIBUTIONS OF BAYESIAN ESTIMATES[a] AND SAMPLING-THEORY ESTIMATES OF γ (TRUE VALUE = 2.0): RUN III

Interval (1)	Sample Size $T = 20$		Sample Size $T = 40$		Sample Size $T = 60$		Sample Size $T = 100$	
	Bayesian Estimates[a] (2)	Sampling-Theory Estimates (3)	Bayesian Estimates[a] (4)	Sampling-Theory Estimates (5)	Bayesian Estimates[a] (6)	Sampling-Theory Estimates (7)	Bayesian Estimates[a] (8)	Sampling-Theory Estimates (9)
0.700–0.899	0	1	0	0	0	0	0	0
1.000–1.099	0	0	0	0	0	0	0	0
1.100–1.299	0	0	0	0	0	0	0	0
1.300–1.499	0	1	0	0	0	0	0	0
1.500–1.699	2	3	0	0	1	1	0	0
1.700–1.899	6	12	5	8	4	9	5	7
1.900–2.099	41	24	31	32	37	32	44	42
2.000–2.299	1	9	14	10	8	8	1	1
2.300–2.499	0	0	0	0	0	0	0	0
Total:	50	50	50	50	50	50	50	50

[a] Posterior modal values.

regard as a "large sample" is critically dependent on the features of the underlying model. For Run I we may say, *somewhat loosely*, that large-sample results take hold for sample sizes in the vicinity of 100 ± 30, whereas for Run III they appear to take hold even in samples of size $T = 40$. However, having the exact posterior distribution in the Bayesian approach makes the issue of "how large is large" less critical than in the sampling-theory approach in which it is usual to rely on asymptotic theory to justify finite sample inferences.

9.7.5 Experimental Results: Confidence Intervals

We now turn to consider the performance of Bayesian and approximate sampling-theory confidence intervals. For each sample the intervals discussed above were computed. Then for each sample size the number of intervals covering the true value of γ was determined. This number, expressed as a percentage of 50, the number of trials, is reported in Table 9.5. For Runs I, II, and III the Bayesian confidence intervals perform remarkably well. With 50 trials it is impossible to get $47\frac{1}{2}$ intervals that cover; thus 95.0% coverage is impossible. The actual figures reported in columns (2) and (3) indicate that the Bayesian intervals have relatively good sampling properties.[53]

With respect to the performance of the approximate sampling-theory confidence intervals the results reported in Table 9.5 indicate that the nominal 95% confidence level is not generally realized. The percentage of coverage in a number of instances was found to be significantly below 95. This indicates that inferences, based on these approximate sampling-theory intervals, may generally be erroneous in "small-sample" situations.

9.7.6 Concluding Remarks on the Monte Carlo Experiments

The experiments reported above indicate that the differences in the sampling properties of Bayesian and sampling theory estimators in "small-sample" situations are quite striking. Under the conditions we have examined Bayesian procedures produced better results than sampling-theory estimation procedures. With respect to point estimation Bayesian estimates tended to be more highly concentrated about the true value of the parameter being estimated than sampling-theory estimates, particularly in small-sample situations. In this connection it appears relevant to note that the sampling-theory estimators considered are usually given a large-sample justification. Thus there is no assurance that these estimators will perform well in small-

[53] For some analysis of the sampling properties of Bayesian intervals, see B. L. Welch and H. W. Peers, "On Formulae for Confidence Points Based on Integrals of Weighted Likelihoods," *Journal of the Royal Statistical Society*, **B25**, 318–324 (1963), and D. J. Bartholomew, "A Comparison of Some Bayesian and Frequentist Inferences," *Biometrika*, **52**, 19–35 (1965).

Table 9.5 PERFORMANCE OF BAYESIAN AND SAMPLING THEORY NOMINAL 95% CONFIDENCE INTERVALS FOR γ

Experiment (1)	Bayesian "Exact Central" Interval[a] (2)	Bayesian "Exact Shortest" Interval[a] (3)	Sampling-Theory "Normal" Interval[a] (4)	Sampling-Theory t Interval[a] (5)
	(Percent of intervals covering true value)			
Run I:				
$T = 20$	96.0	96.0	84.0	88.0
$T = 40$	96.0	98.0	82.0	82.0
$T = 60$	96.0	98.0	76.0	76.0
$T = 100$	96.0	96.0	92.0	96.0
Run II:				
$T = 20$	96.0	98.0	82.0	82.0
$T = 40$	96.0	96.0	86.0	86.0
$T = 60$	90.0	96.0	90.0	94.0
$T = 100$	100.0	100.0	94.0	98.0
Run III:				
$T = 20$	92.0	96.0	84.0	84.0
$T = 40$	100.0	100.0	94.0	94.0
$T = 60$	96.0	96.0	98.0	100.0
$T = 100$	94.0	96.0	88.0	90.0

[a] These intervals are defined in the text.

sample situations.[54] With respect to interval estimation intervals computed from posterior pdf's were found to have rather good sampling properties. On the other hand, *approximate* sampling-theory confidence intervals,[55] often used in practice, were found to be deficient in small-sample situations. In large-sample situations, as theory predicts, Bayesian and sampling-theory procedures performed about equally well in terms of the criterion of performance in repeated samples.

QUESTIONS AND PROBLEMS

1. Consider the following two-equation supply and demand model:

Demand: $p_t = \alpha_1 q_t + \alpha_2 x_{1t} + u_{1t}$
Supply: $q_t = \beta_1 p_{t-1} + \beta_2 x_{2t} + u_{2t},$ $\quad t = 1, 2, \ldots, T$

where the subscript t denotes the value of a variable in the tth period, p_t and

[54] In the text of this chapter we have seen that certain sampling theory estimates can be viewed as approximations to the mean or modal value of posterior pdf's.
[55] The finite sample properties of these approximate procedures have not been analyzed to the best of our knowledge. However, some Monte Carlo experiments investigating this problem have been reported in J. G. Cragg, *loc. cit.*

q_t, price and quantity, respectively, are endogenous variables, x_{1t} and x_{2t}, income and a cost variable, respectively, are assumed exogenous, the α's and β's are structural parameters, and u_{1t} and u_{2t} are serially uncorrelated random disturbance terms. Assume that u_{1t} and u_{2t} are normally distributed with zero means, variances σ_{11} and σ_{22}, and zero covariance, that is $\sigma_{12} = 0$, for all t. Given the initial observation p_0, write down the likelihood function and derive maximum likelihood estimates for the parameters, given sample data for p, q, and the x's. What is the information matrix for this system?

2. Analyze the supply and demand model in Problem 1 with the following diffuse prior pdf for the parameters; $p(\alpha_1, \alpha_2, \beta_1, \beta_2, \sigma_{11}, \sigma_{22}) \propto 1/\sigma_{11}\sigma_{22}$, with $-\infty < \alpha_i, \beta_i < \infty$ and $0 < \sigma_{ii} < \infty$, $i = 1, 2$.
3. Suppose in Problem 1 that the u_{1t}'s are autocorrelated and satisfy $u_{1t} = \eta u_{1t-1} + \epsilon_t$ for all t, where η is an unknown parameter and the ϵ_t's are NID(0, σ^2). Show how maximum likelihood and Bayesian estimates of the parameters α_1 and α_2 can be obtained under the assumption regarding the u_{1t}'s.
4. If in Problem 1 $\sigma_{12} \neq 0$, write down the likelihood function in terms of the α's, β's, σ_{11}, σ_{22}, and $\rho = \sigma_{12}/\sqrt{\sigma_{11}\sigma_{22}}$. Then, using the prior assumptions of Problem 2 and the assumption that $\rho = \rho_0$, a given value, show how to obtain conditional posterior pdf's for the α's and β's, given that $\rho = \rho_0$.
5. In connection with the simple Haavelmo model shown in (9.36) to (9.37), formulate an informative beta-prior pdf for α, the marginal propensity to consume, and show how it can be employed, along with other prior assumptions, to obtain a posterior pdf for α.
6. In Problem 5 from the beta-prior pdf for α deduce the implied prior pdf for the multiplier $\pi = 1/(1 - \alpha)$ and examine how its properties depend on the parameters of the prior pdf for α.
7. Use Haavelmo's data for the United States, 1929 to 1941, shown below, to compute the posterior pdf for α derived in Problem 5:

Year	c_t	y_t	Year	c_t	y_t
1929	\$474	\$534	1936	\$463	\$511
1930	439	478	1937	469	520
1931	399	440	1938	444	477
1932	350	372	1939	471	517
1933	364	381	1940	494	548
1934	392	419	1941	529	629
1935	416	449			

(Figures are per-capita price-deflated personal consumption expenditures and disposable income.)

8. Formulate a relatively diffuse prior pdf for the parameters of the following slightly expanded Haavelmo model:

$$\left.\begin{aligned} c_t &= \beta + \alpha y_t + \gamma c_{t-1} + u_t \\ y_t &= c_t + z_t \end{aligned}\right\} \qquad t = 1, 2, \ldots, T,$$

where z_t is assumed to be exogenous and the u_t's are NID(0, σ^2).

9. Using the prior pdf formulated in Problem 8 and Haavelmo's data, shown in Problem 7, derive and compute posterior pdf's for the parameters of the model in Problem 8, given the initial value for c_t, $c_0 = 466$, the 1928 value. In particular, examine the posterior pdf for γ to decide whether the information in the data and prior assumptions suggest a zero value for this parameter.

10. In Problem 9 from the joint posterior pdf for α and γ explain how to obtain the marginal posterior pdf for $\alpha/(1 - \gamma)$, the "long-run" marginal propensity to consume. Then compute the marginal posterior pdf for $\alpha/(1 - \gamma)$.

11. For the ith firm $i = 1, 2, \ldots, n$ let y_{0i} = log of output, y_{1i} = log of labor input, and y_{2i} = log of capital input. If the production function is of the Cobb-Douglas (CD) form and the firm is assumed to maximize the mathematical expectation of profit, it has been shown[56] that the following model for the observations (y_{0i}, y_{1i}, y_{2i}) results:

$$\begin{aligned}
\text{Production relation:}\quad & y_{0i} - \alpha_1 y_{1i} - \alpha_2 y_{2i} - \alpha_0 = u_{0i}, \\
\text{Condition on labor input:}\quad & (\alpha_1 - 1)y_{1i} + \alpha_2 y_{2i} - \beta_1 = u_{1i}, \\
\text{Condition on capital input:}\quad & \alpha_1 y_{1i} + (\alpha_2 - 1)y_{2i} - \beta_2 = u_{2i},
\end{aligned}$$

for $i = 1, 2, \ldots, n$, where α_1 and α_2 are the labor and capital coefficients in the CD function, respectively, α_0, β_1, and β_2 are parameters, and the triplets (u_{0i}, u_{1i}, u_{2i}) are assumed to be independently and normally distributed, with zero mean vector and common 3×3 pds covariance matrix Σ, given by

$$\Sigma = \begin{pmatrix} \sigma_{00} & \mathbf{0}' \\ \mathbf{0} & \Sigma_* \end{pmatrix}.$$

In this last expression σ_{00} is the variance of the u_{0i}'s, Σ_* is a 2×2 covariance matrix for u_{1i} and u_{2i}, and $\mathbf{0}' = (0, 0)$. Write down the likelihood function for the system and derive maximum likelihood estimates for the parameters.

12. Use the likelihood function in Problem 11 along with the following diffuse prior pdf for the parameters $\boldsymbol{\alpha}' = (\alpha_1, \alpha_2, \alpha_0)$, $\boldsymbol{\beta}' = (\beta_1, \beta_2)$, σ_{00} and Σ_*, $p(\boldsymbol{\alpha}, \boldsymbol{\beta}, \sigma_{00}, \Sigma_*) \propto p_1(\boldsymbol{\alpha}, \boldsymbol{\beta})\, p_2(\sigma_{00})\, p_3(\Sigma_*)$, with $p_1(\boldsymbol{\alpha}, \boldsymbol{\beta}) \propto |1 - \alpha_1 - \alpha_2|^{-1}$, $p_2(\sigma_{00}) \propto 1/\sigma_{00}$, and $p_3(\Sigma_*) \propto |\Sigma_*|^{-3/2}$. Here we are assuming that the location parameters $\boldsymbol{\alpha}$ and $\boldsymbol{\beta}$ are a priori independent of the scale parameters σ_{00} and Σ_*. Further, application of Jeffreys' invariance theory[57] provides the form for $p_1(\boldsymbol{\alpha}, \boldsymbol{\beta})$, shown above. Assuming that $0 < \sigma_{00} < \infty$, $0 < |\Sigma_*|$, and α_0, β_1, and β_2 range from $-\infty$ to ∞, write down the joint posterior pdf for the parameters and then integrate to obtain the marginal posterior pdf for α_0, α_1, and α_2, which is in the trivariate Student-t form centered at the maximum likelihood estimates if it is assumed that α_1 and α_2 both range from $-\infty$ to ∞.

13. In Problem 12 provide an alternative prior pdf for α_1 and α_2 which reflects information suggested by economic theory, namely $\alpha_1, \alpha_2 > 0$, and show how

[56] See A. Zellner, J. Kmenta, and J. Drèze, "Specification and Estimation of Cobb-Douglas Production Function Models," *Econometrica*, **34**, 784–795 (1966), for details.
[57] See Appendix to Chapter 2.

it can be employed in the analysis of the CD model given in Problem 11. How will such prior information affect the posterior pdf's for α_1 and α_2 and optimal Bayesian point estimates when the sample size n is small and when it is large?

14. Let $y_{1t} = y_{2t}\gamma + u_{1t}$ and $y_{2t} = \mathbf{x}_t'\boldsymbol{\pi} + u_{2t}$, $t = 1, 2, \ldots, T$, where y_{1t} and y_{2t} are endogenous variables, γ is a scalar parameter, $\mathbf{x}_t'$ is a $1 \times k$ vector of given quantities with $X' = (\mathbf{x}_1, \mathbf{x}_2, \ldots, \mathbf{x}_T)$ a $k \times T$ matrix of rank k, $\boldsymbol{\pi}$ is a $k \times 1$ vector of parameters, and u_{1t} and u_{2t} are disturbance terms. Assume that the pairs (u_{1t}, u_{2t}) are normally and independently distributed, each with zero mean vector and 2×2 pds covariance matrix Σ. Is the above system observationally equivalent to the following system? $y_{1t} = \mathbf{x}_t'\boldsymbol{\pi}\gamma + \epsilon_{1t}$ and $y_{2t} = \mathbf{x}_t'\boldsymbol{\pi} + \epsilon_{2t}$, where the pairs $(\epsilon_{1t}, \epsilon_{2t})$ are assumed to be normally and independently distributed, each with zero mean vector and 2×2 pds covariance matrix Ω.

15. Suppose that we formulate the following prior pdf for parameters appearing in the second system of Problem 14:

$$p(\gamma, \boldsymbol{\pi}, \Omega) \propto p_1(\gamma)|\Omega|^{-3/2},$$

with $-\infty < \pi_i < \infty$, $i = 1, 2, \ldots, k$, $|\Omega| > 0$, and $p_1(\gamma)$, the marginal prior pdf for γ. What does this prior pdf imply about the prior pdf for Σ, the disturbance covariance matrix for the first system in Problem 14?

16. Using the prior pdf shown in Problem 15, analyze the system in Problem 14 to obtain a marginal posterior pdf for γ.

17. As in Section 9.5, analyze the first equation of a simultaneous equation system but with the assumption that the structural disturbance covariance has the following form:

$$\Sigma = \begin{pmatrix} \sigma_{11} & \mathbf{0}' \\ \mathbf{0} & \Sigma_* \end{pmatrix},$$

where Σ is an $m \times m$ pds matrix, σ_{11} is the common variance of the elements of $\mathbf{u}_1$, Σ_* is an $(m - 1) \times (m - 1)$ full matrix, and $\mathbf{0}'$ is a $1 \times (m - 1)$ vector of zeros.

18. If in connection with the system shown in (9.1) Γ and Σ are known to be block diagonal,

$$\Gamma = \begin{bmatrix} \Gamma_{11} & 0 & \cdots & 0 \\ 0 & \Gamma_{22} & \cdots & 0 \\ \vdots & \vdots & \ddots & \vdots \\ 0 & 0 & \cdots & \Gamma_{GG} \end{bmatrix} \quad \text{and} \quad \Sigma = \begin{pmatrix} \Sigma_{11} & 0 & \cdots & 0 \\ 0 & \Sigma_{22} & \cdots & 0 \\ \vdots & \vdots & \ddots & \vdots \\ 0 & 0 & \cdots & \Sigma_{GG} \end{pmatrix},$$

with Σ_{ii} the covariance matrix for the disturbance terms appearing in the subset of equations with endogenous variable coefficient matrix Γ_{ii}, $i, = 1, 2, \ldots, G$, show how this leads to a factorization of the likelihood function.

CHAPTER X

On Comparing and Testing Hypotheses

In many circumstances there is the problem of comparing alternative hypotheses; for example, we may be interested in comparing Friedman's permanent income hypothesis (PIH) and the absolute income hypothesis (AIH), or we may wish to compare the hypothesis that a parameter is zero with the hypothesis that it is different from zero. Given a *precise* statement of the hypotheses to be compared, the way in which we actually make the comparison will depend on the purpose to be served by our analysis, the state of our prior information, and whether we have an explicitly formulated loss function. Each of these points is considered below.

As regards purpose of an analysis, a body of data may be analyzed just to provide a revision of prior probabilities associated with alternative hypotheses[1]; for example, if, initially, our prior probabilities for the PIH and AIH are each equal to $\frac{1}{2}$, sample information, used in a way to be described below, may change these probabilities to $\frac{3}{4}$ for the PIH and $\frac{1}{4}$ for the AIH. An investigator may stop here with the conclusion that on the basis of initial prior probabilities ($= \frac{1}{2}$) and of the information in the data the posterior odds in favor of the PIH are three to one; that is $\frac{3}{4} \div \frac{1}{4} = 3$. Clearly, this result is useful and important. Further, it may be that the investigator has no idea how his research findings are to be used. Thus he cannot be or is not interested in formulating a decision problem with an explicit loss function. However, by obtaining posterior probabilities associated with alternative hypotheses, he has provided an important ingredient for those interested in solving decision problems with explicitly given loss functions. Further, analysis of additional sample data will result in a revision of posterior probabilities. After much data have been analyzed it may be that the posterior probability associated with one of the hypotheses is close to one. In this situation we would say that this hypothesis is probably true.[2] The process of revising prior probabilities associated with alternative hypotheses does not necessarily involve a decision to reject or accept these hypotheses, which is

[1] Note that in the Bayesian approach it is considered meaningful to introduce probabilities associated with hypotheses.

[2] The use of the term "probably true" allows for the possibility of error.

why we have used the term "comparing hypotheses" rather than "testing hypotheses."

On the other hand, we recognize that on many occasions the objective of an analysis is to reach a decision, say accept or reject, with respect to alternative hypotheses; that is, we may wish to state that on the basis of the available information we can reject the AIH and accept the PIH. If we have an explicitly given loss function, the prescription act to minimize expected loss is usually employed to reach a decision.[3] If, however, we have no explicitly given loss function, there is some unavoidable degree of arbitrariness in our decision; that is, we may decide to accept the PIH and reject the AIH if the posterior odds in favor of the PIH are 20 to 1 or greater. One may well ask why 20 to 1 rather than 30 to 1? No satisfactory answer to this question can be given when the consequences of the acts, reject or accept, are not or cannot be spelled out explicitly.[4] Thus, when no explicit statement of the consequences of the acts, accept or reject, is given, going beyond the reporting of posterior probabilities associated with alternative hypotheses to reach the conclusion, reject or accept, will involve some element of arbitrariness.

The last general point to be considered is the role of prior information in comparing hypotheses. Just as in estimation problems, it will be seen that prior information can be incorporated in analyses involving the comparison of hypotheses. As with estimation, the amount and kind of prior information to be employed in an analysis will depend on what we know and what we judge appropriate to incorporate into the analysis. We recognize that there are situations in which we know very little and thus want procedures for comparing hypotheses with the use of little prior information. There are other circumstances, however, when we have prior information, say from analyses of past samples of data, and wish to incorporate it in our comparison of hypotheses. We shall see how this can be done in the Bayesian approach.

10.1 POSTERIOR PROBABILITIES ASSOCIATED WITH HYPOTHESES

We now turn to the problem of using data to revise prior probabilities associated with hypotheses.[5] As a first example let us consider two mutually exclusive and exhaustive hypotheses, H_0 and H_1. Initially, we assume that under H_0 our observation vector $\mathbf{y}$ has a pdf $p(\mathbf{y}|\boldsymbol{\theta} = \boldsymbol{\theta}_0)$ and under H_1,

[3] This is in accord with the expected utility hypothesis. For an interesting discussion of this precept for consistent behavior see R. D. Luce and H. Raiffa, *Games and Decisions*. New York: Wiley, 1958.

[4] The same arbitrariness arises in the choice of a significance level in sampling theory tests. That is, why use the 5 per cent rather than the 1 per cent level of significance?

[5] Much of what follows is due to Harold Jeffreys. See his *Theory of Probability* (3rd ed.). Oxford: Clarendon, 1961, Chapters 5 and 6.

$p(\mathbf{y}|\boldsymbol{\phi} = \boldsymbol{\phi}_1)$, where $\boldsymbol{\theta}_0$ and $\boldsymbol{\phi}_1$ are specific values for the parameter vectors $\boldsymbol{\theta}$ and $\boldsymbol{\phi}$, respectively.[6] Further, let w be a dichotomous random variable such that

$$w = \begin{cases} 0 & \text{if } H_0 \text{ is true,} \\ 1 & \text{if } H_1 \text{ is true.} \end{cases} \tag{10.1}$$

Our prior probabilities associated with the hypotheses are $p(H_0) = p(w = 0)$ and $p(H_1) = p(w = 1)$ with $p(H_0) + p(H_1) = p(w = 0) + p(w = 1) = 1$.

Now consider the joint pdf for $\mathbf{y}$ and w,

$$\begin{aligned} p(\mathbf{y}, w) &= p(w)\, p(\mathbf{y}|w) \\ &= p(\mathbf{y})\, p(w|\mathbf{y}), \end{aligned} \tag{10.2}$$

from which we obtain

$$p(w|\mathbf{y}) = \frac{p(w)\, p(\mathbf{y}|w)}{p(\mathbf{y})}, \tag{10.3}$$

where $p(w|\mathbf{y})$ is the discrete posterior pdf for w, given the sample information, $p(w)$ is the discrete prior pdf for w, $p(\mathbf{y}|w)$ is the conditional pdf for $\mathbf{y}$, given w, and $p(\mathbf{y}) = p(\mathbf{y}|w = 0)\, p(w = 0) + p(\mathbf{y}|w = 1)\, p(w = 1)$ is the marginal pdf for $\mathbf{y}$, assumed nonzero. Then from (10.3) we have for the posterior probability associated with H_0

$$\begin{aligned} p(H_0|\mathbf{y}) = p(w = 0|\mathbf{y}) &= \frac{p(w = 0)\, p(\mathbf{y}|w = 0)}{p(\mathbf{y})} \\ &= \frac{p(H_0)\, p(\mathbf{y}|\boldsymbol{\theta} = \boldsymbol{\theta}_0)}{p(\mathbf{y})}, \end{aligned} \tag{10.4}$$

and similarly for H_1

$$\begin{aligned} p(H_1|\mathbf{y}) = p(w = 1|\mathbf{y}) &= \frac{p(w = 1)\, p(\mathbf{y}|w = 1)}{p(\mathbf{y})} \\ &= \frac{p(H_1)\, p(\mathbf{y}|\boldsymbol{\phi} = \boldsymbol{\phi}_1)}{p(\mathbf{y})}. \end{aligned} \tag{10.5}$$

The expressions in (10.4) and (10.5) can be employed to compute posterior probabilities, given that we have prior probabilities and explicit functional forms for $p(\mathbf{y}|\boldsymbol{\theta} = \boldsymbol{\theta}_0)$ and $p(\mathbf{y}|\boldsymbol{\phi} = \boldsymbol{\phi}_1)$. Also, the posterior odds in favor of H_0, denoted by K_{01}, are given by[7]

$$K_{01} = \frac{p(H_0|\mathbf{y})}{p(H_1|\mathbf{y})} = \frac{p(H_0)}{p(H_1)} \frac{p(\mathbf{y}|\boldsymbol{\theta} = \boldsymbol{\theta}_0)}{p(\mathbf{y}|\boldsymbol{\phi} = \boldsymbol{\phi}_1)}. \tag{10.6}$$

[6] In some problems $\boldsymbol{\theta} = \boldsymbol{\phi}$, and we consider $H_0\colon \boldsymbol{\theta} = \boldsymbol{\theta}_0$ and $H_1\colon \boldsymbol{\theta} = \boldsymbol{\theta}_1$ as our two hypotheses.

[7] Note that K_{01} in (10.6) can be computed and would remain unchanged if we had more than two mutually exclusive hypotheses.

In (10.6) we see that the posterior odds are the product of the prior odds, $p(H_0)/p(H_1)$, and the likelihood ratio, $p(\mathbf{y}|\boldsymbol{\theta} = \boldsymbol{\theta}_0)/p(\mathbf{y}|\boldsymbol{\phi} = \boldsymbol{\phi}_1)$.

As an example illustrating application of these concepts and operations let us assume that a coin has been fairly and independently tossed three times and that we observe two heads and one tail. Under hypothesis H_0 we assume that the probability of a head is $\frac{1}{2}$. Under the hypothesis H_1 we assume that the probability of a head is $\frac{1}{4}$. If the prior probabilities are $p(H_0) = p(H_1) = \frac{1}{2}$, then the data, two heads in three tries, yields the following posterior odds from (10.6):

$$K_{01} = \frac{p(H_0|\mathbf{y})}{p(H_1|\mathbf{y})} = \frac{(\frac{1}{2})(\frac{1}{2})^2(\frac{1}{2})}{(\frac{1}{2})(\frac{1}{4})^2(\frac{3}{4})}$$
$$= 8/3.$$

Thus the sample evidence changes our prior odds, 1/1, to 8/3 in favor of the hypothesis H_0, namely, that the probability of a head is $\frac{1}{2}$. Equivalently, the data have changed our prior probabilities from $\frac{1}{2}$ to 8/11 and 3/11 for H_0 and H_1, respectively. To indicate how prior odds of 1/1 would be modified by other possible outcomes we show in Table 10.1 posterior odds associated with a range of possible outcomes.

We see that for just two tosses the posterior odds are just 4/3 for H_0, given that one head has appeared. Given two heads in two tosses, however, the odds are 4/1 in favor of H_0. Note, too, how increasing the sample size affects the posterior odds, given various outcomes; for example, having two of four tosses result in a head appearing leads to posterior odds 16/9, which is somewhat greater than those associated with getting one head in two tosses, namely 4/3.

Being able to state posterior odds and probabilities may be all that we may want to do. There are other circumstances, however, in which we may want to take an *action*; that is, accept H_0 or reject H_0.[8] This is a *two-action* problem.

Table 10.1 POSTERIOR ODDS FOR H_0 RELATIVE TO H_1[a]

Number of Trials	Number of Heads Appearing					
	0	1	2	3	4	5
2	4/9	4/3	4	—	—	—
3	8/27	8/9	8/3	8	—	—
4	16/81	16/27	16/9	16/3	16	—
5	32/243	32/81	32/27	32/9	32/3	32

[a] H_0 is the hypothesis that the probability of a head on a single toss is $\frac{1}{2}$, whereas H_1 asserts that this probability is $\frac{1}{4}$.

[8] Here we specifically assume that the action "go on taking data" is not an option.

Further, we recognize that, by assumption, there are two possible states of the world—H_0 true or H_1 true. Thus we have a "two-state–two-action" problem. Let us assume that the *consequences* of our actions are given by the following *loss structure*:

		State of World	
		H_0 true	H_1 true
Action	Accept H_0	$L(H_0, \hat{H}_0) = 0$	$L(H_1, \hat{H}_0)$
	Accept H_1	$L(H_0, \hat{H}_1)$	$L(H_1, \hat{H}_1) = 0$

This particular loss structure is such that we experience zero loss if our actions are in agreement with states of the world. However, if we accept H_0 when H_1 is true, we experience a positive loss $L(H_1, \hat{H}_0)$, where the first argument of L refers to the state of the world and the second to our action; that is, $\hat{H}_0$ is short-hand notation for accept H_0.

With the above loss structure we can evaluate the consequences of our actions, given that we have posterior probabilities for the hypotheses H_0 and H_1; that is, the expected loss associated with the action accept H_0 is

$$E(L|\hat{H}_0) = p(H_0|\mathbf{y})\, L(H_0, \hat{H}_0) + p(H_1|\mathbf{y})\, L(H_1, \hat{H}_0) = p(H_1|\mathbf{y})\, L(H_1, \hat{H}_0), \tag{10.7}$$

since we have assumed $L(H_0, \hat{H}_0) = 0$. Similarly,

$$E(L|\hat{H}_1) = p(H_0|\mathbf{y})\, L(H_0, \hat{H}_1) + p(H_1|\mathbf{y})\, L(H_1, \hat{H}_1) = p(H_0|\mathbf{y})\, L(H_0, \hat{H}_1), \tag{10.8}$$

since we have assumed $L(H_1, \hat{H}_1) = 0$.

Having computed the expected losses in (10.7) and (10.8), we can compare them and choose an action:

$$\text{If } E(L|\hat{H}_0) < E(L|\hat{H}_1), \text{ accept } H_0, \text{ action } \hat{H}_0, \tag{10.9}$$

$$\text{If } E(L|\hat{H}_1) < E(L|\hat{H}_0), \text{ accept } H_1, \text{ action } \hat{H}_1. \tag{10.10}$$

This provides a basis for action[9] in accord with the implications of the expected utility hypothesis, namely, that a decision maker should maximize expected utility (or, equivalently, minimize expected loss) in order to be consistent.

[9] If the expected losses are equal, we could take either action and experience the same consequences.

To see what acting on the basis of a comparison of expected losses implies in terms of sample information, we see from (10.7) and (10.8) that $E(L|\hat{H}_0)$ will be less than $E(L|\hat{H}_1)$ if, and only if,

$$p(H_1|\mathbf{y})\,L(H_1, \hat{H}_0) < p(H_0|\mathbf{y})\,L(H_0, \hat{H}_1). \tag{10.11}$$

Then on substituting for $p(H_0|\mathbf{y})$ and $p(H_1|\mathbf{y})$ from (10.4) and (10.5) we have

$$\frac{p(\mathbf{y}|H_0)}{p(\mathbf{y}|H_1)} > \frac{p(H_1)\,L(H_1, \hat{H}_0)}{p(H_0)\,L(H_0, \hat{H}_1)}. \tag{10.12}$$

Thus we find that (10.12) is logically implied by the condition $E(L|\hat{H}_0) < E(L|\hat{H}_1)$ and is an alternative way of stating the expected loss criterion for taking the action of accepting H_0. In (10.12) we have a likelihood ratio $p(\mathbf{y}|H_0)/p(\mathbf{y}|H_1)$ which is compared with the ratio of *prior* expected losses. The higher the prior expected loss associated with accepting H_0, $p(H_1)\,L(H_1, \hat{H}_0)$, in relation to that associated with accepting H_1, $p(H_0)\,L(H_0, \hat{H}_1)$, the greater the sample evidence in favor of H_0, as reflected by the likelihood ratio on the lhs of (10.12). This indeed appears to be a sensible procedure for determining the "critical value" in a likelihood-ratio test procedure; that is, a likelihood-ratio test procedure states accept H_0 if $p(\mathbf{y}|H_0)/p(\mathbf{y}|H_1) > \lambda$, where λ's value is determined by choice of the significance level for the test. Often the significance level and the associated value of λ are chosen with *implicit* consideration of the relative costs or losses associated with errors of the first and second kind.[10] In the Bayesian approach *explicit* consideration is given to the loss structure. That this leads to a likelihood-ratio test procedure is indeed an interesting justification for use of this procedure in testing hypotheses. Further, that explicit consideration of the loss structure provides a natural choice of the critical value λ is certainly satisfying.

It is extremely interesting to investigate the implications of various loss structures. One particularly simple one is the "symmetric" loss structure:

$$L(H_0, \hat{H}_1) = L(H_1, \hat{H}_0); \qquad L(H_0, \hat{H}_0) = L(H_1, \hat{H}_1) = 0. \tag{10.13}$$

For this loss structure errors of the first and second kind have equal associated losses. When it is appropriate to use such a loss structure, we see from (10.12) that the critical point in the likelihood-ratio test is just the prior odds for H_1 in relation to H_0. If the likelihood ratio is greater than the prior odds for H_1, we accept H_0, an action in accord with the expected utility hypothesis, given the symmetric loss function in (10.13). Equivalently, we can make a

[10] An error of the first kind is accepting H_1, given that H_0 is true, while an error of the second kind is accepting H_0 when H_1 is true; $L(H_0, \hat{H}_1)$ and $L(H_1, \hat{H}_0)$ are the losses associated with errors of the first and second kind, respectively.

comparison of the expected losses in (10.5) and (10.6) under the symmetric loss structure. This shows that $E(L|\hat{H}_0) < E(L|\hat{H}_1)$ if, and only if,

$$p(H_0|\mathbf{y}) > p(H_1|\mathbf{y}) \quad \text{or} \quad \frac{p(H_0|\mathbf{y})}{p(H_1|\mathbf{y})} > 1. \tag{10.14}$$

Thus *under the symmetric loss structure* a comparison of the posterior probabilities will provide a basis for choosing between H_0 and H_1. Obviously, with other loss structures the specific prescription for action will be different but the principles will be the same.

To bring out salient points we have considered alternative hypotheses that assign specific values to all parameters of the pdf for the observations $p(\mathbf{y}|H)$, as in the example involving coin tossing. This comparison of *simple* hypotheses is sometimes encountered in practice. More frequently, we encounter so-called composite hypotheses; for example, the hypothesis that the probability of a head is not equal to $\frac{1}{2}$. No specific value is suggested. Rather, all possible values other than $\frac{1}{2}$ are consistent with the hypothesis. To compute posterior probabilities for these hypotheses involves some extension of the above analysis, which we now undertake.

Again we consider two mutually exclusive and exhaustive hypotheses, H_0 and H_1, with prior probabilities $p(H_0) = p(w = 0)$ and $p(H_1) = p(w = 1)$, where w is the random variable defined in (10.1). Let $\boldsymbol{\theta}$ denote the parameter vector associated with hypothesis H_0 under which the pdf for the observation vector $\mathbf{y}$ is $p(\mathbf{y}|\boldsymbol{\theta}) \equiv p(\mathbf{y}|w = 0, \Theta)$ and $\boldsymbol{\phi}$, the parameter vector associated with hypothesis H_1 under which the pdf for $\mathbf{y}$ is $p(\mathbf{y}|\boldsymbol{\phi}) \equiv p(\mathbf{y}|w = 1, \Theta)$. We have for the joint pdf for $\mathbf{y}$, w, and Θ

$$\begin{aligned} p(\mathbf{y}, w, \Theta) &= p(\mathbf{y})\, p(w, \Theta|\mathbf{y}) \\ &= p(w, \Theta)\, p(\mathbf{y}|w, \Theta) \end{aligned} \tag{10.15}$$

or

$$\begin{aligned} p(w, \Theta|\mathbf{y}) &= \frac{p(w, \Theta)\, p(\mathbf{y}|w, \Theta)}{p(\mathbf{y})} \\ &= \frac{p(w)\, p(\Theta|w)\, p(\mathbf{y}|w, \Theta)}{p(\mathbf{y})}, \end{aligned} \tag{10.16}$$

where $p(w)$ is the prior pdf for w and $p(\Theta|w)$ is the conditional prior pdf for Θ, given w.

We have $p(\Theta|w = 0) = p(\boldsymbol{\theta})$ and $p(\Theta|w = 1) = p(\boldsymbol{\phi})$, the prior pdf's for $\boldsymbol{\theta}$ and $\boldsymbol{\phi}$, respectively. Then the posterior probability associated with H_0 can be obtained from (10.16) by inserting $w = 0$ and integrating with respect to $\boldsymbol{\theta}$; that is

$$p(H_0|\mathbf{y}) = \frac{p(w = 0) \int p(\boldsymbol{\theta})\, p(\mathbf{y}|\boldsymbol{\theta})\, d\boldsymbol{\theta}}{p(\mathbf{y})} \tag{10.17}$$

and similarly

$$p(H_1|\mathbf{y}) = \frac{p(w = 1)\int p(\boldsymbol{\phi})\, p(\mathbf{y}|\boldsymbol{\phi})\, d\boldsymbol{\phi}}{p(\mathbf{y})}. \tag{10.18}$$

These expressions can be employed to compute the posterior probabilities associated with the hypotheses, provided that the integrals converge. Also, the posterior odds in favor of H_0 are given by

$$K_{01} = \frac{p(H_0|\mathbf{y})}{p(H_1|\mathbf{y})} = \frac{p(H_0)}{p(H_1)} \cdot \frac{\int p(\boldsymbol{\theta})\, p(\mathbf{y}|\boldsymbol{\theta})\, d\boldsymbol{\theta}}{\int p(\boldsymbol{\phi})\, p(\mathbf{y}|\boldsymbol{\phi})\, d\boldsymbol{\phi}}, \tag{10.19}$$

where we have written $p(H_0) = p(w = 0)$ and $p(H_1) = p(w = 1)$ as the prior probabilities associated with H_0 and H_1, respectively.

From (10.19) we see that the posterior odds are equal to the prior odds $p(H_0)/p(H_1)$ times the ratio of *averaged* likelihoods with the prior pdf's $p(\boldsymbol{\theta})$ and $p(\boldsymbol{\phi})$ serving as the weighting functions. This contrasts with the usual likelihood-ratio testing procedure which involves taking the ratio of *maximized* likelihood functions under H_0 and H_1, a procedure that amounts to using maximum likelihood estimates *as if they were true values of the unknown parameters* in forming a likelihood ratio appropriate for two simple hypotheses.

10.2 ANALYZING HYPOTHESES WITH DIFFUSE PRIOR PDF'S FOR PARAMETERS

In this section we consider Lindley's procedure[11] for Bayesian tests of significance. As Lindley emphasizes, his procedure is appropriate only when prior information is vague or diffuse; that is, if we are concerned with the hypothesis that a scalar parameter θ is equal to θ_0, the value under the "null hypothesis," and the alternative hypothesis is $\theta \neq \theta_0$, the prior distribution in the neighborhood of the null value θ_0 ". . . must be reasonably smooth for the tests to be sensible."[12] This means that the situation is assumed to be such that we have no reason to believe that $\theta = \theta_0$ more strongly than $\theta = \theta_1$, where θ_1 is any value for θ near θ_0. We shall see that for many, but not all, problems Lindley's procedure leads to tests that are computationally equivalent to sampling-theory tests. However, the interpretation of Lindley's Bayesian tests of significance is fundamentally different from that of sampling-theory tests. It must also be recognized that when our prior information is neither vague nor diffuse results will usually be importantly influenced by prior information, as demonstrated below.

[11] D. V. Lindley, *Introduction to Probability and Statistics from a Bayesian Viewpoint, Part 2. Inference*. Cambridge: University Press, 1965, p. 58 ff.
[12] *Ibid.*, p. 61.

In Lindley's procedure the posterior pdf for the parameter (or parameters) of a model is derived by using a diffuse prior pdf. Let the posterior pdf for the parameter θ be denoted by $p(\theta|\mathbf{y})$, where $\mathbf{y}$ denotes the sample information. Further assume that $p(\theta|\mathbf{y})$ is unimodal. To perform a Lindley significance test of the hypothesis $\theta = \theta_0$, where θ_0 is a suggested value for θ, at the α (say 0.05) level of significance, we construct an interval such that $\Pr\{a < \theta < b|\mathbf{y}\} = 1 - \alpha$.[13] (We referred to such an interval as a Bayesian confidence interval in Chapter 2.) If θ_0 falls in this interval, that is, $a < \theta_0 < b$, we accept the hypothesis $\theta = \theta_0$ at the α level of significance; if θ_0 falls outside the interval a to b, that is, $\theta_0 < a$ or $\theta_0 > b$, we reject the hypothesis $\theta = \theta_0$ at the α level of significance.

The rationale for the Lindley procedure is that the posterior pdf for a parameter, say θ, gives us a basis for expressing beliefs about possible values of θ. If the value $\theta = \theta_0$ is in a region in which the posterior probability density is not high, this fact leads one to suspect that this value for θ is not believable and thus he rejects the hypothesis $\theta = \theta_0$.

Some particular features of Lindley's procedure deserve to be emphasized. First, since it is based on the use of the likelihood function, all the sample information is employed. This contrasts with certain test procedures based on the distribution of a sample statistic which is not a sufficient statistic. Second, we are judging the hypothesis $\theta = \theta_0$ on the basis of posterior beliefs, as represented by the posterior pdf for θ. We are not judging it on the fact that some test statistic may assume a usual or unusual value, given that $\theta = \theta_0$, as in usual sampling-theory test procedures. The level of significance for sampling-theory tests should not be interpreted as measuring the degree of belief that the hypothesis $\theta = \theta_0$ is valid, even though this interpretation is frequently encountered. Sampling theorists do not consider it meaningful to assign probabilities to hypotheses. Third, Lindley points out that "... the significance level is ... an incomplete expression of posterior beliefs." Usually an investigator will study properties of a posterior pdf in detail and not rely solely on a significance test to summarize his beliefs. Finally, it is worth emphasizing again that Lindley suggests use of his procedure only when prior information is vague or diffuse. When prior information is not vague, we shall use posterior odds in comparing and testing hypotheses.

Example 10.1. Suppose that our observations $\mathbf{y}_1' = (y_1, y_2, \ldots, y_n)$ are independently drawn from a normal population with unknown mean μ and known standard deviation $\sigma = \sigma_0$. Under what condition will we reject the hypothesis $\mu = 5$ at the 0.05 level of significance, given diffuse prior beliefs about μ?

[13] As discussed in Chapter 2, we determine a and b such that $b - a$ is minimized subject to the condition that $\Pr\{a < \theta < b|\mathbf{y}\} = 1 - \alpha$.

Applying Bayes' theorem, the posterior pdf for μ is given by

$$p(\mu|\mathbf{y}, \sigma = \sigma_0) \propto \exp\left[-\frac{1}{2\sigma_0^2}\sum_{i=1}^{n}(y_i - \mu)^2\right]$$

$$\propto \exp\left[-\frac{n}{2\sigma_0^2}(\mu - \hat{\mu})^2\right],$$

where $\hat{\mu} = \sum_{i=1}^{n} y_i/n$, the sample mean. Thus, a posteriori, μ is normally distributed with mean $\hat{\mu}$ and variance σ_0^2/n. Then $z = (\mu - \hat{\mu})\sqrt{n}/\sigma_0$ is a standardized normally distributed variable; that is, $Ez = 0$ and $\text{Var}\, z = 1$. By consulting tables of the normal distribution we find $\Pr\{-1.96 < z < 1.96\} = 0.95$. A logically equivalent statement is

$$\text{(10.20)} \qquad \Pr\left\{\hat{\mu} - \frac{1.96\sigma_0}{\sqrt{n}} < \mu < \hat{\mu} + \frac{1.96\sigma_0}{\sqrt{n}} \,|\hat{\mu}, \sigma_0, n\right\} = 0.95.$$

Then, if the value 5 lies in the interval $\hat{\mu} \pm 1.96\sigma_0/\sqrt{n}$, we accept the hypothesis $\mu = 5$ at the 0.05 significance level. If not, we reject.

In (10.20) it is important to emphasize that μ is considered random, whereas $\hat{\mu}$, which depends on the given sample information, is given with σ_0 and n. This contrasts with the sampling-theory approach in which $z' = (\hat{\mu} - 5)\sqrt{n}/\sigma_0$ is considered random and the hypothesis $\mu = 5$ is accepted at the 0.05 significance level if $-1.96 < z' < 1.96$ and rejected if $|z'| > 1.96$; that is, accept if $\hat{\mu}$ is in the interval $5 \pm 1.96\sigma_0/\sqrt{n}$ and reject if $\hat{\mu}$ lies outside this interval. Note that this action is precisely equivalent to those taken on the basis of (10.20) when we follow Lindley's procedure. However, the interpretation and justification for the sampling-theory procedure are quite different from those for the Bayesian procedure.

Example 10.2. In Chapter 3 it was found that with a diffuse prior pdf for regression coefficients and the error terms' common standard deviation in the linear normal regression model the marginal posterior pdf for a single regression coefficient, say β_1 was in the univariate Student-t form; that is, a posteriori $t_\nu = (\beta_1 - \hat{\beta}_1)/s\sqrt{m^{11}}$ has a univariate Student-t pdf with $\nu = n - k$ degrees of freedom [see (3.39), Chapter 3]. If we wish to test the hypothesis $\beta_1 = 0$ by using Lindley's procedure[14] at, say, the 0.20 level of significance, we consult tables of the Student-t distribution for $\nu = n - k$ degrees of freedom to find c such that $\Pr\{-c < t_\nu < c\} = 0.80$. Given the

[14] Note that to be in accord with Lindley's assumptions the value $\beta_1 = 0$ must not be in any sense an unusual value for β_1. If, for example, a theory predicts that $\beta_1 = 0$, it may very well be the case that an investigator will want to incorporate this information in his prior pdf, and thus Lindley's approach with a diffuse prior pdf would not be appropriate.

definition of t_ν above, we have $\Pr\{\hat{\beta}_1 - cs\sqrt{m^{11}} < \beta_1 < \hat{\beta}_1 + cs\sqrt{m^{11}}\} = 0.80$. Then, if the value $\beta_1 = 0$ falls in the interval $\hat{\beta}_1 \pm cs\sqrt{m^{11}}$ we accept; otherwise we reject. As in Example 10.1, this leads to the same action as a sampling-theory test based on the random variable $\hat{\beta}_1/s\sqrt{m^{11}}$.

Example 10.3. Consider the posterior pdf for the autocorrelation coefficient ρ in (4.19). Using numerical integration procedures, we can find the shortest interval such that $\Pr\{a < \rho < b\} = 1 - \alpha$. To test the hypothesis that $\rho = \rho_0$, where ρ_0 is a given value, we accept if $a < \rho_0 < b$ and reject otherwise at the α level of significance. In this instance there does not appear to be a simple sampling-theory test which yields comparable results.

In the above examples we have used Lindley's approach to test hypotheses about a single scalar parameter. If we have a joint hypothesis about two or more parameters, say $\boldsymbol{\theta}$, a Bayesian "highest posterior density" confidence region for $\boldsymbol{\theta}$ is first obtained with probability content $1 - \alpha$. If our hypothesis is $\boldsymbol{\theta} = \boldsymbol{\theta}_0$, where $\boldsymbol{\theta}_0$ is a given vector, we accept if $\boldsymbol{\theta}_0$ is contained in the confidence region and reject otherwise at the α level of significance.

Example 10.4. Consider the simple linear normal regression model in Section 3.1. Equation 3.18 and the associated distributional result provide us with what is needed to construct a Bayesian confidence region (ellipse) with any specified probability content, say $1 - \alpha$. Then, if our hypothesis is $\beta_1 = \beta_{10}$ and $\beta_2 = \beta_{20}$, where β_{10} and β_{20} are given values, we accept if the point (β_{10}, β_{20}) is within the Bayesian confidence region and reject otherwise at the α level of significance. Equivalently, we compute ψ in (3.18) with $\beta_1 = \beta_{10}$ and $\beta_2 = \beta_{20}$ and designate this computed value by ψ_0. Since ψ a posteriori is distributed as $F_{2,\nu}$, where $\nu = n - 2$, we can consult tables of the F-distribution to find c such that $\Pr\{\psi < c\} = 1 - \alpha$. If $\psi_0 < c$, we accept the hypothesis that $\beta_1 = \beta_{10}$ and $\beta_2 = \beta_{20}$ at the α significance level. If $\psi_0 > c$, we reject the hypothesis at the α significance level.

In summary, Lindley's procedure for tests of significance are appropriate when prior information is vague or diffuse. Basically, the rationale for such tests is accept when the suggested value for a parameter under the null hypothesis is in an interval in which the posterior probability density is high and reject otherwise. No decision theoretic justification appears available for this procedure. Rather it is based entirely on what is reasonable a posteriori. For a number of problems Lindley's procedure leads to actions, accept or reject, which are identical to those flowing from sampling-theory test procedures. In some problems, however, for example, testing a hypothesis about an autocorrelation parameter, Lindley's procedure leads to easily computable results, whereas comparable sampling theory results are not available. Finally in large samples, when the likelihood function assumes a normal form, with mean equal to the maximum likelihood estimate for a parameter vector $\boldsymbol{\theta}$ and covariance matrix equal to the inverse of the estimated information matrix (see Chapter 2, Section 2.10), Lindley's testing procedure yields results that are computationally equivalent to large-sample sampling-theory tests which utilize the large-sample normality of maximum likelihood estimators centered at the true parameter vector with approximate covariance matrix given by the inverse of the estimated information matrix.

10.3 COMPARING AND TESTING HYPOTHESES WITH NONDIFFUSE PRIOR INFORMATION

In situations in which prior information is not diffuse, it is important to take this fact into account in comparing and testing hypotheses. If we consider a simple hypothesis $\theta = \theta_0$, where θ_0 is a value suggested by a theory, we may well believe that θ_0 is a more probable value for θ than are other possible values for θ. If this is the situation, then it is important to have test procedures that allow us to incorporate nondiffuse prior information.[15]

To illustrate how nondiffuse prior information can be introduced in comparing and testing hypotheses, assume that we have n observations $\mathbf{y}' = (y_1, y_2, \ldots, y_n)$ independently drawn from a normal population with unknown mean μ and known variance $\sigma^2 = 1$. Assume that our null hypothesis is $H_0: \mu = 0$ and that we have some reason for thinking that $\mu = 0$ with probability π_1 and $\mu \neq 0$ with probability $1 - \pi_1$. Further assume that the prior probability, $1 - \pi_1$, that $\mu \neq 0$ is spread out uniformly over the interval $-M/2$ to $+M/2$; that is, our prior beliefs about μ are as shown in Figure 10.1. Under these assumptions the prior odds ratio pertaining to the hypotheses $H_0: \mu = 0$ and $H_1: \mu \neq 0$ is $\pi_1/(1 - \pi_1)$. Under H_0, the pdf for $\mathbf{y}$, given $\mu = 0$ and $\sigma = 1$, is

$$(10.21) \qquad \begin{aligned} p(\mathbf{y}|\mu = 0, \sigma = 1) &= (\sqrt{2\pi})^{-n} \exp\left(-\tfrac{1}{2} \sum_{i=1}^{n} y_i^2\right) \\ &= (\sqrt{2\pi})^{-n} \exp\left[-\tfrac{1}{2}(\nu s^2 + n\bar{y}^2)\right], \end{aligned}$$

where $\nu = n - 1$, $\nu s^2 = \sum_{i=1}^{n} (y_i - \bar{y})^2$, and $\bar{y} = \sum_{i=1}^{n} y_i/n$. Under the hypothesis $H_1: \mu \neq 0$, we have

$$(10.22) \qquad \begin{aligned} p(\mathbf{y}|\mu \neq 0, \sigma = 1) &= (\sqrt{2\pi})^{-n} \exp\left[-\tfrac{1}{2} \sum_{i=1}^{n} (y_i - \mu)^2\right] \\ &= (\sqrt{2\pi})^{-n} \exp\{-\tfrac{1}{2}[\nu s^2 + n(\mu - \bar{y})^2]\}. \end{aligned}$$

[15] Jeffreys, *op. cit.*, p. 251, points out that often when the hypothesis $\theta = 0$ is being tested the mere fact that it has been suggested that the parameter's value is zero "... corresponds to some presumption that it is fairly small." Thus use of a diffuse prior pdf for θ would be inappropriate in this case.

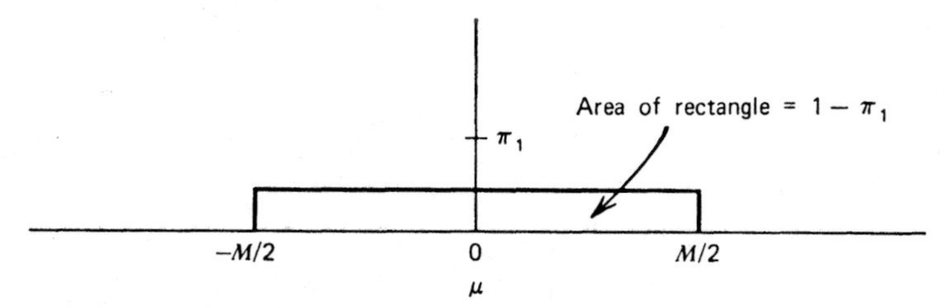

Figure 10.1 Prior pdf for μ.

Now, as explained in Section 10.1, the posterior odds ratio K_{01}, relating to the hypotheses $H_0: \mu = 0$ and $H_1: \mu \neq 0$, is given by (10.19), which specializes in the present instance to[16]

$$(10.23) \qquad K_{01} = \frac{\pi_1}{1 - \pi_1} \frac{\exp[-(n/2)\bar{y}^2]}{\int_{-M/2}^{M/2} \exp[-(n/2)(\mu - \bar{y})^2]\, d\mu/M}.$$

The quantity K_{01} can be readily computed, given values for π_1, n, $\bar{y}$, and M.

Several points regarding (10.23) should be appreciated. First it gives us a basis for comparing the hypotheses $H_0: \mu = 0$ and $H_1: \mu \neq 0$ which incorporate nondiffuse prior information[17] as well as sample information. Second, as explained in Section 10.1, we can choose between H_0 and H_1 to minimize expected loss if we are given losses associated with possible actions and states of the world. Third, as Lindley has demonstrated,[18] the result of using K_{01} in (10.23) to compare the hypotheses H_0 and H_1 can differ from what is obtained in a sampling-theory approach. It is to this point that we now turn.

Assume that $\bar{y}$ is well within the interval $-M/2$ to $M/2$ so that

$$(10.24) \qquad \int_{-M/2}^{M/2} \exp\left[-\frac{n}{2}(\mu - \bar{y})^2\right] d\mu \doteq \left(\frac{2\pi}{n}\right)^{1/2}.$$

This approximation amounts to neglecting the area under the normal pdf to the right of $M/2$ and to the left of $-M/2$ which will be small if $\bar{y}$ is well within the interval $-M/2$ to $M/2$. On substituting from (10.24) into (10.23),

[16] Since the prior pdf in the present problem is part continuous and part discrete, to be formally consistent with (10.19), (10.19) should be expressed in terms of Stieltjes integrals.

[17] Prior information in forms other than that shown in Figure 10.1 can easily be introduced.

[18] See D. V. Lindley, "A Statistical Paradox," *Biometrika*, **44** (1957), 187–192. In M. S. Bartlett's comment on Lindley's paper, *Biometrika*, **44**, 533–534 (1957), a minor error in Lindley's paper is reported.

we have

$$(10.25)\qquad \begin{aligned} K_{01} &= \frac{\pi_1}{1-\pi_1} M\left(\frac{n}{2\pi}\right)^{1/2} \exp\left(-\frac{n}{2}\bar{y}^2\right) \\ &= \frac{\pi_1}{1-\pi_1} M\left(\frac{n}{2\pi}\right)^{1/2} \exp\left(-\frac{z_\alpha^2}{2}\right), \end{aligned}$$

where $z_\alpha = \sqrt{n}\,\bar{y}$. Now, if $z_\alpha \geq 1.96$, a sampling theorist would reject the hypothesis $\mu = 0$ at the $\alpha = 0.05$ significance level. On inserting $z_\alpha = 1.96$ in (10.25), we see that the resulting expression for K_{01}, the posterior odds, depends on the quantities π_1, M, and n. For certain values of these quantities the odds in favor of H_0 can be high, even though $z_\alpha = \sqrt{n}\,\bar{y} = 1.96$. To illustrate this point assume that $\pi_1 = \frac{1}{2}$ and $M = 1$. Then $K_{01} = \sqrt{n/2\pi}\exp(-z_\alpha^2/2)$ and we can table values of n, K_{01}, and the posterior probability that $\mu = 0$, given by $K_{01}/(1 + K_{01})$. For large n the posterior probability that $\mu = 0$ is close to one, even though $\sqrt{n}\,\bar{y} = 1.96$, a value that would lead to *rejection* of the hypothesis $\mu = 0$ at the $\alpha = 0.05$ significance level. This is Lindley's paradox which illustrates very nicely that a sampling-theory test of significance can give results differing markedly from those obtained from a calculation of posterior probabilities which takes account of nondiffuse prior and sample information. In Table 10.2 the results diverge most markedly as n grows large.[19] Basically, the point to appreciate is that in general one minus a significance level in a sampling-theory test cannot be equated with a degree of belief in a hypothesis represented by a posterior probability.

We shall now consider Lindley's problem by using a more general prior pdf for μ following the analysis put forward by Jeffreys.[20] Let us assume that our prior probabilities on the hypotheses $H_0: \mu = 0$ and $H_1: \mu \neq 0$ are π_1 and $1 - \pi_1$, respectively. Given that H_1 is assumed to be true, let $p(\mu)$ be a

Table 10.2

n	K_{01}	$\Pr\{\mu = 0 \mid \sqrt{n}\,\bar{y} = 1.96, \pi_1 = \frac{1}{2}, M = 1\}$
1	0.058	0.055
10	0.185	0.156
100	0.584	0.369
300	1.012	0.503
10,000	5.843	0.854
100,000	18.477	0.949

[19] If a sampling theorist were to adjust his significance level upward as n grows larger, which seems reasonable, z_α would grow with n and tend to counteract somewhat the influence of the $\sqrt{n}$ factor in the expression for K_{01}.

[20] Jeffreys, *op. cit.*, p. 246 ff.

continuous proper prior pdf for μ; that is, $\int p(\mu)\, d\mu = 1$ under H_1. Let $\Pr\{w = 0\}$ denote the prior probability that H_0 is true and $\Pr\{w = 1\}$, the prior probability that H_1 is true, where w is the discrete random variable introduced in connection with (10.2). Using (10.16), we have

$$p(w, \mu|\mathbf{y}) = \frac{p(w)\, p(\mu|w)\, p(\mathbf{y}|\mu, w)}{p(\mathbf{y})}. \tag{10.26}$$

Then the posterior probability that H_0 is true, that is, $w = 0$, is from (10.17):

$$\begin{aligned} p(w = 0|\mathbf{y}) &= \frac{p(w = 0)\, p(\mathbf{y}|\mu = 0, w = 0)}{p(\mathbf{y})} \\ &= \frac{\pi_1 p(\mathbf{y}|\mu = 0, w = 0)}{p(\mathbf{y})}. \end{aligned} \tag{10.27}$$

Note that, given H_0 (or $w = 0$), $p(\mu|w = 0) = 1$ for $\mu = 0$ and 0 for $\mu \neq 0$. The posterior probability that H_1 is true, that is, $w = 1$, is given by

$$\begin{aligned} p(w = 1|\mathbf{y}) &= \int p(w = 1, \mu|\mathbf{y})\, d\mu \\ &= \frac{p(w = 1) \int p(\mu)\, p(\mathbf{y}|\mu, w = 1)\, d\mu}{p(\mathbf{y})} \\ &= \frac{(1 - \pi_1) \int p(\mu)\, p(\mathbf{y}|\mu, w = 1)\, d\mu}{p(\mathbf{y})}. \end{aligned} \tag{10.28}$$

Then the posterior odds ratio K_{01} is

$$K_{01} = \frac{p(w = 0|\mathbf{y})}{p(w = 1|\mathbf{y})} = \frac{\pi_1}{1 - \pi_1} \frac{p(\mathbf{y}|\mu = 0, w = 0)}{\int p(\mu)\, p(\mathbf{y}|\mu, w = 1)\, d\mu}. \tag{10.29}$$

Under the above assumptions that the elements of $\mathbf{y}' = (y_1, y_2, \ldots, y_n)$ are independently drawn from a normal population with standard deviation $\sigma = \sigma_0$, with σ_0 known, (10.29) becomes

$$K_{01} = \left(\frac{\pi_1}{1 - \pi_1}\right) \frac{\exp\left[-(n/2\sigma_0^2)\bar{y}^2\right]}{\int p(\mu) \exp\left[-(n/2\sigma_0^2)(\mu - \bar{y})^2\right] d\mu}. \tag{10.30}$$

This expression can be evaluated if π_1 and $p(\mu)$ are given.

Let us assume that in (10.30) π_1 equals $\frac{1}{2}$; that is, the probability that $H_0\colon \mu = 0$ is true is $\frac{1}{2}$. Then $\pi_1/(1 - \pi_1) = 1$ and (10.30) becomes

$$K_{01} = \frac{\exp\left[-(n/2\sigma_0^2)\bar{y}^2\right]}{\int p(\mu) \exp\left[-(n/2\sigma_0^2)(\mu - \bar{y})^2\right] d\mu}. \tag{10.31}$$

Along with Jeffreys, consider two extreme cases. First assume that there is a finite interval, say $-a$ to a, such that $\int_{-a}^{a} p(\mu)\, d\mu = 1$. If $\bar{y}$ lies within the interval $-a$ to a and $\sigma_0/\sqrt{n}$ is so large that $n(\mu - \bar{y})^2/2\sigma_0^2$ is small for $-a < \mu < a$, we have both the numerator and the denominator of K_{01} in (10.31), each approximately equal to one, and under these conditions $K_{01} \doteq 1$. Thus there is no discrimination between the hypotheses when the standard deviation $\sigma_0/\sqrt{n}$ associated with the mean $\bar{y}$ is much greater than the range for μ, $-a$ to a, permitted under the hypothesis H_1. The result $K_{01} \doteq 1$ under these conditions appears to be reasonable.

As another extreme case, assume that $\sigma_0/\sqrt{n}$ is very small so that $\exp[-(n/2\sigma_0^2)(\mu - \bar{y})^2]$ can take very large values in the interval $-a$ to a. The integral in the denominator of (10.31) is approximately equal to $(\sqrt{2\pi}\,\sigma_0/\sqrt{n})\,p(\bar{y})$, where $p(\bar{y})$ is the prior pdf for μ evaluated at $\mu = \bar{y}$. Then

$$K_{01} \doteq \frac{1}{\sqrt{2\pi}\,p(\bar{y})}\frac{\sqrt{n}}{\sigma_0}\exp\left(-\frac{n}{2}\frac{\bar{y}^2}{\sigma_0^2}\right). \tag{10.32}$$

As Jeffreys points out, if $\bar{y} = 0$, K_{01} is proportional to $\sqrt{n}$ and, as n grows, K_{01} does also, which indicates support for the hypothesis $H_0\colon \mu = 0$. If $|\bar{y}| \gg \sigma_0/\sqrt{n}$, the exponential factor in (10.32) will be small and thus the observations tend to support $H_1\colon \mu \neq 0$; that is, K_{01} tends to be small under this condition. Further, for given n there will be a value for $\sqrt{n}\,\bar{y}/\sigma_0$ such that $K_{01} = 1$. Last, for values of $\bar{y}$ such that $|\bar{y}| < \sigma_0/\sqrt{n}$, as n grows, K_{01} grows. This is reasonable, since having $|\bar{y}|$ less than its standard deviation, $\sigma_0/\sqrt{n}$, should carry more weight for believing that $\mu = 0$ when n is large than when n is small.[21]

Since nondiffuse prior information affects posterior odds in both small and large samples, it is obvious that care must be exercised in representing the prior information to be employed in an analysis. That this information exerts an influence on posterior odds, even in large samples, contrasts with the situation in estimation in which the influence of nondogmatic prior information on the shape of a posterior pdf generally diminishes as the sample size grows.

10.4 COMPARING REGRESSION MODELS

Often in econometrics and elsewhere, we have the problem of comparing alternative regression models designed to explain the variation of a particular dependent variable; for example, in the work of Friedman and Meiselman[22] two models were studied:

$$\left.\begin{aligned} C_t &= v_0 + v_1 M_t + u_{1t} \quad (10.33)\\ C_t &= m_0 + m_1 A_t + u_{2t} \quad (10.34)\end{aligned}\right\} \qquad t = 1, 2, \ldots, T,$$

[21] See Jeffreys, *op. cit.*, Chapters 5 and 6, for further consideration of tests of significance.

[22] M. Friedman and D. Meiselman, "The Relative Stability of Velocity and the Investment Multiplier," in the Commission on Money and Credit volume, *Stabilization Policies*. Englewood Cliffs, N.J.: Prentice-Hall, 1963.

where the subscript t denotes the value of a variable in the tth time period, C_t, consumption, M_t, the money supply, A_t, autonomous spending, and u_{1t} and u_{2t}, error terms. The v's and m's are unknown regression parameters. To choose between (10.33) and (10.34) Friedman and Meiselman used a goodness of fit measure, R^2. It is of interest to determine conditions under which this criterion, namely, the choice of the model with the higher R^2, is compatible with minimizing expected loss in a Bayesian decision theoretic approach.[23] In addition, we provide general Bayesian procedures for comparing and choosing models.

Assume that there are just two possible models for explaining the variation of a dependent variable; that is, we assume that the observation vector $\mathbf{y}' = (y_1, y_2, \ldots, y_n)$ is generated by

$$M_1: \mathbf{y} = X_1\boldsymbol{\beta}_1 + \mathbf{u}_1 \tag{10.35}$$

or by

$$M_2: \mathbf{y} = X_2\boldsymbol{\beta}_2 + \mathbf{u}_2, \tag{10.36}$$

where M_1 denotes the first model and M_2, the second, X_1 and X_2 are each $n \times k$ matrices of given quantities, each with rank k, $\boldsymbol{\beta}_1$ and $\boldsymbol{\beta}_2$ are each $k \times 1$ coefficient vectors with no common elements, and $\mathbf{u}_1$ and $\mathbf{u}_2$ are each $n \times 1$ vectors of error terms. Given that M_1 is correct, the elements of $\mathbf{u}_1$ are assumed to be normally and independently distributed, each with mean zero and common variance σ_1^2. Similarly, in the case that M_2 is correct, the elements of $\mathbf{u}_2$ are assumed to be normally and independently distributed, each with mean zero and common variance σ_2^2.

As regards prior pdf's for the parameters $\boldsymbol{\beta}_i$ and σ_i, $i = 1, 2$, we employ the following natural conjugate forms for $i = 1, 2$:

$$p(\boldsymbol{\beta}_i, \sigma_i) = p(\boldsymbol{\beta}_i|\sigma_i)\, p(\sigma_i), \tag{10.37}$$

with

$$p(\boldsymbol{\beta}_i|\sigma_i) = \frac{|C_i|^{1/2}}{(2\pi)^{k/2}\sigma_i^{\,k}} \exp\left[-\frac{1}{2\sigma_i^{\,2}}(\boldsymbol{\beta}_i - \bar{\boldsymbol{\beta}}_i)'C_i(\boldsymbol{\beta}_i - \bar{\boldsymbol{\beta}}_i)\right] \tag{10.38}$$

and

$$p(\sigma_i) = \frac{K_i}{\sigma_i^{q_i+1}} \exp\left(-\frac{q_i\bar{s}_i^{\,2}}{2\sigma_i^{\,2}}\right), \tag{10.39}$$

[23] This problem has been analyzed in H. Thornber, "Applications of Decision Theory to Econometrics," unpublished doctoral dissertation, University of Chicago, 1966, and in M. S. Geisel, "Comparing and Choosing Among Parametric Statistical Models: A Bayesian Analysis with Macroeconomic Applications," unpublished doctoral dissertation, University of Chicago, 1970. See also G. E. P. Box and W. J. Hill, "Discrimination Among Mechanistic Models," *Technometrics*, **9**, 57–71 (1967).

where the normalizing constant $K_i = 2(q_i\bar{s}_i^2/2)^{q_i/2}/\Gamma(q_i/2)$. In (10.38) we have assumed a proper normal prior pdf for the elements of $\boldsymbol{\beta}_i$, given σ_i, with prior mean vector $\bar{\boldsymbol{\beta}}_i$ and covariance matrix $\sigma_i^2 C_i^{-1}$; C_i, a $k \times k$ matrix to be assigned by the investigator, is assumed to be a positive definite symmetric matrix. In (10.39) the prior pdf for σ_i is in the inverted gamma form with parameters q_i and $\bar{s}_i^2$ to be assigned values by the investigator. For (10.39) to be proper, we need $0 < q_i, \bar{s}_i^2 < \infty$, $i = 1, 2$.

With the above assumptions about the models M_1 and M_2 and the prior pdf's for the parameters, we can apply (10.19) to obtain posterior odds, K_{12}, pertaining to the two models. If we let $p(M_1)/p(M_2)$ denote the prior odds, then[24]

$$K_{12} = \frac{p(M_1)}{p(M_2)} \frac{\int p(\boldsymbol{\beta}_1|\sigma_1)\, p(\sigma_1)\, p(\mathbf{y}|\boldsymbol{\beta}_1, \sigma_1, M_1)\, d\boldsymbol{\beta}_1\, d\sigma_1}{\int p(\boldsymbol{\beta}_2|\sigma_2)\, p(\sigma_2)\, p(\mathbf{y}|\boldsymbol{\beta}_2, \sigma_2, M_2)\, d\boldsymbol{\beta}_2\, d\sigma_2}, \tag{10.40}$$

where, for $i = 1, 2$, $p(\mathbf{y}|\boldsymbol{\beta}_i, \sigma_i, M_i)$, viewed as a function of $\boldsymbol{\beta}_i$ and σ_i, is the likelihood function, given that M_i is assumed to be true; that is, for $i = 1, 2$,

(10.41)
$$p(\mathbf{y}|\boldsymbol{\beta}_i, \sigma_i, M_i) = \frac{1}{(2\pi)^{n/2}} \frac{1}{\sigma_i^n} \exp\left\{-\frac{1}{2\sigma_i^2}[\nu s_i^2 + (\boldsymbol{\beta}_i - \hat{\boldsymbol{\beta}}_i)' X_i' X_i (\boldsymbol{\beta}_i - \hat{\boldsymbol{\beta}}_i)]\right\},$$

where $\hat{\boldsymbol{\beta}}_i = (X_i'X_i)^{-1}X_i'\mathbf{y}$, $\nu = n - k$ and $\nu s_i^2 = (\mathbf{y} - X_i\hat{\boldsymbol{\beta}}_i)'(\mathbf{y} - X_i\hat{\boldsymbol{\beta}}_i)$.

To evaluate K_{12} in (10.40), we must perform the indicated integrations; that is, we have

(10.42)
$$\begin{aligned}
&\int p(\boldsymbol{\beta}_1|\sigma_1)\, p(\sigma_1)\, p(\mathbf{y}|\boldsymbol{\beta}_1, \sigma_1, M_1)\, d\boldsymbol{\beta}_1\, d\sigma_1 \\
&= \frac{K_1|C_1|^{1/2}}{(2\pi)^{(n+k)/2}} \int \frac{1}{\sigma_1^{n+q_1+k+1}} \exp\left\{-\frac{1}{2\sigma_1^2}[q_1\bar{s}_1^2 + \nu s_1^2 + (\boldsymbol{\beta}_1 - \bar{\boldsymbol{\beta}}_1)' C_1 (\boldsymbol{\beta}_1 - \bar{\boldsymbol{\beta}}_1)\right. \\
&\qquad \left. + (\boldsymbol{\beta}_1 - \hat{\boldsymbol{\beta}}_1)' X_1' X_1 (\boldsymbol{\beta}_1 - \hat{\boldsymbol{\beta}}_1)]\right\} d\boldsymbol{\beta}_1\, d\sigma_1.
\end{aligned}$$

To integrate with respect to the elements of $\boldsymbol{\beta}_1$, complete the square on $\boldsymbol{\beta}_1$ in the exponent as follows:

$$\begin{aligned}
&(\boldsymbol{\beta}_1 - \bar{\boldsymbol{\beta}}_1)' C_1 (\boldsymbol{\beta}_1 - \bar{\boldsymbol{\beta}}_1) + (\boldsymbol{\beta}_1 - \hat{\boldsymbol{\beta}}_1)' X_1' X_1 (\boldsymbol{\beta}_1 - \hat{\boldsymbol{\beta}}_1) \\
&\quad = \boldsymbol{\beta}_1' A_1 \boldsymbol{\beta}_1 - 2\boldsymbol{\beta}_1'(C_1\bar{\boldsymbol{\beta}}_1 + X_1'X_1\hat{\boldsymbol{\beta}}_1) + \bar{\boldsymbol{\beta}}_1' C_1 \bar{\boldsymbol{\beta}}_1 + \hat{\boldsymbol{\beta}}_1' X_1' X_1 \hat{\boldsymbol{\beta}}_1 \\
&\quad = (\boldsymbol{\beta}_1 - \tilde{\boldsymbol{\beta}}_1)' A_1 (\boldsymbol{\beta}_1 - \tilde{\boldsymbol{\beta}}_1) + \bar{\boldsymbol{\beta}}_1' C_1 \bar{\boldsymbol{\beta}}_1 + \hat{\boldsymbol{\beta}}_1' X_1' X_1 \hat{\boldsymbol{\beta}}_1 - \tilde{\boldsymbol{\beta}}_1' A_1 \tilde{\boldsymbol{\beta}}_1 \\
&\quad = (\boldsymbol{\beta}_1 - \tilde{\boldsymbol{\beta}}_1)' A_1 (\boldsymbol{\beta}_1 - \tilde{\boldsymbol{\beta}}_1) + (\bar{\boldsymbol{\beta}}_1 - \tilde{\boldsymbol{\beta}}_1)' C_1 (\bar{\boldsymbol{\beta}}_1 - \tilde{\boldsymbol{\beta}}_1) \\
&\qquad + (\hat{\boldsymbol{\beta}}_1 - \tilde{\boldsymbol{\beta}}_1)' X_1' X_1 (\hat{\boldsymbol{\beta}}_1 - \tilde{\boldsymbol{\beta}}_1),
\end{aligned}$$

[24] Note that (10.40) can be expressed as $K_{12} = [p(M_1)/p(M_2)][p(\mathbf{y}|M_1)/p(\mathbf{y}|M_2)]$, where $p(\mathbf{y}|M_i)$ is the marginal pdf for the observations $i = 1, 2$, given that M_i is assumed to be true.

where

(10.43) $A_1 = C_1 + X_1'X_1$ and $\tilde{\boldsymbol{\beta}}_1 = A_1^{-1}(C_1\bar{\boldsymbol{\beta}}_1 + X_1'X_1\hat{\boldsymbol{\beta}}_1)$.

On substituting in (10.42) and integrating with respect to the elements of $\boldsymbol{\beta}_1$, using properties of the multivariate normal pdf, we obtain

(10.44)

$$\frac{K_1}{(2\pi)^{n/2}} |C_1|^{1/2}|A_1|^{-1/2} \int \frac{1}{\sigma_1^{n+q_1+1}} \exp\left\{-\frac{1}{2\sigma_1^2}[q_1\bar{s}_1^2 + \nu s_1^2 + Q_{1a} + Q_{1b}]\right\} d\sigma_1,$$

where we have introduced the following definitions:

(10.45) $Q_{1a} = (\bar{\boldsymbol{\beta}}_1 - \tilde{\boldsymbol{\beta}}_1)'C_1(\bar{\boldsymbol{\beta}}_1 - \tilde{\boldsymbol{\beta}}_1)$ and $Q_{1b} = (\hat{\boldsymbol{\beta}}_1 - \tilde{\boldsymbol{\beta}}_1)'X_1'X_1(\hat{\boldsymbol{\beta}}_1 - \tilde{\boldsymbol{\beta}}_1)$.

Then on integrating (10.44) with respect to σ_1, the result is[25]

$$\frac{1}{2}\frac{K_1}{(2\pi)^{n/2}}\left(\frac{|C_1|}{|A_1|}\right)^{1/2} \frac{2^{(n+q_1)/2}\Gamma[(n+q_1)/2]}{(q_1\bar{s}_1^2 + \nu s_1^2 + Q_{1a} + Q_{1b})^{(n+q_1)/2}}.$$

When similar operations are performed to evaluate the integral in the denominator of the expression for K_{12} in (10.40), the result for K_{12} is

(10.46)

$$\begin{aligned} K_{12} &= \frac{p(M_1)}{p(M_2)}\frac{K_1}{K_2}\left[\frac{|C_1|/|A_1|}{|C_2|/|A_2|}\right]^{1/2} \frac{2^{(n+q_1)/2}}{2^{(n+q_2)/2}}\frac{\Gamma[(n+q_1)/2]}{\Gamma[(n+q_2)/2]} \\ &\quad\times \frac{(q_1\bar{s}_1^2 + \nu s_1^2 + Q_{1a} + Q_{1b})^{-(n+q_1)/2}}{(q_2\bar{s}_2^2 + \nu s_2^2 + Q_{2a} + Q_{2b})^{-(n+q_2)/2}} \\ &= \frac{p(M_1)}{p(M_2)}\left[\frac{|C_1|/|A_1|}{|C_2|/|A_2|}\right]^{1/2} \frac{\Gamma[(n+q_1)/2]/\Gamma(q_1/2)}{\Gamma[(n+q_2)/2]/\Gamma(q_2/2)}\frac{(q_1\bar{s}_1^2)^{q_1/2}}{(q_2\bar{s}_2^2)^{q_2/2}} \\ &\quad\times \frac{(q_1\bar{s}_1^2 + \nu s_1^2 + Q_{1a} + Q_{1b})^{-(n+q_1)/2}}{(q_2\bar{s}_2^2 + \nu s_2^2 + Q_{2a} + Q_{2b})^{-(n+q_2)/2}}, \end{aligned}$$

where $A_2 = C_2 + X_2'X_2$ and Q_{2a} and Q_{2b} are defined similarly to Q_{1a} and Q_{2b} in (10.45); that is

(10.47) $Q_{2a} = (\bar{\boldsymbol{\beta}}_2 - \tilde{\boldsymbol{\beta}}_2)'C_2(\bar{\boldsymbol{\beta}}_2 - \tilde{\boldsymbol{\beta}}_2)$ and $Q_{2b} = (\hat{\boldsymbol{\beta}}_2 - \tilde{\boldsymbol{\beta}}_2)'X_2'X_2(\hat{\boldsymbol{\beta}}_2 - \tilde{\boldsymbol{\beta}}_2)$,

with $\hat{\boldsymbol{\beta}}_2 = (X_2'X_2)^{-1}X_2'\mathbf{y}$, $\tilde{\boldsymbol{\beta}}_2 = A_2^{-1}(C_2\bar{\boldsymbol{\beta}}_2 + X_2'X_2\hat{\boldsymbol{\beta}}_2)$, and $\bar{\boldsymbol{\beta}}_2$ is the prior mean vector for $\boldsymbol{\beta}_2$. The expression for K_{12} in (10.46) can also be expressed as

(10.48)
$$K_{12} = \frac{p(M_1)}{p(M_2)}\left[\frac{|C_1|/|A_1|}{|C_2|/|A_2|}\right]^{1/2}\left(\frac{\delta_1}{\delta_2}\right)^{-n/2}\frac{\bar{s}_1^2/\delta_1}{\bar{s}_2^2/\delta_2}\frac{f_{q_1,n}(\bar{s}_1^2/\delta_1)}{f_{q_2,n}(\bar{s}_2^2/\delta_2)},$$

where, for $i = 1, 2$, $\delta_i = (\nu s_i^2 + Q_{ia} + Q_{ib})/n$ and $f_{q_i,n}(\bar{s}_1^2/\delta_i)$ denotes the ordinate of the F pdf with q_i and n degrees of freedom.

We now turn to the interpretation of the expression for K_{12} in (10.48).

[25] Note for a, $b > 0$, $\int_0^\infty (\sigma^{a+1})^{-1} \exp(-b/2\sigma^2)\, d\sigma = \frac{1}{2}(2/b)^{a/2} \int_0^\infty x^{a/2-1}e^{-x}\, dx = \frac{1}{2}(2/b)^{a/2}\Gamma(a/2)$, where Γ denotes the gamma function.

1. The first factor $p(M_1)/p(M_2)$ is the prior odds ratio. If, for example, we had no reason to believe more in one model than the other, we would take $p(M_1) = p(M_2) = \frac{1}{2}$ and the odds ratio $p(M_1)/p(M_2) = 1$.

2. The second factor involves the ratios $|C_1|/|A_1|$ and $|C_2|/|A_2|$. Note from (10.38) and (10.43) that $|C_i|/|A_i|$ is a measure of the precision (or information) in the prior pdf for $\boldsymbol{\beta}_i$, given σ_i in relation to the posterior precision (or information), given σ_i for $i = 1, 2$. The posterior precision, given σ_i, is proportional to $|A_i| = |C_i + X_i'X_i|$ and thus depends on C_i and the "design" matrix $X_i'X_i$. The posterior odds ratio K_{12} will be larger (smaller), the larger (smaller) $|C_1|/|A_1|$ is relative to $|C_2|/|A_2|$. This is reasonable, since, *ceteris paribus*, we should favor the model with more prior information, as measured by $|C_i|/|A_i|$, provided, of course, that the prior information is in accord with sample information (on this point see below).

3. The third factor in (10.48), namely, $(\delta_1/\delta_2)^{-n/2}$, involves quantities δ_1 and δ_2 which reflect what the sample information has to say about the goodness of fit of the models and the extent to which the prior information about the coefficient vectors is in accord with the sample information; for example, $\delta_1 = (\nu s_1^2 + Q_{1a} + Q_{1b})/n$. Now $\nu s_1^2 = (\mathbf{y} - X_1\hat{\boldsymbol{\beta}}_1)'(\mathbf{y} - X_1\hat{\boldsymbol{\beta}}_1)$ is just the residual sum of squares, a goodness of fit measure, and

$$Q_{1a} = (\bar{\boldsymbol{\beta}}_1 - \tilde{\boldsymbol{\beta}}_1)'C_1(\bar{\boldsymbol{\beta}}_1 - \tilde{\boldsymbol{\beta}}_1)$$

involves the difference between the prior mean vector $\bar{\boldsymbol{\beta}}_1$ and $\tilde{\boldsymbol{\beta}}_1$, which is a "matrix" weighted average of $\bar{\boldsymbol{\beta}}_1$ and $\hat{\boldsymbol{\beta}}_1$, where $\hat{\boldsymbol{\beta}}_1$ is $(X_1'X_1)^{-1}X_1'\mathbf{y}_1$, the sample estimate. Thus the closer the agreement between $\bar{\boldsymbol{\beta}}_1$ and $\tilde{\boldsymbol{\beta}}_1$, the smaller Q_{1a}; $Q_{1b} = (\hat{\boldsymbol{\beta}}_1 - \tilde{\boldsymbol{\beta}}_1)'X_1'X_1(\hat{\boldsymbol{\beta}}_1 - \tilde{\boldsymbol{\beta}}_1)$ depends on the difference between the sample estimate $\hat{\boldsymbol{\beta}}_1$ and the average of $\hat{\boldsymbol{\beta}}_1$ and $\bar{\boldsymbol{\beta}}_1$, namely $\tilde{\boldsymbol{\beta}}_1$. Similar considerations relate to δ_2. Thus, for given n, the larger δ_1 is relative to δ_2, due perhaps to poor fit (high νs_1^2) and/or incompatibility of prior and sample information about $\boldsymbol{\beta}_1$ (high Q_{1a} and Q_{1b}), the lower will be the posterior odds, K_{12}, in favor of M_1.

4. The final factor in (10.48),

$$\frac{w_1 f_{q_1,n}(w_1)}{w_2 f_{q_2,n}(w_2)}, \tag{10.49}$$

where $w_i = \bar{s}_i^2/\delta_i$, $i = 1, 2$, reveals the dependence of K_{12} on the prior information regarding σ_1 and σ_2 [see (10.39) for the prior pdf's for these parameters.][26] Since w_i depends on $\bar{s}_i^2$, a prior parameter connected with the location of the prior pdf for σ_i^2, and s_i^2, a sample estimate of σ_i^2, K_{12} will be affected by divergences between these two quantities. Further assume that $\bar{s}_1^2 = \bar{s}_2^2$ and $q_1 = q_2$; that is, we assign the same prior pdf's for σ_1 and σ_2.

[26] Note that the inverted gamma pdf for σ_i, $i = 1, 2$, in (10.39) has a modal value at $\bar{s}_i\sqrt{q_i/(1 + q_i)}$. Also, the prior mean of σ_i^2 is $q_i\bar{s}_i^2/(q_i - 2)$, for $q_i > 2$.

Then, should $\delta_1 = \delta_2$, the factor in (10.49) assumes a value of one, which appears reasonable. Its value is also one under the more general conditions $q_1 = q_2$ and $\bar{s}_1^2/\delta_1 = \bar{s}_2^2/\delta_2$.

Next, it is interesting to determine the behavior of the posterior odds K_{12}, given in (10.48), when n grows large. As n grows large, $\delta_i \to s_i^2$, since Q_{ia}/n and Q_{ib}/n both approach zero.[27] Assume further that $[|C_1|/|A_1| \div |C_2|/|A_2|]^{1/2} \to 1$ as n grows large, $q_1 = q_2 = q$, and $\bar{s}_1^2 = \bar{s}_2^2 = \bar{s}^2$. Under these conditions and noting that $(1/q)f_{q,n}(w)$ approaches the χ^2 pdf with q degrees of freedom when n is large,[28] we obtain the following expression for the posterior odds:

$$K_{12} = \left(\frac{s_1^2}{s_2^2}\right)^{-(n+q)/2} \frac{\exp(-q\bar{s}^2/2s_1^2)}{\exp(-q\bar{s}^2/2s_2^2)} \tag{10.50}$$

if we take $p(M_1) = p(M_2)$. The value of K_{12} is vitally dependent on the relative sizes of the sample quantities s_1^2 and s_2^2.[29] If $s_1^2 = s_2^2$, $K_{12} = 1$; if $s_1^2 > s_2^2$, $K_{12} < 1$; and if $s_1^2 < s_2^2$, $K_{12} > 1$. These results are intuitively pleasing. They relate, however, to situations in which n is large and the other assumptions introduced above are satisfied.

Last it is useful to determine what happens as we let our prior information become diffuse; that is $|C_i| \to 0$[30] and $q = q_1 = q_2 \to 0$ with the assumption that $p(M_1) = p(M_2)$. Under these conditions $\tilde{\boldsymbol{\beta}}_i \to \hat{\boldsymbol{\beta}}_i$ and thus $Q_{ib} \to 0$ and $Q_{ia} \to 0$.[31] Thus $\delta_i \to \nu s_i^2/n$ as our prior information about $\boldsymbol{\beta}_1$ and $\boldsymbol{\beta}_2$ becomes diffuse. Further we assume that, as $|C_i| \to 0$, $i = 1, 2$, $[|C_1|/|A_1| \div |C_2|/|A_2|]^{1/2} \to 1$. Last we assume that $\bar{s}_1^2 = \bar{s}_2^2$. Under these assumptions the

[27] We assume that $\lim_{n\to\infty} X_i'X_i/n$ is finite, $i = 1, 2$.

[28] That is, the pdf of qw, which is $(1/q)f_{q,n}(w)$, approaches $[2^m(qw)^{m-1}]/\Gamma(m) \exp(-qw/2)$ as n grows large.

[29] If q is very small so that the ratio of exponentials in (10.50) is close to one, then $K_{12} \doteq (s_1^2/s_2^2)^{-n/2}$.

[30] Since $|C_i|$ equals the product of the roots of C_i, $|C_i| \to 0$ can result from having all the roots of $C_i \to 0$, which is what we assume.

[31] $Q_{ia} = (\bar{\boldsymbol{\beta}}_i - \tilde{\boldsymbol{\beta}}_i)'C_i(\bar{\boldsymbol{\beta}}_i - \tilde{\boldsymbol{\beta}}_i) = \mathbf{z}_i'P_i'C_iP_i\mathbf{z}_i = \mathbf{z}_i'D_i\mathbf{z}_i$, where $\bar{\boldsymbol{\beta}}_i - \tilde{\boldsymbol{\beta}}_i = P_i\mathbf{z}_i$ with P_i an orthogonal matrix such that $P_i'C_iP_i = D_i$, a diagonal matrix with the roots of C_i on the diagonal. If these roots $\to 0$, $Q_{ia} \to 0$. With respect to $\tilde{\boldsymbol{\beta}}_i - \hat{\boldsymbol{\beta}}_i$ as the roots of $C_i \to 0$, we have

$$\tilde{\boldsymbol{\beta}}_i = (X_i'X_i + C_i)^{-1}(C_i\bar{\boldsymbol{\beta}}_i + X_i'X_i\hat{\boldsymbol{\beta}}_i) = [I + (X_i'X_i)^{-1}C_i]^{-1}(X_i'X_i)^{-1}(C_i\bar{\boldsymbol{\beta}}_i + X_i'X_i\hat{\boldsymbol{\beta}}_i).$$

Now let H_i be an orthogonal matrix such that $H_i'(X_i'X_i)^{-1}C_iH_i$ is a diagonal matrix, say G_i, with diagonal elements of less than one. Then $[I + (X_i'X_i)^{-1}C_i]^{-1} = H_i(I + G_i)^{-1}H_i' = H_i(I - G_i + G_i^2 - \cdots) H_i' = I$ as the roots of $C_i \to 0$. This leaves us with $\tilde{\boldsymbol{\beta}}_i = \hat{\boldsymbol{\beta}}_i + (X_i'X_i)^{-1}C_i\bar{\boldsymbol{\beta}}_i$. Now write $(X_i'X_i)^{-1}C_i\bar{\boldsymbol{\beta}}_i = H_iH_i'(X_i'X_i)^{-1}C_iH_iH_i'\bar{\boldsymbol{\beta}}_i = H_iG_iH_i'\bar{\boldsymbol{\beta}}_i$, which approaches zero as the elements of the diagonal matrix $G_i \to 0$, as they will do if the roots of $C_i \to 0$; that is, $|G_i| = |(X_i'X_i)^{-1}C_i| = \prod_{i=1}^k \phi_i\psi_i$, where the ϕ_i's are the roots of $(X_i'X_i)^{-1}$ and the ψ_i's are the roots of C_i. More simply, if all the roots of $C_i \to 0$, then $C_i \to a$ null matrix and the results above follow immediately.

expression for K_{12} in (10.48) becomes

$$K_{12} = \left(\frac{s_1^2}{s_2^2}\right)^{-n/2}, \tag{10.51}$$

which is a function[32] of the ratio s_1^2/s_2^2. If $s_1^2/s_2^2 = 1$, $K_{12} = 1$, if $s_1^2/s_2^2 > 1$, $K_{12} < 1$, and, if $s_1^2/s_2^2 < 1$, $K_{12} > 1$. Thus, as pointed out in Section 10.1, if our loss function is symmetric, acting to minimize expected loss in the choice between M_1 and M_2 involves choosing the model with the higher posterior probability. Under the present assumptions this is compatible with the rule of choosing the model with the smaller s^2 (or higher R^2). Of course, this rule is appropriate only in the case of a symmetric loss function and diffuse prior information or when n is large and other conditions are satisfied, as explained above.

10.5 COMPARING DISTRIBUTED LAG MODELS

In Chapter 7 the estimation of distributed-lag consumption function models was considered. In this section we take up the problem of computing posterior odds pertaining to alternative formulations with nondiffuse prior information about the parameters incorporated.[33] As in Chapter 7, we consider the following equation:

$$C_t = \lambda C_{t-1} + (1 - \lambda)kY_t + u_t - \lambda u_{t-1}, \qquad t = 1, 2, \ldots, T. \tag{10.52}$$

In (10.52) the subscript t denotes the value of a variable in the tth time period, C_t and Y_t are price-deflated seasonally adjusted personal-consumption expenditures and disposable income, respectively, λ and k are parameters, and u_t is a disturbance term. Our alternative models for the observations, the C_t's, taking C_0 as given, are described here:

M_1: Equation 10.52, with $u_t - \lambda u_{t-1} = \epsilon_{1t}$, $t = 1, 2, \ldots, T$, where the ϵ_{1t}'s are normally and independently distributed, each with mean zero and variance σ_1^2.

M_2: Equation 10.52, with $u_t = \epsilon_{2t}$, $t = 1, 2, \ldots, T$, where the ϵ_{2t}'s are normally and independently distributed, each with mean zero and variance σ_2^2.

Note that, as long as $\lambda \neq 0$, M_1 and M_2 are mutually exclusive.

As prior pdf's for the parameters of (10.52) we employ the prior pdf's

[32] Thornber, *loc. cit.*, and Geisel, *loc. cit.*, arrive at the same result with slightly different assumptions.

[33] These results are taken from A. Zellner and M. S. Geisel, "Analysis of Distributed Lag Models with Applications to Consumption Function Estimation," invited paper presented to the Econometric Society, Amsterdam, September 1968, and published in *Econometrica*, **38**, 865–888 (1971).

discussed in Section 7.5. For convenience they are reproduced below, with, in each instance, $0 < \lambda < 1$, $0 < k < 1$, and $0 < \sigma_i < \infty$:

First Prior Pdf's:

$$p(\lambda, k, \sigma_i) \propto \frac{1}{\sigma_i}, \qquad i = 1, 2. \tag{10.53}$$

Second Prior Pdf's:

$$p(\lambda, k, \sigma_i) \propto \frac{\lambda^{34}(1-\lambda)^{14}k^{89}(1-k)^{9}}{\sigma_i}, \qquad i = 1, 2. \tag{10.54}$$

Third Prior Pdf's:

$$p(\lambda, k, \sigma_i) \propto \frac{\lambda(1-\lambda)^{7}k^{89}(1-k)^{9}}{\sigma_i}, \qquad i = 1, 2. \tag{10.55}$$

In each instance we assume that λ, k, and σ_i are independently distributed and we take a diffuse prior pdf for σ_i. With respect to λ and k in (10.53), these parameters are assumed to be uniformly distributed in the interval zero to one. In (10.54) and (10.55) independent beta prior pdf's are employed for λ and k. The beta prior for k has mean 0.9 and variance 0.00089 in both (10.54) and (10.55). The prior pdf for λ has mean 0.7 and variance 0.0041 in (10.54) and mean 0.2 and variance 0.0146 in (10.55).

The posterior odds ratio is given by

$$K_{12} = \frac{p(M_1)\int\int\int p(\lambda, k, \sigma_1)\, p_1(\mathbf{C}|\lambda, k, \sigma_1)\, d\lambda\, dk\, d\sigma_1}{p(M_2)\int\int\int p(\lambda, k, \sigma_2) p_2\,(\mathbf{C}|\lambda, k, \sigma_2)\, d\lambda\, dk\, d\sigma_2}, \tag{10.56}$$

where $p(M_1)/p(M_2)$ is the prior odds ratio, $\mathbf{C}' = (C_1, C_2, \ldots, C_T)$, and $p_i(\mathbf{C}|\lambda, k, \sigma_i)$ is the likelihood function, $i = 1, 2$, given C_0. The likelihood functions are given by

$$\begin{aligned} p_1(\mathbf{C}|\lambda, k, \sigma_1) \propto \frac{1}{\sigma_1{}^T} \exp\Big\{-\frac{1}{2\sigma_1{}^2}&[\mathbf{C} - \lambda\mathbf{C}_{-1} - (1-\lambda)k\mathbf{Y}]' \\ &\times [\mathbf{C} - \lambda\mathbf{C}_{-1} - (1-\lambda)k\mathbf{Y}]\Big\} \end{aligned} \tag{10.57}$$

and

$$\begin{aligned} p_2(\mathbf{C}|\lambda, k, \sigma_2) \propto \frac{|G|^{-1/2}}{\sigma_2{}^T} \exp\Big\{-\frac{1}{2\sigma_2{}^2}&[\mathbf{C} - \lambda\mathbf{C}_{-1} - (1-\lambda)k\mathbf{Y}]'G^{-1} \\ &\times [\mathbf{C} - \lambda\mathbf{C}_{-1} - (1-\lambda)k\mathbf{Y}]\Big\}, \end{aligned} \tag{10.58}$$

where $\mathbf{C}_{-1}' = (C_0, C_1, \ldots, C_{T-1})$, $\mathbf{Y}' = (Y_1, Y_2, \ldots, Y_T)$ and

$$G = \begin{bmatrix} 1+\lambda^2 & -\lambda & & \\ -\lambda & 1+\lambda^2 & \ddots & \\ & \ddots & \ddots & -\lambda \\ & & -\lambda & 1+\lambda^2 \end{bmatrix},$$

a $T \times T$ matrix with nonzero elements just in the three bands shown.

On inserting (10.57) and (10.58) in (10.56) and performing the integrations with respect to σ_1 and σ_2, we have

(10.59)

$$K_{12} = \frac{p(M_1)}{p(M_2)} \frac{\iint p(\lambda, k)\{[\mathbf{C} - \lambda\mathbf{C}_{-1} - (1-\lambda)k\mathbf{Y}]' \times [\mathbf{C} - \lambda\mathbf{C}_{-1} - (1-\lambda)k\mathbf{Y}]\}^{-T/2}\, d\lambda\, dk}{\iint p(\lambda, k)|G|^{-1/2}\{[\mathbf{C} - \lambda\mathbf{C}_{-1} - (1-\lambda)k\mathbf{Y}]'G^{-1} \times [\mathbf{C} - \lambda\mathbf{C}_{-1} - (1-\lambda)k\mathbf{Y}]\}^{-T/2}\, d\lambda\, dk},$$

where $p(\lambda, k)$ denotes the prior pdf for λ and k. Bivariate numerical integration was performed to evaluate the double integrals in (10.59) for each of the prior pdf's shown above, with the following results:

Prior Odds	Prior Pdf for λ and k	Posterior Odds
$p(M_1)/p(M_2) = 1$	(10.53)	$K_{12} = 1.62 \times 10^8$
$p(M_1)/p(M_2) = 1$	(10.54)	$K_{12} = 6.32 \times 10^8$
$p(M_1)/p(M_2) = 1$	(10.55)	$K_{12} = 1.49 \times 10^6$

It is seen that in each instance the information in the data[34] moves us from a prior odds ratio of one to a posterior odds ratio overwhelmingly in favor of M_1. Note also that although the K_{12} values are high, no matter which of the three prior pdf's is employed, the K_{12} values show substantial variation as we change our prior assumptions about λ and k.

As a second example of the comparison of alternative distributed lag schemes by computation of posterior odds, we take up the analysis of Solow's distributed lag model[35] which has a flexible two-parameter weighting scheme. Solow's model for the observations $\mathbf{y}' = (y_1, y_2, \ldots, y_T)$ is

$$y_t = \sum_{i=0}^{\infty} \alpha_i x_{t-i} + u_t, \qquad t = 1, 2, \ldots, T, \tag{10.60}$$

[34] These are U.S. quarterly data for the period 1947I–1960IV taken from Z. Griliches, G. S. Maddala, R. Lucas, and N. Wallace, "Notes on Estimated Aggregate Quarterly Consumption Functions," *Econometrica*, **30**, 491–500, pp. 499–500 (1965).

[35] R. M. Solow, "On a Family of Lag Distributions," *Econometrica*, **28**, 393–406 (1960). Analyses of Solow's distributed lag model from the Bayesian point of view have been reported in V. K. Chetty, "Discrimination. Estimation, and Aggregation of Distributed Lag Models," manuscript, Columbia University, 1968, and in M. S. Geisel, *loc. cit.*

where x_{t-i} is the value of an exogenous variable in period $t - i$ and the weights, the α_i's, are given by

(10.61)
$$\alpha_i = k\binom{r+i-1}{i}(1-\lambda)^r\lambda^i, \qquad 0 < \lambda < 1, r > 0 \text{ and } i = 0, 1, 2, \ldots,$$

where k, r, and λ are unknown parameters. It is seen that α_i is k times a factor that is precisely in the form of probabilities associated with the Pascal distribution (or with r not an integer, the negative binomial distribution). Since Solow considers positive integral values of r adequate for his purposes, we shall do so also. He notes that the mean and variance of the Pascal distribution are $r\lambda/(1 - \lambda)$ and $r\lambda/(1 - \lambda)^2$, respectively. Thus both the mean and variance increase with r and λ separately. Also, the mode is always less than the mean, and in this sense the distribution is skewed to the right. As Solow points out, the larger the λ and the smaller the r, the greater the skewness. Note that if $r = 1$ we have the special case of geometrically declining α_i's.

On introducing a lag operator L such that $Lx_t = x_{t-1}$ and $L^ix_t = x_{t-i}$ and noting that

$$\sum_{i=0}^{\infty}\binom{r+i-1}{i}L^i = (1-L)^{-r},$$

we can rewrite (10.60) as

$$(1 - \lambda L)^r y_t = k(1-\lambda)^r x_t + (1-\lambda L)^r u_t. \tag{10.62}$$

In analyzing this equation, we shall follow Solow's first case in which he assumes that $(1 - \lambda L)^r u_t = \epsilon_t$, $t = 1, 2, \ldots, T$, and that the ϵ_t's have zero means and common variance σ^2.[36] Further, we assume that the ϵ_t's are normally and independently distributed.

The problem we pose is the following one. Given our observations and prior assumptions about the parameters, what are the posterior probabilities associated with these hypotheses: $H_1{:}r = 1$, $H_2{:}r = 2$, $H_3{:}r = 3$, and $H_4{:}r = 4$? These four hypotheses are mutually exclusive and are assumed to be the entire set of possibilities. The direct solution to this problem is simply to compute the posterior pdf for r, a discrete pdf that contains the information we require.

Our four models are

$$M_1{:}r = 1, \qquad y_t = \lambda y_{t-1} + k(1-\lambda)x_t + \epsilon_{1t}, \tag{10.63}$$
$$M_2{:}r = 2, \qquad y_t = 2\lambda y_{t-1} - \lambda^2 y_{t-2} + k(1-\lambda)^2 x_t + \epsilon_{2t}, \tag{10.64}$$
$$M_3{:}r = 3, \qquad y_t = 3\lambda y_{t-1} - 3\lambda^2 y_{t-2} + \lambda^3 y_{t-3} + k(1-\lambda)^3 x_t + \epsilon_{3t}, \tag{10.65}$$

[36] Solow rationalizes this assumption as follows: it is possible that an investigator will formulate the model $(1 - \lambda L)^r y_t = \beta(1 - \lambda)^r x_t + \epsilon_t$ as his basic model without reference to (10.60) and the subsequent steps leading to (10.62). He suggests an analysis of residuals to determine whether the ϵ_t's are nonautocorrelated.

and

$$(10.66)\quad M_4: r = 4, \qquad y_t = 4\lambda y_{t-1} - 6\lambda^2 y_{t-2} + 4\lambda^3 y_{t-3} - \lambda^4 y_{t-4} + k(1-\lambda)^4 x_t + \epsilon_{4t}.$$

For given r, say $r = i$, our prior pdf for the parameters is denoted by $p(\lambda, k, \sigma_i | r = i)$, where σ_i is the standard deviation of ϵ_{it}, $i = 1, 2, 3, 4$. Then the posterior odds ratio relating to, say, models M_1 and M_2 is given by

$$(10.67)\qquad K_{12} = \frac{\Pr\{r = 1\} \int\int\int p(\lambda, k, \sigma_1 | r = 1) l_1 \, d\lambda \, dk \, d\sigma_1}{\Pr\{r = 2\} \int\int\int p(\lambda, k, \sigma_2 | r = 2) l_2 \, d\lambda \, dk \, d\sigma_2},$$

where $\Pr\{r = 1\}/\Pr\{r = 2\}$ is the prior odds ratio and l_i denotes the likelihood function for the parameters of M_i, $i = 1, 2$. From the posterior odds so computed it is straightforward to obtain posterior probabilities.[37]

Geisel has applied[38] this approach to the analysis of U.S. quarterly data in which y_t equals price-deflated seasonally adjusted personal-consumption expenditures and x_t equals price-deflated seasonally adjusted personal-disposable income. His prior pdf's for the parameters were taken as follows for $i = 1, 2, 3$ and 4:

$$(10.68)\qquad p(\lambda, k, \sigma_i | r = i) \propto \frac{1}{\sigma_i} \qquad \begin{array}{l} 0 < \sigma_i < \infty, \\ 0 < \lambda, k < 1. \end{array}$$

Also, he assumed $\Pr\{r = i\}/\Pr\{r = j\} = 1$ for $i, j = 1, 2, 3, 4$. On substituting from (10.68) in (10.67), the integration with respect to the σ_i's can be done analytically. To integrate with respect to λ and k, bivariate numerical integration techniques were employed. The results of his analysis, based on quarterly data, 1948 to 1967, are shown below:

Value of r	Posterior Probability
1	0.762
2	0.177
3	0.043
4	0.018

It is seen that the posterior probability associated with $r = 1$ is quite a bit larger than those associated with other values of r. However, the probability that $r = 2$, 0.177, is not negligible and thus the analysis points to a possible departure from the model with geometrically declining weights ($r = 1$).[39]

[37] If the posterior probabilities are denoted π_i, $i = 1, 2, 3, 4$, we have $1 = \pi_1 + \pi_2 + \pi_3 + \pi_4 = \pi_1(1 + \pi_2/\pi_1 + \pi_3/\pi_1 + \pi_4/\pi_1)$. Thus $\pi_1 = 1/(1 + K_{21} + K_{31} + K_{41})$, and so forth, where $K_{ij} = \pi_i/\pi_j$.

[38] M. S. Geisel, *loc. cit.* He assumed given initial values in formulating his likelihood functions.

[39] See Geisel's and Chetty's work for results bearing on this issue.

In concluding this chapter, it is pertinent to emphasize that Bayesian methods for comparing and testing hypotheses constitute a unified set of principles that is operational and applicable to a broad range of problems. These methods permit the incorporation of prior information when comparing and testing hypotheses and, more important, yield posterior probabilities associated with hypotheses, which are useful in a wide range of circumstances. Last, when a choice has to be made, the Bayesian-decision theoretic approach to testing is one that involves acting to maximize expected utility in accord with some basic results in the economic theory of choice.

QUESTIONS AND PROBLEMS

1. Suppose that we have $n = 10$ observations drawn independently from a normal population with unknown mean μ and standard deviation $\sigma = 1$. If the sample mean is 1.52, compute the posterior odds relating to the hypotheses $H_1: \mu = 2.0$ and $H_2: \mu = 1.0$ with each of the prior probabilities on these hypotheses taken to be $\frac{1}{2}$.
2. In connection with the hypotheses in Problem 1, assume the following loss structure:

		State of World	
		H_1 true	H_2 true
Action	Accept H_1	0	4
	Accept H_2	2	0

   and compute and compare expected losses associated with the actions Accept H_1 and Accept H_2, by using posterior probabilities obtained from results in Problem 1.
3. If we have $n = 15$ observations drawn independently from a normal population with $\mu = 1.0$ and unknown variance σ^2 and $s^2 = \sum_{i=1}^{15} (y_i - \mu)^2/n = 2.2$, obtain posterior odds relating to the hypotheses $H_1: \sigma^2 = 1.0$ and $H_2: \sigma^2 = 2.5$ by using equal prior probabilities for the hypotheses.
4. If, in Problem 3, we considered a third hypothesis, $H_3: \sigma^2 = 1.9$, and assigned equal prior probabilities to H_1, H_2, and H_3, would the posterior odds associated with H_1 and H_2 be changed from what was obtained in Problem 3? Compare the posterior probabilities relating to H_1, H_2, and H_3 with those obtained in Problem 3. What is the effect of increasing the number of mutually exclusive hypotheses considered on the value of a posterior probability for one of them, given a particular sample of data and equal prior probabilities?
5. Show how the posterior odds for hypotheses H_1 and H_2 of Problem 1 can be computed when the population standard deviation σ has an unknown value and we use the following prior pdf: $p(\sigma|\nu_0, s_0^2) = k\sigma^{-(\nu_0+1)} \exp(-\nu_0 s_0^2/2\sigma^2)$,

$0 < \sigma < \infty$, where ν_0 and s_0^2 are positive parameters whose values are assigned by an investigator and $k = 2(\nu_0 s^2/2)^{\nu_0/2}/\Gamma(\nu_0/2)$.

6. Discuss Lindley's paradox, presented in Section 10.3, under the assumption that the sampling-theory significance level for the test is increased as n grows in size. Why would it be reasonable to change the significance level as n increases?

7. Suppose that we have n independent observations $\mathbf{y}' = (y_1, y_2, \ldots, y_n)$ from a normal population with unknown mean μ and standard deviation $\sigma = 1$. Consider the hypotheses $\mu = 0$ and $\mu \neq 0$. If, under $\mu \neq 0$, we were to use the diffuse prior pdf $p(\mu) \propto$ constant, $-\infty < \mu < \infty$, what would be the value of the posterior odds associated with the two hypotheses? In what sense is the result obtained in agreement with the fact that the maximum likelihood estimate for μ will have zero probability of assuming the value zero?

8. Use Lindley's approach to testing for situations in which prior information is vague or diffuse to construct a test of the hypothesis $\mathbf{l}'\boldsymbol{\beta} = c$, where $\mathbf{l}$ is a $k \times 1$ vector of given constants, c is a given scalar constant, and $\boldsymbol{\beta}$ is a $k \times 1$ vector of regression parameters appearing in a standard, linear, normal multiple-regression equation, $\mathbf{y} = X\boldsymbol{\beta} + \mathbf{u}$.

9. In connection with (10.40), it was noted that the posterior odds pertaining to the regression models M_1 and M_2, shown in (10.35) to (10.36) can be expressed as

$$K_{12} = \frac{p(M_1)\, p(\mathbf{y}|M_1)}{p(M_2)\, p(\mathbf{y}|M_2)}.$$

Under the assumption that M_1 and M_2 are the only models under consideration, obtain posterior probabilities for M_1 and M_2 from the expression for K_{12}.

10. Explain how the posterior probabilities for M_1 and M_2 in Problem 9 can be employed to obtain a weighted average of the mean of a future observation assumed to be generated from M_1 in (10.35) with probability equal to the posterior probability for M_1 or from M_2 with probability equal to the posterior probability for M_2.

11. Express K_{12} in (10.51) in terms of R_1^2 and R_2^2, the usual squared multiple-correlation coefficients and note how K_{12} depends on these quantities, noting also that K_{12} is a *relative* measure of belief. Then, under the assumption that the set of models under consideration includes just M_1 and M_2, obtain posterior probabilities for M_1 and for M_2 in terms of R_1^2 and R_2^2 and examine their dependence on these latter quantities.

12. Explain how informative prior pdf's for σ_i, $i = 1, 2$, in the inverted gamma form can be incorporated in the analysis of the posterior odds ratio shown in (10.56).

CHAPTER XI

Analysis of Some Control Problems

In control problems we usually have the following elements: (a) a criterion function, (b) a model containing some variables appearing in the criterion function, and (c) a subset of the model's variables which can be controlled; for example, in the economic theory of the firm the criterion function is usually taken to be the profit function. This function depends on output and the prices of outputs and inputs, which are usually incorporated in a model of the markets for output and inputs. The variables assumed to be under control of the firm are the levels of the inputs, say labor and capital services. Generally, we wish to determine values of the control variables, consistent with possible restrictions imposed by the model, to maximize (or minimize) the criterion function. In the firm example values of the input variables, consistent with the non-negativity requirements of the model and technological constraints, are sought to maximize profits. This problem can usually be solved without much difficulty for nonstochastic models. When, however, the model is stochastic, and has unknown parameters which have to be estimated, the problem becomes more difficult. In what follows we analyze several problems of this kind.[1] It will be seen that the Bayesian approach is convenient for obtaining solutions, since it treats stochastic elements and uncertainty about parameter values in a systematic and unified manner.

When control problems involve optimization over several time periods with stochastic elements and uncertainty about parameter values present, another basic complication appears. The setting of control variables in one period importantly affects the information regarding parameter values we have for other time periods. Thus in multiperiod control problems involving random variables and uncertainty about parameter values a full solution to the optimization problem will have to take account of the flow of information about parameter values as we proceed through time. Bayesian solutions to problems of this kind, called "adaptive control" problems, provide a sequence of optimizing actions, settings of the control variables through time,

[1] See, for example, M. Aoki, *Optimization of Stochastic Systems.* New York: Academic, 1967, for a more extensive coverage of problems and methods.

which not only take account of what we learn from new data but also achieve the *combined and interdependent* goals of control and learning about parameters' values most effectively. Thus a Bayesian solution to an adaptive control problem is a simultaneous solution to a combined control and sequential design of experiments problem.

The plan of this chapter is as follows. In Section 11.1 we analyze several one-period control problems. Emphasis is placed on how uncertainty about parameter values and the cost of changing control variables affect solutions. In the next two sections we generalize some of the one-period results to apply to the problems of controlling the outputs of multiple and multivariate regression models. Sensitivity of results to the form of the criterion function is explored in Section 11.4. In Section 11.5 we take up the analysis of a two-period problem, and in Section 11.6 two multiperiod problems are considered.

11.1 SOME SIMPLE ONE PERIOD CONTROL PROBLEMS

Our first problem involves control of a simple regression model; that is, we assume that the model generating our observations is the simple regression model considered in Chapter 3[2]:

$$y_t = \beta x_t + u_t, \qquad t = 1, 2, \ldots, T. \tag{11.1}$$

In (11.1) β is an unknown parameter, y_t and x_t are the tth observations on the dependent and independent variables, respectively, and u_t is the tth random unobservable disturbance term. We assume that the values of the independent variable can be controlled and that the u_t's are normally and independently distributed, each with zero mean and unknown variance σ^2.

In the first future period $t = T + 1$ we have, with $z \equiv y_{T+1}$ and $w \equiv x_{T+1}$,

$$z = \beta w + u_{T+1}, \tag{11.2}$$

where u_{T+1} is normally distributed with mean zero and variance σ^2 and independent of all previous disturbance terms. It is assumed that we have not yet observed $z \equiv y_{T+1}$ nor determined a value for $w \equiv x_{T+1}$.

To form our criterion function we assume that we wish z to be close to a given target value, denoted by a. The loss associated with being off target is assumed to be given by the following quadratic loss function[3]:

$$L(z, a) = (z - a)^2. \tag{11.3}$$

[2] To keep the analysis as simple as possible, we initially suppress the intercept term. The analysis of the problem of controlling a simple regression process, presented below, draws on the work presented in A. Zellner and M. S. Geisel, "Sensitivity of Control to Uncertainty and Form of the Criterion Function," in D. G. Watts (Ed.), *The Future of Statistics*. New York: Academic, 1968, pp. 269–283.

[3] A quadratic utility function $U = c_0 + 2c_1 z - c_2 z^2$ with c_1 and c_2 positive constants can be brought into the following form by completing the square on z: $U = a_0 -$

Since z is random, $L(z, a)$ is random, and it is impossible to minimize a random function. Rather, we pose our control problem to be minimization of the mathematical expectation of $L(z, a)$ with respect to the choice of w; that is

$$\min_w EL(z, a) = \min_w E(z - a)^2 \tag{11.4}$$

is our problem in which z depends on the control variable w, as shown in (11.2).

The mathematical expectation of the loss function is given by

$$\begin{aligned} EL(z, a) &= \int_{-\infty}^{\infty} L(z, a)\, p(z|\mathbf{y}, w)\, dz \\ &= \int_{-\infty}^{\infty} (z - a)^2\, p(z|\mathbf{y}, w)\, dz, \end{aligned} \tag{11.5}$$

where $p(z|\mathbf{y}, w)$ is the predictive pdf for $z \equiv y_{T+1}$, given the sample information $\mathbf{y}$ and the value of the control variable $w \equiv x_{T+1}$. From the results in Chapter 3 $p(z|\mathbf{y}, w)$ is known to be in the following Student-t form when we use a diffuse prior pdf[4] for β and σ:

$$p(z|\mathbf{y}, w) = \frac{\Gamma[(\nu + 1)/2]}{\Gamma(1/2)\Gamma(\nu/2)} \frac{g^{1/2}}{\nu^{1/2}} \left[1 + \frac{g}{\nu}(z - w\hat{\beta})^2\right]^{-(\nu+1)/2}, \tag{11.6}$$

where $\nu = T - 1$, $\hat{\beta} = \sum_{t=1}^T x_t y_t / \sum_{t=1}^T x_t^2$, $g = [s^2(1 + w^2/\sum_{t=1}^T x_t^2)]^{-1}$, and $\nu s^2 = \sum_{t=1}^T (y_t - \hat{\beta} x_t)^2$. The mean of this pdf is $w\hat{\beta}$ which clearly depends on w. Also, since g depends on w, the "spread" of the pdf, as well as its mean, depends on the value of the control variable w.

Since $E(z - a)^2 = E[z - Ez - (a - Ez)]^2 = \text{Var } z + (a - Ez)^2$, we have[5] for $\nu > 2$

$$E(z - a)^2 = \frac{\nu s^2}{\nu - 2}\left(1 + \frac{w^2}{\sum_{t=1}^T x_t^2}\right) + (a - w\hat{\beta})^2. \tag{11.7}$$

Since (11.7) is quadratic in w, it is easy to establish that the value of w, which minimizes expected loss, denoted by w^*, is[6]

$$w^* = \frac{a}{\hat{\beta}}\left(\frac{1}{1 + 1/t_0^2}\right), \tag{11.8}$$

$c_2(z - a)^2$, where $a_0 = c_0 + c_1^2/c_2$ and $a = c_1/c_2$. Since U is a decreasing function of $(z - a)^2$, minimizing the expectation of $(z - a)^2$, our loss function in (11.3), will maximize expected utility.

[4] That is, $p(\beta, \sigma) \propto 1/\sigma$ with $-\infty < \beta < \infty$ and $0 < \sigma < \infty$. The above analysis can also be performed easily with a natural conjugate prior pdf for β and σ.

[5] From (11.6) $t = \sqrt{g}\,(z - w\hat{\beta})$ has mean zero and variance $\nu/(\nu - 2)$. Thus z has variance $\nu/g(\nu - 2)$.

[6] This is essentially the result obtained by W. D. Fisher, "Estimation in the Linear Decision Model," *Intern. Econ. Rev.*, **3**, 1–29 (1962), except that we have considered σ to be an unknown parameter.

where $t_0^2 = \hat{\beta}^2 m_{xx}/\bar{s}^2$ with $m_{xx} = \sum_{t=1}^T x_t^2$ and $\bar{s}^2 = \nu s^2/(\nu - 2)$. It is seen that w^* is equal to the product of two factors. The first, $a/\hat{\beta}$, is the target value divided by the mean of the posterior pdf for β, namely, $\hat{\beta}$. The second factor is a function of $t_0^{-2} = \hat{\beta}^2 m_{xx}/\bar{s}^2$. Since $\bar{s}^2/m_{xx}$ is the posterior variance for β, t_0^2 is the square of the coefficient of variation associated with the posterior pdf for β. As the precision of estimation, measured by $m_{xx}/\bar{s}^2$, increases, t_0^2 increases and the second factor in (11.8) approaches one[7]; that is, if we have a very precise estimate for β, (11.8) is approximately $a/\hat{\beta}$. To appreciate how well $a/\hat{\beta}$ approximates w^*, the following table is helpful:

t_0:	1.0	2.0	3.0	4.0	5.0
$(1 + 1/t_0^2)^{-1}$:	0.50	0.80	0.90	0.94	0.96.

It is seen that, even for $t_0 = 3.0$, w^* is 0.90 times $a/\hat{\beta}$. Thus, when the precision of estimation is not too high, use of the approximation $a/\hat{\beta}$ results in suboptimal settings for w and, of course, higher expected loss (see below).

At this point it is pertinent to point out that if we went ahead conditional on $\beta = \hat{\beta}$, as appears to be the procedure suggested by the "certainty equivalence" approach,[8] our loss function would be approximated by $(w\hat{\beta} - a)^2$ and the value of w, which minimizes this expression, denoted by w_{ce}, would be

$$w_{ce} = \frac{a}{\hat{\beta}}. \tag{11.9}$$

The certainty equivalence solution is seen to be just the first factor of (11.8). The second factor of (11.8), which reflects the precision with which β has been estimated, does not appear in (11.9).

To demonstrate how use of the suboptimal certainty equivalence value for w leads to higher expected loss than the use of w^* in (11.8), we calculate $E(z - a)^2$ for $w = w^*$ and for $w = w_{ce}$ from (11.7). The results are

$$E(L|w = w^*) = \bar{s}^2 + \frac{a^2}{1 + t_0^2} \tag{11.10}$$

and

$$E(L|w = w_{ce}) = \bar{s}^2 + \frac{a^2}{t_0^2}. \tag{11.11}$$

[7] Note, too, that $t_0 = \hat{\beta} m_{xx}^{1/2}/\bar{s}$ is very nearly the same as the usual sampling theory statistic, $\hat{\beta} m_{xx}^{1/2}/s$, which is used to test the hypothesis $\beta = 0$.

[8] The "certainty equivalence" approach is described in H. A. Simon, "Dynamic Programming under Uncertainty with a Quadratic Criterion Function," *Econometrica*, **24**, 74–81 (1956); H. Theil, "A Note on Certainty Equivalence in Dynamic Planning," *Econometrica*, **25**, 346–349 (1957); and C. Holt, J. F. Muth, F. Modigliani, and H. A. Simon, *Planning Production, Inventories and Work Force.* Englewood Cliffs, N.J.: Prentice-Hall, 1960.

The absolute increase in expected loss associated with using $w = w_{ce}$ rather than $w = w^*$, the optimal value, is

$$E(L|w = w_{ce}) - E(L|w = w^*) = \frac{a^2}{t_0^2(1 + t_0^2)}. \tag{11.12}$$

It is seen that this increase in expected loss depends not only on the precision of estimation, as measured by t_0^2, but also on the squared value of the target a^2. Only when $a = 0$ is there no increase in expected loss associated with use of $w = w_{ce}$.[9] For target values far from the origin the contribution of a^2 to the expression in (11.12) can be substantial.

To provide further information on this last point Table 11.1 lists the relative expected losses *REL* given by the following expression:

$$REL = \frac{E(L|w = w^*)}{E(L|w = w_{ce})} = \frac{[1 + (a/\bar{s})^2/(1 + t_0^2)]}{[1 + (a/\bar{s})^2/t_0^2]}. \tag{11.13}$$

It is seen from the figures in Table 11.1 that the use of the optimal setting for w, namely, w^* given in (11.8), results in a reduction in expected loss *vis à vis* use of the approximate certainty equivalence solution. The reduction in expected loss is greater, the smaller t_0^2, a measure of the precision with which β has been estimated, and the greater $(a/\bar{s})^2$.

To illustrate application of this analysis let us take $y_t = Y_t - Y_{t-1}$, the annual change in aggregate U.S. income and $x_t = M_t - M_{t-1}$, the annual change in the U.S. money supply. Under the assumptions made in connection with (11.1), our problem is to use data[10] for 1921 to 1929, along with a

Table 11.1 TABULATION OF RELATIVE EXPECTED LOSS (11.13) AS A FUNCTION OF $(a/\bar{s})^2$ AND t_0^2

Values of $(a/\bar{s})^2$	Values of t_0^2							
	1.0	2.0	3.0	4.0	5.0	6.0	...	∞
0	1.00	1.00	1.00	1.00	1.00	1.00	...	1.00
2	0.75	0.83	0.90	0.94	0.95	0.96	...	1.00
4	0.60	0.78	0.86	0.91	0.93	0.94	...	1.00
6	0.57	0.75	0.83	0.90	0.92	0.93	...	1.00
8	0.56	0.73	0.82	0.88	0.91	0.92	...	1.00
10	0.54	0.72	0.81	0.87	0.90	0.91	...	1.00
⋮	⋮	⋮	⋮	⋮	⋮	⋮	⋮	⋮
∞	0.50	0.67	0.75	0.80	0.85	0.86	...	1.00

[9] This is because we have assumed that the regression goes through the origin [see (11.1)].

[10] The data are taken from M. Friedman and D. Meiselman, "The Relative Stability of Monetary Velocity and the Investment Multiplier in the United States, 1897–1958,"

diffuse prior pdf for β and σ, to find the change in the money supply for 1930, $w \equiv x_{T+1}$, to have $z = y_{T+1}$, income change for 1930, close to a target value of 10 billion dollars. The symmetry of the quadratic loss function $(z - 10)^2$ implies that we consider overshoots involving inflation as serious as undershoots involving deflation. Based on the data for 1921 to 1929, we have

$$\hat{y}_t = \underset{(0.8813)}{2.0676} x_t, \tag{11.14}$$

where the figure in parentheses is the conventional standard error, $\hat{\beta} = 2.0676$ and $s^2 = \sum_{t=1}^{T} (y_t - \hat{\beta}x_t)^2/\nu = 0.2681 \times 10^8$ with $\nu = T - 1 = 8$. Given these sample quantities, we can readily compute the optimal setting for the 1930 change in the money supply and the associated expected loss. For comparative purposes we also compute the certainty equivalence approximate solution and its associated loss. The results are

$$w^* = 2.893; \qquad E(L|w = w^*) = 55.29,$$

$$w_{ce} = 4.837; \qquad E(L|w = w_{ce}) = 60.03.$$

In this example w^* and w_{ce} are quite different and use of w^* results in about an 8% decrease in expected loss.

Next, it is of interest to elaborate slightly the simple quadratic loss function employed above to take account of possible costs of changing the control variable. This will be done in the following manner:

$$L = (z - a)^2 + c(w - x_T)^2, \tag{11.15}$$

where c is a known non-negative constant, $z \equiv y_{T+1}$, $w \equiv x_{T+1}$, a is the target value, and x_T, the setting for the control variable in period T. Since $w = x_{T+1}$, $w - x_T$ is the change in the control variable in period $T + 1$. We assume that the cost of change is proportional to $(w - x_T)^2$. Using the pdf for z in (11.6), we obtain the expected loss:

$$\begin{aligned} EL &= \operatorname{Var} z + (a - Ez)^2 + c(w - x_T)^2 \\ &= \bar{s}^2\left(1 + \frac{w^2}{m_{xx}}\right) + \hat{\beta}^2\left(\frac{a}{\hat{\beta}} - w\right)^2 + c(w - x_T)^2. \end{aligned} \tag{11.16}$$

Clearly the first term on the rhs of (11.16) will be minimized if $w = 0$, the second, if $w = a/\hat{\beta}$, and the third, if $w = x_T$. On determining the value of w, which minimizes EL in (11.16), we find

$$w^{**} = \frac{\hat{\beta}^2(a/\hat{\beta}) + cx_T}{\hat{\beta}^2 + c + \bar{s}^2/m_{xx}}, \tag{11.17}$$

in the Commission on Money and Credit Research Study, Stabilization Policies, Englewood Cliffs, N.J.: Prentice-Hall, 1963. Their money variable is currency in the hands of the public plus all commercial bank deposits, and their income variable is consumption expenditures plus their autonomous expenditures A.

a weighted average of the values 0, $a/\hat{\beta}$, and x_T. Of course, if $c = 0$, (11.17) yields the result in (11.8) which can be interpreted as a weighted average of 0 and $a/\hat{\beta}$, namely, $w^* = \hat{\beta}^2(a/\hat{\beta})/(\hat{\beta}^2 + \bar{s}^2/m_{xx})$. From (11.17), as c grows large, $w^{**} \to x_T$; that is, as change becomes more costly, w^{**} assumes a value closer to its initial value x_T, which results in a smaller change and less of a contribution to expected loss. In general, for finite positive c the solution w^{**} in (11.17) is always between the solution for $c = 0$, w^* shown in (11.8), and x_T; that is, for $x_T < w^*$, $x_T < w^{**} < w^*$ and for $x_T > w^*$, $w^* < w^{**} < x_T$ are inequalities satisfied by the solutions if c has a positive value. Thus, with the cost of change introduced, there is a tendency toward conservatism introduced in the sense that the optimal change in the control variable, $w^{**} - x_T$, will be smaller in absolute size with $c > 0$ than with $c = 0$, that is, with no cost of changing the control variable.

As another example of a one period control problem, let us consider the problem of a profit-maximizing monopolist under the assumption that he is uncertain about the parameters of his total cost function and of his demand function. With π = profits, p = price, the control variable, q = quantity produced, and C total costs, we have

$$\pi = pq - C, \tag{11.18}$$

where all variables' values are for the first future period, $T + 1$. Further assume that the monopolist's demand function and total cost function are given by

$$q_t = \beta_0 + \beta_1 p_t + u_t, \qquad t = 1, 2, \ldots, T, T + 1 \tag{11.19}$$

and

$$C_t = \alpha_0 + \alpha_1 q_t + \alpha_2 q_t^2 + v_t, \qquad t = 1, 2, \ldots, T, T + 1, \tag{11.20}$$

where the α's and β's are unknown parameters, assumed random, and the u_t's and v_t's are normal, independent, random-error terms with zero means and constant variances, σ_u^2 and σ_v^2, respectively.

Given past data on q_t, c_t, and p_t and a prior pdf for the parameters, it is possible to obtain posterior pdf's for the parameters and predictive pdf's for $C \equiv C_{T+1}$ and $q \equiv q_{T+1}$. The latter can be employed to obtain the mathematical expectation of $\pi_{T+1} = \pi$, which is random because of its dependence on q and C, both random. Then the optimal value of $p \equiv p_{T+1}$, price for period $T + 1$, can be determined. An alternative simpler and equivalent procedure for the present problem is to substitute from (11.19) and (11.20) in (11.18) to obtain

(11.21)
$$\pi = p(\beta_0 + \beta_1 p + u) - \alpha_0 - \alpha_1(\beta_0 + \beta_1 p + u) - \alpha_2(\beta_0 + \beta_1 p + u)^2 - v,$$

where the α's, β's, u, and v are random. The expectation of π is given by[11]

$$(11.22)\qquad \begin{aligned} E\pi = p(\bar{\beta}_0 + \bar{\beta}_1 p) - \bar{\alpha}_0 - \bar{\alpha}_1(\bar{\beta}_0 + \bar{\beta}_1 p) - \bar{\alpha}_2 \\ \times [(\bar{\beta}_0 + \bar{\beta}_1 p)^2 + \sigma_0^2 + p^2\sigma_1^2 + 2p\sigma_{01} + \bar{\sigma}_u^2], \end{aligned}$$

where the $\bar{\beta}$'s and $\bar{\alpha}$'s are posterior means, σ_0^2, σ_1^2, and σ_{01} are the posterior variances and covariance, respectively, of β_0 and β_1, and $\bar{\sigma}_u^2$ is the posterior mean of σ_u^2. In taking the expectation of π in (11.22), it is important to note that the parameters of the demand function, the β's, are a posteriori distributed independently of the parameters of the cost function, the α's, since we have assumed that the u_t's and v_t's in (11.19) and (11.20) are independently distributed and further that our prior pdf's incorporate the assumption that these two sets of parameters are independent.

On differentiating (11.22) with respect to the control variable, p, we obtain

$$(11.23)\qquad \frac{dE\pi}{dp} = \bar{\beta}_0 + 2\bar{\beta}_1 p - \bar{\alpha}_1\bar{\beta}_1 - 2\bar{\alpha}_2[\bar{\beta}_1(\bar{\beta}_0 + \bar{\beta}_1 p) + p\sigma_1^2 + \sigma_{01}].$$

The value of p which sets this derivative equal to zero, that is, the price that maximizes expected profits,[12] is

$$(11.24)\qquad p^* = \frac{1}{2\bar{\beta}_1}\left[\frac{\bar{\alpha}_1\bar{\beta}_1 - \bar{\beta}_0 + 2\bar{\alpha}_2\bar{\beta}_0\bar{\beta}_1(1 + \phi_{01})}{1 - \bar{\alpha}_2\bar{\beta}_1(1 + \phi_1)}\right],$$

where $\phi_1 = \sigma_1^2/\bar{\beta}_1^2$ and $\phi_{01} = \sigma_{01}/\bar{\beta}_0\bar{\beta}_1$. The quantity in (11.24) must be positive to be an acceptable solution.[13]

Note that if we had substituted the mean values for the α's, β's, u, and v in (11.21) and maximized with respect to p, the result would be

$$(11.25)\qquad \tilde{p} = \frac{1}{2\bar{\beta}_1}\left(\frac{\bar{\alpha}_1\bar{\beta}_1 - \bar{\beta}_0 + 2\bar{\alpha}_2\bar{\beta}_0\bar{\beta}_1}{1 - \bar{\alpha}_2\bar{\beta}_1}\right),$$

exactly what is obtained from (11.24) by setting ϕ_0 and ϕ_1 both equal to zero. Alternatively, ϕ_1 will approach zero as information about β_1 grows, whereas σ_{01} could be zero if β_0 and β_1 are independent parameters. Under these special conditions $\tilde{p}$ in (11.25) will approximate the optimal p^* in (11.24).

[11] Note that $E(\beta_0 + \beta_1 p + u)^2 = E[\beta_0 - \bar{\beta}_0 + (\beta_1 - \bar{\beta}_1)p + \bar{\beta}_0 + \bar{\beta}_1 p + u]^2 = E(\beta_0 - \bar{\beta}_0)^2 + p^2E(\beta_1 - \bar{\beta}_1)^2 + 2pE(\beta_0 - \bar{\beta}_0)(\beta_1 - \bar{\beta}_1) + (\bar{\beta}_0 + \bar{\beta}_1 p)^2 + E\sigma_u^2$.

[12] For a maximum, $d^2E\pi/dp^2 = 2(\bar{\beta}_1 - \bar{\alpha}_2\bar{\beta}_1^2 - \bar{\alpha}_2\sigma_1^2)$ must be negative. Usually, $\bar{\beta}_1$ and $\bar{\alpha}_2$ are negative. If this is so, this second derivative will be negative, given that $\bar{\beta}_1/\bar{\alpha}_2 > \sigma_1^2 + \bar{\beta}_1^2$. Since $|\bar{\alpha}_2|$ is often quite small, this condition is reasonable. The dependence of the second-order condition on the size of σ_1^2, a measure of uncertainty about β_1, is surprising.

[13] $\bar{\beta}_0$ will usually be large and positive, whereas $\bar{\beta}_1$ is negative and often not too large in absolute value. Thus $-\bar{\beta}_0/\bar{\beta}_1$ will be large and positive. Note, too, that $\bar{\alpha}_2$ will be small and negative in most cases.

From (11.24) we see that as uncertainty about β_1, measured by σ_1, decreases, p^* will fall, given $\bar{\beta}_1$, $\bar{\alpha}_2 < 0$. Thus, as the monopolist learns about the value of β_1 in the sense that he has a posterior pdf for this parameter with a progressively declining variance, his profit-maximizing price falls. Also (11.24) indicates an interesting dependence of p^* on the posterior covariance of β_0 and β_1. Given that $\bar{\alpha}_2\bar{\beta}_0 < 0$, the greater the value of σ_{10}, the lower is p^*. As the above example indicates, allowing for imperfect knowledge and stochastic elements modifies and enriches the results of traditional economic theory.[14]

11.2 SINGLE-PERIOD CONTROL OF MULTIPLE REGRESSION PROCESSES

Assume that our observation vector $\mathbf{y}' = (y_1, y_2, \ldots, y_T)$ is generated by a multiple-regression process

$$\mathbf{y} = X\boldsymbol{\beta} + \mathbf{u}, \tag{11.26}$$

where $X = (\mathbf{x}_1, \mathbf{x}_2, \ldots, \mathbf{x}_k)$ is a $T \times k$ matrix with rank k of observations of the past settings of k control variables,[15] $\boldsymbol{\beta}$ is a $k \times 1$ vector of unknown regression parameters, and $\mathbf{u}$ is a $T \times 1$ vector of normal and independent disturbance terms, each with mean zero and unknown variance σ^2. As prior pdf for the unknown parameters, we shall employ the diffuse pdf[16] introduced in Chapter 3:

$$p(\boldsymbol{\beta}, \sigma) \propto \frac{1}{\sigma}, \quad 0 < \sigma < \infty, \quad \text{and} \quad -\infty < \beta_i < \infty, \quad i = 1, 2, \ldots, k. \tag{11.27}$$

Let z be a future value of the dependent variable, say $z \equiv y_{T+1}$, the first future value assumed to satisfy

$$z = \mathbf{w}'\boldsymbol{\beta} + u_{T+1}, \tag{11.28}$$

where u_{T+1} is a normal disturbance term, distributed independently of $\mathbf{u}$, with zero mean and variance σ^2, and $\mathbf{w}'$, a $1 \times k$ vector, denotes the

[14] See also R. M. Cyert and M. H. De Groot, "Bayesian Analysis and Duopoly Theory," manuscript, Carnegie-Mellon University, April 1968; M. S. Feldstein, "Production with Uncertain Technology: Some Econometric Implications," manuscript, Harvard University, 1969; and A. Zellner, J. Kmenta, and J. Drèze, "Specification and Estimation of Cobb-Douglas Production Function Models," *Econometrica*, **34**, 784–795 (1966).

[15] Here we assume that all k independent variables can be controlled; see below for a relaxation of this condition.

[16] A natural conjugate prior pdf could easily be employed instead of a diffuse prior in working the problems of this section.

settings of the control variables in the $(T + 1)$st period; that is, $\mathbf{w}' \equiv (x_{1,T+1} \cdots x_{k,T+1})$.

Now suppose that we wish to choose the control vector $\mathbf{w}$ to keep $z \equiv y_{T+1}$ close to a target value, denoted by a, and that our loss function is

$$L(z, a) = (z - a)^2, \tag{11.29}$$

the same squared error loss function employed in Section 11.1.

From results in Chapter 3 the predictive pdf for $z \equiv y_{T+1}$, under the above assumptions, is

$$p(z|\mathbf{y}, \mathbf{w}) \propto [\nu + (z - \mathbf{w}'\hat{\boldsymbol{\beta}})^2 H]^{-(\nu+1)/2}, \tag{11.30}$$

where $\hat{\boldsymbol{\beta}} = (X'X)^{-1}X'\mathbf{y}$, $H = (1/s^2)[1 + \mathbf{w}'(X'X)^{-1}\mathbf{w}]^{-1}$, $\nu = T - k$, and $\nu s^2 = (\mathbf{y} - X\hat{\boldsymbol{\beta}})'(\mathbf{y} - X\hat{\boldsymbol{\beta}})$. The mean and variance of z are given by[17]

$$Ez = \mathbf{w}'\hat{\boldsymbol{\beta}} \tag{11.31}$$

and

$$\begin{aligned} E(z - Ez)^2 &= \frac{H^{-1}\nu}{(\nu - 2)} \\ &= \bar{s}^2[1 + \mathbf{w}'(X'X)^{-1}\mathbf{w}], \end{aligned} \tag{11.32}$$

where $\bar{s}^2 = \nu s^2/(\nu - 2)$.

Now we shall take the expectation of the loss function in (11.29) and determine the value of $\mathbf{w}$, the control vector, which minimizes expected loss. We have

$$\begin{aligned} E(z - a)^2 &= E[(z - Ez) - (a - Ez)]^2 \\ &= E(z - Ez)^2 + (a - Ez)^2 \\ &= \bar{s}^2(1 + \mathbf{w}'(X'X)^{-1}\mathbf{w}) + (a - \mathbf{w}'\hat{\boldsymbol{\beta}})^2. \end{aligned} \tag{11.33}$$

Differentiating $E(z - a)^2$ with respect to the elements of $\mathbf{w}$ yields

$$\frac{\partial E(z - a)^2}{\partial \mathbf{w}} = 2\bar{s}^2(X'X)^{-1}\mathbf{w} - 2a\hat{\boldsymbol{\beta}} + 2\hat{\boldsymbol{\beta}}\hat{\boldsymbol{\beta}}'\mathbf{w}.$$

The value of $\mathbf{w}$, say $\mathbf{w}^*$, which sets this derivative equal to zero satisfies

$$[\bar{s}^2(X'X)^{-1} + \hat{\boldsymbol{\beta}}\hat{\boldsymbol{\beta}}']\mathbf{w}^* = a\hat{\boldsymbol{\beta}} \tag{11.34}$$

[17] For the variance to exist we need $\nu = T - k > 2$.

and thus[18]

$$\mathbf{w}^* = \frac{X'X\hat{\boldsymbol{\beta}}a}{\bar{s}^2 + \hat{\boldsymbol{\beta}}'X'X\hat{\boldsymbol{\beta}}} \tag{11.35}$$

is the setting for $\mathbf{w}$ which minimizes[19] expected loss. Multiplying both sides of (11.35) by $\hat{\boldsymbol{\beta}}'$ and rearranging terms, we have

$$\begin{aligned} a &= \mathbf{w}^{*\prime}\hat{\boldsymbol{\beta}}\left(1 + \frac{1}{\hat{\boldsymbol{\beta}}'X'X\hat{\boldsymbol{\beta}}/\bar{s}^2}\right) \\ &= \mathbf{w}^{*\prime}\hat{\boldsymbol{\beta}}^*; \end{aligned} \tag{11.36}$$

$\hat{\boldsymbol{\beta}}^*$ in (11.36) can be interpreted as the implied estimate for $\boldsymbol{\beta}$.[20]

Substituting from (11.35) in (11.33), we find

$$E(L|\mathbf{w} = \mathbf{w}^*) = \bar{s}^2\left(1 + \frac{a^2}{\bar{s}^2 + \hat{\boldsymbol{\beta}}'X'X\hat{\boldsymbol{\beta}}}\right) \tag{11.37}$$

for the expected loss associated with the optimal setting for $\mathbf{w}$, namely, $\mathbf{w} = \mathbf{w}^*$. The first term in the expression for expected loss, $\bar{s}^2$, does not disappear as the sample size grows, whereas the second term, $\bar{s}^2a^2/(\bar{s}^2 + \hat{\boldsymbol{\beta}}'X'X\hat{\boldsymbol{\beta}})$ approaches zero as the sample size grows.[21] From (11.36) we see also that $\hat{\boldsymbol{\beta}}^*$, the implied optimal estimate, approaches $\hat{\boldsymbol{\beta}}$ as the sample size grows.

Since in many problems not all independent variables are under control, we extend the above analysis to the case in which $X = (X_1 \vdots X_2)$, where X_1 denotes observations on control variables and X_2 denotes observations on variables not under control. Our regression model for the sample observations is

$$\begin{aligned} \mathbf{y} &= X\boldsymbol{\beta} + \mathbf{u} \\ &= X_1\boldsymbol{\beta}_1 + X_2\boldsymbol{\beta}_2 + \mathbf{u}, \end{aligned} \tag{11.38}$$

[18] Note that $[\bar{s}^2(X'X)^{-1} + \hat{\boldsymbol{\beta}}\hat{\boldsymbol{\beta}}']^{-1} = (1/\bar{s}^2)[X'X - X'X\hat{\boldsymbol{\beta}}\hat{\boldsymbol{\beta}}'X'X/(\bar{s}^2 + \hat{\boldsymbol{\beta}}'X'X\hat{\boldsymbol{\beta}})]$ and can be verified as follows with $Q \equiv \hat{\boldsymbol{\beta}}'X'X\hat{\boldsymbol{\beta}}$: $[\bar{s}^2(X'X)^{-1} + \hat{\boldsymbol{\beta}}\hat{\boldsymbol{\beta}}'](1/\bar{s}^2)[X'X - X'X\hat{\boldsymbol{\beta}}\hat{\boldsymbol{\beta}}'X'X/(\bar{s}^2 + Q)] = I - \hat{\boldsymbol{\beta}}\hat{\boldsymbol{\beta}}'X'X/(\bar{s}^2 + Q) + \hat{\boldsymbol{\beta}}\hat{\boldsymbol{\beta}}'X'X/\bar{s}^2 - \hat{\boldsymbol{\beta}}\hat{\boldsymbol{\beta}}'X'X\hat{\boldsymbol{\beta}}\hat{\boldsymbol{\beta}}'X'X/(\bar{s}^2 + Q) = I - \hat{\boldsymbol{\beta}}\hat{\boldsymbol{\beta}}'X'X[1/(\bar{s}^2 + Q) - 1/\bar{s}^2 + Q/\bar{s}^2(\bar{s}^2 + Q)] = I$. The formula for the inverse of a matrix in the form encountered in (11.34), given in C. R. Rao, *Linear Statistical Inference and Its Applications*. New York: Wiley, 1965, p. 29, Example 2.8, is slightly in error. It should read:

$$(A + UV')^{-1} = A^{-1} - \frac{(A^{-1}U)(V'A^{-1})}{1 + V'A^{-1}U},$$

where A is a nonsingular matrix and U and V are two column vectors.

[19] Note that $\partial^2 E(z - a)^2/\partial\mathbf{w}^2$ is a positive definite matrix under our assumptions.

[20] See W. D. Fisher, *loc. cit.*, for some related results.

[21] Note that $\hat{\boldsymbol{\beta}}'X'X\hat{\boldsymbol{\beta}} = \hat{\mathbf{y}}'\hat{\mathbf{y}}$ grows in size as T grows.

and for the future observation, $z \equiv y_{T+1}$,

$$\begin{aligned} z &= \mathbf{w}'\boldsymbol{\beta} + v \\ &= \mathbf{w}_1'\boldsymbol{\beta}_1 + \mathbf{w}_2'\boldsymbol{\beta}_2 + v, \end{aligned} \tag{11.39}$$

where $\boldsymbol{\beta}' = (\boldsymbol{\beta}_1' \vdots \boldsymbol{\beta}_2')$ and all other variables are defined above with $\mathbf{w}' = (\mathbf{w}_1' \vdots \mathbf{w}_2')$ and $\mathbf{w}_2$ is given. We make the same stochastic and prior assumptions made in connection with (11.26) and (11.27).

Let our loss function be

$$L = (z - a)^2 + (\mathbf{w}_1 - \mathbf{x}_{1T})'G(\mathbf{w}_1 - \mathbf{x}_{1T}), \tag{11.40}$$

where $\mathbf{x}_{1T}$ denotes the setting for the control variables (the variables in X_1) in period T and G is a positive definite symmetric matrix. The loss function in (11.40) incorporates losses associated with being off the target value a and costs of changing the control variables.[22] Given that (11.30) is the predictive pdf for z, we readily obtain the following expression for expected loss:

$$\begin{aligned} EL &= \bar{s}^2[1 + \mathbf{w}'(X'X)^{-1}\mathbf{w}] + (a - \mathbf{w}'\hat{\boldsymbol{\beta}})^2 + (\mathbf{w}_1 - \mathbf{x}_{1T})'G(\mathbf{w}_1 - \mathbf{x}_{1T}) \\ &= \bar{s}^2\left[1 + (\mathbf{w}_1'\mathbf{w}_2')\begin{pmatrix} M^{11} & M^{12} \\ M^{21} & M^{22} \end{pmatrix}\begin{pmatrix} \mathbf{w}_1 \\ \mathbf{w}_2 \end{pmatrix}\right] + (a - \mathbf{w}_1'\hat{\boldsymbol{\beta}}_1 - \mathbf{w}_2'\hat{\boldsymbol{\beta}}_2)^2 \\ &\quad + (\mathbf{w}_1 - \mathbf{x}_{1T})'G(\mathbf{w}_1 - \mathbf{x}_{1T}), \end{aligned} \tag{11.41}$$

where $\hat{\boldsymbol{\beta}}' = (\hat{\boldsymbol{\beta}}_1' \vdots \hat{\boldsymbol{\beta}}_2')$ has been partitioned to correspond to the partitioning of $X = (X_1 \vdots X_2)$ and similarly for

$$(X'X)^{-1} = \begin{pmatrix} M^{11} & M^{12} \\ M^{21} & M^{22} \end{pmatrix}.$$

On differentiating EL with respect to the elements of $\mathbf{w}_1$ and solving for the minimizing value $\mathbf{w}_1 = \mathbf{w}_1^*$, we find

$$\begin{aligned} \mathbf{w}_1^* &= [\bar{s}^2 M^{11} + G + \hat{\boldsymbol{\beta}}_1\hat{\boldsymbol{\beta}}_1']^{-1}[\hat{\boldsymbol{\beta}}_1(a - \mathbf{w}_2'\hat{\boldsymbol{\beta}}_2) - \bar{s}^2 M^{12}\mathbf{w}_2 + G\mathbf{x}_{1T}] \\ &= \left(P^{-1} - \frac{P^{-1}\hat{\boldsymbol{\beta}}_1\hat{\boldsymbol{\beta}}_1'P^{-1}}{1 + \hat{\boldsymbol{\beta}}_1'P^{-1}\hat{\boldsymbol{\beta}}_1}\right)[\hat{\boldsymbol{\beta}}_1(a - \mathbf{w}_2'\hat{\boldsymbol{\beta}}_2) - \bar{s}^2 M^{12}\mathbf{w}_2 + G\mathbf{x}_{1T}] \\ &= P^{-1}\hat{\boldsymbol{\beta}}_1\frac{(a - w_2'\hat{\boldsymbol{\beta}}_2)}{1 + Q_1} - \left(P^{-1} - \frac{P^{-1}\hat{\boldsymbol{\beta}}_1\hat{\boldsymbol{\beta}}_1'P^{-1}}{1 + Q_1}\right)(\bar{s}^2 M^{12}\mathbf{w}_2 - G\mathbf{x}_{1T}), \end{aligned} \tag{11.42}$$

where $P = \bar{s}^2 M^{11} + G$ and $Q_1 = \hat{\boldsymbol{\beta}}_1'P^{-1}\hat{\boldsymbol{\beta}}_1$. The first term on the rhs of the third line of (11.42) is similar in form to that encountered in (11.35) except that P^{-1} replaces $X'X/\bar{s}^2$, $a - \mathbf{w}_2'\boldsymbol{\beta}_2$ replaces a, and Q_1 replaces $\hat{\boldsymbol{\beta}}'X'X\hat{\boldsymbol{\beta}}/\bar{s}^2$.

[22] Cost of changing the control variables has usually been incorporated in loss functions. See, for example, W. D. Fisher, *loc. cit.*, and S. J. Press, "On Control of Bayesian Regression Models," manuscript (undated).

The other term in the third line of (11.42) reflects both the interdependence of $\mathbf{w}_1$ and $\mathbf{w}_2$ in determining the variance of z and the cost of moving the control variables. If X_2 were orthogonal to X_1 so that $M^{12} = 0$ and, further, if $G = 0$, then (11.42) reduces to

$$\mathbf{w}_1^* = \frac{X_1'X_1\hat{\boldsymbol{\beta}}_1(a - \mathbf{w}_2'\hat{\boldsymbol{\beta}}_2)}{\bar{s}^2 + \hat{\boldsymbol{\beta}}_1'X_1'X_1\hat{\boldsymbol{\beta}}_1}, \tag{11.43}$$

which is quite similar to (11.35) in form.

11.3 CONTROL OF MULTIVARIATE NORMAL REGRESSION PROCESSES[23]

In this section we extend the analysis of the preceding section to apply to the traditional multivariate regression model[24] considered in Chapter 8. Our model for the observations, $Y = (\mathbf{y}_1, \mathbf{y}_2, \ldots, \mathbf{y}_m)$, a $T \times m$ matrix, is given by

$$\begin{aligned} Y &= XB + U \\ &= X_1B_1 + X_2B_2 + U, \end{aligned} \tag{11.44}$$

where X is a $T \times k$ matrix of rank k, B is a $k \times m$ matrix of unknown parameters, and U is a $T \times m$ matrix of random disturbance terms. We have partitioned $X = (X_1 \vdots X_2)$, with X_1 containing observations on variables under control and X_2, those on variables not under control. The B matrix has been partitioned to correspond to the partitioning of X.

The first future *vector* of observations, $\mathbf{z}' = (y_{1,T+1}, y_{2,T+1}, \ldots, y_{m,T+1})$, is assumed to be generated by the same process generating Y; that is,

$$\begin{aligned} \mathbf{z}' &= \mathbf{w}'B + \mathbf{v}' \\ &= \mathbf{w}_1'B_1 + \mathbf{w}_2'B_2 + \mathbf{v}', \end{aligned} \tag{11.45}$$

where $\mathbf{w}' = (\mathbf{w}_1' \vdots \mathbf{w}_2')$ is a $1 \times k$ vector of values for the independent variables in the future period $T + 1$ and $\mathbf{v}' = (u_{1,T+1}, u_{2,T+1}, \ldots, u_{m,T+1})$ is the disturbance vector for period $T + 1$.

In Chapter 8 (8.57) we have the predictive pdf for observations generated by a traditional multivariate normal regression model when we employ

[23] For some previous work on this problem from the Bayesian point of view, cf. W. D. Fisher, *loc. cit.*, S. J. Press, *loc. cit.*, and A. Zellner and V. K. Chetty, "Prediction and Decision Problems in Regression Models from the Bayesian Point of View," *J. Am. Statist. Assoc.*, **60**, 608–616 (1965).

[24] This model can also be considered to be the "unrestricted" reduced form equations of a "simultaneous equation" econometric model. If lagged endogenous variables appear in the system, we assume that initial values for these variables are given in forming the likelihood function. Also, such variables are obviously not control variables.

diffuse prior pdf's for the unknown parameters [see (8.8) and (8.9)]. In the case of one future period (8.57) specializes to a pdf in the multivariate Student t form:

$$p(\mathbf{z}|Y) \propto [\nu + h(\mathbf{z} - \hat{B}'\mathbf{w})'\bar{S}^{-1}(\mathbf{z} - \hat{B}'\mathbf{w})]^{-(\nu+m)/2}, \tag{11.46}$$

where $\nu = T - (k-1) - m$, $\hat{B} = (X'X)^{-1}X'Y$, $\bar{S} = (Y - X\hat{B})'(Y - X\hat{B})/\nu$ and $h = [1 - \mathbf{w}'(X'X + \mathbf{w}\mathbf{w}')^{-1}\mathbf{w}]$. Then we note[25] that

$$E(\mathbf{z}'|Y) = \mathbf{w}'\hat{B} \tag{11.47}$$

and

$$\begin{aligned} \operatorname{Var}(\mathbf{z}|Y) &= \frac{\nu}{\nu - 2} h^{-1}\bar{S} \\ &= \frac{\nu}{\nu - 2}[1 + \mathbf{w}'(X'X)^{-1}\mathbf{w}]\bar{S}. \end{aligned} \tag{11.48}$$

Now let our loss function be given by[26]

$$L = (\mathbf{z} - \mathbf{a})'C(\mathbf{z} - \mathbf{a}), \tag{11.49}$$

where $\mathbf{a}' = (a_1, a_2, \ldots, a_m)$ is a vector of given target values, one for each of the m dependent variables,[27] and C is a positive definite symmetric matrix. Then the expectation of the loss function is

$$\begin{aligned} EL &= E(\mathbf{z} - \mathbf{a})'C(\mathbf{z} - \mathbf{a}) \\ &= E(\mathbf{z} - E\mathbf{z})'C(\mathbf{z} - E\mathbf{z}) + (\mathbf{a} - E\mathbf{z})'C(\mathbf{a} - E\mathbf{z}) \\ &= \sum \{\tilde{s}_{\alpha l}[1 + \mathbf{w}'(X'X)^{-1}\mathbf{w}] + (a_\alpha - \mathbf{w}'\hat{\boldsymbol{\beta}}_\alpha)(a_l - \mathbf{w}'\hat{\boldsymbol{\beta}}_l)\} c_{\alpha l}, \end{aligned} \tag{11.50}$$

where the summation extends over α, $l = 1, 2, \ldots, m$, $\tilde{s}_{\alpha l}$ is the (α, l)th element of $\nu\bar{S}/(\nu - 2)$, $Ez_\alpha = \mathbf{w}'\hat{\boldsymbol{\beta}}_\alpha$, with $\hat{\boldsymbol{\beta}}_\alpha$ the αth column of $\hat{B}$, and $c_{\alpha l}$ is the (α, l)th element of C. In (11.50) we partition $\mathbf{w}' = (\mathbf{w}_1' \vdots \mathbf{w}_2')$ and $(X'X)^{-1}$, $\hat{\boldsymbol{\beta}}_\alpha$, and $\hat{\boldsymbol{\beta}}_l$ correspondingly; for example, $\hat{\boldsymbol{\beta}}_\alpha' = (\hat{\boldsymbol{\beta}}_{\alpha_1}' \vdots \hat{\boldsymbol{\beta}}_{\alpha_2}')$, where $\hat{\boldsymbol{\beta}}_{\alpha_1}$ is a column vector with the dimensions of $\mathbf{w}_1$. Then (11.50) becomes

$$\begin{aligned} EL = \sum [&\tilde{s}_{\alpha l}(1 + \mathbf{w}_1'M^{11}\mathbf{w}_1 + 2\mathbf{w}_1'M^{12}\mathbf{w}_2 + \mathbf{w}_2'M^{22}\mathbf{w}_2) \\ &+ (a_\alpha - \mathbf{w}_1'\hat{\boldsymbol{\beta}}_{\alpha_1} - \mathbf{w}_2'\hat{\boldsymbol{\beta}}_{\alpha_2})(a_l - \mathbf{w}_1'\hat{\boldsymbol{\beta}}_{l_1} - \mathbf{w}_2'\hat{\boldsymbol{\beta}}_{l_2})]c_{\alpha l}, \end{aligned} \tag{11.51}$$

[25] For both (11.47) and (11.48) to exist we need $\nu > 2$.

[26] Since it is straightforward to include a quadratic term for the cost of changing the control variables, we do not add this complication in the present analysis. Also, if, as Fisher, *loc. cit.*, does, we employ a cardinal utility function of the form $U = 2b'\mathbf{z} - \mathbf{z}'C\mathbf{z}$, where $\mathbf{b}$ is a given vector and C, a positive definite symmetric matrix, it is well known that we can complete the square on $\mathbf{z}$ and write $U = \mathbf{b}'C^{-1}\mathbf{b} - (\mathbf{z} - C^{-1}\mathbf{b})'C(\mathbf{z} - C^{-1}\mathbf{b})$. If we let $\mathbf{a} = C^{-1}\mathbf{b}$, minimizing $E(\mathbf{z} - \mathbf{a})'C(\mathbf{z} - \mathbf{a})$ is equivalent to maximizing EU.

[27] In some problems we may have target values just for a subset of the m dependent variables, say m', with $1 \leq m' < m$.

with $(X'X)^{-1} = \{M^{ij}\}$, $i, j = 1, 2$. We can now differentiate (11.51) with respect to the elements of $\mathbf{w}_1$ and find the minimizing[28] value $\mathbf{w}_1 = \mathbf{w}_1^*$. This operation yields

$$\text{(11.52)} \quad \begin{aligned} \mathbf{w}_1^* &= [\sum (2\tilde{s}_{\alpha l}M^{11} + \hat{\boldsymbol{\beta}}_{\alpha_1}\hat{\boldsymbol{\beta}}_{l_1}' + \hat{\boldsymbol{\beta}}_{l_1}\hat{\boldsymbol{\beta}}_{\alpha_1}')c_{\alpha l}]^{-1} \\ &\quad \times \{\sum [(a_\alpha - \mathbf{w}_2'\hat{\boldsymbol{\beta}}_{\alpha_2})\boldsymbol{\beta}_{l_1} + (a_l - \mathbf{w}_2'\hat{\boldsymbol{\beta}}_{l_2})\hat{\boldsymbol{\beta}}_{\alpha_1} - 2\tilde{s}_{\alpha l}M^{12}\mathbf{w}_2]c_{\alpha l}\}; \end{aligned}$$

again all summations extend over $\alpha, l = 1, 2, \ldots, m$. Given the sample quantities and the elements of $C = \{c_{\alpha l}\}$, the optimal setting for $\mathbf{w}_1^*$ can readily be computed.

11.4 SENSITIVITY OF CONTROL TO FORM OF LOSS FUNCTION

In the preceding sections we have analyzed several control problems using symmetric quadratic loss functions. Now we present the results of calculations[29] to illustrate how incorrect assumptions about the form of the loss function affect results, an area that may be called "robustness under changes of the loss function."[30]

To provide some numerical results $T = 15$ observations were generated from the following simple regression model:

$$\text{(11.53)} \qquad y_t = 2.0x_t + u_t, \qquad t = 1, 2, \ldots, 15;$$

the u_t's were independently drawn from a normal distribution with zero mean and variance $= 9.0$, and the x_t's were independently drawn from a normal distribution with mean zero and variance 0.64. The data so generated are shown in Table 11.2. Using the data in Table 11.2, we compute the following sample quantities[31]:

$$\hat{\beta} = \frac{\sum x_t y_t}{\sum x_t^2} = 1.5885, \qquad s^2 = \left(\frac{1}{\nu}\right) \sum (y_t - \hat{\beta}x_t)^2 = 5.7351,$$

with $\nu = T - 1 = 14$.

As regards loss functions, we first consider the following symmetric functions:

$$\text{(11.54)} \qquad L(z, a) = |z - a|^\alpha, \qquad \alpha = 0.5, 1, 2, \text{ and } 4,$$

[28] Note that the matrix of second derivatives of (11.51) with respect to the elements of $\mathbf{w}_1$ is positive definite; hence the value of $\mathbf{w}_1$ which sets the first matrix derivative equal to zero is a minimizing value.

[29] These are taken from A. Zellner and M. S. Geisel, "Sensitivity of Control to Uncertainty and Form of the Criterion Function," in D. G. Watts (editor), *The Future of Statistics*. New York: Academic, 1968, pp. 269–283.

[30] This phrase was suggested by J. C. Kiefer.

[31] The conventional standard error associated with $\hat{\beta}$ is $s/(\sum x_t^2)^{1/2} = 0.8259$.

Table 11.2 DATA GENERATED FROM SIMPLE REGRESSION MODEL (11.53)

t	y_t	x_t	t	y_t	x_t
1	−0.039	−1.026	9	0.348	−0.300
2	2.730	1.542	10	4.428	0.924
3	0.997	−0.532	11	−0.723	0.699
4	−2.990	−0.253	12	−0.632	0.182
5	2.660	0.040	13	−2.434	1.078
6	−0.624	−0.882	14	1.864	0.512
7	6.598	0.875	15	−1.620	−0.438
8	−0.669	−0.066			

where $z = y_{T+1}$ and $a = 4$. In (11.54) we have included square-root error ($\alpha = 0.5$), absolute error ($\alpha = 1$), and squared error ($\alpha = 2$) and quartic error ($\alpha = 4$) loss functions. For each of them the following integral was evaluated[32]:

$$\int_{-\infty}^{\infty} L(z, a)\, p(z|\mathbf{y}, w)\, dz \tag{11.55}$$

for various values of $w = x_{T+1}$, where $p(z|\mathbf{y}, w)$ is the predictive pdf shown in (11.6). The results of these calculations are shown in Table 11.3. From Table 11.3 we see that the optimal setting for w in the squared-error loss function is $w = 1.9144$ and the associated loss is 10.533. Suppose, now, that we were mistaken in assuming the loss function to be $|z - 4|^2$, for in fact it is $|z - a|$. With $w = 1.9144$, our expected loss with the absolute-error loss function is 2.5475. This is close to the minimal expected loss, 2.5465, associated with the optimal w value for the absolute-error loss function, namely $w = 1.9584$. Similar results are summarized in Table 11.4. It is seen that for this problem the optimal solution for the squared-error loss function is remarkably robust under changes in the form of the loss function[33] as long as we restrict ourselves to symmetric loss functions.

To study robustness of solutions, when there is a departure from symmetry in the loss function, calculations similar to those reported above have been carried through by using the following loss functions:

$$L(z, a) = \begin{cases} k|z - a| & \text{for } z \geq a \\ |z - a| & \text{for } z < a \end{cases} \qquad k = \begin{matrix} 0.25, 0.50, 0.75, 1.0, \\ 1.5, 2.0, 3.0. \end{matrix} \tag{11.56}$$

In (11.56) we have a class of linear loss functions which is asymmetric when $k \neq 1$. As above, we take $a = 4$.

[32] Numerical integration was employed.

[33] As can be seen from the figures in Table 11.3, this conclusion does not hold for the approximate certainty equivalence solution, $w = 4/\hat{\beta} = 2.5181$.

Table 11.3 EXPECTED LOSS AS A FUNCTION OF THE CONTROL VARIABLE SETTING AND FORM OF THE LOSS FUNCTION

Value of Control Variable, w	Form of Loss Function			
	$\|z - 4\|^{0.5}$	$\|z - 4\|$	$\|z - 4\|^2$	$\|z - 4\|^4$
3.6780	1.7432	3.6031	20.853	1457.7
2.9895	1.5789	2.9680	14.370	731.98
2.5181[a]	1.4992	2.6783	11.743	496.01
2.1751	1.4681	2.5666	10.759	412.52
2.1035	1.4651	2.5554	10.652	402.44
2.0364	1.4636	2.5490	10.582	395.05
2.0045	1.4632	2.5473	...	...
1.9812	1.4632[c]	2.5467	...	...
1.9735	1.4632	2.5466	10.544	389.85
1.9584	1.4633	2.5465[c]	...	...
1.9144[b]	1.4638	2.5475	10.533[c]	386.42
1.8587	1.4653	2.5513	10.543	384.44
1.8061	1.4675	2.5576	10.572	383.67
1.7960	...	...	...	383.64[c]
1.7565	1.4704	2.5658	10.615	383.89
1.7095	1.4737	2.5757	10.672	384.93
1.5442	1.4906	2.6271	10.987	394.90
1.4080	1.5104	2.6887	11.383	410.36

[a] $w = 4/\hat{\beta}$, certainty equivalence solution for $|z - 4|^2$.
[b] $w = (4/\hat{\beta})[t_0^2/(1 + t_0^2)]$, optimal solution for $|z - 4|^2$ where $t_0^2 = \hat{\beta}^2 m_{xx}/\bar{s}^2$.
[c] Minimal expected loss for given loss function.

For each of these loss functions the value of w, which minimizes (11.55), was found by successive numerical integrations. The results of these calculations are shown in Table 11.5. We note that in contrast to what was found in the case of symmetric loss functions the optimal solution for the quadratic loss function, denoted by w_q, is not very robust to marked departures from

Table 11.4 EXPECTED LOSSES ASSOCIATED WITH OPTIMAL CONTROL AND USE OF SOLUTION FOR SQUARED ERROR LOSS FUNCTION

Item	Form of Loss Function			
	$\|z - 4\|^{0.5}$	$\|z - 4\|$	$\|z - 4\|^2$	$\|z - 4\|^4$
Optimal value for w	1.9812	1.9584	1.9144	1.7960
Expected loss for optimal w	1.4632	2.5465	10.533	383.64
Expected loss for $w = 1.9144$	1.4638	2.5475	10.533	386.42

Table 11.5 COMPARISON OF OPTIMAL SOLUTIONS FOR ASYMMETRIC LINEAR LOSS FUNCTIONS WITH RESULTS OF EMPLOYING CERTAINTY EQUIVALENCE AND QUADRATIC CONTROL SETTINGS[a]

Slope Parameter of Loss Function[b]	Optimal Setting $w = w^*$	Expected Loss for $w = w^*$	Quadratic Setting $w = w_q$	Expected Loss for $w = w_q$	Certainty Equivalence Setting $w = w_{ce}$	Expected Loss for $w = w_{ce}$
$k = 0.25$	3.4778	1.5518	1.9144	1.9519	2.5181	1.6739
$k = 0.50$	2.6214	2.0059	1.9144	2.1504	2.5181	2.0087
$k = 0.75$	2.2321	2.3131	1.9144	2.3489	2.5181	2.3435
$k = 1.0$	1.9584	2.5465	1.9144	2.5475	2.5181	2.6783
$k = 1.5$	1.6226	2.8937	1.9144	2.9446	2.5181	3.3478
$k = 2.0$	1.3928	3.1511	1.9144	3.3416	2.5181	4.0174
$k = 3.0$	1.0944	3.5272	1.9144	4.1358	2.5181	5.3565

[a] Computations based on generated data in Table 11.2.
[b] The loss function employed is $L = |z - 4|$ for $-\infty < z \leq 4$; $k|z - 4|$ for $4 < z < \infty$.

symmetry; for example, when $k = 3.0$, the optimal setting for w is 1.0944, with an associated expected loss of 3.5272, whereas for $w_q = 1.9144$ the expected loss is 4.1358. However, for values of k in the range 0.50 to 2.0 use of the quadratic solution does not result in raising the expected loss very much. Use of the approximate certainty equivalence solution $w_{ce} = 2.5181$ gives rise to quite an increase in expected loss in relation to minimal expected loss when $k = 0.25$ and $k = 1.5$, 2.0, and 3.0. For certain values of k use of w_{ce} leads to smaller expected loss than does w_q.

In summary, for the range of loss functions, problem and data considered, it is found that the optimal solution for the quadratic case is quite robust to changes in the form of the loss function except in the case of considerable asymmetry.

11.5 TWO-PERIOD CONTROL OF THE MULTIPLE REGRESSION MODEL[34]

Here we are concerned with the problem of setting the values of independent variables in such a way that the dependent variable is kept close to a target value in not just one future but several future periods. That we seek to achieve optimal control for several future periods adds new dimensions to our analysis. As has been recognized in the literature, learning and design considerations are involved in the multiperiod control problem; that is, as

[34] The material in this section is based in part on A. Zellner, "On Controlling and Learning about a Normal Regression Model," manuscript, 1966, presented to the Information, Decision and Control Workshop, University of Chicago.

we proceed into the future, we get more sample information that permits us to learn more about the values of unknown parameters. Also, how much we learn about unknown parameter values will depend on the settings of control variables, a design-of-experiments consideration. We shall see that the solution to a multiperiod control problem provides an optimal sequence of actions that takes explicit account of control, learning, and design considerations. Since the future actions embodied in the solution involve adapting our actions to new data as they appear, the multiperiod problem is often referred to as an *adaptive control problem.*

To make these considerations more concrete, let us consider a two-period problem. Our model for the observations is

$$\mathbf{y} = X\boldsymbol{\beta} + \mathbf{u}, \tag{11.57}$$

where $\mathbf{y}' = (y_1, y_2, \ldots, y_T)$, X is a $T \times k$ matrix with rank k of observations on k control variables, $\boldsymbol{\beta}$ is a $k \times 1$ vector of unknown coefficients, and $\mathbf{u}$ is a $T \times 1$ normal disturbance vector with mean zero and covariance matrix $\sigma^2 I_T$.

The future observation vector $\mathbf{z}' = (z_1, z_2) \equiv (y_{T+1}, y_{T+2})$ is assumed to be generated as follows:

$$\mathbf{z} = W\boldsymbol{\beta} + \mathbf{v}, \tag{11.58}$$

where $W = (\mathbf{w}_1, \mathbf{w}_2)'$ is a $2 \times k$ matrix of future settings for the control variables and $\mathbf{v}' = (u_{T+1}, u_{T+2})$ denotes a vector of future disturbance terms, assumed to be distributed normally and independently with zero means and common variances σ^2 and independent of $\mathbf{u}$.

If we employ a diffuse prior pdf for the elements of $\boldsymbol{\beta}$ and σ, it has been shown in Chapter 3 that the predictive pdf for $\mathbf{z}$ is in the following bivariate Student t form:

$$p(\mathbf{z}|\mathbf{y}) \propto [\nu + (\mathbf{z} - W\hat{\boldsymbol{\beta}})'H(\mathbf{z} - W\hat{\boldsymbol{\beta}})]^{-(\nu+2)/2}, \tag{11.59}$$

where $\hat{\boldsymbol{\beta}} = (X'X)^{-1}X'\mathbf{y}$, $\nu = T - k$, and $H = [I + W(X'X)^{-1}W']^{-1}/s^2$, with $s^2 = (\mathbf{y} - X\hat{\boldsymbol{\beta}})'(\mathbf{y} - X\hat{\boldsymbol{\beta}})/\nu$.

Let us assume that our loss function is given by[35]

$$L(\mathbf{z}, \mathbf{a}) = (z_1 - a_1)^2 + (z_2 - a_2)^2, \tag{11.60}$$

where a_1 and a_2 are given target values for future periods 1 and 2, respectively. Then our problem is to formulate a procedure for determining the values of $\mathbf{w}_1$ and $\mathbf{w}_2$, where $W = (\mathbf{w}_1, \mathbf{w}_2)'$, to minimize expected loss. Let us consider several possible approaches for solving this problem.

[35] The analysis can be extended to apply to a loss function of the following form: $L(\mathbf{z}, \mathbf{a}) = (\mathbf{z} - \mathbf{a})'A(\mathbf{z} - \mathbf{a})$, where A is any positive definite symmetric matrix.

Approach I. "Here and Now" Solution

Use the pdf in (11.59) to evaluate the expectation of the loss function in (11.60) and then minimize with respect to the elements of $\mathbf{w}_1$ and $\mathbf{w}_2$.[36] The solution to this problem, termed a "here and now" solution, is appropriate when for some reason we must announce our actual settings for *both* $\mathbf{w}_1$ and $\mathbf{w}_2$ at the beginning of the first future period. The costs associated with being required to announce settings for $\mathbf{w}_1$ and $\mathbf{w}_2$ at the beginning of period $T + 1$ are the following:

1. We cannot use the information provided by $z_1 \equiv y_{T+1}$ to obtain an optimal setting for $\mathbf{w}_2$.
2. We cannot take account of how our choice of $\mathbf{w}_1$ will affect our determination of an optimal setting for $\mathbf{w}_2$.

However, since there are situations in which a "here and now" solution is needed and also for comparative purposes, it is of interest to present the "here and now" solution. We have

$$\begin{aligned} EL(\mathbf{z}, \mathbf{a}) &= \underset{z_1}{E}(z_1 - a_1)^2 + \underset{z_2}{E}(z_2 - a_2)^2 \\ &= \bar{s}^2[1 + \mathbf{w}_1'(X'X)^{-1}\mathbf{w}_1] + (a_1 - \mathbf{w}_1'\hat{\boldsymbol{\beta}})^2 \\ &\quad + \bar{s}^2[1 + \mathbf{w}_2'(X'X)^{-1}\mathbf{w}_2] + (a_2 - \mathbf{w}_2'\hat{\boldsymbol{\beta}})^2. \end{aligned} \tag{11.61}$$

Then the optimal settings for $\mathbf{w}_1$ and $\mathbf{w}_2$ are given by

$$\mathbf{w}_1^* = \frac{X'X\hat{\boldsymbol{\beta}}a_1}{\bar{s}^2 + \hat{\boldsymbol{\beta}}'X'X\hat{\boldsymbol{\beta}}} \quad \text{and} \quad \mathbf{w}_2^* = \frac{X'X\hat{\boldsymbol{\beta}}a_2}{\bar{s}^2 + \hat{\boldsymbol{\beta}}'X'X\hat{\boldsymbol{\beta}}}. \tag{11.62}$$

On inserting these values in (11.61), the expected loss is

$$EL(\mathbf{z}, \mathbf{a} | W = W^*) = \bar{s}^2\left(1 + \frac{a_1^2}{\bar{s}^2 + Q}\right) + \bar{s}^2\left(1 + \frac{a_2^2}{\bar{s}^2 + Q}\right), \tag{11.63}$$

where $Q = \hat{\boldsymbol{\beta}}'X'X\hat{\boldsymbol{\beta}}$. As mentioned above, the solution for each future period is completely analogous to that presented in Section 11.2 for a one-period problem.

Approach II. "Sequential Updating" Solution

Use the pdf in (11.59) to evaluate $E_{z_1}(z_1 - a_1)^2$ and minimize with respect to the elements of $\mathbf{w}_1$. This provides a setting for $\mathbf{w}_1$. Then *after* the first future period has passed and $z_1 \equiv y_{T+1}$ has been observed use $p(z_2|z_1, \mathbf{y})$ to

[36] Since z_1 and z_2 have marginal pdf's in the univariate Student t form and the marginal pdf for z_1 does not involve $\mathbf{w}_2$ and that for z_2 does not involve $\mathbf{w}_1$, the present problem reduces to two one-period problems, each identical to one considered in Section 11.2 [see (11.26) and the results that follow it].

evaluate $E(z_2 - a_2)^2$ and minimize with respect to the elements of $\mathbf{w}_2$. In this procedure the value of $\mathbf{w}_1$ is determined without taking account of how it affects the determination of $\mathbf{w}_2$ and on this count the solution is not optimal. However, in contrast to the "here and now" solution, the "sequential updating" solution does take account of information in period $T + 1$ to arrive at the setting for $\mathbf{w}_2$ and, as will be seen, is often a good approximation to the optimal solution for the present problem.

In the sequential updating solution for the first future period we have

$$\underset{z_1}{E}(z_1 - a_1)^2 = \bar{s}^2[1 + \mathbf{w}_1'(X'X)^{-1}\mathbf{w}_1] + (a_1 - \mathbf{w}_1'\hat{\boldsymbol{\beta}})^2 \tag{11.64}$$

and

$$\mathbf{w}_1^* = \frac{X'X\hat{\boldsymbol{\beta}}a_1}{\bar{s}^2 + \hat{\boldsymbol{\beta}}'X'X\hat{\boldsymbol{\beta}}}. \tag{11.65}$$

The value for $\mathbf{w}_1$ in (11.65) is precisely the same as in the "here and now" solution. For period $T + 2$, we take account of $\mathbf{w}_1 = \mathbf{w}_1^*$ and the new observation $z_1 \equiv y_{T+1}$; that is, we take the expectation of $(z_2 - a)^2$ with respect to z_2, given z_1. This yields[37]

$$\begin{aligned}\underset{z_2}{E}[(z_2 - a_2)^2|z_1, \mathbf{w}_1] &= \underset{z_2}{E}\{[z_2 - E(z_2|z_1)]^2|z_1\} + (a_2 - Ez_2|z_1)^2 \\ &= \bar{s}_2^2(1 + \mathbf{w}_2'M_2^{-1}\mathbf{w}_2) + (a_2 - \mathbf{w}_2'\hat{\boldsymbol{\beta}}_2)^2,\end{aligned} \tag{11.66}$$

where

$$M_2 = X'X + \mathbf{w}_1\mathbf{w}_1', \tag{11.67}$$

$$\begin{aligned}\hat{\boldsymbol{\beta}}_2 &= M_2^{-1}(X'\mathbf{y} + \mathbf{w}_1 z_1) \\ &= \hat{\boldsymbol{\beta}} + (X'X)^{-1}\mathbf{w}_1[1 + \mathbf{w}_1'(X'X)^{-1}\mathbf{w}_1]^{-1}(z_1 - \mathbf{w}_1'\hat{\boldsymbol{\beta}}),\end{aligned} \tag{11.68}$$

with $\hat{\boldsymbol{\beta}} = (X'X)^{-1}X'\mathbf{y}$ and

$$\begin{aligned}\bar{s}_2^2 &= \frac{\nu_2 s_2^2}{\nu_2 - 2} \\ &= \frac{[(\mathbf{y} - X\hat{\boldsymbol{\beta}})'(\mathbf{y} - X\hat{\boldsymbol{\beta}}) + (z_1 - \mathbf{w}_1'\hat{\boldsymbol{\beta}})^2(1 - \mathbf{w}_1'M_2^{-1}\mathbf{w}_1)]}{\nu_2 - 2} \\ &= \frac{[(\mathbf{y} - X\hat{\boldsymbol{\beta}}_2)'(\mathbf{y} - X\hat{\boldsymbol{\beta}}_2) + (z_1 - \mathbf{w}_1'\hat{\boldsymbol{\beta}}_2)^2]}{\nu_2 - 2},\end{aligned} \tag{11.69}$$

with $\nu_2 = \nu + 1$, where $\nu = T - k$. In (11.67) M_2 is the moment matrix for the independent variables at the beginning of $T + 2$, $\hat{\boldsymbol{\beta}}_2$ in (11.68) is the least squares quantity computed at the beginning of $T + 2$, and $\bar{s}_2^2$ involves the sum of squared residuals at the beginning of $T + 2$.

[37] See Appendix 1 of this chapter for the derivation of the result shown in (11.66).

On minimizing (11.66) with respect to the elements of $\mathbf{w}_2$, we obtain

$$\mathbf{w}_2^{+} = \frac{M_2\hat{\boldsymbol{\beta}}_2 a_2}{\bar{s}_2^{\,2} + \hat{\boldsymbol{\beta}}_2' M_2 \hat{\boldsymbol{\beta}}_2}, \tag{11.70}$$

which differs from the "here and now" setting shown in (11.62) in that (11.70) incorporates information pertaining to period $T + 1$ whereas w_2^* in (11.62) does not. Further, since $\mathbf{w}_2^{+}$ depends on z_1, it cannot be evaluated until after the first future period has passed and z_1 has been observed.

On substituting from (11.70) in (11.66), we find that the expected loss in period $T + 2$, *conditional on* z_1, $\mathbf{w}_1$, and $\mathbf{w}_2 = \mathbf{w}_2^{+}$, is

$$E[(z_2 - a_2)^2 | z_1, \mathbf{w}_1, \mathbf{w}_2 = \mathbf{w}_2^{+}] = \bar{s}_2^{\,2}\left(1 + \frac{a_2^{\,2}}{\bar{s}_2^{\,2} + \hat{\boldsymbol{\beta}}_2' M_2 \hat{\boldsymbol{\beta}}_2}\right). \tag{11.71}$$

Since the rhs of (11.71) depends on z_1, we can obtain its mean approximately by using the predictive pdf for z_1. The result is[38]

$$\begin{aligned} \underset{z_1}{E}\left[\bar{s}_2^{\,2}\left(1 + \frac{a_2^{\,2}}{\bar{s}_2^{\,2} + \hat{\boldsymbol{\beta}}_2' M_2 \hat{\boldsymbol{\beta}}_2}\right)\right] \\ \doteq \bar{s}^2\left[1 + \frac{a_2^{\,2}}{\bar{s}^2 + \hat{\boldsymbol{\beta}}' X' X \hat{\boldsymbol{\beta}}} - \frac{\nu - 2}{\nu - 1} \frac{a_2^{\,2}(\mathbf{w}_1'\hat{\boldsymbol{\beta}})^2}{(\bar{s}^2 + \hat{\boldsymbol{\beta}}' X' X \hat{\boldsymbol{\beta}})^2}\right]. \end{aligned} \tag{11.72}$$

If we now substitute $\mathbf{w}_1 = \mathbf{w}_1^*$, with $\mathbf{w}_1^*$ given in (11.65), we have expected loss for the second future period. With this substitution made in both (11.64) and (11.72), the total expected loss for the two periods for the sequential updating solution is approximately

$$EL \doteq \bar{s}^2\left(1 + \frac{a_1^{\,2}}{\bar{s}^2 + Q}\right) + \bar{s}^2\left(1 + \frac{a_2^{\,2}}{\bar{s}^2 + Q}\right) - \frac{\nu - 2}{\nu - 1}\frac{(a_1 a_2 \bar{s} Q)^2}{(\bar{s}^2 + Q)^4}, \tag{11.73}$$

where $Q = \hat{\boldsymbol{\beta}}' X' X \hat{\boldsymbol{\beta}}$. On comparing (11.73) with (11.63), the expression for expected loss associated with the "here and now" solution, we see that the first two terms of (11.73) are identical to the two terms of (11.63). The last term in (11.73) represents a reduction of expected loss associated with using the information for period $T + 1$ in determining the setting for $\mathbf{w}_2$.

Approach III. Adaptive Control Solution

Here we describe the adaptive control procedure for obtaining a solution to our two-period problem.[39]

Step 1. We consider ourselves to be at the beginning of the second future period, $T + 2$, and use the conditional predictive pdf for z_2, given z_1, $\mathbf{w}_1$, and

[38] See Appendix 2 of this chapter for the derivation of (11.72).

[39] Although we specialize it to our two-period regression problem, it should be recognized that it is more generally applicable.

the given sample information, $\mathbf{y}$ and X, to evaluate[40]

$$E_{z_2}(L|z_1, \mathbf{w}_1, \mathbf{y}, X) = g(z_1, \mathbf{w}_1, \mathbf{w}_2, \mathbf{y}, X). \tag{11.74}$$

It is to be emphasized that (11.74) is valid for whatever value z_1 takes and for whatever value is given to $\mathbf{w}_1$.

Step 2. Minimize the expression in (11.74) with respect to the elements of $\mathbf{w}_2$. If $\tilde{\mathbf{w}}_2$ denotes the solution to this problem, we have

$$\tilde{\mathbf{w}}_2 = h(z_1, \mathbf{w}_1, \mathbf{y}, X). \tag{11.75}$$

From (11.75), we see that $\tilde{\mathbf{w}}_2$ depends on the as yet unobserved value of $z_1 = y_{T+1}$, the as yet undetermined value of $\mathbf{w}_1$, and on the given sample information.[41]

Step 3. Substitute $\mathbf{w}_2 = \tilde{\mathbf{w}}_2$ in (11.74) with $\tilde{\mathbf{w}}_2$ as given in (11.75). This leads to

$$E_{z_2}(L|z_1, \mathbf{w}_1, \mathbf{y}, X) = g(z_1, \mathbf{w}_1, \tilde{\mathbf{w}}_2, \mathbf{y}, X), \tag{11.76}$$

which is a function of z_1, $\mathbf{w}_1$, and the given sample information.

Step 4. Take the expectation of (11.76) with respect to z_1 to obtain

$$E_{z_1} g(z_1, \mathbf{w}_1, \tilde{\mathbf{w}}_2, \mathbf{y}, X) = f(\mathbf{w}_1, \mathbf{y}, X). \tag{11.77}$$

Step 5. Minimize (11.77) with respect to the elements of $\mathbf{w}_1$ to obtain the optimal setting for $\mathbf{w}_1$, say $\tilde{\mathbf{w}}_1$, given by

$$\tilde{\mathbf{w}}_1 = \tilde{\mathbf{w}}_1(\mathbf{y}, X), \tag{11.78}$$

which is seen to depend on just the given sample information,[42] hence can be computed at the beginning of the first future period.

The optimal setting for $\mathbf{w}_1$, shown in (11.78), takes account of how the future information z_1 will be employed in optimizing in the second future period and how the first period setting for $\mathbf{w}_1$ will affect actions in the second future period. Further, the second period setting, $\mathbf{w}_2 = \tilde{\mathbf{w}}_2$, shown in (11.75), incorporates all the sample information available at the beginning of $T + 2$; that is, $\mathbf{y}$, X, $z_1 \equiv y_{T+1}$ and $\mathbf{w}_1 = \tilde{\mathbf{w}}_1$, the optimal setting for $\mathbf{w}_1$.

We now turn to the problem of obtaining the adaptive control solution to the two-period regression control problem with loss function given in (11.60)

[40] Also, (11.74) is conditioned by our given prior information about $\boldsymbol{\beta}$ and σ, diffuse in the present problem.

[41] It should be emphasized that $\tilde{\mathbf{w}}_2$ cannot be given a specific value until we observe z_1 and are given a setting for w_1.

[42] If we employ an informative prior pdf for $\boldsymbol{\beta}$ and σ, $\tilde{\mathbf{w}}_1$ would depend on parameters of the prior pdf as well as on sample information.

and predictive pdf given in (11.59). Specializing Step 1, shown in (11.74), we have[43]

(11.79)

$$\begin{aligned}\underset{z_2}{E}[L(\mathbf{z}, \mathbf{a})|z_1, \mathbf{w}_1] &= (z_1 - a_1)^2 + \underset{z_2}{E}[(z_2 - a_2)^2|z_1, \mathbf{w}_1] \\ &= (z_1 - a_1)^2 + \bar{s}_2^2(1 + \mathbf{w}_2' M_2^{-1}\mathbf{w}_2) + (a_2 - \mathbf{w}_2'\hat{\boldsymbol{\beta}}_2)^2,\end{aligned}$$

wherein (11.66) has been utilized and the quantities $\bar{s}_2^2$, M_2, and $\hat{\boldsymbol{\beta}}_2$ have been defined in (11.67) to (11.69).

Step 2 involves minimization of (11.79) with respect to the elements of $\mathbf{w}_2$. This yields[44]

$$\tilde{\mathbf{w}}_2 = \frac{M_2\hat{\boldsymbol{\beta}}_2 a_2}{\bar{s}_2^2 + \hat{\boldsymbol{\beta}}_2' M_2 \hat{\boldsymbol{\beta}}_2}. \tag{11.80}$$

Since M_2, $\hat{\boldsymbol{\beta}}_2$, and $\bar{s}_2^2$ depend on $\mathbf{w}_1$ and z_1, $\tilde{\mathbf{w}}_2$ is a function of these quantities and cannot be computed until values for them are available.

In Step 3 we substitute from (11.80) in (11.79) to obtain

$$\underset{z_2}{E}[L(\mathbf{z}, \mathbf{a})|z_1, \mathbf{w}_1, \mathbf{w}_2 = \tilde{\mathbf{w}}_2] = (z_1 - a_1)^2 + \bar{s}_2^2\left(1 + \frac{a_2^2}{\bar{s}_2^2 + \hat{\boldsymbol{\beta}}_2' M_2 \hat{\boldsymbol{\beta}}_2}\right), \tag{11.81}$$

which is a function of z_1 and $\mathbf{w}_1$.

Step 4 involves computation of the mean of (11.81) with respect to z_1. This computation yields the following approximate result[45]:

$$\begin{aligned}&\underset{z_1}{E}\left[(z_1 - a_1)^2 + \bar{s}_2^2\left(1 + \frac{a_2^2}{\bar{s}_2^2 + \hat{\boldsymbol{\beta}}_2' M_2 \hat{\boldsymbol{\beta}}_2}\right)\right] \\ &\doteq \bar{s}^2[1 + \mathbf{w}_1'(X'X)^{-1}\mathbf{w}_1] + (a_1 - \mathbf{w}_1'\hat{\boldsymbol{\beta}})^2 \\ &\quad + \bar{s}^2\left[1 + \frac{a_2^2}{\bar{s}^2 + Q} - \frac{\nu - 2}{\nu - 1}\frac{a_2^2(\mathbf{w}_1'\hat{\boldsymbol{\beta}})^2}{(\bar{s}^2 + Q)^2}\right]\end{aligned} \tag{11.82}$$

where $Q = \hat{\boldsymbol{\beta}}' X' X \hat{\boldsymbol{\beta}}$.

In Step 5 we minimize (11.82) with respect to the elements of $\mathbf{w}_1$ to get our first-period setting. This calculation produces[46]

$$\tilde{\mathbf{w}}_1 = \frac{X'X\hat{\boldsymbol{\beta}}a_1}{\bar{s}^2 + Q(1 - K)}, \quad \text{for} \quad 0 < K < 1, \tag{11.83}$$

[43] In what follows it is understood that we are taking our sample information $\mathbf{y}$ and X as given.

[44] Note that $\tilde{\mathbf{w}}_2$ in (11.80) is in the same form as $\mathbf{w}_2^+$ in (11.70). In evaluating $\tilde{\mathbf{w}}_2$, we employ a setting for $\mathbf{w}_1$ which differs from that employed in evaluating $\mathbf{w}_2^+$.

[45] The second term on the rhs of (11.81) is in exactly the same form as (11.71). Thus we can employ the results in Appendix 11.2 to get its approximate mean.

[46] In (11.83) the condition $0 < K < 1$ is needed in connection with the second-order condition for a minimum.

Table 11.6 FIRST-PERIOD SETTINGS AND TWO-PERIOD EXPECTED LOSSES FOR VARIOUS CONTROL SOLUTIONS[a]

Control Solution	First Period Setting for $\mathbf{w}_1$[b]	Expected Two-Period Loss[c]
I. "Here and now"	$\mathbf{w}_1 = \dfrac{X'X\hat{\boldsymbol{\beta}}a_1}{\bar{s}^2 + Q}$	$\bar{s}^2\left(2 + \dfrac{a_1^2 + a_2^2}{\bar{s}^2 + Q}\right)$
II. "Sequential updating"	$\mathbf{w}_1 = \dfrac{X'X\hat{\boldsymbol{\beta}}a_1}{\bar{s}^2 + Q}$	$\bar{s}^2\left(2 + \dfrac{a_1^2 + a_2^2}{\bar{s}^2 + Q} - \dfrac{\nu - 2}{\nu - 1}\dfrac{a_1^2a_2^2Q^2}{(\bar{s}^2 + Q)^4}\right)$
III. Adaptive control	$\mathbf{w}_1 = \dfrac{X'X\hat{\boldsymbol{\beta}}a_1}{\bar{s}^2 + Q(1 - K)}$	$\bar{s}^2\left(2 + \dfrac{a_1^2 + a_2^2}{\bar{s}^2 + Q} - \dfrac{\nu - 2}{\nu - 1}\dfrac{a_1^2a_2^2Q}{(\bar{s}^2 + Q)^3}\right)$

[a] In the table $Q = \hat{\boldsymbol{\beta}}'X'X\hat{\boldsymbol{\beta}}$ and $K = (\nu - 2)\bar{s}^2a_2^2/(\nu - 1)(\bar{s}^2 + Q)^2$.
[b] Approximate in III.
[c] Approximate in II and III. Terms of $0(T^{-3})$ and higher order of smallness have been dropped.

with $K = [(\nu - 2)/(\nu - 1)]\bar{s}^2a_2^2/(\bar{s}^2 + Q)^2$. It is seen that as $K \to 0$ (11.83) approaches $\mathbf{w}_1^*$, the solution value for the first period for both the "here and now" approach and the "sequential updating" approach [see (11.62) and (11.65)]. The factor $1 - K$ in the rhs of (11.83) provides a modification[47] to take account of how the first-period setting affects information about unknown parameters in solving the second-period problem.

Finally, we can substitute from (11.83) in (11.82) to obtain the following expression for expected loss, given the approximately optimal setting for $\mathbf{w}_1$:

(11.84)
$$E(L|\mathbf{w}_1 = \tilde{\mathbf{w}}_1) \doteq \bar{s}^2\left(1 + \frac{a_1^2}{\bar{s}^2 + Q}\right) + \bar{s}^2\left(1 + \frac{a_2^2}{\bar{s}^2 + Q}\right) - \frac{\nu - 2}{\nu - 1}\frac{(a_1a_2\bar{s})^2Q}{(\bar{s}^2 + Q)^3},$$

where $Q = \hat{\boldsymbol{\beta}}'X'X\hat{\boldsymbol{\beta}}$ and terms of order T^{-q}, $q \geq 3$, have been dropped. On comparing the expected loss in (11.84) with that of the "sequential updating" solution in (11.73), we see that the first two terms are identical. The third terms differ and the expected loss in (11.84) is smaller.[48] The difference in the present instance, however, involves a difference in terms of $0(T^{-2})$ which will be small in many cases.

In Table 11.6 some of the results obtained above are brought together. The main points to be appreciated are that the solutions presented can readily be applied in practice. Also, as already mentioned, there is a reduction in expected loss associated with the "sequential updating" and adaptive control solutions *vis à vis* the "here and now" solution. Next, a comparison of approximate expected losses for the "sequential updating"

[47] Note that K is of $0(T^{-2})$ and thus may often not be very different from zero.
[48] Note that the ratio of the third term of (11.84) to that of (11.73) is $(\bar{s}^2 + Q)/Q > 1$. Thus the expected loss in (11.84) is smaller than that in (11.73).

and adaptive control solutions indicates that the latter is smaller; however, the difference will often be small when T is not small and target values are not large. Last, these conclusions are presented in relation to a particular model and two-period problem and thus should not be crudely generalized to other models and problems.

11.6 SOME MULTIPERIOD CONTROL PROBLEMS

First it is rather straightforward to generalize the "sequential updating" solution of Section 11.5 to the case of controlling a regression model for $q(q > 2)$ future periods rather than two. Assume that our sample data $\mathbf{y}$ and the future observations $\mathbf{z}' = (z_1, z_2, \ldots, z_q)$, with $z_i = y_{T+i}$ $(i = 1, 2, \ldots, q)$, are generated by a standard normal linear-regression model.[49] Further, assume that our prior information about parameters is represented by (11.27) and that our loss function is given by

$$L(\mathbf{z}, \mathbf{a}) = \sum_{i=1}^{q} (z_i - a_i)^2, \tag{11.85}$$

where $\mathbf{a}' = (a_1, a_2, \ldots, a_q)$ is a vector whose elements are given target values. The future values $z_1, z_2, \ldots, z_q$ satisfy

$$z_i = \mathbf{w}_i'\boldsymbol{\beta} + v_i, \qquad i = 1, 2, \ldots, q, \tag{11.86}$$

where the $\mathbf{w}_i$'s are future settings of the control variables to be determined and the v_i's are normal and independent error terms, each with zero mean and common variance σ^2.

The application of the "sequential updating" approach in the present instance leads to the setting for $\mathbf{w}_1$ shown in (11.65) for the first future period:

$$\mathbf{w}_1^* = \frac{M_1\hat{\boldsymbol{\beta}}_1 a_1}{\bar{s}_1^2 + \hat{\boldsymbol{\beta}}_1'M_1\hat{\boldsymbol{\beta}}_1}, \tag{11.87}$$

where the subscript 1 denotes quantities available and known at the beginning of the first future period; that is, $M_1 = X'X$, $\hat{\boldsymbol{\beta}}_1 = M_1^{-1}X'\mathbf{y} = M_1^{-1}m_1$ and $\bar{s}_1^2 = (\mathbf{y} - X\hat{\boldsymbol{\beta}}_1)'(\mathbf{y} - X\hat{\boldsymbol{\beta}}_1)/(\nu_1 - 2) = (g_1 - \hat{\boldsymbol{\beta}}_1'M_1\hat{\boldsymbol{\beta}}_1)/(\nu_1 - 2)$, where $g_1 = \mathbf{y}'\mathbf{y}$ and $\nu_1 = \nu = T - k$.

For the second future period we have observed z_1 and know the setting for $\mathbf{w}_1$. Thus, on computing $E_{z_2}[(z_2 - a_2)^2|z_1]$ and optimizing with respect to the elements of $\mathbf{w}_2$ as in (11.64) to (11.65), we are led to

$$\mathbf{w}_2^* = \frac{M_2\hat{\boldsymbol{\beta}}_2 a_2}{\bar{s}_2^2 + \hat{\boldsymbol{\beta}}_2'M_2\hat{\boldsymbol{\beta}}_2}, \tag{11.88}$$

[49] See (11.26) and the surrounding text for a description of the model under consideration. Below it is assumed that all independent variables can be controlled. Generalization to the case in which just a subset can be controlled is direct and thus is not presented.

where the subscript 2 denotes quantities available and known at the beginning of the second future period; that is, $M_2 = X'X + \mathbf{w}_1^*\mathbf{w}_1^{*\prime}$, $\hat{\boldsymbol{\beta}}_2 = M_2^{-1}(X'\mathbf{y} + \mathbf{w}_1^* z_1) = M_2^{-1}\mathbf{m}_2$, and $\bar{s}_2^2 = (g_2 - \hat{\boldsymbol{\beta}}_2' M_2 \hat{\boldsymbol{\beta}}_2)/(\nu_2 - 2)$, where $g_2 = y'y + z_1^2$ and $\nu_2 = \nu_1 + 1 = \nu + 1$.

For the jth future period we have observed $z_1, z_2, \ldots, z_{j-1}$ and know the settings for $\mathbf{w}_1, \mathbf{w}_2, \ldots, \mathbf{w}_{j-1}$. On minimizing $E_{z_j}(z_j - a_j)^2$ with respect to the elements of $\mathbf{w}_j$, we obtain

$$\mathbf{w}_j^* = \frac{M_j \hat{\boldsymbol{\beta}}_j a_j}{\bar{s}_j^2 + \hat{\boldsymbol{\beta}}_j' M_j \hat{\boldsymbol{\beta}}_j}, \tag{11.89}$$

where

$$M_j = X'X + \sum_{i=1}^{j-1} \mathbf{w}_i^*\mathbf{w}_i^{*\prime}, \qquad \hat{\boldsymbol{\beta}}_j = M_j^{-1}\mathbf{m}_j \quad \text{and} \quad \bar{s}_j^2 = \frac{g_j - \hat{\boldsymbol{\beta}}_j' M_j \hat{\boldsymbol{\beta}}_j}{\nu_j - 2}, \tag{11.90}$$

with

$$\mathbf{m}_j = X'\mathbf{y} + \sum_{i=1}^{j-1} \mathbf{w}_i^* z_i, \qquad g_j = \mathbf{y}'\mathbf{y} + \sum_{i=1}^{j-1} z_i^2, \quad \text{and} \quad \nu_j = \nu + j - 1. \tag{11.91}$$

Thus (11.89), for $j = 1, 2, \ldots, q$, yields the sequence of "sequential updating" settings[50] for the q control vectors, $\mathbf{w}_1, \mathbf{w}_2, \ldots, \mathbf{w}_q$.

Now we turn to one of several multiperiod problems considered by Prescott.[51] He discusses control of a multivariate system with autoregressive features and obtains approximations to the adaptive control solution. His approximate solutions take account of uncertainty about parameter values and new information as it becomes available, considerations which were seen to be important above. In addition, for a particular system he uses Monte Carlo techniques to evaluate the average losses associated with his approximate solutions and with several "certainty equivalence" or "linear decision rule" solutions which neglect uncertainty about parameter values. The average losses associated with Prescott's approximate adaptive control solutions are found to be quite a bit smaller than those associated with solutions that neglect uncertainty about parameter values. This finding attests to the importance of allowing for uncertainty about the values of parameters in solving control problems, particularly when parameter estimates are not very precise.

[50] For computational convenience relations in the form of (11.68) and (11.69) can be employed to update the $\hat{\boldsymbol{\beta}}_i$'s and $\bar{s}_i^2$'s. In the second line of (11.68) note that $[1 + \mathbf{w}_1'(X'X)^{-1}\mathbf{w}_1]^{-1} = 1 - \mathbf{w}_1' M_2^{-1}\mathbf{w}_1$, where $M_2 = X'X + \mathbf{w}_1\mathbf{w}_1'$, and thus $\hat{\boldsymbol{\beta}}_2 = \hat{\boldsymbol{\beta}}_1 + M_1^{-1}\mathbf{w}_1(1 - \mathbf{w}_1' M_2^{-1}\mathbf{w}_1)(z_1 - \mathbf{w}_1'\hat{\boldsymbol{\beta}}_1)$, where $M_1 = X'X$ and $\hat{\boldsymbol{\beta}}_1 = M_1^{-1}\mathbf{m}_1$ with $\mathbf{m}_1 = X'\mathbf{y}$.

[51] E. C. Prescott, *Adaptive Decision Rules for Macro Economic Planning*, doctoral dissertation, Graduate School of Industrial Administration, Carnegie-Mellon University, 1967.

The multivariate system considered by Prescott is given by

$$\mathbf{y}_t = A_y\mathbf{y}_{t-1} + A_w\mathbf{w}_t + A_x\mathbf{x}_t + \mathbf{u}_t, \qquad t = 1, 2, \ldots, q, \tag{11.92}$$

where $\mathbf{y}_t$ is an $m \times 1$ dependent or endogenous variable vector, $\mathbf{y}_{t-1}$ is the dependent variable vector lagged one period,[52] $\mathbf{w}_t$ is a $k \times 1$ vector of control variables, $\mathbf{x}_t$ is a $p \times 1$ vector of independent (or exogenous) variables not under control, and $\mathbf{u}_t$ is an $m \times 1$ vector of random disturbance terms; A_y, A_w and A_x are coefficient matrices. Time is measured with $t = 0$ denoting the current period; thus $t = 1$ denotes the first future period, $t = 2$, the second, and so on.[53] The disturbance vectors are assumed to be normally and independently distributed, each with zero mean vector and common positive definite symmetric covariance matrix.

The loss function for the q future periods is assumed to have the following form:

$$L = \sum_{t=1}^{q} L_t(\mathbf{y}_t), \tag{11.93}$$

where each $L_t(\mathbf{y}_t)$ is a quadratic in the elements of $\mathbf{y}_t$.[54]

As new information becomes available, we learn more about the parameters of (11.92). To represent this learning process it is convenient to rewrite (11.92) as

$$\mathbf{y}_t = A\mathbf{d}_t + \mathbf{u}_t, \qquad t = 1, 2, \ldots, q, \tag{11.94}$$

where $A = (A_y \vdots A_w \vdots A_x)$ and $\mathbf{d}_t' = (\mathbf{y}_{t-1}' \vdots \mathbf{w}_t' \vdots \mathbf{x}_t')$. As prior pdf for the elements of A at $t = 0$, Prescott considers, among others, independent multivariate normal pdf's for the rows of A; that is, if A_i is the transpose of the ith row of A,

$$p_{i1}(A_i|\mathbf{m}_{i1}, H_{i1}) = \text{MVN}(\mathbf{m}_{i1}, H_{i1}), \qquad i = 1, 2, \ldots, m, \tag{11.95}$$

where $\mathbf{m}_{i1}$ is the prior mean vector and H_{i1}, the prior precision matrix[55] of the multivariate normal (MVN) prior pdf at the beginning of the first future period. The pdf in (11.95) would have the form of the posterior pdf based on T sample observations ($t = -1, -2, \ldots, -T$) if our initial ($t = -T$) prior

[52] By suitable definition of $\mathbf{y}_t$ it is well known that higher order difference equation systems can be written in the form of (11.92). In this case, as well as others, some of the coefficients of the system will be known with certainty.

[53] With time so measured the sample period extends from $t = -T + 1$ to $t = 0$, T observations in all.

[54] Prescott, *op. cit.*, p. 17, notes that by suitable definition of $\mathbf{y}_t$ the loss functions $L_t(\mathbf{y}_t)$ can include lagged and unlagged endogenous, exogenous, and control variables. Of course, if $\mathbf{y}_t$ is so redefined, the system in (11.92) should be correspondingly redefined.

[55] The prior precision matrix is just the inverse of the prior covariance matrix.

pdf on the elements of A were diffuse and the covariance matrix for $\mathbf{u}_t$ were diagonal with known elements.[56]

After the first future period ($t = 1$) has passed we can use the new information to update the prior pdf in (11.95) by an application of Bayes' theorem, which yields[57]

$$
\begin{aligned}
p_{i2}(A_i|I_2) &\propto \exp\left[-\tfrac{1}{2}(A_i - \mathbf{m}_{i1})'H_{i1}(A_i - \mathbf{m}_{i1}) - \tfrac{1}{2}(y_{i1} - \mathbf{d}_1'A_i)^2\right] \\
&\propto \exp\left[-\tfrac{1}{2}(A_i - \mathbf{m}_{i2})'H_{i2}(A_i - \mathbf{m}_{i2})\right],
\end{aligned}
\tag{11.96}
$$

where $p_{i2}(A_i|I_2)$ is the posterior pdf for the elements of A_i, given the information denoted by I_2, available at the beginning of $t = 2$. In the second line of (11.96).

$$H_{i2} = H_{i1} + \mathbf{d}_1\mathbf{d}_1', \tag{11.97}$$

and

$$\mathbf{m}_{i2} = H_{i2}^{-1}(H_{i1}\mathbf{m}_{i1} + y_{i1}\mathbf{d}_1); \tag{11.98}$$

H_{i2} and $\mathbf{m}_{i2}$ are the precision matrix and mean vector, respectively, for $p_{i2}(A_i|I_2)$. Since the posterior pdf's $p_{it}(A_i|I_t)$, $t = 1, 2, \ldots, q$, are all normal, (11.97) to (11.98) can be generalized by induction to

$$
\left.
\begin{aligned}
&(11.99) & H_{i,t+1} &= H_{it} + \mathbf{d}_t\mathbf{d}_t' \\
&(11.100) & \mathbf{m}_{i,t+1} &= H_{i,t+1}^{-1}(H_{it}\mathbf{m}_{it} + y_{it}\mathbf{d}_t)
\end{aligned}
\right\} \quad t = 0, 1, 2, \ldots, q - 1,
$$

where $H_{i,t}$ and $\mathbf{m}_{it}$ are the precision matrix and mean vector, respectively, of $p_{it}(A_i|I_t)$, a multivariate normal pdf. The relations (11.99) to (11.100) are useful for updating our pdf's for the elements of A_i as new observations become available.[58]

With this said about updating our prior information, we turn to our basic problem of selecting values sequentially for the control vector $\mathbf{w}_t$ in (11.92) for $t = 1, 2, \ldots, q$, to minimize the expected value of the loss function shown in (11.93).[59] Formally, let f_t denote the *minimum* expected value of the loss

[56] Other initial assumptions can lead to a posterior pdf in the form of (11.95); for example, we could initially ($t = -T$) have independent MVN prior pdf's for the elements of the A_i's, given the elements of a diagonal disturbance covariance matrix. Generalization to the case of an unknown and nondiagonal disturbance matrix is considered by Prescott.

[57] Below, for convenience, we take the known diagonal disturbance covariance matrix to be the identity matrix; that is, $E\mathbf{u}_t\mathbf{u}_t' = I$ for all t. With known variance, this can be achieved by rescaling the data.

[58] Note that the predictive pdf for y_{it}, $p_{it}(y_{it}|I_t)$, is given by

$$p_{it}(y_{it}|I_t) = \int p_{it}(y_{it}|A_i)p_{it}(A_i|I_t)\, dA_i,$$

where $p_{it}(y_{it}|A_i) \propto \exp\left[-\tfrac{1}{2}(y_{it} - \mathbf{d}_t'A_i)^2\right]$ and $p_{it}(A_i|I_t)$ is the multivariate normal pdf for the elements of A_i, given I_t, information available at the beginning of period t.

[59] We shall assume that a minimum exists.

for periods t through q, inclusive; that is,[60]

$$f_t(I_t, \mathbf{y}_{t-1}) = \min E\left[\sum_{i=t}^{q} L_i(\mathbf{y}_i) | I_t, \mathbf{y}_{t-1}\right]. \tag{11.101}$$

Now Bellman's principle of optimality states:

An optimal policy [here sequential rules for selecting values of the control vector, the $\mathbf{w}_t$'s] has the property that whatever the initial state and initial decision are the remaining decisions must constitute an optimal policy with regard to the state resulting from the first decision.[61]

Thus by applying this principle we can write for the present problem

$$f_t(I_t, \mathbf{y}_{t-1}) = \min_{\mathbf{w}_t} E[f_{t+1}(I_{t+1}, \mathbf{y}_t) + L_t(\mathbf{y}_t) | I_t, \mathbf{y}_{t-1}] \tag{11.102}$$

for $t = 1, 2, \ldots, q$, with $f_{q+1} \equiv 0$. As we can see from (11.102), the optimal setting for $\mathbf{w}_t$ and optimal settings for future periods are determined by taking past settings and outcomes as given, whatever they may have been; thus the formal solution procedure, shown in (11.102), reflects the principle of optimality.

In (11.102) we have a system of q functional equations which in theory can be solved[62]; for example, if

$$E[f_{t+1}(I_{t+1}, \mathbf{y}_t) + L_t(\mathbf{y}_t) | I_t, \mathbf{y}_{t-1})] \tag{11.103}$$

is a quadratic function in $\mathbf{y}_{t-1}$ and $\mathbf{w}_t$, the $\mathbf{w}_t$ which minimizes (11.103) will be linear in $\mathbf{y}_{t-1}$ and the function's minimal value is quadratic in $\mathbf{y}_{t-1}$.[63] Then, using (11.102), we determine the function $f_t(I_t, \mathbf{y}_{t-1})$ explicitly. If the values of the elements in the coefficient matrix A in (11.94) were all known with certainty, (11.103) would be quadratic in $\mathbf{y}_{t-1}$ and $\mathbf{w}_t$ and the approach described above could be employed. However, when some or all elements of A are unknown and have to be estimated, (11.103) is not quadratic in $\mathbf{y}_{t-1}$ and $\mathbf{w}_t$ and a different solution procedure is required.[64]

[60] That f_t is a function of I_t, the information available at the beginning of period t, and $\mathbf{y}_{t-1}$ is due to the fact that (11.92) is a Markov process in the vector $\mathbf{y}_t$.

[61] Quoted from R. Bellman and R. Kalaba, "Dynamic Programming and Adaptive Processes: Mathematical Foundations," reprinted from *IRE Trans. Automatic Control*, **AC-5**, No. 1 (January 1960), in R. Bellman and R. Kalaba (Eds.), *Selected Papers on Mathematical Trends in Control Theory*. New York: Dover, 1964, pp. 195–200, p. 197.

[62] See, for example, Bellman and Kalaba, *loc. cit.*, M. Freimer, "A Dynamic Programming Approach to Adaptive Control Processes," *IRE Trans.*, AC-4, **2**, No. 2, 10–15 (1959), and M. Aoki, *loc. cit.*, for solutions for particular problems.

[63] Of course, (11.103) depends also on quantities that describe the available information I_t. Note, too, that I_{t+1} will depend on $\mathbf{y}_t$ as well as other quantities pertaining to period t.

[64] Note, for example, that the lhs of (11.82) is not quadratic in z_1 and $\mathbf{w}_1$.

In developing his approximate solution to the present problem, Prescott considers the first equation of (11.102):

$$f_1(I_1, \mathbf{y}_0) = \min_{\mathbf{w}_1} E[f_2(I_2, \mathbf{y}_1) + L_1(\mathbf{y}_1)|I_1, \mathbf{y}_0]. \tag{11.104}$$

If $f_2(I_2, \mathbf{y}_1)$ had a known functional form, we could evaluate the expectation on the rhs of (11.104) and find the minimizing value for $\mathbf{w}_1$. Indeed, this was done above for a two-period problem [see Section 11.5, particularly (11.79) and the subsequent analysis]. For three or more future periods, however, the determination of an exact form for $f_2(I_2, \mathbf{y}_1)$ does not appear to be possible. Since this is the case, Prescott introduces a function that approximates $f_2(I_2, \mathbf{y}_1)$. Given that I_2 represents our information about the unknown parameters for $t = 2$ and that this information is not updated for subsequent periods, we can define the minimum value of the expected loss for periods t through q as $h_t(I_2, \mathbf{y}_{t-1})$, $t = 2, 3, \ldots, q$. Then, applying the optimality principle, we have

$$h_t(I_2, \mathbf{y}_{t-1}) = \min_{\mathbf{w}_t} E[h_{t+1}(I_2, \mathbf{y}_t) + L_t(\mathbf{y}_t)|I_2, y_{t-1}] \tag{11.105}$$

for $t = 2, 3, \ldots, q$ with $h_{q+1} \equiv 0$. Given I_2, the system of equations in (11.105) can be solved because the function $h_t(I_2, \mathbf{y}_{t-1})$ is quadratic in $\mathbf{y}_{t-1}$. In terms of (11.104), $h_2(I_2, \mathbf{y}_1)$ is taken as an approximation for $f_2(I_2, \mathbf{y}_1)$.

The setting for $\mathbf{w}_1$ is obtained by minimizing

$$\underset{y_1}{E}[h_2(I_2, \mathbf{y}_1) + L_1(\mathbf{y}_1)], \tag{11.106}$$

where I_2 depends on I_1, $\mathbf{w}_1$, $\mathbf{y}_1$, and $\mathbf{x}_1$. Since I_2 depends on $\mathbf{w}_1$, the first term of (11.106) reflects the influence of the first period setting of $\mathbf{w}_1$ on the losses for subsequent periods.[65] To provide a practical computation procedure for medium to large systems, a final approximation is introduced; that is, (11.106) is approximated by

$$h_2(I_2{}^a, E\mathbf{y}_1) + \underset{y_1}{E}L_1(\mathbf{y}_1), \tag{11.107}$$

where $E\mathbf{y}_1$ is the mean of the predictive pdf for $\mathbf{y}_1$ and $I_2{}^a$ denotes the prior information for $t = 2$, with $E\mathbf{y}_1$ replacing $\mathbf{y}_1$. Then (11.107) can be minimized with respect to the elements of $\mathbf{w}_1$ by using computer search procedures. As a convenient initial value for $\mathbf{w}_1$, Prescott suggests the value of $\mathbf{w}_1$ which minimizes expected loss if the prior pdf for the first future period were never updated.

Since Prescott's solution is an approximate one that incorporates an allowance for uncertainty about parameters and shows how current control

[65] It should be noted that this consideration is not taken into account in the "sequential updating" approach described and applied above.

variable settings affect future losses, it is of interest to consider its performance in relation to other approaches. In this connection Prescott generated data from the following model[66]:

$$
\begin{aligned}
\Delta C &= 0.308\Delta Y_1 + 0.194\Delta C_{-1} + 0.408\Delta M \\
&\quad + 0.078\Delta G + 87.7P_{-1} - 4797 + u_1, \\
\Delta I_1 &= 0.278\Delta Y_1 - 0.663 I_{1,-1} + 0.009 Y_{-1}^* \\
&\quad + 0.168\Delta G + 159P_{-1} - 6125 + u_2, \\
\Delta I_2 &= 0.105\Delta Y_1 - 0.220\Delta R - 0.510 I_{2,-1} \\
&\quad + 0.041 Y_{-1}^* + 92.8P_{-1} - 5996 + u_3, \\
\Delta R &= 0.111\Delta Y_1 - 0.739\Delta M + 0.318\Delta M_{-1} \\
&\quad + 0.187\Delta G - 937 + u_4, \\
\Delta T &= 0.21\Delta Y,
\end{aligned}
\tag{11.108}
$$

where the time subscripts have been suppressed, Δ denotes the first difference operator, for example, $\Delta C \equiv C_t - C_{t-1}$, and a subscript -1 denotes a variable lagged one period, for example, $C_{-1} = C_{t-1}$. Definitions of the variables follow:

$C =$ personal consumption expenditures,
$I_1 =$ gross private investment, less new construction expenditures,
$I_2 =$ new construction expenditures,
$I = I_1 + I_2$, gross private investment,
$G =$ government purchase of goods and budgetary surplus,
$T - G =$ budgetary government surplus (or deficit if $T - G$ is negative) on income and product account,
$T =$ taxes,
$Y^* = Y_1 + G - T$,
$Y_1 = C + I$,
$M =$ currency and demand deposits adjusted in the middle of the year,
$P =$ GNP deflator,
$R =$ yield of 20-year corporate bonds, annual percentage rate multiplied by 10,
$Y = C + I + G$, with variables measured in billions of current dollars.
$u_i =$ random disturbance term ($i = 1, 2, 3, 4$).

The variables assumed under control are G and M. All other variables are endogenous, except for P, which is assumed to be exogenous. The random disturbance terms were generated by independent drawings from normal distributions with mean zero and a different variance for each of the four

[66] The model used is one developed and estimated by G. Chow in "Multiplier, Accelerator and Liquidity Preference in the Determination of National Income in the United States," *IBM Record Report*, **RC 1455** (1966).

disturbance terms.[67] For purposes of generating the data the exogenous price level was assumed to have grown at 2% a year in a 12-observation "sample period"[68] and at 1.5% rate for the "future planning period" of eight years. Further, since information regarding M_t and G_t is needed to generate the data, Prescott used the following relations:

$$\Delta M_t = 2.1 + 2.2u_{5t},$$

$$\Delta G_t = 4.1 + 4.0u_{6t},$$

where u_{5t} and u_{6t} are independently and normally distributed disturbance terms with zero means and unit variances.

The loss function, reflecting the goals of full employment, rapid economic growth, and price stability, is assumed to have the following form:

$$L = \sum_{t=1965}^{1972} [2(Y_t - Y_t^0)^2 + (C_t - C_t^0)^2 + (G_t - G_t^0)^2 + (I_t - I_t^0)^2], \tag{11.109}$$

where Y_t^0, C_t^0, G_t^0, and I_t^0 are given optimal or target values for the variables in period t.[69] In (11.109) we have an eight-year planning horizon for which the first future period is 1965. Some calculations were performed by using a four-year horizon, 1965 to 1968.

Given data generated as explained above for the "sample" period, 1953 to 1964, the policy maker must decide on values of M and G for 1965 and then, on the basis of his augmented sample after 1965's data are available, values for 1966, and so on. The following are several of the decision rules considered by Prescott:

1. Linear Decision Rules (LDR)

The reduced form equations of the system in (11.108) are estimated from the generated data by using classical least squares. The parameter estimates are regarded as being equal to the true parameter values and the expectation of (11.109) is minimized to provide settings for M and G.

[67] The variances, $\text{Var}(u_i)$, $i = 1, 2, 3, 4$ were given values equal to their respective sample estimates.

[68] The parameter values used to generate the data are equal to estimates obtained by G. Chow, *loc. cit.*, based on actual data. In other respects the experimental set-up is designed to approximate conditions of the U.S. economy in the late 1950's and early 1960's. Observed values of variables in 1951 and 1952 were used as initial conditions in generating the data for 1953 to 1964, the "sample" period, and subsequent future periods.

[69] The target values assigned were selected with an eye toward providing realism. An estimate of "capacity" GNP for 1964 was assumed to grow at 3.5% per year. Given that P grows at 1.5% per year, Y_t^0 was assumed to increase 5% per year. Other target values were obtained as follows: $C_t^0 = 0.64\,Y_t^0$, $I_t^0 = 0.16\,Y_t^0$, and $G_t^0 = 0.20\,Y_t^0$.

2. Linear Decision Rules I and II (LDR-I and LDR-II)

The two-stage least squares method is used to estimate parameters in the structural equations.[70] If the coefficient estimate for ΔM in the consumption equation has the "wrong" a priori algebraic sign, the corresponding variable is deleted from the equation.[71] Reduced-form coefficient estimates are obtained from the structural coefficient estimates and regarded as being the true values of the reduced-form coefficients. The procedure described above in (1) is then applied to obtain settings for M and G. LDR-II is the same as LDR-I except that the "sign test" is not used.

3. Adaptive Decision Rules (ADR)

The system in (11.108) has the form of (11.92) and the control solution ADR is the approximate one developed by Prescott. The initial (1953) prior pdf for the parameters is diffuse. Further, the variances of disturbance terms are assumed to be equal to their respective sample estimates. This is an approximation that can be relaxed in applications.

4. Adaptive Decision Rules (ADR-0)

Here the adaptive decision-rule procedure is one in which the decision maker sets the values of M and G for period t by using information available at that time but *not* allowing for the future updating of prior pdf's in subsequent periods. ADR-0 is what we have called the "sequential updating" approach.

5. Perfect Information Decision Rule (PIDR)

This decision rule involves minimization of the expected value of the loss function (11.109) under the assumption that the decision maker knows the *true* values of the model's parameters used in the generation of the data. Of course, in practice these values are unknown and have to be estimated. However, it is of interest to compare the resulting loss associated with a solution under this assumption with those associated with an application of decision rules when the parameters are unknown and have to be estimated.

In Table 11.7 the losses actually experienced with different decision rules and several sets of data, generated as explained above, are presented.[72] As Prescott concludes, "In every case, the procedure using adaptive decision rules performed better than LDR, clearly indicating the superiority of our analysis in this example."[73] In many cases the margin of superiority is

[70] These are not shown in the text.

[71] This a priori "sign test" is frequently employed in practice and appears to have been used by Chow in constructing his model.

[72] E. Prescott, *loc. cit.*, presents additional results generally similar to those shown in Table 11.7.

[73] E. Prescott, *op. cit.*, p. 63.

Table 11.7 LOSSES ASSOCIATED WITH DIFFERENT DECISION RULES FOR VARIOUS GENERATED SETS OF DATA[a]

		Four-Year Planning Horizon (1965–1968)				
Data Set	PIDR	Adaptive Decision Rules ADR	ADR-0	Linear Decision Rules LDR-I	LDR-II	LDR
1.1	4.1	5.2	5.2			5.5
1.2	4.0	6.0	6.0			8.7
1.3	6.0	4.3	6.6			8.7
1.4	2.8	5.6	5.9			10.9
1.5	5.9	10.0	10.0			10.6
Average for 1.1–1.5	4.6	6.2	6.7			8.9
1.6			4.2	10.0	7.1	5.9
1.7			4.4	3.8	25.1	9.5
1.8			3.9	2.8	4.4	5.7
1.9			6.0	16.3	7.4	6.5
1.10			6.6	12.8	8.6	9.0
Average for 1.6–1.10			5.0	11.1	10.5	7.3
		Eight-Year Planning Horizon (1965–1972)				
2.1	8.4	15.2	15.2			
2.2	8.0	8.1	8.1			
2.3	9.7	9.9	7.4			
2.4	11.1	12.9	13.0			
2.5	7.6	9.1	9.8			
Average for 2.1–2.5	9.0	11.0	10.7			
2.6			15.0	11.3	14.7	18.0
2.7			20.1	33.1	65.1	20.9
2.8			11.0	71.0	32.0	16.4
2.9			16.1	21.6	13.9	19.3
2.10			20.0	26.7	18.1	21.1
Average for 2.6–2.10			16.4	32.5	28.7	19.1

[a] Taken from E. Prescott, *op. cit.*, pp. 64–65. Where entries are not shown, they were not calculated.

substantial. Prescott points out that ADR performs better for the four-year horizon and ADR-0 does slightly better for the eight-year horizon. In discussing this difference he points to excessive experimentation when ADR is employed and when the horizon is long.[74] He suggests possible use of a moving horizon of three or four periods in connection with his approximate solution procedure.

The results in Table 11.7 point up the need to be extremely careful in determining "optimal" settings for control variables, particularly when sample information is not extensive. In addition, it must be realized that models representing economies and loss functions, such as that shown in (11.109), are approximations. Further research and experience with applications appear to be needed before it is possible to state whether adaptive control solutions will be "good enough" to aid in macroeconomic policy making.

APPENDIX 1 The Conditional Predictive Pdf for $\mathbf{z}_2$ Given $\mathbf{z}_1$

In this appendix we derive the conditional predictive pdf for z_2, given z_1, from the joint predictive pdf for these variables, which, as shown in (11.59), is in the following bivariate Student t form when we employ a diffuse prior pdf for $\boldsymbol{\beta}$ and σ:

$$p(\mathbf{z}|\mathbf{y}) \propto [\nu + (\mathbf{z} - W\hat{\boldsymbol{\beta}})'H(\mathbf{z} - W\hat{\boldsymbol{\beta}}]^{-(\nu+2)/2}, \tag{1}$$

with all quantities as defined in connection with (11.59). Now we partition $\mathbf{z} - W\hat{\boldsymbol{\beta}}$ as

$$\mathbf{z} - W\hat{\boldsymbol{\beta}} = \begin{pmatrix} z_1 - \mathbf{w}_1'\hat{\boldsymbol{\beta}} \\ z_2 - \mathbf{w}_2'\hat{\boldsymbol{\beta}} \end{pmatrix} = \begin{pmatrix} e_1 \\ e_2 \end{pmatrix} \tag{2}$$

and $H = [I + W(X'X)^{-1}W']^{-1}/s^2$ as

$$H = \begin{pmatrix} h_{11} & h_{12} \\ h_{21} & h_{22} \end{pmatrix}, \tag{3}$$

a 2×2 symmetric positive definite matrix. Then the quadratic form in (1) can be written as

$$\begin{aligned}(\mathbf{z} - W\hat{\boldsymbol{\beta}})'H(\mathbf{z} - W\hat{\boldsymbol{\beta}}) &= h_{22}e_2^2 + 2h_{12}e_1e_2 + h_{11}e_1^2 \\ &= h_{22}\left(e_2 + \frac{h_{12}}{h_{22}}e_1\right)^2 + \left(\frac{h_{11}h_{22} - h_{12}^2}{h_{22}}\right)e_1^2.\end{aligned}$$

[74] As he points out, *ibid.*, pp. 69–70, his approximate "... solution selected the current period's decision on the basis that all learning would occur from that observation and that the prior would not be updated subsequently. Therefore, ADR placed too great an emphasis on experimenting when the planning horizon was long."

On introducing this result in (1), we have

$$p(z_1, z_2|\mathbf{y}) \propto \left[\nu + \frac{e_1^2}{h^{11}} + h_{22}\left(e_2 - \frac{h^{12}}{h^{11}} e_1\right)^2\right]^{-(\nu+2)/2}, \tag{4}$$

where e_1 and e_2 are as defined in (2) and h^{ij} denotes the (i, j)th element of H^{-1}; that is,

$$H^{-1} = [h^{ij}] = s^2\begin{bmatrix} 1 + \mathbf{w}_1'(X'X)^{-1}\mathbf{w}_1 & \mathbf{w}_1'(X'X)^{-1}\mathbf{w}_2 \\ \mathbf{w}_2'(X'X)^{-1}\mathbf{w}_1 & 1 + \mathbf{w}_2'(X'X)^{-1}\mathbf{w}_2 \end{bmatrix}.$$

To put (4) in a more appealing form note that

$$\begin{aligned} e_2 - \frac{h^{12}}{h^{11}} e_1 &= z_2 - \mathbf{w}_2'\hat{\boldsymbol{\beta}} - \frac{\mathbf{w}_2'(X'X)^{-1}\mathbf{w}_1}{1 + \mathbf{w}_1'(X'X)^{-1}\mathbf{w}_1}(z_1 - \mathbf{w}_1'\hat{\boldsymbol{\beta}}) \\ &= z_2 - \mathbf{w}_2'\left[\hat{\boldsymbol{\beta}} + \frac{(X'X)^{-1}\mathbf{w}_1}{1 + \mathbf{w}_1'(X'X)^{-1}\mathbf{w}_1}(z_1 - \mathbf{w}_1'\hat{\boldsymbol{\beta}})\right]. \end{aligned} \tag{5}$$

The quantity in square brackets is equal to

$$\hat{\boldsymbol{\beta}}_2 = (X'X + \mathbf{w}_1\mathbf{w}_1')^{-1}(X'\mathbf{y} + \mathbf{w}_1 z_1), \tag{6}$$

the least squares quantity computed on the basis of information available at the beginning of period $T + 2$. This is shown as follows[75]:

$$\begin{aligned} \hat{\boldsymbol{\beta}}_2 &= \left[(X'X)^{-1} - \frac{(X'X)^{-1}\mathbf{w}_1\mathbf{w}_1'(X'X)^{-1}}{1 + \mathbf{w}_1'(X'X)^{-1}\mathbf{w}_1}\right](X'\mathbf{y} + \mathbf{w}_1 z_1) \\ &= \hat{\boldsymbol{\beta}} + \frac{(X'X)^{-1}\mathbf{w}_1(z_1 - \mathbf{w}_1'\hat{\boldsymbol{\beta}})}{1 + \mathbf{w}_1'(X'X)^{-1}\mathbf{w}_1}, \end{aligned} \tag{7}$$

which is just the quantity in brackets in the second line of (5). Thus

$$e_2 - \frac{h^{12}}{h^{11}} e_1 = z_2 - \mathbf{w}_2'\hat{\boldsymbol{\beta}}_2 \tag{8}$$

and, *given* z_1, (4) can be written as

$$p(z_2|z_1, \mathbf{y}) \propto \left[1 + \frac{h_{22}}{\nu + e_1^2/h^{11}}(z_2 - \mathbf{w}_2'\hat{\boldsymbol{\beta}}_2)^2\right]^{-(\nu+2)/2}. \tag{9}$$

We now obtain a more intelligible expression for the quantity $h_{22}/(\nu + e_1^2/h^{11})$ appearing in (9). From $H = [I + W(X'X)^{-1}W']^{-1}/s^2 = [I - W(X'X + W'W)^{-1}W']/s^2$ we have, with $M_2 = X'X + \mathbf{w}_1\mathbf{w}_1'$,

$$\begin{aligned} h_{22} &= \frac{1 - \mathbf{w}_2'(M_2 + \mathbf{w}_2\mathbf{w}_2')^{-1}\mathbf{w}_2}{s^2} \\ &= \frac{1}{(1 + \mathbf{w}_2'M_2^{-1}\mathbf{w}_2)s^2}. \end{aligned}$$

[75] The expression in brackets in the first line of (7) is $(X'X + \mathbf{w}_1\mathbf{w}_1')^{-1}$.

Then

(10)
$$\frac{h_{22}}{\nu + e_1^2/h^{11}} = \left(\frac{1}{1 + \mathbf{w}_2'M_2^{-1}\mathbf{w}_2}\right)\left(\frac{1}{\nu s^2 + (z_1 - \mathbf{w}_1'\hat{\boldsymbol{\beta}})^2/[1 + \mathbf{w}_1'(X'X)^{-1}\mathbf{w}_1]}\right).$$

From $\nu s^2 = (\mathbf{y} - X\hat{\boldsymbol{\beta}})'(\mathbf{y} - X\hat{\boldsymbol{\beta}})$ and the expression for $\hat{\boldsymbol{\beta}}_2$ in the second line of (7), it is straightforward to show that

$$(11)\quad (\mathbf{y} - X\hat{\boldsymbol{\beta}}_2)'(\mathbf{y} - X\hat{\boldsymbol{\beta}}_2) + (z_1 - \mathbf{w}_1'\hat{\boldsymbol{\beta}}_2)^2 = \nu s^2 + \frac{(z_1 - \mathbf{w}_1'\hat{\boldsymbol{\beta}})^2}{1 + \mathbf{w}_1'(X'X)^{-1}\mathbf{w}_1}.$$

Writing the lhs of (11), the residual sum of squares at the beginning of period $T + 2$, as $\nu_2 s_2^2$, with $\nu_2 = \nu + 1$, we find that the expression in (10) becomes $1/\nu_2 s_2^2(1 + \mathbf{w}_2'M_2^{-1}\mathbf{w}_2)$ and the conditional pdf for z_2, given z_1, in (9) can be expressed as[76]

$$(12)\quad p(z_2|z_1, \mathbf{y}) \propto \left[\nu_2 + \frac{1}{s_2^2(1 + \mathbf{w}_2'M_2^{-1}\mathbf{w}_2)}(z_2 - \mathbf{w}_2'\hat{\boldsymbol{\beta}}_2)^2\right]^{-(\nu_2+1)/2}.$$

This pdf is in the form of a univariate Student t pdf with $\nu_2 = \nu + 1$ degrees of freedom. The conditional mean of z_2, given z_1, is $\mathbf{w}_2'\hat{\boldsymbol{\beta}}_2$, whereas its variance[77] is $\nu_2 s_2^2(1 + \mathbf{w}_2'M_2\mathbf{w}_2)/(\nu_2 - 2)$ as stated in the text in connection with (11.66).

APPENDIX 2 Derivation of Approximate Mean Given in (11.72)

In (11.72) we have the following expectation to evaluate: $E_{z_1}\bar{s}_2^2[1 + a_2^2/(\bar{s}_2^2 + \hat{\boldsymbol{\beta}}_2'M_2\hat{\boldsymbol{\beta}}_2)]$, where $\bar{s}_2^2$, $\hat{\boldsymbol{\beta}}_2$, and M_2 have been defined in (11.67) to (11.69) and a_2 is the given target value for period $T + 2$. For the first term we have[78]

$$(1)\quad \begin{aligned} \underset{z_1}{E}\bar{s}_2^2 &= \underset{z_1}{E}\frac{(\mathbf{y} - X\hat{\boldsymbol{\beta}})'(\mathbf{y} - X\hat{\boldsymbol{\beta}}) + (z_1 - \mathbf{w}_1'\hat{\boldsymbol{\beta}})^2(1 - \mathbf{w}_1'M_2^{-1}\mathbf{w}_1)}{\nu_2 - 2} \\ &= \frac{(\mathbf{y} - X\hat{\boldsymbol{\beta}})'(\mathbf{y} - X\hat{\boldsymbol{\beta}}) + \bar{s}^2[1 + \mathbf{w}_1'(X'X)^{-1}\mathbf{w}_1][1 - \mathbf{w}_1'(X'X + \mathbf{w}_1\mathbf{w}_1')^{-1}\mathbf{w}_1]}{\nu_2 - 2} \\ &= \frac{\nu s^2 + \bar{s}^2(1 + Q_1)[1 - Q_1/(1 + Q_1)]}{\nu_2 - 2} \\ &= \frac{\nu s^2 + \bar{s}^2}{\nu_2 - 2} \\ &= \bar{s}^2, \end{aligned}$$

[76] For the pdf to be proper we need $\nu_2 > 0$.

[77] For the mean and variance to exist we need $\nu_2 > 1$ and $\nu_2 > 2$, respectively.

[78] In going from the second to the third line of (1), we use $(X'X + \mathbf{w}_1\mathbf{w}_1')^{-1} = (X'X)^{-1} - (X'X)^{-1}\mathbf{w}_1\mathbf{w}_1'(X'X)^{-1}/[1 + \mathbf{w}_1'(X'X)^{-1}\mathbf{w}_1]$. Thus $\mathbf{w}_1'(X'X + \mathbf{w}_1\mathbf{w}_1')^{-1}\mathbf{w}_1 = Q_1 - Q_1^2/(1 + Q_1) = Q_1/(1 + Q_1)$, where $Q_1 = \mathbf{w}_1'(X'X)^{-1}\mathbf{w}_1$.

where $\nu_2 = \nu + 1$, $\bar{s}^2 = \nu s^2/(\nu - 2)$, $\nu s^2 = (\mathbf{y} - X\hat{\boldsymbol{\beta}})'(\mathbf{y} - X\hat{\boldsymbol{\beta}})$, and $Q_1 = \mathbf{w}_1'(X'X)^{-1}\mathbf{w}_1$.

Next we have to evaluate $E_{z_1} a_2^2 \bar{s}_2^2/(\bar{s}_2^2 + \hat{\boldsymbol{\beta}}_2' M_2 \hat{\boldsymbol{\beta}}_2)$.[79] Using (11.69) and (11.70), we can rewrite this last expression as

$$(2)\quad E_{z_1}\left\{\frac{a_2^2[s_0^2 + e_1^2/(1 + Q_1)(\nu - 1)]}{a_0 + \hat{\boldsymbol{\beta}}'\mathbf{w}_1\mathbf{w}_1'\hat{\boldsymbol{\beta}} + 2\mathbf{w}_1'\hat{\boldsymbol{\beta}}e_1 + [Q_1 + 1/(\nu - 1)]e_1^2/(1 + Q_1)]}\right\},$$

where $e_1 = z_1 - \mathbf{w}_1\hat{\boldsymbol{\beta}}$, $a_0 = s_0^2 + \hat{\boldsymbol{\beta}}'X'X\hat{\boldsymbol{\beta}}$, $s_0^2 = \nu s^2/(\nu - 1)$, and $Q_1 = \mathbf{w}_1'(X'X)^{-1}\mathbf{w}_1$. Unfortunately, it does not appear possible to evaluate the expectation in (2) exactly.[80] Therefore we approximate it by

$$(3)\quad \frac{(a_2^2/a_0)[s_0^2 + \bar{s}^2/(\nu - 1)]}{1 + [\hat{\boldsymbol{\beta}}'\mathbf{w}_1\mathbf{w}_1'\hat{\boldsymbol{\beta}} + Q_1 + 1/(\nu - 1)]/a_0}.$$

The expression in (3) is obtained by noting that (2) can be written in terms of e_1/a_0 and e_1^2/a_0, random variables with finite means and variances, the latter being of $0(T^{-2})$, since $a_0 = s_0^2 + \hat{\boldsymbol{\beta}}'X'X\hat{\boldsymbol{\beta}}$ is $0(T)$. Then an approximation to the expectation in (2) is obtained by replacing the random variables with their mean values,[81] an approximation that here involves neglecting terms of $0(T^{-3})$ and higher order of smallness.

Since the denominator of (3) is one plus a term of $0(T^{-1})$, we can expand it to obtain[82]

$$(4)\quad \frac{a_2^2}{a_0}\left[s_0^2 + \frac{\bar{s}^2}{\nu - 1} - \frac{s_0^2}{a_0}(\mathbf{w}_1'\hat{\boldsymbol{\beta}})^2\right],$$

where terms of $0(T^{-2})$ have been retained and those of higher order of smallness have been dropped.

We can now add (1) and (4) to get our final result:

$$(5)\quad \begin{aligned} \underset{z_1}{E}\bar{s}_2^2\left(1 + \frac{a_2^2}{\bar{s}_2^2 + \hat{\boldsymbol{\beta}}_2'M_2\hat{\boldsymbol{\beta}}_2}\right) &\doteq \bar{s}^2 + \frac{a_2^2}{a_0}\left[s_0^2 + \frac{\bar{s}^2}{\nu - 1} - \frac{s_0^2}{a_0}(\hat{\boldsymbol{\beta}}'\mathbf{w}_1\mathbf{w}_1'\hat{\boldsymbol{\beta}} + \bar{s}^2 Q_1)\right] \\ &= \bar{s}^2\left[1 + \frac{a_2^2}{a_0} - \left(\frac{\nu - 2}{\nu - 1}\right)\frac{a_2^2}{a_0^2}(\mathbf{w}_1'\hat{\boldsymbol{\beta}})^2\right], \end{aligned}$$

where in going from the first to the second line $s_0^2 = (\nu - 2)/(\nu - 1)\bar{s}^2$ has been used. Since $a_0 = \bar{s}^2 + \hat{\boldsymbol{\beta}}'X'X\hat{\boldsymbol{\beta}}$ and $Q_1 = \mathbf{w}_1'(X'X)^{-1}\mathbf{w}_1$, it is the case that (5) is identical to the expression shown in (11.72).

[79] Note that the denominator $\bar{s}_2^2 + \hat{\boldsymbol{\beta}}_2'M_2\hat{\boldsymbol{\beta}}_2$ is positive with probability one.

[80] This problem is similar to the problem of taking the expectation of the reciprocal of a positive constant plus a quadratic in a normal variable discussed by M. Aoki who states that the problem cannot be solved exactly; *op. cit.*, p. 113 ff.

[81] See M. G. Kendall and A. Stuart, *The Advanced Theory of Statistics*, Vol. I. London: Griffin, 1958, p. 231 ff., for a discussion of this type of approximation to the mean of a function of random variables.

[82] The expansion is of the form $(1 + x)^{-1} = 1 - x + \cdots$, which requires $0 \leq |x| < 1$.

QUESTIONS AND PROBLEMS

1. Explain why and how the expression for expected loss in equation (11.7) depends on $\sum_{t=1}^{T} x_t^2$.
2. Consider the loss function in (11.3): $L(z, a) = (z - a)^2$. With $\hat{z} = w\hat{\beta}$, the mean of the predictive pdf for z, show that $E_z(z - a)^2 = \text{const} + (a - \hat{z})^2 + 0(T^{-1})$, where $0(T^{-1})$ denotes a term of order T^{-1}.
3. In the analysis of the one-period control of the simple regression process in (11.1), with the loss function shown in (11.3), we employed a diffuse prior pdf for the parameters β and σ. Use the same loss function and an informative prior pdf for β and σ in the normal-gamma form; that is, $p_1(\beta|\sigma)\, p_2(\sigma)$, where $p_1(\beta|\sigma)$ is normal with prior mean $\bar{\beta}$ and prior variance $c^2\sigma^2$ and $p_2(\sigma) \propto \sigma^{-(\nu_0+1)} \exp(-\nu_0 s_0^2/2\sigma^2)$, where $\bar{\beta}$, c, ν_0, and s_0 are assigned values to reflect prior information about β and σ to obtain the optimal setting for the control variable $w = x_{T+1}$. Compare this value for w and the associated expected loss with the corresponding expressions (11.8) and (11.10), obtained by using a diffuse prior pdf.
4. Under what circumstances will t_0^2 in (11.8) tend to be large so that w^* will be approximately equal to $a/\hat{\beta}$?
5. Would it be meaningful to consider the quantities $\bar{s}^2$ and t_0^2 on the rhs of (11.10) as random and to evaluate the mean of (11.10) by using the pdf for $\bar{s}^2$ and t_0^2?
6. Discuss the extent to which the loss function $(z - 10)^2$, where z is the nominal annual change in U.S. aggregate income and 10 is a target value for z, is a satisfactory loss function for use in economic policy making.
7. If we have a loss function of unspecified form, say $L(z, a)$, where a is a given quantity and z has the pdf shown in (11.6), explain how the mean of L $E_zL(z, a)$, assumed to exist, can be approximated by expanding $L(z, a)$ in a Taylor series about the mean of z.
8. Use the expansion technique of Problem 7 to provide an approximation to $E_z(z - a)^2 - E_zL(z, a)$, where $L(z, a)$ is a "true" loss function, a is a given constant, and z has pdf given in (11.6) of the text. Evaluate terms in the series when $L(z, a) = (z - a)^4$, indicating the order in T of each term retained and of terms omitted.
9. Suppose that a "true" loss function is given by $L(z, a) = (z - a)^2$, where a is the target value and $z = y_{T+1}$, with predictive pdf given by (11.6). If, instead of the true loss function, we use $L(z, a_0) = (z - a_0)^2$, where $a_0 \neq a$, compare the optimizing values of $w = x_{T+1}$ and the expected losses associated with the use of $L(z, a)$ and $L(z, a_0)$.
10. In Problem 9 explore the consequences of decisions based on $L(z, a_0)$ where $a_0 = (1 + b)a$ with those based on $L(z, a_0)$ where $a_0 = (1 - b)a$, and b is a given constant satisfying $0 < b < 1$.

11. In connection with the loss function shown in (11.15) explore the consequences of using a loss function that misrepresents the cost of change; that is, $L = (z - a)^2 + c_0(w - x_T)^2$, where $c_0 \neq c$.

12. Assume that the model in (11.1) has an intercept term; that is, $y_t = \beta_0 + \beta x_t + u_t$. With a diffuse prior pdf for the parameters, derive the predictive pdf for $z = y_{T+1}$, use it to obtain the mathematical expectation of the loss function in (11.15), and examine its three components. Then, analogous to the result in (11.17), show that the value of $w = x_{T+1}$, which minimizes expected loss, can be expressed as a weighted average of the values of w, which minimize the three components of expected loss individually.

13. In the profit maximization problem analyzed in (11.18) and following the control variable was assumed to be price. Analyze the problem under the assumption that q, quantity produced, is the control variable and the demand equation is $p_t = \gamma_0 + \gamma_1 q_t + \epsilon_t$, where γ_0 and γ_1 are parameters with unknown values and that ϵ_t is a random-error term. In this analysis supply needed assumptions and use (11.20) as the cost function.

14. In Problem 13 assume that there are costs associated with changing output q and that these costs can be approximated by a quadratic cost function. How is the solution to Problem 13 modified by the introduction of costs of changing q?

15. In connection with the model in (11.38), if some of the variables not under control are stochastic (e.g., one might be a measure of rainfall), explain how the problem of determining a value for $\mathbf{w}_1$ in (11.39) to minimize the expectation of the loss function in (11.40) is affected. Then provide assumptions about the stochastic uncontrolled variables that permit determination of an optimizing value for $\mathbf{w}_1$.

16. The expression in (11.73) provides a basis for comparing a sequential updating solution with a here and now solution. What is the order in T of the last term on the rhs of (11.73) which is the approximate reduction in loss associated with the sequential updating solution *vis à vis* the here and now solution?

17. Provide a critical appraisal of the economic considerations and implications of the loss function shown in (11.109).

CHAPTER XII

Conclusion

In the preceding chapters the Bayesian approach was applied in analyses of a broad range of models and problems. At this point it is useful to summarize some of the major attributes of the Bayesian approach to inference in econometrics.

1. As Jeffreys and others have emphasized, the Bayesian approach to inference complements very nicely the activities of researchers. A researcher is often concerned with how information in data modifies his beliefs about empirical phenomena. In the Bayesian approach to inference an investigator has operational techniques for determining how information in data modifies his beliefs; that is, initial beliefs represented by prior probabilities are combined by means of Bayes' theorem with information in data, incorporated in the likelihood function, to yield posterior probabilities relating to parameters or hypotheses. In a fundamental sense the Bayesian procedure for changing initial beliefs is a learning model of great value in accomplishing a major objective of science—learning from experience.

2. As regards estimation, we have seen that Bayes' theorem can be applied in analyses of all kinds of models to yield exact finite sample posterior pdf's for parameters. That one simple principle has such wide applicability is indeed appealing, since it obviates the need for the use of *ad hoc* procedures which are often needed in other systems of inference to get reasonable results. Further, in the area of point estimation, the Bayesian prescription (choice of the point estimate that minimizes expected loss) is a general operational principle in accord with the expected utility hypothesis. The theoretical appeal, generality, and practical aspects of this solution to the problem of point estimation stand in marked contrast to the many and varied rules for generating point estimates in sampling-theory approaches to inference, some of which have only a large-sample justification. That Bayesian estimators are admissible, consistent, and minimize average risk, when it exists, are additional features that commend their use in practice.

3. Bayesian methods for analyzing prediction problems are simple, operational, and generally applicable. Whatever the model, we derive the predictive

pdf for future observations. This pdf enables us to make probability statements about future observations. For a given loss function involving prediction errors it is generally possible to obtain a point prediction that minimizes expected loss.

4. With respect to control and decision problems, we have seen that the Bayesian approach yields solutions that take account of uncertainty about parameters and the future values of random variables. That these difficulties are dealt with by a straightforward application of basic Bayesian principles is further testimony to their generality and fruitfulness.

5. In the Bayesian approach prior information about parameters or models can be readily and formally incorporated in analyses of estimation, prediction, control, and hypothesis-testing problems. This flexibility contrasts markedly with currently available sampling theory techniques. As seen in the analyses of the errors-in-the-variables and simultaneous equation models, prior information is required to identify parameters of interest. Sampling theorists usually introduce this information in the form of exact restrictions. In such situations Bayesians can introduce less restrictive prior information by choice of a suitable prior pdf and perform an analysis by using less restrictive prior information than the sampling theorist.[1]

6. The problem of nuisance parameters is solved quite straightforwardly and neatly in the Bayesian approach. Parameters not of interest to an investigator, so-called nuisance parameters, can be integrated out of a posterior pdf to obtain the marginal posterior pdf for the parameters of interest. In the sampling theory approach it is often the case that "optimal" estimators or test statistics depend on nuisance parameters whose values are unknown; for example, minimum variance linear unbiased estimators frequently depend on elements of the covariance matrix of disturbance terms. In many instances these unknown parameters are replaced by sample estimates. However, this procedure provides just an approximation to the optimal estimator and is usually justified by large-sample theory. That a Bayesian can generally integrate out nuisance parameters enables him to perform a finite sample analysis without having to rely on a large-sample justification.

7. The Bayesian approach is convenient for the analysis of effects of departures from specifying assumptions; that is, use of conditional posterior pdf's enables an investigator to determine how sensitive his inferences about a particular subset of parameters are to what is assumed about other parameters. In Chapter 4 this procedure was applied in the analysis of the regression model with autocorrelated errors, a departure from the assumption of independent error terms. In Chapter 9 the same approach was pursued to

[1] This point has been emphasized by Jacques Drèze.

determine how inferences about the marginal propensity to consume are affected when the assumption that an investment variable is exogenous is relaxed.

8. In the Bayesian approach inferences about parameters, etc., can be made on the basis of the prior and sample information we have. There is no need to justify inference procedures in terms of their behavior in repeated, as yet unobserved, samples as is usually done in the sampling theory approach. This is not to say that properties of procedures in repeated samples are not of interest; in fact, Bayesian estimators have desirable sampling properties in that they are admissible and constructed to minimize average risk, when it exists. In a given analysis, however, the information currently available is used to make inferences in the Bayesian approach and there is no need to bring in considerations regarding performance in repeated samples.

9. Last, in the area of comparing and testing hypotheses the Bayesian approach is distinguished from sampling-theory approaches in that it associates probabilities with hypotheses and provides simple operational techniques for computing them. These posterior probabilities incorporate prior and sample information and represent degrees of belief. As seen in Chapter 10, such probabilities are useful in comparing non-nested models. Then, too, if one has explicit losses associated with actions such as accept or reject and possible states of the world, he can act to minimize expected loss in testing hypotheses.

From what has been said above and from the material presented in preceding chapters we see that the Bayesian approach to inference is a unified one that works well on a broad range of problems and models, both in small and large samples. The conclusion that emerges is that use of Bayesian methods can result in more fruitful and meaningful econometric analyses of a wide range of problems. This is not to say, however, that all problems associated with the Bayesian approach have been solved. In the preceding chapters several technical problems associated with analyses of posterior pdf's that remain to be solved were encountered. Also, more work is required to formulate, understand, and use a broader range of prior pdf's to represent our prior information, particularly in multiparameter problems. The technical aspects of developing and interpreting procedures for computing posterior odds for a broader range of models must be considered further. Additional results for and experience with adaptive control problems would be welcome. These problems, among others, deserve attention. It is hoped that the present volume will serve as a useful point of departure for research on these and other problems, research that will help to make the Bayesian approach even more useful than it is at present.

APPENDIX A

Properties of Several Important Univariate Pdf's

Here we provide properties of several important univariate pdf's which have appeared at various points in the text.

A.1 UNIVARIATE NORMAL (UN) PDF

A random variable, $\tilde{x}$, is normally distributed if, and only if, its pdf has the following form:

$$\text{(A.1)} \quad p(x|\theta, \sigma) = \frac{1}{\sqrt{2\pi}\,\sigma} \exp\left[-\frac{1}{2\sigma^2}(x - \theta)^2\right], \qquad -\infty < x < \infty.$$

This pdf has two parameters: a location parameter θ, $-\infty < \theta < \infty$, and a scale parameter σ, $0 < \sigma < \infty$. It also has a single mode at $x = \theta$ and is symmetric about the point. Thus θ is the median and the mean of the UN pdf.[1] Given symmetry about $x = \theta$, the odd-order moments about θ are all zero; that is,

(A.2)

$$\mu_{2r-1} \equiv E(\tilde{x} - \theta)^{2r-1} = \int_{-\infty}^{\infty} (x - \theta)^{2r-1} p(x|\theta, \sigma)\, dx = 0, \quad r = 1, 2, 3, \ldots;$$

for example, $\mu_1 = 0$ or $E\tilde{x} = \theta$.

The even-order moments about the mean are given by

$$\text{(A.3)} \quad \begin{aligned} \mu_{2r} \equiv E(\tilde{x} - \theta)^{2r} &= \int_{-\infty}^{\infty} (x - \theta)^{2r} p(x|\theta, \sigma)\, dx \\ &= \frac{2^r \sigma^{2r}}{\sqrt{\pi}} \Gamma(r + \tfrac{1}{2}), \qquad r = 1, 2, \ldots, \end{aligned}$$

where $\Gamma(r + \frac{1}{2})$ denotes the gamma function with argument $r + \frac{1}{2}$ [see (A.6) for a definition of this well-known and important function].

[1] That $x = \theta$ is the modal value of the UN pdf can be seen by inspection from (A.1). That $x = \theta$ is also median follows from the fact that (A.1) is a normalized pdf [i.e., $\int_{-\infty}^{\infty} p(x|\theta, \sigma)\, dx = 1$] and that it is symmetric about the single modal value $x = \theta$.

Further, we note that by a change of variables $z = (x - \theta)/\sigma$, (A.1) can be brought into the standardized UN form[2]:

$$p(z) = \frac{1}{\sqrt{2\pi}} e^{-z^2/2}, \qquad -\infty < z < \infty. \tag{A.4}$$

The moments of this pdf are easily obtained from the expressions for the moments shown in (A.2) and (A.3).

Proofs of Properties of the UN Pdf

First, we establish that (A.1) is a proper normalized pdf; that is $\int_{-\infty}^{\infty} p(x|\theta, \sigma)\, dx = 1$. Note that $p(x|\theta, \sigma) > 0$ for all x such that $-\infty < x < \infty$. Now make the change of variable $z = (x - \theta)/\sigma$ to obtain (A.4) and note that $\int_{-\infty}^{\infty} p(x|\theta, \sigma)\, dx = \int_{-\infty}^{\infty} p(z)\, dz$. Now, letting $u = z^2/2$, we have $0 < u < \infty$ and $du = z\, dz$, or $dz = du/\sqrt{2u}$. Then[3]

$$\frac{1}{\sqrt{2\pi}} \int_{-\infty}^{\infty} e^{-z^2/2}\, dz = \frac{1}{\sqrt{\pi}} \int_0^{\infty} u^{-1/2} e^{-u}\, du. \tag{A.5}$$

The integral on the rhs of (A.5) is the gamma function with argument $\frac{1}{2}$; that is, the gamma function, denoted by $\Gamma(q)$, is defined as

$$\Gamma(q) = \int_0^{\infty} u^{q-1} e^{-u}\, du, \qquad 0 < q < \infty, \tag{A.6}$$

and thus the rhs of (A.5) is $(1/\sqrt{\pi})\, \Gamma(\frac{1}{2})$. Since it is shown in the calculus that $\Gamma(\frac{1}{2}) = \sqrt{\pi}$, the rhs of (A.5) is indeed equal to one.

Second, we show that the odd-order moments of $x - \theta$ are all zero, as shown in (A.2). This is equivalent to showing that $E\tilde{z}^{2r-1} = 0$, $r = 1, 2, \ldots,$ since $\tilde{z} = (\tilde{x} - \theta)/\sigma$. We have

$$E\tilde{z}^{2r-1} = \int_{-\infty}^{0} z^{2r-1} p(z)\, dz + \int_0^{\infty} z^{2r-1} p(z)\, dz.$$

Since z^{2r-1}, with $-\infty < z < 0$, is negative and $p(-z) = p(z)$ from the symmetrical form of (A.4), the first integral on the rhs can be expressed as the negative of the second and thus their sum is zero, as was to be shown.

Third, we derive the expression for the even-order moments shown in (A.3). We shall obtain the even-order moments of $p(z)$ in (A.4) and from

[2] Note from $z = (x - \theta)/\sigma$, $dz = dx/\sigma$ and thus the factor $1/\sigma$ in (A.1) does not appear in (A.4).

[3] Remember that in changing variables by $u = \frac{1}{2}z^2$ we have $\int_{-\infty}^{\infty} p(z)\, dz = 2 \int_0^{\infty} p(u)\, du/\sqrt{2u}$, with the factor 2 appearing to take account of the area under $p(z)$ for *both* positive and negative values of z.

them obtain the expression in (A.3). Let $u = \frac{1}{2}z^2$; then

$$\int_{-\infty}^{\infty} z^{2r}\, p(z)\, dz = 2\int_0^{\infty} (2u)^r\, p(\sqrt{2u})\, \frac{du}{\sqrt{2u}}$$

(A.7)
$$= \frac{2^r}{\sqrt{\pi}}\int_0^{\infty} u^{r-1/2} e^{-u}\, du$$

$$= \frac{2^r}{\sqrt{\pi}}\Gamma(r + \tfrac{1}{2}), \qquad r = 1, 2, \ldots,$$

where $\Gamma(r + \frac{1}{2})$ is the gamma function in (A.6) with argument $q = r + \frac{1}{2}$. Since $z = (x - \mu)/\sigma$, the even-order moments, denoted μ_{2r}, in (A.3), are just σ^{2r} times the expression given in (A.7) and the result shown in (A.3) is established.

For convenience we provide explicit expressions for the second and fourth moments shown in (A.3)[4]:

(A.8)
$$\mu_2 = E(\tilde{x} - \theta)^2 = \frac{2\sigma^2}{\sqrt{\pi}}\Gamma(1 + \tfrac{1}{2}) = \sigma^2$$

and

(A.9)
$$\mu_4 = E(\tilde{x} - \theta)^4 = \frac{2^2\sigma^4}{\sqrt{\pi}}\Gamma(2 + \tfrac{1}{2}) = 3\sigma^4.$$

Explicit expressions for higher order even moments can be obtained in a similar fashion.

As regards measures other than moments to characterize properties of univariate pdf's, we shall just take up measures of skewness and kurtosis. Measures of skewness, that is, of departures from symmetry, include K. Pearson's measure, shown in (A.10), and two others:

(A.10)
$$Sk = \frac{\text{mean-mode}}{\sigma},$$

(A.11)
$$\beta_1 = \frac{{\mu_3}^2}{{\mu_2}^3},$$

and

(A.12)
$$\gamma_1 = \frac{\mu_3}{{\mu_2}^{3/2}}.$$

Of course, for the symmetric UN pdf all of these measures have a zero value. With respect to kurtosis, the following measure, the "excess," is often employed:

(A.13)
$$\gamma_2 \equiv \beta_2 - 3,$$

[4] Use is made of the following two properties of the gamma function: (i) $\Gamma(q + 1) = q\Gamma(q)$, for $q > 0$, and (ii) $\Gamma(\frac{1}{2}) = \sqrt{\pi}$. In (A.9) property (i) is used twice.

where $\beta_2 = \mu_4/\mu_2^2$. The measure γ_2 assumes a value of zero for the UN pdf; pdf's for which $\gamma_2 = 0$ are called mesokurtic, those for which $\gamma_2 > 0$ are leptokurtic, and those for which $\gamma_2 < 0$, platykurtic. As already pointed out,

"... it was thought that leptokurtic curves were more sharply peaked, and platykurtic curves more flat-topped, than the normal curve. This, however, is not necessarily so and although the terms are useful they are best regarded as describing the sign of γ_2 rather than the shape of the curve."[5]

A.2 THE UNIVARIATE STUDENT *t* (US *t*) Pdf

A random variable, $\tilde{x}$, is distributed in the US *t* form if, and only if, it has the following pdf:

(A.14)

$$p(x|\theta, h, \nu) = \frac{\Gamma[(\nu+1)/2]}{\Gamma(1/2)\,\Gamma(\nu/2)}\left(\frac{h}{\nu}\right)^{1/2}\left[1 + \frac{h}{\nu}(x-\theta)^2\right]^{-(\nu+1)/2}, \qquad -\infty < x < \infty,$$

where $-\infty < \theta < \infty$, $0 < h < \infty$, $0 < \nu$ and where Γ denotes the gamma function. This pdf has three parameters, θ, h, and ν. From inspection of (A.14) it is seen that the US *t* pdf has a single mode at $x = \theta$ and is symmetric about the modal value $x = \theta$. Thus $x = \theta$ is the median and mean (which exists for $\nu > 1$—see below) of the US *t* pdf. The following expressions give the odd- and even-order moments about the mean:

(A.15)

$$\mu_{2r-1} \equiv E(\tilde{x}-\theta)^{2r-1} = \int_{-\infty}^{\infty} (x-\theta)^{2r-1}\, p(x|\theta, h, \nu)\, dx = 0\Big\} \qquad \begin{matrix} r = 1, 2, \ldots, \\ \nu > 2r-1, \end{matrix}$$

and

(A.16)

$$\begin{aligned} \mu_{2r} \equiv E(\tilde{x}-\theta)^{2r} &= \int_{-\infty}^{\infty} (x-\theta)^{2r}\, p(x|\theta, h, \nu)\, dx \\ &= \frac{\Gamma(r+1/2)\Gamma(\nu/2-r)}{\Gamma(1/2)\,\Gamma(\nu/2)}\left(\frac{\nu}{h}\right)^r\Big\} \qquad \begin{matrix} r = 1, 2, \ldots, \\ \nu > 2r. \end{matrix} \end{aligned}$$

As shown below, for the existence of the $(2r-1)$st moment about θ we must have $\nu > 2r - 1$. Similarly, for the existence of the $2r$th-order moment about θ, ν must satisfy $\nu > 2r$. Given the symmetry of the US *t* pdf about $x = \theta$, all existing odd-order moments in (A.15) are zero. In particular, $E(\tilde{x} - \theta) = 0$ so that $E\tilde{x} = \theta$ which exists for $\nu > 1$.[6] With respect to the even-order

[5] M. G. Kendall and A. Stuart, *The Advanced Theory of Statistics*, Vol. I. London: Griffin, 1958. New York: Hafner, p. 86. However, for many distributions encountered in practice a positive γ_2 does mean a sharper peak with higher tails than if the distribution were normal.

[6] For $\nu = 1$ the US *t* pdf is identical to the Cauchy pdf for which the first- and higher-order moments do not exist.

moments in (A.16) the second- and fourth-order moments are given by[7]

$$\text{(A.17)} \qquad \mu_2 \equiv E(\tilde{x} - \theta)^2 = \frac{1}{\nu - 2}\frac{\nu}{h}, \qquad \text{for } \nu > 2,$$

and

$$\text{(A.18)} \qquad \mu_4 \equiv E(\tilde{x} - \theta)^4 = \frac{3}{(\nu - 2)(\nu - 4)}\left(\frac{\nu}{h}\right)^2, \qquad \text{for } \nu > 4.$$

Given that $\nu > 2$, the variance μ_2 exists and is seen to depend on ν and h. When $\nu > 4$, the fourth-order moment μ_4 exists and, as with μ_2, depends on just ν and h.

Since the US t pdf is symmetric about $x = \theta$, the measures of skewness, discussed in connection with the UN pdf, are all zero, provided, of course, that moments on which they depend exist. With respect to kurtosis, we have from (A.17) and (A.18)

$$\text{(A.19)} \qquad \begin{aligned} \gamma_2 &= \frac{\mu_4}{{\mu_2}^2} - 3 \\ &= 3\left(\frac{\nu - 2}{\nu - 4} - 1\right) = \frac{6}{\nu - 4}, \qquad \nu > 4. \end{aligned}$$

Thus, for finite ν the US t pdf is leptokurtic ($\gamma_2 > 0$), probably because it has fatter tails than a UN pdf with mean θ and variance $1/h$. As ν gets large, the US t pdf assumes the shape of a UN pdf with mean θ and variance $\mu_2 = 1/h$.[8]

We can obtain the standardized form of the US t pdf from (A.14) by the following change of variable

$$\text{(A.20)} \qquad t = \sqrt{h}\,(x - \theta), \qquad -\infty < t < \infty.$$

Using (A.20), we obtain

$$\text{(A.21)} \quad p(t|\nu) = \frac{\Gamma[(\nu + 1)/2]}{\sqrt{\nu}\,\Gamma(1/2)\,\Gamma(\nu/2)}\left(1 + \frac{t^2}{\nu}\right)^{-(\nu+1)/2}, \qquad -\infty < t < \infty,$$

which, with $\nu > 0$, is the proper normalized standard US t pdf. This pdf has a single mode at $t = 0$ and is symmetric about this point. From (A.20) the moments of $\tilde{t}$ can be obtained from those of $\tilde{x} - \theta$ when they exist.

Proofs of Properties of the US t Pdf

To establish properties of the US t pdf we need the following results from the calculus:

[7] In obtaining these results, repeated use is made of the relationship $\Gamma(q + 1) = q\Gamma(q)$, $q > 0$, a fundamental property of the gamma function.

[8] For ν about 30 these two pdf's are similar; (A.17) and (A.18) show explicitly how μ_2 and μ_4 for the US t are related to the corresponding moments of the limiting normal pdf; that is, $\mu_2 = 1/h$ and $\mu_4 = 3/h^2$.

1. If $f(v)$ is continuous for $a \le v < \infty$ and $\lim_{v\to\infty} v^r f(v) = A$, a finite constant, for $r > 1$, then $\int_a^\infty |f(v)|\, dv < \infty$; that is, the integral converges absolutely.[9]

2. If $g(v)$ is continuous for $-\infty < v \le b$ and $\lim_{v\to-\infty} (-v)^r g(v) = c$, a constant, for $r > 1$, then $\int_{-\infty}^b |g(v)|\, dv < \infty$.

3. The relation connecting the beta function, denoted $B(u, v)$, and the gamma function is[10]

$$\text{(A.22)} \qquad B(u, v) = \frac{\Gamma(u)\,\Gamma(v)}{\Gamma(u + v)}, \qquad 0 < u, v < \infty,$$

where

$$\text{(A.23)} \qquad B(u, v) \equiv \int_0^1 x^{u-1}(1 - x)^{v-1}\, dx, \qquad 0 < u, v < \infty.$$

With these results stated, we first show that for $\nu > 0$ the US t pdf in (A.21) is a proper normalized pdf.[11] We note that $p(t|\nu) > 0$ for $-\infty < t < \infty$, and letting $t' = t/\sqrt{\nu}$ we can write (A.21) as

$$\text{(A.24)} \quad p(t'|\nu) = \left[B\left(\frac{1}{2}, \frac{\nu}{2}\right)\right]^{-1}(1 + t'^2)^{-(\nu+1)/2}, \qquad -\infty < t' < \infty,$$

with

$$B\left(\frac{1}{2}, \frac{\nu}{2}\right) = \Gamma\left(\frac{1}{2}\right)\Gamma\left(\frac{\nu}{2}\right)\Big/\Gamma\left(\frac{\nu + 1}{2}\right)$$

from the result given in (A.22). Now make the change of variable $z = 1/(1 + t'^2)$, $0 < z < 1$, to obtain[12]

$$\text{(A.25)} \quad p(z|\nu) = \left[B\left(\frac{1}{2}, \frac{\nu}{2}\right)\right]^{-1} z^{\nu/2-1}(1 - z)^{-1/2}\frac{dz}{2}, \qquad 0 < z < 1.$$

Noting that $\int_{-\infty}^\infty p(t'|\nu)\, dt' = 2\int_0^1 p(z|\nu)\, dz$, we have

$$\text{(A.26)} \quad \int_{-\infty}^\infty p(t'|\nu)\, dt' = \left[B\left(\frac{1}{2}, \frac{\nu}{2}\right)\right]^{-1}\int_0^1 z^{\nu/2-1}(1 - z)^{-1/2}\, dz = 1,$$

since the integral on the rhs of (A.26) is just $B(1/2, \nu/2)$, *provided that* $\nu > 0$. This condition is required in order to have $\int_{-\infty}^\infty p(t'|\nu)\, dt' < \infty$ [see (1) and (2)].

[9] Note that $\int_a^\infty |f(v)|\, dv < \infty$ implies $\int_a^\infty f(v)\, dv < \infty$; that is, absolute convergence implies simple convergence. See, for example, D. V. Widder, *Advanced Calculus*. New York: Prentice-Hall, 1947, p. 271. See also p. 273 ff. for proofs of (1) and (2).

[10] Equation A.22 implies $B(u, v) = B(v, u)$.

[11] Since (A.21) is obtained from (A.14) by the change of variable in (A.20), showing that (A.21) is a proper normalized pdf will imply that (A.14) also has this property.

[12] From $z = 1/(1 + t'^2)$, $|dz/dt'| = 2t'/(1 + t'^2)^2$ and $t'^2 = (1 - z)/z$; thus $|dt'/dz| = z^{-3/2}(1 - z)^{-1/2}/2$.

The results for the odd-order moments in (A.15) can be established easily by considering

$$\int_{-\infty}^{\infty} t'^{2r-1}\, p(t'|\nu)\, dt', \qquad -\infty < t' < \infty, \tag{A.27}$$

with $t' = t/\sqrt{\nu} = \sqrt{h/\nu}\,(x - \theta)$. For the integral in (A.27) to converge we need $2r - 1 < \nu$ by application of (1) and (2). Thus, if $2r - 1 < \nu$, the $(2r - 1)$st odd-order moment exists and is zero by virtue of the symmetry of $p(t'|\nu)$ about $t' = 0$.

The expression for the even-order moments in (A.16) is most easily obtained by evaluating

$$\int_{-\infty}^{\infty} t'^{2r}\, p(t'|\nu)\, dt', \qquad -\infty < t' < \infty, \tag{A.28}$$

with $t' = \sqrt{h/\nu}\,(x - \theta)$. For the integral in (A.28) to converge we need $\nu > 2r$. If this condition is satisfied, we use the transformation $z = 1/(1 + t'^2)$, or $t'^2 = (1 - z)/z$ [see (A.24)] to obtain

$$\begin{aligned}\int_{-\infty}^{\infty} t'^{2r} p(t'|\nu)\, dt' &= \left[B\left(\frac{1}{2}, \frac{\nu}{2}\right)\right]^{-1} \int_0^1 \left(\frac{1 - z}{z}\right)^r z^{\nu/2-1}(1 - z)^{-1/2}\, dz \\ &= \left[B\left(\frac{1}{2}, \frac{\nu}{2}\right)\right]^{-1} \int_0^1 z^{\nu/2-r-1}(1 - z)^{r-1/2}\, dz \\ &= \frac{B(\nu/2 - r, r + 1/2)}{B(1/2, \nu/2)} = \frac{\Gamma(r + 1/2)\,\Gamma(\nu/2 - r)}{\Gamma(1/2)\,\Gamma(\nu/2)}, \qquad \nu > 2r.\end{aligned} \tag{A.29}$$

This gives the $2r$th moment of $p(t'|\nu)$. From $x - \theta = \sqrt{\nu/h}\, t'$ we obtain the $2r$th moment of $p(x|\theta, h, \nu)$,

$$\mu_{2r} \equiv E(\tilde{x} - \theta)^{2r} = \frac{\Gamma(r + 1/2)\,\Gamma(\nu/2 - r)}{\Gamma(1/2)\,\Gamma(\nu/2)} \left(\frac{\nu}{h}\right)^r, \qquad \nu > 2r,$$

which is just (A.16).

A.3 THE GAMMA (G) AND χ^2 Pdf's

As the name implies, the G pdf is closely linked to the gamma function. A random variable, $\tilde{x}$, is distributed according to the G distribution if, and only if, its pdf is given by

$$p(x|\gamma, \alpha) = \frac{x^{\alpha-1}}{\Gamma(\alpha)\gamma^{\alpha}}\, e^{-x/\gamma}, \qquad 0 < x < \infty, \tag{A.30}$$

where α and γ are strictly positive parameters; that is, $\alpha, \gamma > 0$. From the form of (A.30) it is seen that γ is a scale parameter. When $\alpha \geq 1$, the pdf has a single mode[13] at $x = \gamma(\alpha - 1)$. For small values of α the pdf has a long tail to the right. As α grows in size for any given value of γ, the pdf becomes more symmetric and approaches a normal form.

The G pdf can be brought into a standardized form by the change of variable $z = x/\gamma$, which results in

$$p(z|\alpha) = \frac{1}{\Gamma(\alpha)} z^{\alpha-1}e^{-z}, \qquad 0 < z < \infty. \tag{A.31}$$

From the definition of the gamma function it is obvious that (A.31) is a proper normalized pdf and that moments of all orders exist. The moments about zero, denoted by μ_r', are given by

$$\mu_r' = \int_0^\infty z^r\, p(z|\alpha)\, dz = \frac{\Gamma(r + \alpha)}{\Gamma(\alpha)}, \qquad r = 1, 2, \ldots. \tag{A.32}$$

From (A.32) we have for the first four moments about zero[14]

$$\mu_1' = \alpha; \qquad \mu_2' = (1 + \alpha)\alpha; \qquad \mu_3' = (2 + \alpha)(1 + \alpha)\alpha; \qquad \text{and} \quad \mu_4' = (3 + \alpha)(2 + \alpha)(1 + \alpha)\alpha. \tag{A.33}$$

From these results it is seen that α is the mean of the G pdf and, surprisingly, also its variance.[15] Further, for the third and fourth moments about the mean we have $\mu_3 = 2\alpha$ and $\mu_4 = 3\alpha(2 + \alpha)$. Collecting these results, we have

$$\mu_1' = \alpha, \qquad \mu_2 = \alpha, \qquad \mu_3 = 2\alpha \quad \text{and} \quad \mu_4 = 3\alpha(2 + \alpha). \tag{A.34}$$

Given that $0 < \alpha < \infty$, the skewness is always positive. Since the mode is located at $z = \alpha - 1$ for $\alpha \geq 1$, we have for Pearson's measure of skewness $Sk = (\text{mean-mode})/\sqrt{\mu_2} = 1/\sqrt{\alpha}$. Clearly, as α grows in size, this measure of skewness approaches zero. As regards kurtosis, $\gamma_2 = \mu_4/\mu_2^2 - 3 = 6/\alpha$, which also approaches zero as α grows large. That $Sk \to 0$ and $\gamma_2 \to 0$ as $\alpha \to \infty$ is, of course, connected with the fact that the G pdf assumes a normal form as $\alpha \to \infty$.[16]

The χ^2 pdf is a special case of the G pdf (A.30) in which $\alpha = \nu/2$ and $\gamma = 2$; that is, the χ^2 pdf has the following form[17]:

$$p(x|\nu) = \frac{x^{\nu/2-1}e^{-x/2}}{2^{\nu/2}\,\Gamma(\nu/2)}, \qquad 0 < x < \infty, \tag{A.35}$$

[13] For $0 < \alpha < 1$ the pdf has no mode.

[14] In obtaining the expressions for the moments below, we use the relation $\Gamma(1 + q) = q\,\Gamma(q)$ repeatedly.

[15] The variance μ_2 is in general related to moments about zero by the following relationship: $\mu_2 = \mu_2' - \mu_1'^2$. For the G pdf $\mu_2 = (1 + \alpha)\alpha - \alpha^2 = \alpha$.

[16] Moments, etc., for the unstandardized G pdf in (A.30) are readily obtained from those for the standardized pdf (A.31) by taking note of $z = x/\gamma$.

[17] Often (A.35) is written with $x = \chi^2$.

where $0 < \nu$. The parameter ν is usually referred to as the "number of degrees of freedom." Since (A.35) is a special case of (A.30), the standardized form is given by (A.31) with α replaced by $\nu/2$. Also, the moments of the standardized form are given by (A.33) and (A.34), again with $\alpha = \nu/2$. Since the standardized variable z is related to the unstandardized variable x by $z = x/\gamma = x/2$, moments of the unstandardized χ^2 pdf in (A.35) can be obtained easily from moments of the standardized χ^2 pdf. For the reader's convenience we present the moments associated with (A.35)[18]:

$$\text{(A.36)} \quad \mu_1' = \nu; \qquad \mu_2 = 2\nu; \qquad \mu_3 = 8\nu; \quad \text{and} \quad \mu_4 = 24\nu\left(2 + \frac{\nu}{2}\right).$$

A very important property of the χ^2 pdf is that any sum of squared independent, standardized normal random variables has a pdf in the χ^2 form; that is, if $\tilde{x} = \tilde{z}_1^2 + \tilde{z}_2^2 + \cdots + \tilde{z}_n^2$, where the $\tilde{z}_i$'s are independent, standardized normal random variables, then $\tilde{x}$ has a pdf in the form of (A.35) with $\nu = n$.[19]

A.4 THE INVERTED GAMMA (IG) Pdf's

The IG pdf is obtained from the G pdf in (A.30) by letting y equal the positive square root of $1/x$; that is, $y = |\sqrt{1/x}|$ and thus $y^2 = 1/x$. With this change of variable the IG pdf is[20]

$$\text{(A.37}a\text{)} \qquad p(y|\gamma, \alpha) = \frac{2}{\Gamma(\alpha)\gamma^\alpha y^{2\alpha+1}} e^{-1/\gamma y^2}, \qquad 0 < y < \infty,$$

where $\gamma, \alpha > 0$. Since this pdf is encountered frequently in connection with prior and posterior pdf's for a standard deviation, we rewrite (A.37a), letting $\sigma = y$, $\alpha = \nu/2$, and $\gamma = 2/\nu s^2$ to obtain

$$\text{(A.37}b\text{)} \quad p(\sigma|\nu, s) = \frac{2}{\Gamma(\nu/2)} \left(\frac{\nu s^2}{2}\right)^{\nu/2} \frac{1}{\sigma^{\nu+1}} e^{-\nu s^2/2\sigma^2}, \qquad 0 < \sigma < \infty,$$

where $\nu, s > 0$. The pdf in (A.37b) has a single mode at the following value[21] of σ:

$$\text{(A.38)} \qquad \sigma_{\text{mod}} = s\left(\frac{\nu}{\nu + 1}\right)^{1/2}.$$

Clearly, as ν gets large, $\sigma_{\text{mod}} \to s$.

[18] These are obtained from (A.34) with $\alpha = \nu/2$ and $z = x/2$, where x is the χ^2 variable in (A.35), z is the standardized variable in (A.31), and α is the parameter in (A.31).

[19] See M. G. Kendall and A. Stuart, *op. cit.*, pp. 246–247, for a proof of this result.

[20] Since (A.37a) is obtained from a proper normalized pdf by a simple one-to-one differentiable change of variable, it is a proper normalized pdf.

[21] This is easily established by taking the log of both sides of (A.37b) and then finding the value of σ for which $\log p(\sigma|\nu, s)$ achieves its largest value.

The moments of (A.37*b*), when they exist, are obtained by evaluating the following integral:

$$\mu_r' = c\int_0^\infty \sigma^{r-(\nu+1)}e^{-\nu s^2/2\sigma^2}\,d\sigma, \tag{A.39}$$

where

$$c = \frac{2}{\Gamma(\nu/2)}\left(\frac{\nu s^2}{2}\right)^{\nu/2}.$$

On letting $y = \nu s^2/2\sigma^2$, (A.39) can be expressed as

$$\mu_r' = \tfrac{1}{2}c\left(\frac{\nu s^2}{2}\right)^{(r-\nu)/2}\int_0^\infty y^{(\nu-r)/2-1}e^{-y}\,dy. \tag{A.40}$$

The integral in (A.40) is the gamma function. For it to converge we must have

$$\nu - r > 0, \tag{A.41}$$

which is the condition for existence of a moment of order r. Inserting the definition of c in (A.40), we have

$$\mu_r' = \frac{\Gamma[(\nu-r)/2]}{\Gamma(\nu/2)}\left(\frac{\nu s^2}{2}\right)^{r/2}, \qquad \nu > r, \tag{A.42}$$

a convenient expression for the moments about zero. The first four moments are

$$\mu_1' = \frac{\Gamma[(\nu-1)/2]}{\Gamma(\nu/2)}\left(\frac{\nu}{2}\right)^{1/2}s, \qquad \nu > 1, \tag{A.43}$$

$$\mu_2' = \frac{\Gamma(\nu/2-1)}{\Gamma(\nu/2)}\left(\frac{\nu}{2}\right)s^2 = \frac{\nu s^2}{\nu-2}, \qquad \nu > 2, \tag{A.44}$$

$$\mu_3' = \frac{\Gamma[(\nu-3)/2]}{\Gamma(\nu/2)}\left(\frac{\nu}{2}\right)^{3/2}s^3, \qquad \nu > 3, \tag{A.45}$$

and

$$\mu_4' = \frac{\Gamma(\nu/2-2)}{\Gamma(\nu/2)}\left(\frac{\nu}{2}\right)^2 s^4 = \frac{\nu^2}{(\nu-2)(\nu-4)}s^4, \qquad \nu > 4. \tag{A.46}$$

It is seen from (A.43) that the mean μ_1' is intimately related to the parameter s. As ν gets large, $\mu_1' \to s$,[22] which is also approximately the modal value for large ν (see above).

[22] In the calculus it is shown that with $0 < q < \infty$, as $q \to \infty$, $q^{b-a}\,\Gamma(q+a)/\Gamma(q+b) \to 1$ for a and b finite.

As regards moments about the mean,[23] we have

$$\mu_2 = \mu_2' - \mu_1'^2, \qquad \nu > 2,$$

(A.47)
$$= \frac{\nu s^2}{\nu - 2} - \mu_1'^2,$$

(A.48) $$\mu_3 = \mu_3' - 3\mu_1'\mu_2 - \mu_1'^3, \qquad \nu > 3,$$

and

(A.49) $$\mu_4 = \mu_4' - 4\mu_1'\mu_3 - 6\mu_1'^2\mu_2 - \mu_1'^4, \qquad \nu > 4.$$

These formulas are useful when we have to evaluate higher order moments of the IG pdf.

The Pearson measure of skewness for the IG pdf is given by

(A.50)
$$Sk = \frac{\text{mean} - \text{mode}}{\sqrt{\mu_2}}$$
$$= \left[\frac{\Gamma[(\nu - 1)/2]}{\Gamma(\nu/2)}\left(\frac{\nu}{2}\right)^{1/2} - \left(\frac{\nu}{\nu + 1}\right)^{1/2}\right]\frac{s}{\sqrt{\mu_2}}, \qquad \nu > 2.$$

Since this measure is generally positive, the IG pdf is skewed to the right. Clearly, as ν gets large $Sk \to 0$. For small to moderate ν the IG pdf has a rather long tail to the right.[24]

A.5 THE BETA (B) Pdf

A random variable, $\tilde{x}$, is said to be distributed according to beta distribution if, and only if, its pdf has the following form:

(A.51) $$p(x|a, b, c) = \frac{1}{cB(a, b)}\left(\frac{x}{c}\right)^{a-1}\left(1 - \frac{x}{c}\right)^{b-1}, \qquad 0 \leq x \leq c,$$

where $a, b, c > 0$ and $B(a, b)$ denotes the beta function, shown in (A.23), with arguments a and b. It is seen that the range of the B pdf is 0 to c.

By a change of variable, $z = x/c$, we can obtain the standardized B pdf,

(A.52) $$p(z|a, b) = \frac{1}{B(a, b)} z^{a-1}(1 - z)^{b-1}, \qquad 0 \leq z \leq 1,$$

which has a range zero to one. Some properties of (A.52) follow.

That (A.52) is a proper normalized pdf is established by observing that the pdf is non-negative in $0 \leq z \leq 1$ and that $B(a, b) = \int_0^1 z^{a-1}(1 - z)^{b-1}\, dz$,

[23] See M. G. Kendall and A. Stuart, *op. cit.*, p. 56, for formulas connecting moments about zero with moments about the mean.

[24] Since it is rather straightforward to obtain the pdf for σ^n, $n = 2, 3, \ldots$, from that for σ in (A.37*b*) and to establish its properties, we do not provide these results.

which converges for all $a, b > 0$. If $a > 2$, (A.52) is tangential to the abscissa at $z = 0$, and if $b > 2$ it is tangential at $z = 1$. For $a, b > 1$,[25] the mode is

$$z_{\text{mod}} = \frac{a - 1}{a + b - 2}. \tag{A.53}$$

The first and higher moments about zero for the standardized B pdf in (A.52) are

$$\begin{aligned} \mu_r' &= \frac{1}{B(a, b)} \int_0^1 z^{r+a-1}(1 - z)^{b-1}\, dz \\ &= \frac{B(r + a, b)}{B(a, b)} = \frac{\Gamma(r + a)}{\Gamma(r + a + b)} \frac{\Gamma(a + b)}{\Gamma(a)} \\ &= \frac{a(a + 1)(a + 2)\cdots(a + r - 1)}{(a + b)(a + b + 1)\cdots(a + b + r - 1)}, \qquad r = 1, 2, \ldots, \end{aligned} \tag{A.54}$$

where (A.22) and the recurrence relation for the gamma function, $\Gamma(q + 1) = q\,\Gamma(q)$, have been employed. Thus the first three moments from (A.54) are

$$\mu_1' = \frac{a}{a + b}, \tag{A.55}$$

$$\mu_2' = \frac{a(a + 1)}{(a + b)(a + b + 1)}, \tag{A.56}$$

$$\mu_3' = \frac{a(a + 1)(a + 2)}{(a + b)(a + b + 1)(a + b + 2)}, \tag{A.57}$$

and so forth. The mean and higher moments are seen to depend simply on the parameters a and b. Further, the variance is given by

$$\mu_2 = \frac{ab}{(a + b)^2(a + b + 1)}. \tag{A.58}$$

As regards skewness, Pearson's measure for $a, b > 1$ is

$$\begin{aligned} Sk &= \frac{a/(a + b) - (a - 1)/(a + b - 2)}{\sqrt{\mu_2}} \\ &= \frac{(b - a)/(a + b)(a + b - 2)}{\sqrt{\mu_2}}. \end{aligned} \tag{A.59}$$

Thus, if $b = a$, $Sk = 0$ and the pdf is symmetric. If $b > a$, there is positive skewness, whereas if $b < a$ there is negative skewness.[26]

[25] For $0 < a, b < 1$ the pdf approaches ∞ as $z \to 0$ or 1.

[26] Since the variable of the unstandardized B pdf in (A.51) is related to that of the standardized B pdf in (A.52) by $x = cz$, the moments, etc., associated with (A.51) are easily obtained.

A useful and important result connecting the standardized gamma (G) and beta (B) pdf's follows. Let $\tilde{z}_1$ and $\tilde{z}_2$ be two independent random variables, each with a standardized G pdf with parameters α_1 and α_2, respectively [see (A.31)]. Then the random variable $\tilde{z} = \tilde{z}_1/(\tilde{z}_1 + \tilde{z}_2)$ has a standardized B pdf with parameters α_1 and α_2; that is, the pdf for $\tilde{z} = \tilde{z}_1/(\tilde{z}_1 + \tilde{z}_2)$ is $z^{\alpha_1 - 1}(1 - z)^{\alpha_2 - 1}/B(\alpha_1, \alpha_2)$.[27] This result is often used when $\tilde{z}_1$ and $\tilde{z}_2$ are independent sums of squares of independent standardized normal random variables. Further, since the B pdf can be transformed to the F pdf, as shown below, we can state on this basis that $\tilde{z} = \tilde{z}_1/(\tilde{z}_1 + \tilde{z}_2)$ has a pdf transformable to an F pdf.

A pdf closely associated with the B pdf is the beta prime or inverted beta (IB) pdf.[28] Its standardized form is obtained from the standardized B pdf in (A.52) by letting $z = 1/(1 + u)$ to obtain the IB pdf:

$$\text{(A.60)} \qquad p(u|a, b) = \frac{1}{B(a, b)} \frac{u^{b-1}}{(1 + u)^{a+b}}, \qquad 0 \le u < \infty,$$

with $a, b > 0$. The moments of this pdf are given by

$$\text{(A.61)} \qquad \begin{aligned} \mu_r' &= \frac{1}{B(a, b)} \int_0^\infty \frac{u^{r+b-1}}{(1 + u)^{a+b}} du \\ &= \frac{B(b + r, a - r)}{B(a, b)}, \qquad r < a, \end{aligned}$$

a result that is obtained by a change of variable $u = 1/(1 + z)$ in (A.61) and noting that the result is in the form of a standardized unnormalized B pdf. Then, from (A.61),

$$\text{(A.62)} \qquad \mu_1' = \frac{b}{a - 1}, \qquad a > 1,$$

$$\text{(A.63)} \qquad \mu_2' = \frac{b(b + 1)}{(a - 1)(a - 2)}, \qquad a > 2,$$

and so on. The variance is

$$\text{(A.64)} \qquad \mu_2 = \frac{b(a + b - 1)}{(a - 1)^2(a - 2)}, \qquad a > 2.$$

[27] The joint pdf for $\tilde{z}_1$ and $\tilde{z}_2$ is the product of their individual pdf's, since they are assumed to be independent; that is, $p(z_1, z_2|\alpha_1, \alpha_2)\, dz_1\, dz_2 = [\Gamma(\alpha_1)\,\Gamma(\alpha_2)]^{-1} z_1^{\alpha_1 - 1} \times z_2^{\alpha_2 - 1} e^{-(z_1 + z_2)}\, dz_1\, dz_2$. Now change variables to $v = z_1 + z_2$ and $z = z_1/(z_1 + z_2)$, with $dz_1\, dz_2 = v\, dz\, dv$, and integrate with respect to v from zero to infinity to obtain the result above.

[28] See, for example, J. F. Kenney and E. S. Keeping, *Mathematics of Statistics: Part Two* (2nd ed.). Princeton, N.J.: Van Nostrand, 1951, pp. 95–96, and H. Raiffa and R. Schlaifer, *Applied Statistical Decision Theory*, Boston: Graduate School of Business Administration, Harvard University, 1961, pp. 220–221.

The IB pdf has a single mode if $b > 1$ at

$$u_{\text{mod}} = \frac{b - 1}{a + 1}. \tag{A.65}$$

Then Pearson's measure of skewness for the IB pdf is

$$\begin{aligned} Sk &= \frac{b/(a - 1) - (b - 1)/(a + 1)}{\sqrt{\mu_2}} \\ &= \frac{2b + a - 1}{a + 1}\left[\frac{a - 2}{b(a + b - 1)}\right]^{1/2}, \qquad b > 1 \quad \text{and} \quad a > 2, \end{aligned} \tag{A.66}$$

which is positive and shows that the IB pdf usually has a long tail to the right.

Last, we can obtain an important alternative form of (A.60) by letting $u = y/c$, with $0 < c$, to yield

$$p(y|a, b, c) = \frac{1}{cB(a, b)}\frac{(y/c)^{b-1}}{(1 + y/c)^{a+b}}, \qquad 0 \leq y < \infty, \tag{A.67}$$

where $a, b, c > 0$. Since $y = cu$ and we have already found the moments associated with (A.60), the moments for (A.67) are directly available from (A.61) to (A.64), that is the rth moment about zero is $c^r\mu_r'$ with μ_r' given in (A.61). It will be seen in the next section that the F and US t pdf's are special cases of (A.67).

A.6 THE FISHER-SNEDECOR F PDF

A random variable, $\tilde{x}$, is said to have an F distribution if, and only if, it has a pdf in the following form:

(A.68)

$$p(x|\nu_1, \nu_2) = \left[B\left(\frac{\nu_1}{2}, \frac{\nu_2}{2}\right)\right]^{-1}\left(\frac{\nu_1}{\nu_2}\right)^{\nu_1/2}\frac{x^{\nu_1/2-1}}{(1 + (\nu_1/\nu_2)x)^{(\nu_1+\nu_2)/2}}, \qquad 0 < x < \infty,$$

where $\nu_1, \nu_2 > 0$. It is seen that (A.68) is a special case of the IB pdf in (A.67), where $a = \nu_2/2$, $b = \nu_1/2$, and $c = \nu_2/\nu_1$. The parameters ν_1 and ν_2 are usually referred to as degrees of freedom and (A.68) is called the F pdf with ν_1 and ν_2 degrees of freedom.

If $\nu_1/2 > 1$, the F pdf has a single mode at

$$x_{\text{mod}} = \frac{\nu_2}{\nu_1}\frac{\nu_1/2 - 1}{\nu_2/2 + 1}. \tag{A.69}$$

The moments of the F pdf can, of course, be obtained directly from those associated with the IB pdf shown in (A.62) to (A.64). For easy reference we list the moments of the F pdf:

$$\mu_1' = \frac{\nu_2}{\nu_1}\frac{\nu_1/2}{\nu_2/2 - 1}, \qquad \frac{\nu_2}{2} > 1, \tag{A.70}$$

$$\mu_2' = \left(\frac{\nu_2}{\nu_1}\right)^2 \frac{(\nu_1/2)(\nu_1/2 + 1)}{(\nu_2/2 - 1)(\nu_2/2 - 2)}, \qquad \frac{\nu_2}{2} > 2, \tag{A.71}$$

and so on. The variance of the F pdf is

$$\mu_2 = \left(\frac{\nu_2}{\nu_1}\right)^2 \frac{(\nu_1/2)[(\nu_1 + \nu_2)/2 - 1]}{(\nu_2/2 - 1)^2(\nu_2/2 - 2)}, \qquad \frac{\nu_2}{2} > 2. \tag{A.72}$$

We now review relations of the F pdf to several other well-known pdf's.

1. If in the F pdf in (A.68) $\nu_1 = 1$ and we let $t^2 = x$, the F pdf is transformed to a standardized US t pdf with ν_2 degrees of freedom.
2. If $\tilde{z}_1$ and $\tilde{z}_2$ are independent random variables with χ^2 pdf's that have ν_1 and ν_2 degrees of freedom, respectively, then $\tilde{x} = (\tilde{z}_1/\nu_1)/(\tilde{z}_2/\nu_2)$ has an F pdf with ν_1 and ν_2 degrees of freedom, provided $\nu_1, \nu_2 > 0$.
3. If $\tilde{x}$ has the F pdf in (A.68), then, when $\nu_2 \to \infty$, the random variable $\nu_1\tilde{x}$ will have a χ^2 pdf with ν_1 degrees of freedom.
4. If $\tilde{x}$ has the F pdf in (A.68), then, when $\nu_1 \to \infty$, the random variable $\nu_2/\tilde{x}$ will have a χ^2 pdf with ν_2 degrees of freedom.
5. If $\tilde{x}$ has the F pdf in (A.68), then, when $\nu_2 \to \infty$ with $\nu_1 = 1$, the random variable $\sqrt{\tilde{x}}$ will have a standardized UN pdf.
6. If $\tilde{\sigma}_1$ and $\tilde{\sigma}_2$ are independent random variables with IG pdf's in the form of (A.37*b*) and parameters ν_1, s_1, and ν_2, s_2, respectively, the random variable $\tilde{x} = (\tilde{\sigma}_1{}^2/s_1{}^2)/(\tilde{\sigma}_2{}^2/s_2{}^2)$ will have an F pdf with ν_2 and ν_1 degrees of freedom.

Proposition (1) is established by making the change of variable $t^2 = x$ in (A.68) and noting that with $\nu_1 = 1$ the resulting pdf is precisely the standardized US t pdf with ν_2 degrees of freedom.

Proposition (2) is established by noting that the joint pdf for $\tilde{z}_1$ and $\tilde{z}_2$ is

$$p(z_1, z_2|\nu_1, \nu_2) = kz_1^{\nu_1/2-1}z_2^{\nu_2/2-1}e^{-(z_1+z_2)/2}, \qquad 0 < z_1, z_2 < \infty,$$

with $1/k = 2^{(\nu_1+\nu_2)/2}\Gamma(\nu_1/2)\Gamma(\nu_2/2)$. Now let $v = z_1/z_2$ and $y = (z_1 + z_2)/2$, which implies that $z_1 = 2vy/(v + 1)$ and $z_2 = 2y/(v + 1)$. The Jacobian of this transformation is $4y/(v + 1)^2$ and thus the pdf for v and y is

$$p(v, y|\nu_1, \nu_2) = 2^{(\nu_1+\nu_2)/2}k \frac{v^{\nu_1/2-1}}{(1 + v)^{(\nu_1+\nu_2)/2}} y^{(\nu_1+\nu_2)/2-1}e^{-y}, \qquad 0 < v, y, <\infty.$$

On integrating with respect to y, the result is

$$p(v|\nu_1, \nu_2) = \frac{\Gamma[(\nu_1 + \nu_2)/2]}{\Gamma(\nu_1/2, \nu_2/2)} \frac{v^{\nu_1/2-1}}{(1 + v)^{(\nu_1+\nu_2)/2}}, \qquad 0 < v < \infty.$$

Finally, on letting $x = (\nu_2/\nu_1)v$, we obtain the F pdf in (A.68).

To prove Proposition (3) let $z = \nu_1 x$ in (A.68) to obtain

$$p(z|\nu_1, \nu_2) = \frac{\Gamma[(\nu_1 + \nu_2)/2]}{\Gamma(\nu_1/2)\Gamma(\nu_2/2)} \frac{1}{\nu_2^{\nu_1/2}} \frac{z^{\nu_1/2-1}}{(1 + z/\nu_2)^{(\nu_1+\nu_2)/2}}, \qquad 0 < z < \infty,$$

and as $\nu_2 \to \infty$ with ν_1 fixed the limit is

$$\lim_{\nu_2 \to \infty} p(z|\nu_1, \nu_2) = \left[2^{\nu_1/2}\,\Gamma\left(\frac{\nu_1}{2}\right)\right]^{-1} z^{\nu_1/2-1}e^{-z/2}, \qquad 0 < z < \infty,$$

which is a χ^2 pdf with ν_1 degrees of freedom. Propositions (4) and (5) can be established by using similar methods.

Proposition (6) is established by writing the joint pdf for $\tilde{\sigma}_1$ and $\tilde{\sigma}_2$:

$$p(\sigma_1, \sigma_2|\nu_1, \nu_2, s_1, s_2) = k(\sigma_1^{\nu_1+1}\sigma_2^{\nu_2+1})^{-1} \exp\left(-\frac{\nu_1 s_1^2}{2\sigma_1^2} - \frac{\nu_2 s_2^2}{2\sigma_2^2}\right), \qquad 0 < \sigma_1, \; \sigma_2 < \infty,$$

with

$$k = \frac{4}{\Gamma(\nu_1/2)\Gamma(\nu_2/2)}\left(\frac{\nu_1 s_1^2}{2}\right)^{\nu_1/2}\left(\frac{\nu_2 s_2^2}{2}\right)^{\nu_2/2}.$$

Now make the following change of variables $\lambda = \sigma_1^2/\sigma_2^2$ and $\phi = \sigma_1$ to obtain

$$p(\lambda, \phi|\nu_1, \nu_2, s_1, s_2) = \frac{k}{2}\frac{\lambda^{\nu_2/2-1}}{\phi^{\nu_1+\nu_2+1}} \exp\left(-\frac{1}{2\phi^2}(\nu_1 s_1^2 + \lambda\nu_2 s_2^2)\right), \qquad 0 < \lambda, \phi, < \infty.$$

On integrating with respect to ϕ, the result is

$$\begin{aligned} p(\lambda|\nu_1, \nu_2, s_1, s_2) &= \frac{k}{4}\Gamma\left(\frac{\nu_1 + \nu_2}{2}\right)\frac{2^{(\nu_1+\nu_2)/2}\lambda^{\nu_2/2-1}}{(\nu_1 s_1^2 + \lambda\nu_2 s_2^2)^{(\nu_1+\nu_2)/2}} \\ &= \frac{\Gamma[(\nu_1 + \nu_2)/2]}{\Gamma(\nu_1/2)\Gamma(\nu_2/2)}\left(\frac{\nu_2 s_2^2}{\nu_1 s_1^2}\right)^{\nu_2/2} \frac{\lambda^{\nu_2/2-1}}{(1 + \nu_2 s_2^2/\nu_1 s_1^2 \lambda)^{(\nu_1+\nu_2)/2}}. \end{aligned}$$

If we change variables by $x = s_2^2\lambda/s_1^2$, the pdf for x will be precisely the F pdf in (A.68).

APPENDIX B

Properties of Some Multivariate Pdf's

B.1 THE MULTIVARIATE NORMAL (MN) Pdf

The elements of a random vector $\tilde{\mathbf{x}}' = (\tilde{x}_1, \tilde{x}_2, \ldots, \tilde{x}_m)$ are said to be jointly normally distributed if, and only if, they have a pdf in the following form:

$$\text{(B.1)} \qquad p(\mathbf{x}|\boldsymbol{\theta}, \Sigma) = \frac{|\Sigma|^{-1/2}}{(2\pi)^{m/2}} \exp\{-\tfrac{1}{2}(\mathbf{x} - \boldsymbol{\theta})' \Sigma^{-1} (\mathbf{x} - \boldsymbol{\theta})\}, \qquad -\infty < x_i < \infty, \ i = 1, 2, \ldots, m,$$

where $\mathbf{x}' = (x_1, x_2, \ldots, x_m)$, $\boldsymbol{\theta}' = (\theta_1, \theta_2, \ldots, \theta_m)$, with $-\infty < \theta_i < \infty$, $i = 1, 2, \ldots, m$, and Σ is an $m \times m$ positive definite symmetric (PDS) matrix. For convenience the MN pdf is often written

$$\text{(B.2)} \qquad p(\mathbf{x}|\boldsymbol{\theta}, V) = \frac{|V|^{1/2}}{(2\pi)^{m/2}} \exp\{-\tfrac{1}{2}(\mathbf{x} - \boldsymbol{\theta})' V (\mathbf{x} - \boldsymbol{\theta})\}, \qquad -\infty < x_i < \infty, \ i = 1, 2, \ldots, m,$$

where

$$\text{(B.3)} \qquad V \equiv \Sigma^{-1}$$

is PDS.

That (B.2) is a proper normalized pdf can be shown by observing that the pdf is positive over the region for which it is defined. Also,

$$\text{(B.4)} \qquad \int_{-\infty}^{\infty} \cdots \int_{-\infty}^{\infty} p(\mathbf{x}|\boldsymbol{\theta}, V)\, d\mathbf{x} = 1,$$

where $d\mathbf{x} = dx_1\, dx_2 \cdots dx_m$; (B.4) can be shown easily by employing the following change of variables:

$$\text{(B.5)} \qquad \mathbf{x} - \boldsymbol{\theta} = C\mathbf{z}$$

with C an $m \times m$ nonsingular symmetric matrix such that $C'VC = I_m$.[1]

[1] Since V is PDS, there exists an orthogonal matrix, say P, such that $P'VP = D$, where D is an $m \times m$ diagonal matrix with the positive roots of V on the diagonal. Then $C = PD^{-1/2}$ is a nonsingular symmetric matrix such that $C'VC = I_m$.

The Jacobian of the transformation in (B.5) is $|C|$ and thus (B.2) can be expressed as[2]

$$
\text{(B.6)}\qquad
\begin{aligned}
p(\mathbf{z}) &= \frac{|C|\,|V|^{1/2}}{(2\pi)^{m/2}} \exp\{-\tfrac{1}{2}\mathbf{z}'C'VC\mathbf{z}\} \\
&= \frac{1}{(2\pi)^{m/2}} \exp\{-\tfrac{1}{2}\mathbf{z}'\mathbf{z}\}, \qquad -\infty < z_i < \infty,\ i = 1, 2, \ldots, m.
\end{aligned}
$$

The pdf in (B.6) is in the form of a product of m standardized, proper normalized UN pdf's. Thus (B.6), integrated with respect to z_i, $-\infty < z_i < \infty$, $i = 1, 2, \ldots, m$, is equal to 1, which establishes (B.4).

The pdf in (B.6) is usually referred to as a standardized multivariate normal (SMN) pdf.[3] If the elements of an $m \times 1$ random vector $\tilde{\mathbf{z}}$ have a SMN pdf, it is clear from the form of (B.6) that they are independently distributed with

$$
\text{(B.7)}\qquad E\tilde{\mathbf{z}} = \mathbf{0} \quad \text{and} \quad E\tilde{\mathbf{z}}\tilde{\mathbf{z}}' = I_m;
$$

that is, each $\tilde{z}_i$ has a zero mean and unit variance and all covariances, $E\tilde{z}_i\tilde{z}_j$, $i, j = 1, 2, \ldots, m$, $i \neq j$, are zero.

From $\tilde{\mathbf{x}} - \boldsymbol{\theta} = C\tilde{\mathbf{z}}$ and the results in (B.7) we have

$$
\text{(B.8)}\qquad E(\tilde{\mathbf{x}} - \boldsymbol{\theta}) = CE\tilde{\mathbf{z}} = \mathbf{0}
$$

and

$$
\text{(B.9)}\qquad
\begin{aligned}
E(\tilde{\mathbf{x}} - \boldsymbol{\theta})(\tilde{\mathbf{x}} - \boldsymbol{\theta})' &= CE\tilde{\mathbf{z}}\tilde{\mathbf{z}}'C' \\
&= CC' = V^{-1} = \Sigma.^{4}
\end{aligned}
$$

The result in (B.8) gives $E\tilde{\mathbf{x}} = \boldsymbol{\theta}$, as the mean vector of the MN pdf, whereas (B.9) yields V^{-1} (or Σ) as its covariance matrix.

We now derive the conditional pdf for $\tilde{\mathbf{x}}_1$, given $\tilde{\mathbf{x}}_2$, where $\tilde{\mathbf{x}}' = (\tilde{\mathbf{x}}_1'\tilde{\mathbf{x}}_2')$ has a MN pdf given by (B.2). Partitioning $\mathbf{x} - \boldsymbol{\theta}$ and V to correspond to the partitioning of $\tilde{\mathbf{x}}$, that is

$$
\mathbf{x} - \boldsymbol{\theta} = \begin{pmatrix} \mathbf{x}_1 - \boldsymbol{\theta}_1 \\ \mathbf{x}_2 - \boldsymbol{\theta}_2 \end{pmatrix} \quad \text{and} \quad V = \begin{pmatrix} V_{11} & V_{12} \\ V_{21} & V_{22} \end{pmatrix},
$$

[2] Note that $|C|\,|V|^{1/2} = |C'VC|^{1/2} = 1$.

[3] Often the term "spherical" is used rather than "standardized" to emphasize the fact that the contours of (B.6) are spherical (or, with $m = 2$, circular).

[4] From $C'VC = I_m$ we have $V = (C')^{-1}C^{-1}$ and thus $V^{-1} = CC'$.

we can express the quadratic form in the exponent of (B.2) as

$$(\mathbf{x} - \boldsymbol{\theta})'V(\mathbf{x} - \boldsymbol{\theta}) = (\mathbf{x}_1 - \boldsymbol{\theta}_1)'V_{11}(\mathbf{x}_1 - \boldsymbol{\theta}_1) + 2(\mathbf{x}_1 - \boldsymbol{\theta}_1)'V_{12}(\mathbf{x}_2 - \boldsymbol{\theta}_2) + (\mathbf{x}_2 - \boldsymbol{\theta}_2)'V_{22}(\mathbf{x}_2 - \boldsymbol{\theta}_2)$$

$$\text{(B.10)} \qquad = [\mathbf{x}_1 - \boldsymbol{\theta}_1 + V_{11}^{-1}V_{12}(\mathbf{x}_2 - \boldsymbol{\theta}_2)]'V_{11} \times [\mathbf{x}_1 - \boldsymbol{\theta}_1 + V_{11}^{-1}V_{12}(\mathbf{x}_2 - \boldsymbol{\theta}_2)] + (\mathbf{x}_2 - \boldsymbol{\theta}_2)'(V_{22} - V_{21}V_{11}^{-1}V_{12})(\mathbf{x}_2 - \boldsymbol{\theta}_2)$$

by completing the square on $\mathbf{x}_1$. Further, we have

$$\text{(B.11)} \qquad |V| = \begin{vmatrix} V_{11} & V_{12} \\ V_{21} & V_{22} \end{vmatrix} = |V_{11}|\,|V_{22} - V_{21}V_{11}^{-1}V_{12}|.$$

On substituting from (B.10) and (B.11) into (B.2), we can write (B.2) as the product of two factors

$$p(\mathbf{x}_1, \mathbf{x}_2|\boldsymbol{\theta}, V) = \left(\frac{|V_{11}|^{1/2}}{(2\pi)^{m_1/2}} \exp\{-\tfrac{1}{2}[\mathbf{x}_1 - \boldsymbol{\theta}_1 + V_{11}^{-1}V_{12}(\mathbf{x}_2 - \boldsymbol{\theta}_2)]'V_{11} \times [\mathbf{x}_1 - \boldsymbol{\theta}_1 + V_{11}^{-1}V_{12}(\mathbf{x}_2 - \boldsymbol{\theta}_2)]\}\right)$$

$$\text{(B.12)} \qquad \times \left\{\frac{|V_{22} - V_{21}V_{11}^{-1}V_{12}|^{1/2}}{(2\pi)^{m_2/2}} \times \exp[-\tfrac{1}{2}(\mathbf{x}_2 - \boldsymbol{\theta}_2)'(V_{22} - V_{21}V_{11}^{-1}V_{12})(\mathbf{x}_2 - \boldsymbol{\theta}_2)]\right\},$$

where m_1 and m_2 are the number of elements in $\mathbf{x}_1$ and $\mathbf{x}_2$, respectively, with $m_1 + m_2 = m$. Both factors in (B.12) are in the form of normalized MN pdf's. The first factor on the rhs of (B.12) is the conditional pdf for $\mathbf{x}_1$, given $\mathbf{x}_2$, since in general we can write $p(\mathbf{x}_1, \mathbf{x}_2|\boldsymbol{\theta}, V) = p(\mathbf{x}_1|\mathbf{x}_2, \boldsymbol{\theta}, V)\, p(\mathbf{x}_2|\boldsymbol{\theta}, V)$. It is a MN pdf with mean vector

$$\text{(B.13}a\text{)} \qquad E(\tilde{\mathbf{x}}_1|\tilde{\mathbf{x}}_2) = \boldsymbol{\theta}_1 - V_{11}^{-1}V_{12}(\mathbf{x}_2 - \boldsymbol{\theta}_2)$$

and covariance matrix

$$\text{(B.14}a\text{)} \qquad \operatorname{Cov}(\tilde{\mathbf{x}}_1|\tilde{\mathbf{x}}_2) = V_{11}^{-1}.$$

Since $V = \Sigma^{-1}$, we can express (B.13) and (B.14) in terms of submatrices of Σ[5]:

$$\text{(B.13}b\text{)} \qquad E(\tilde{\mathbf{x}}_1|\tilde{\mathbf{x}}_2) = \boldsymbol{\theta}_1 + \Sigma_{12}\Sigma_{22}^{-1}(\mathbf{x}_2 - \boldsymbol{\theta}_2)$$

[5] That is, we partition Σ^{-1} to correspond to the partitioning of V,

$$\begin{pmatrix} V_{11} & V_{12} \\ V_{21} & V_{22} \end{pmatrix} = \begin{pmatrix} \Sigma^{11} & \Sigma^{12} \\ \Sigma^{21} & \Sigma^{22} \end{pmatrix}.$$

Then $V_{11}^{-1} = (\Sigma^{11})^{-1}$ and $V_{12} = \Sigma^{12}$. If we partition Σ correspondingly as

$$\Sigma = \begin{pmatrix} \Sigma_{11} & \Sigma_{12} \\ \Sigma_{21} & \Sigma_{22} \end{pmatrix},$$

and

$$\text{(B.14}b\text{)} \qquad \operatorname{Cov}(\tilde{\mathbf{x}}_1|\tilde{\mathbf{x}}_2) = \Sigma_{11} - \Sigma_{12}\Sigma_{22}^{-1}\Sigma_{21}.$$

The marginal pdf for $\mathbf{x}_2$ can be obtained from (B.12) by integrating with respect to the elements of $\mathbf{x}_1$. Since $\mathbf{x}_1$ appears just in the first factor on the rhs of (B.12), and this is in the form of a normalized MN pdf, integrating the first factor with respect to the elements of $\mathbf{x}_1$ yields 1. Thus the second factor on the rhs of (B.12) is the marginal pdf for $\mathbf{x}_2$. It is a MN pdf with mean vector $\boldsymbol{\theta}_2$ and covariance matrix[6]

$$\text{(B.15)} \qquad \begin{aligned} \operatorname{Cov}(\tilde{\mathbf{x}}_2) &= (V_{22} - V_{21}V_{11}^{-1}V_{12})^{-1} \\ &= \Sigma_{22}. \end{aligned}$$

By similar operations the marginal pdf for $\tilde{\mathbf{x}}_1$ is found to be MN, with mean $\boldsymbol{\theta}_1$ and covariance matrix $(V_{11} - V_{12}V_{22}^{-1}V_{21})^{-1} = \Sigma_{11}$.

Last, consider linear combinations of the elements of $\tilde{\mathbf{x}}$; that is

$$\text{(B.16)} \qquad \tilde{\mathbf{w}}_1 = L_1\tilde{\mathbf{x}},$$

where $\tilde{\mathbf{x}}$ is an $m \times 1$ vector of normal random variables with a MN pdf, as shown in (B.2), and L_1 is a $n \times m$ matrix of given quantities with $n \leq m$ of rank n. Then $\tilde{\mathbf{w}}_1$ is an $n \times 1$ vector whose elements are linear combinations of the elements of $\tilde{\mathbf{x}}$. If $n < m$, write

$$\text{(B.17)} \qquad \tilde{\mathbf{w}} = \begin{pmatrix} \tilde{\mathbf{w}}_1 \\ \tilde{\mathbf{w}}_2 \end{pmatrix} = \begin{pmatrix} L_1 \\ L_2 \end{pmatrix}\tilde{\mathbf{x}} = L\tilde{\mathbf{x}}$$

with the matrix L an $m \times m$ nonsingular matrix. Then $E\tilde{\mathbf{w}} = LE\tilde{\mathbf{x}} = L\boldsymbol{\theta}$, and we can write

$$\text{(B.18)} \qquad \tilde{\mathbf{w}} - L\boldsymbol{\theta} = L(\tilde{\mathbf{x}} - \boldsymbol{\theta}).$$

Then, on noting that the Jacobian of the transformation from $\tilde{\mathbf{x}}$ to $\tilde{\mathbf{w}}$ in (B.18) is $|L^{-1}|$, we can obtain the pdf for $\tilde{\mathbf{w}}$:

$$\text{(B.19)} \quad p(\mathbf{w}|\boldsymbol{\theta}, \Sigma, L) = \frac{|L'\Sigma L|^{-1/2}}{(2\pi)^{m/2}} \exp\left[-\tfrac{1}{2}(\mathbf{w} - L\boldsymbol{\theta})'L^{-1\prime}\Sigma^{-1}L^{-1}(\mathbf{w} - L\boldsymbol{\theta})\right],$$

---

we have $\Sigma^{12} = -\Sigma^{11}\Sigma_{12}\Sigma_{22}^{-1}$ and $(\Sigma^{11})^{-1} = \Sigma_{11} - \Sigma_{12}\Sigma_{22}^{-1}\Sigma_{21}$. Thus $V_{11}^{-1} = \Sigma_{11} - \Sigma_{12}\Sigma_{22}^{-1}\Sigma_{21}$ and $V_{11}^{-1}V_{12} = -\Sigma_{12}\Sigma_{22}^{-1}$.

[6] Since $V = \Sigma^{-1}$, we have

$$\begin{pmatrix} \Sigma_{11} & \Sigma_{12} \\ \Sigma_{21} & \Sigma_{22} \end{pmatrix}\begin{pmatrix} V_{11} & V_{12} \\ V_{21} & V_{22} \end{pmatrix} = \begin{pmatrix} I & 0 \\ 0 & I \end{pmatrix},$$

and thus $\Sigma_{21}V_{12} + \Sigma_{22}V_{22} = I$ and $\Sigma_{21}V_{11} + \Sigma_{22}V_{21} = 0$. The second of these relations yields $\Sigma_{21} = -\Sigma_{22}V_{21}V_{11}^{-1}$ which, when substituted in the first, yields $\Sigma_{22} = (V_{22} - V_{21}V_{11}^{-1}V_{12})^{-1}$.

which is a MN pdf with mean vector $L\boldsymbol{\theta}$ and covariance matrix $L\Sigma L'$. Thus $\tilde{\mathbf{w}} = L\tilde{\mathbf{x}}$ has a MN pdf. If we partition $\mathbf{w}$, as shown in (B.17), the marginal pdf for $\mathbf{w}_1$ will be multivariate normal with mean vector $L_1\boldsymbol{\theta}$ and covariance matrix $L_1\Sigma L_1'$, an application of the general result associated with (B.12) which gives the marginal pdf for a MN pdf.

B.2 THE MULTIVARIATE STUDENT (MS) *t* Pdf

A random vector $\tilde{\mathbf{x}}' = (\tilde{x}_1, \tilde{x}_2, \ldots, \tilde{x}_m)$ has elements distributed according to the MS t distribution if, and only if, they have the following pdf:

$$\text{(B.20)}\quad p(\mathbf{x}|\boldsymbol{\theta}, V, \nu, m) = \frac{\nu^{\nu/2}\Gamma[(\nu + m)/2]|V|^{1/2}}{\pi^{m/2}\Gamma(\nu/2)}[\nu + (\mathbf{x} - \boldsymbol{\theta})'V(\mathbf{x} - \boldsymbol{\theta})]^{-(m+\nu)/2},$$
$$-\infty < x_i < \infty,\ i = 1, 2, \ldots, m,$$

where $\nu > 0$, V is an $m \times m$ PDS matrix, and $\boldsymbol{\theta}' = (\theta_1, \theta_2, \ldots, \theta_m)$, with $-\infty < \theta_i < \infty$, $i = 1, 2, \ldots, m$. Since the quadratic form $(\mathbf{x} - \boldsymbol{\theta})'V(\mathbf{x} - \boldsymbol{\theta})$ is PD, the MS t pdf has a single mode at $\mathbf{x} = \boldsymbol{\theta}$. Further, since the pdf is symmetric about $\mathbf{x} = \boldsymbol{\theta}$, $\boldsymbol{\theta}$ is the mean of the MS t pdf which exists for $\nu > 1$, as shown below. The symmetry about $\boldsymbol{\theta}$ implies that odd-order moments about $\boldsymbol{\theta}$, when they exist, will all be zero. The matrix of second-order moments about the mean exists for $\nu > 2$ and is given by $V^{-1}[\nu/(\nu - 2)]$.

To establish that (B.20) is a proper normalized pdf we note that it is positive in the region for which it is defined. If we let

$$\text{(B.21)}\qquad \mathbf{x} - \boldsymbol{\theta} = C\mathbf{z},$$

where C is an $m \times m$ nonsingular matrix such that $C'VC = I_m$, the pdf for $\mathbf{z}$, an $m \times 1$ vector, is[7]

$$\text{(B.22)}\quad p(\mathbf{z}|\nu, m) = \frac{\nu^{\nu/2}\Gamma[(\nu + m)/2]}{\pi^{m/2}\Gamma(\nu/2)}(\nu + \mathbf{z}'\mathbf{z})^{-(m+\nu)/2}, \qquad -\infty < z_i < \infty,\ i = 1, 2, \ldots, m,$$

the standardized form of the MS t pdf. We show that (B.22) is a normalized pdf[8] by making the following change of variables from $z_1, z_2, \ldots, z_m$ to

[7] Note from $C'VC = I_m$, $|V|^{1/2} = |C|^{-1}$. The Jacobian associated with the transformation in (B.21) is $|C|$ and thus $|V|^{1/2}$ times the Jacobian is equal to 1.

[8] Another way of showing this is to observe that $p(\mathbf{z}|\nu)$ can be written as $p(\mathbf{z}|\nu) = p(z_m|\mathbf{z}_{(m-1)}, \nu)\, p(z_{m-1}|\mathbf{z}_{(m-2)}, \nu) \cdots p(z_1|\nu)$, where $\mathbf{z}_{m-j}' = (z_1, z_2, \ldots, z_{m-j})$, $j = 1, 2, \ldots, m - 1$. Each factor is in the form of a US t pdf and can be integrated by using the results of Appendix A.

$u, \alpha_1, \alpha_2, \ldots, \alpha_{m-1}$, given by

$$\begin{aligned} z_1 &= u^{1/2} \cos \alpha_1 \cos \alpha_2 \cdots \cos \alpha_{m-1} \\ z_2 &= u^{1/2} \cos \alpha_1 \cos \alpha_2 \cdots \cos \alpha_{m-2} \sin \alpha_{m-1} \\ &\cdots \\ z_j &= u^{1/2} \cos \alpha_1 \cos \alpha_2 \cdots \cos \alpha_{m-j} \sin \alpha_{m-j+1} \\ &\cdots \\ z_m &= u^{1/2} \sin \alpha_1, \end{aligned} \tag{B.23}$$

where $0 < u < \infty$, $-\pi/2 < \alpha_i < \pi/2$ for $i = 1, 2, \ldots, m-2$, and $0 < \alpha_{m-1} < 2\pi$. From trigonometry (B.23) yields

$$u = z_1^2 + z_2^2 + \cdots + z_m^2 = \mathbf{z}'\mathbf{z}. \tag{B.24}$$

Also, the Jacobian of the transformation in (B.23) is[9] $\frac{1}{2}u^{m/2-1} \cos^{m-2} \alpha_1 \times \cos^{m-3} \alpha_2 \cdots \cos \alpha_{m-2}$. Thus the pdf in (B.22) becomes

$$p(u, \alpha_1, \alpha_2, \ldots, \alpha_{m-1}|\nu, m) = \frac{1}{2} \frac{\nu^{\nu/2}\Gamma[(\nu + m)/2]}{\pi^{m/2}\Gamma(\nu/2)} \frac{u^{m/2-1}}{(\nu + u)^{(\nu+m)/2}} \times \cos^{m-2} \alpha_1 \cos^{m-3} \alpha_2 \ldots \cos \alpha_{m-2}. \tag{B.25}$$

Now (B.25) can be integrated with respect to u and the α's by using

$$\int_0^\infty \frac{u^{m/2-1}}{(\nu + u)^{(\nu+m)/2}} \, du = \frac{1}{\nu^{\nu/2}} B\left(\frac{\nu}{2}, \frac{m}{2}\right) = \frac{\Gamma(\nu/2)\Gamma(m/2)}{\nu^{\nu/2} \, \Gamma[(\nu + m)/2]}, \qquad \nu > 0, \tag{B.26}$$

from (A.60),

$$\int_{-\pi/2}^{\pi/2} \cos^{m-j-1} \alpha_j \, d\alpha_j = \pi^{1/2} \frac{\Gamma[(m - j)/2]}{\Gamma[(m - j - 1)/2 + 1]}, \qquad j = 1, 2, \ldots, m - 2, \tag{B.27}$$

and

$$\int_0^{2\pi} d\alpha_{m-1} = 2\pi. \tag{B.28}$$

On substituting from (B.26), (B.27), and (B.28) into

$$\int \cdots \int p(u, \alpha_1, \alpha_2, \ldots, \alpha_{m-1}|\nu, m) \, du \, d\alpha_1 \cdots d\alpha_{m-1}, \tag{B.29}$$

with the integrand given by (B.25), the integral in (B.29) has a value of 1. Thus (B.25) and (B.20) are normalized pdf's.

Note from (B.25) and (B.26) that the normalized marginal pdf for $u = \mathbf{z}'\mathbf{z}$ is

$$p(u|\nu, m) = \frac{\nu^{\nu/2}}{B(\nu/2, m/2)} \frac{u^{m/2-1}}{(\nu + u)^{(\nu+m)/2}}, \qquad 0 < u < \infty. \tag{B.30}$$

[9] See, for example, M. G. Kendall and A. S. Stuart, *The Advanced Theory of Statistics*, Vol. I. London: Griffin, 1958, p. 247.

Letting $u = my$, we have

(B.31) $$p(y|\nu, m) = \left[B\left(\frac{\nu}{2}, \frac{m}{2}\right)\right]^{-1}\left(\frac{m}{\nu}\right)^{m/2} \frac{y^{m/2-1}}{(1 + (m/\nu)y)^{(\nu+m)/2}}, \qquad 0 < y < \infty,$$

which is an F pdf with m and ν degrees of freedom [see (A.68)]. Thus the random variable $\tilde{y} = \tilde{u}/m = \tilde{\mathbf{z}}'\tilde{\mathbf{z}}/m$, with $\tilde{\mathbf{z}}$ having a pdf given by the standardized MS t pdf in (B.22), has an F pdf with m and ν degrees of freedom. Further, by writing (B.21) to connect random variables we have

(B.32) $$\begin{aligned} \tilde{y} = \frac{\tilde{\mathbf{z}}'\tilde{\mathbf{z}}}{m} &= \frac{(\tilde{\mathbf{x}} - \boldsymbol{\theta})'(C^{-1})'C^{-1}(\tilde{\mathbf{x}} - \boldsymbol{\theta})}{m} \\ &= \frac{(\tilde{\mathbf{x}} - \boldsymbol{\theta})'V(\tilde{\mathbf{x}} - \boldsymbol{\theta})}{m}, \end{aligned}$$

and thus the quadratic form $(\tilde{\mathbf{x}} - \boldsymbol{\theta})'V(\tilde{\mathbf{x}} - \boldsymbol{\theta})/m$ has an F pdf with m and ν degrees of freedom when $\tilde{\mathbf{x}}$'s pdf is the MS t pdf shown in (B.20).

To obtain expressions for the first and second moments associated with (B.20) we determine moments associated with (B.22) and then use (B.21) to find moments for (B.20). As regards existence of moments, consider the rth moment about zero. To evaluate this moment we have to consider the integral

(B.33) $$\int_{-\infty}^{\infty} \frac{z_1^r\, dz_1}{(a + z_1^2)^{(\nu+m)/2}}, \qquad m = 1, 2, \ldots,$$

with $a \equiv \nu + \sum_{i=2}^m z_i^2$. Using the tests for convergence described in Appendix A, we see that (B.33) will converge for any m if $r + 1 < \nu + 1$ or $\nu > r$. Thus for the first moment to exist we need $\nu > 1$, for the second, $\nu > 2$, and so on. From the symmetry of (B.22) we have

(B.34) $$E\tilde{\mathbf{z}} = \mathbf{0}, \qquad \nu > 1,$$

and from $\tilde{\mathbf{x}} - \boldsymbol{\theta} = C\tilde{\mathbf{z}}$, $E\tilde{\mathbf{x}} = \boldsymbol{\theta}$ for $\nu > 1$.

To evaluate the second moments associated with (B.22) consider (B.33) with $r = 2$. By letting $u = z_1^2/a$ (B.33) can be brought into the following form:

(B.35) $$\frac{1}{a^{(\nu+m-3)/2}} \int_0^{\infty} \frac{u^{1/2}}{(1 + u)^{(\nu+m)/2}}\, du = \frac{1}{a^{(\nu+m-3)/2}} B\left(\frac{\nu + m - 3}{2}, \frac{3}{2}\right).$$

Now (B.35) has to be integrated with respect to $z_2, z_3, \ldots, z_m$, which appear in the quantity $a = \nu + \sum_{i=2}^m z_i^2$; that is, using (B.35),

(B.36) $$E\tilde{z}_1^2 = kB\left(\frac{\nu + m - 3}{2}, \frac{3}{2}\right) \int_{-\infty}^{\infty} \cdots \int_{-\infty}^{\infty} \frac{dz_2 \cdots dz_m}{(\nu + \sum_{i=2}^m z_i^2)^{(\nu+m-3)/2}},$$

with $k = \nu^{\nu/2}\Gamma[(\nu + m)/2]/\pi^{m/2}\Gamma(\nu/2)$, the normalizing constant of (B.22). The integrand of (B.36) can be brought into the form of a standardized MS t pdf and integrated[10] to yield

$$E\tilde{z}_1^2 = \frac{\nu}{\nu - 2}. \tag{B.37}$$

Since this argument can be applied separately to $\tilde{z}_2, \tilde{z}_3, \ldots, \tilde{z}_m$ and $E\tilde{z}_i\tilde{z}_j = 0$ for $i \neq j$, the covariance matrix for $\tilde{\mathbf{z}}$ is

$$E\tilde{\mathbf{z}}\tilde{\mathbf{z}}' = \frac{\nu}{\nu - 2} I_m, \qquad \nu > 2. \tag{B.38}$$

From $\tilde{\mathbf{x}} - \boldsymbol{\theta} = C\tilde{\mathbf{z}}$ the covariance matrix for $\tilde{\mathbf{x}} - \boldsymbol{\theta}$ is

$$E(\tilde{\mathbf{x}} - \boldsymbol{\theta})(\tilde{\mathbf{x}} - \boldsymbol{\theta})' = CE\tilde{\mathbf{z}}\tilde{\mathbf{z}}C'$$

$$= \frac{\nu}{\nu - 2} CC' = \frac{\nu}{\nu - 2} V^{-1},^{11} \qquad \nu > 2. \tag{B.39}$$

We now consider the marginal and conditional pdf's associated with the MS t pdf in (B.20). To accomplish this conveniently we let[12] $H \equiv V/\nu$ and rewrite (B.20) as

(B.40)

$$p(\mathbf{x}|\boldsymbol{\theta}, H, \nu) = \frac{\Gamma[(\nu + m)/2]}{\pi^{m/2}\Gamma(\nu/2)} |H|^{1/2}[1 + (\mathbf{x} - \boldsymbol{\theta})'H(\mathbf{x} - \boldsymbol{\theta})]^{-(m+\nu)/2}, \qquad \nu > 0.$$

Now partition $(\mathbf{x} - \boldsymbol{\theta})' = [(\mathbf{x}_1 - \boldsymbol{\theta}_1)'(\mathbf{x}_2 - \boldsymbol{\theta}_2)']$, where $\mathbf{x}_1 - \boldsymbol{\theta}_1$ is an $m_1 \times 1$ vector and $\mathbf{x}_2 - \boldsymbol{\theta}_2$ is an $m_2 \times 1$ vector with $m_1 + m_2 = m$. Also let

$$H = \begin{pmatrix} H_{11} & H_{12} \\ H_{21} & H_{22} \end{pmatrix},$$

[10] The integrand of (B.36) can be written as

$$\left[\left(\frac{\nu}{\nu'}\right)^{(\nu+m-3)/2}\left(\nu' + \frac{\nu'}{\nu}\sum_{i=2}^{m} z_i^2\right)^{(\nu'+m-1)/2}\right]^{-1}, \qquad -\infty < z_i < \infty,\ i = 2, 3, \ldots, m,$$

with $\nu' = \nu - 2$. Then letting $w_i = \sqrt{\nu'/\nu}\, z_i$, (B.36) becomes

$$kB\left(\frac{\nu + m - 3}{2}, \frac{3}{2}\right)\left(\frac{\nu'}{\nu}\right)^{\nu/2 - 1} \int\cdots\int \frac{dw_2\cdots dw_m}{(\nu' + \sum_{i=2}^{m} w_i^2)^{(\nu'+m-1)/2}}$$

$$= kB\left(\frac{\nu + m - 3}{2}, \frac{3}{2}\right)\left(\frac{\nu'}{\nu}\right)^{\nu/2 - 1} \frac{\pi^{(m-1)/2}\Gamma(\nu'/2)}{(\nu')^{\nu'/2}\Gamma[(\nu' + m - 1)/2]} = \frac{\nu}{\nu - 2},$$

$$-\infty < w_i < \infty,\ i = 2, 3, \ldots, m.$$

[11] From $C'VC = I_m$, $V = (C')^{-1}C^{-1}$ or $V^{-1} = CC'$.

[12] With $\nu > 0$, H is PDS.

where the partitioning of H corresponds to that of $\mathbf{x} - \boldsymbol{\theta}$. Then

$$\begin{aligned}1 + (\mathbf{x} - \boldsymbol{\theta})'H(\mathbf{x} - \boldsymbol{\theta}) &= 1 + (\mathbf{x}_1 - \boldsymbol{\theta}_1)'H_{11}(\mathbf{x}_1 - \boldsymbol{\theta}_1) \\ &\quad + 2(\mathbf{x}_1 - \boldsymbol{\theta}_1)'H_{12}(\mathbf{x}_2 - \boldsymbol{\theta}_2) \\ &\quad + (\mathbf{x}_2 - \boldsymbol{\theta}_2)'H_{22}(\mathbf{x}_2 - \boldsymbol{\theta}_2) \\ &= 1 + [\mathbf{x}_1 - \boldsymbol{\theta}_1 + H_{11}{}^{-1}H_{12}(\mathbf{x}_2 - \boldsymbol{\theta}_2)]'H_{11} \\ &\quad \times [\mathbf{x}_1 - \boldsymbol{\theta}_1 + H_{11}{}^{-1}H_{12}(\mathbf{x}_2 - \boldsymbol{\theta}_2)] \\ &\quad + (\mathbf{x}_2 - \boldsymbol{\theta}_2)'(H_{22} - H_{21}H_{11}{}^{-1}H_{12})(\mathbf{x}_2 - \boldsymbol{\theta}_2) \\ &= 1 + Q_{1\cdot 2} + Q_2,\end{aligned} \tag{B.41}$$

where $Q_{1\cdot 2}$ and Q_2 denote the first and second quadratic forms, respectively, on the rhs of the second line of (B.41). Then, noting $|H| = |H_{11}|\,|H_{22} - H_{21}H_{11}{}^{-1}H_{12}|$, we can express (B.40) as

$$\begin{aligned}p(\mathbf{x}_1, \mathbf{x}_2|\boldsymbol{\theta}, H, \nu) &= \frac{\Gamma[(\nu + m)/2]|H|^{1/2}}{\pi^{m/2}\Gamma(\nu/2)}[(1 + Q_2 + Q_{1\cdot 2})^{(m+\nu)/2}]^{-1} \\ &= \left[\frac{k_1|H_{22} - H_{21}H_{11}{}^{-1}H_{12}|^{1/2}}{(1 + Q_2)^{(m_2+\nu)/2}}\right]\left[\frac{k_2(1 + Q_2)^{-m_1/2}|H_{11}|^{1/2}}{[1 + Q_{1\cdot 2}/(1 + Q_2)]^{(m+\nu)/2}}\right],\end{aligned} \tag{B.42}$$

where

$$k_1 = \frac{\Gamma[(\nu + m_2)/2]}{\pi^{m_2/2}\Gamma(\nu/2)} \quad \text{and} \quad k_2 = \frac{\Gamma[(m + \nu)/2]}{\pi^{m_1/2}\Gamma[(\nu + m_2)/2]}.$$

The second line of (B.42) gives explicit expressions for the marginal and conditional pdf's; that is

$$p(\mathbf{x}_1, \mathbf{x}_2|\boldsymbol{\theta}, H, \nu) = p(\mathbf{x}_2|\boldsymbol{\theta}, H, \nu)\,p(\mathbf{x}_1|\mathbf{x}_2, \boldsymbol{\theta}, H, \nu), \tag{B.43}$$

with

$$p(\mathbf{x}_2|\boldsymbol{\theta}, H, \nu) = \frac{k_1|H_{22} - H_{21}H_{11}{}^{-1}H_{12}|^{1/2}}{(1 + Q_2)^{(m_2+\nu)/2}}, \tag{B.44}$$

and[13]

$$p(\mathbf{x}_1|\mathbf{x}_2, \boldsymbol{\theta}, H, \nu) = \frac{k_2(1 + Q_2)^{-m_1/2}|H_{11}|^{1/2}}{[1 + Q_{1\cdot 2}/(1 + Q_2)]^{(m+\nu)/2}}, \tag{B.45}$$

with $Q_{1\cdot 2}$ and Q_2 defined in connection with (B.41); (B.44) and (B.45) show that in general the marginal and conditional pdf's associated with a MS t pdf have the forms of MS t pdf's.[14] From (B.44) we have for the mean and covariance matrix of the marginal pdf for $\mathbf{x}_2$

$$E\tilde{\mathbf{x}}_2 = \boldsymbol{\theta}_2, \qquad \nu > 1, \tag{B.46}$$

[13] The expression for the conditional MS t pdf, given in H. Raiffa and R. Schlaifer, *Applied Statistical Decision Theory*, Boston: Graduate School of Business Administration, Harvard University, 1961, p. 258 is erroneous.

[14] If $\mathbf{x}_2$ in (B.44) and $\mathbf{x}_1$ in (B.45) are scalars, these pdf's are US t pdf's.

and

$$\text{(B.47)} \quad \begin{aligned} \operatorname{Var}(\tilde{\mathbf{x}}_2) &= \frac{1}{\nu - 2}(H_{22} - H_{21}H_{11}^{-1}H_{12})^{-1} \\ &= \frac{\nu}{\nu - 2}(V_{22} - V_{21}V_{11}^{-1}V_{12})^{-1}, \qquad \nu > 2, \end{aligned}$$

since $H \equiv V/\nu$. For the conditional pdf in (B.45) we have

$$\text{(B.48)} \quad \begin{aligned} E(\tilde{\mathbf{x}}_1|\tilde{\mathbf{x}}_2) &= \boldsymbol{\theta}_1 - H_{11}^{-1}H_{12}(\mathbf{x}_2 - \boldsymbol{\theta}_2) \\ &= \boldsymbol{\theta}_1 - V_{11}^{-1}V_{12}(\mathbf{x}_2 - \boldsymbol{\theta}_2), \end{aligned} \qquad m_2 + \nu > 1,$$

and

$$\text{(B.49)} \quad \begin{aligned} \operatorname{Var}(\tilde{\mathbf{x}}_1|\tilde{\mathbf{x}}_2) &= \frac{1}{m_2 + \nu - 2}(1 + Q_2)H_{11}^{-1} \\ &= \frac{\nu}{m_2 + \nu - 2}(1 + Q_2)V_{11}^{-1}, \qquad m_2 + \nu > 2, \end{aligned}$$

where $Q_2 = (\mathbf{x}_2 - \boldsymbol{\theta}_2)'(H_{22} - H_{21}H_{11}^{-1}H_{12})(\mathbf{x}_2 - \boldsymbol{\theta}_2)$.

Next consider linear combinations of the elements of a random vector $\tilde{\mathbf{x}}$ with a MS t pdf, as shown in (B.40):

$$\text{(B.50)} \qquad \tilde{\mathbf{w}} = L\tilde{\mathbf{x}},$$

where L is an $m \times m$ nonsingular matrix. Then $E\tilde{\mathbf{w}} = L\boldsymbol{\theta}$. The Jacobian of the transformation in (B.50) is $|L^{-1}|$ and thus from (B.40) the pdf for $\tilde{\mathbf{w}}$ is

$$\text{(B.51)} \quad p(\mathbf{w}|\boldsymbol{\theta}, F, \nu) = \frac{\Gamma[(\nu + m)/2]}{\pi^{m/2}\Gamma(\nu/2)} |F|^{1/2}[1 + (\mathbf{w} - L\boldsymbol{\theta})'F(\mathbf{w} - L\boldsymbol{\theta})]^{-(m+\nu)/2},$$

where $F \equiv L^{-1\prime}HL^{-1}$. Thus the elements of $\tilde{\mathbf{w}}$ have a MS t pdf with mean $L\boldsymbol{\theta}$ and covariance matrix equal to $(\nu - 2)^{-1}LH^{-1}L' = [\nu/(\nu - 2)]LV^{-1}L'$. The marginal and conditional pdf's associated with (B.51) are easily obtained by using (B.44) and (B.45) and, of course, will be in the MS t form. A single linear combination of the elements of $\tilde{\mathbf{x}}$, say $\tilde{w}_1$, the first element of $\tilde{\mathbf{w}}$, will have a marginal US t pdf.

Last, as is apparent from many examples cited, the MS t is related to the MN and IG pdf's. Consider the joint pdf

$$\text{(B.52)} \qquad p(\mathbf{x}, \sigma|\boldsymbol{\theta}, V, \nu) = g(\mathbf{x}|\boldsymbol{\theta}, \sigma, V)\, h(\sigma|\nu),$$

where $g(\mathbf{x}|\boldsymbol{\theta}, \sigma, V)$ denotes an m-dimensional MN pdf with mean $\boldsymbol{\theta}$ and covariance matrix $V^{-1}\sigma^2$ and $h(\sigma|\nu)$ denotes an IG pdf with parameters $\nu > 0$ and $s = 1$ [see (A.37b)]; that is,

$$\text{(B.53)} \quad p(\mathbf{x}, \sigma|\boldsymbol{\theta}, V, \nu) = k\frac{|V|^{1/2}}{\sigma^{m+\nu+1}} \exp\left\{-\frac{1}{2\sigma^2}[\nu + (\mathbf{x} - \boldsymbol{\theta})'V(\mathbf{x} - \boldsymbol{\theta})]\right\},$$

where k is the normalizing constant. Then, on integrating (B.53), with respect to σ, 0 to ∞, we have

(B.54) $$p(\mathbf{x}|\boldsymbol{\theta}, V, \nu) = k'|V|^{1/2}[\nu + (\mathbf{x} - \boldsymbol{\theta})'V(\mathbf{x} - \boldsymbol{\theta})]^{-(m+\nu)/2},$$

which is precisely in the form of (B.20), with k' the normalizing constant.

B.3 THE WISHART (W) Pdf

The $m(m + 1)/2$ distinct elements of an $m \times m$ PDS random matrix $\tilde{A} = \{\tilde{a}_{ij}\}$ are distributed according to the W distribution if, and only if, they have the following pdf:

(B.55) $$p(A|\Sigma, \nu, m) = k\frac{|A|^{(\nu - m - 1)/2}}{|\Sigma|^{\nu/2}} \exp\{-\tfrac{1}{2} tr\, \Sigma^{-1}A\}, \qquad |A| > 0,$$

where $k^{-1} = 2^{\nu m/2}\pi^{m(m-1)/4} \prod_{i=1}^{m} \Gamma[(\nu + 1 - i)/2]$, $m \leq \nu$, and $\Sigma = [\sigma_{ij}]$, an $m \times m$ PDS matrix. The pdf in (B.55) is defined for the region given by $|A| > 0$. We denote the pdf in (B.55) by $W(\Sigma, \nu, m)$. Some properties of the W pdf are listed below.

1. If $\tilde{\mathbf{z}}_1, \tilde{\mathbf{z}}_2, \ldots, \tilde{\mathbf{z}}_\nu$ are $m \times 1$ mutually independent random vectors, each with a MN pdf, zero mean vector, and common PDS $m \times m$ covariance matrix Σ, the distinct elements of $\tilde{A} = \tilde{Z}\tilde{Z}'$, where $\tilde{Z} = (\tilde{\mathbf{z}}_1, \tilde{\mathbf{z}}_2, \ldots, \tilde{\mathbf{z}}_\nu)$, have a $W(\Sigma, \nu, m)$ pdf. Note that $\tilde{Z}\tilde{Z}' = \sum_{i=1}^{\nu} \tilde{\mathbf{z}}_i\tilde{\mathbf{z}}_i'$ has diagonal elements given by $\sum_{i=1}^{\nu} \tilde{z}_{ij}^2$, $j = 1, 2, \ldots, m$, and off-diagonal elements given by $\sum_{i=1}^{\nu} \tilde{z}_{ij}\tilde{z}_{ik}$, $j \neq k = 1, 2, \ldots, m$. Thus $\tilde{Z}\tilde{Z}'/\nu = \tilde{S}$, the sample covariance matrix, and the distinct elements of $\tilde{S}$ have a $W[(1/\nu)\Sigma, \nu, m]$ pdf.
2. The distinct elements of a random matrix $\tilde{A}$, with the $W(\Sigma, \nu, m)$ pdf in (B.55), have the following means, variances, and covariances:

(B.56) $$E\tilde{a}_{ij} = \nu\sigma_{ij},$$

(B.57) $$\text{Var}\ \tilde{a}_{ij} = \nu(\sigma_{ij}^2 + \sigma_{ii}\sigma_{jj}),$$

and

(B.58) $$\text{Cov}(\tilde{a}_{ij}, \tilde{a}_{kl}) = \nu(\sigma_{ik}\sigma_{jl} + \sigma_{il}\sigma_{jk}).$$

Let us partition A and Σ correspondingly as

$$A = \begin{matrix} & m_1 & m - m_1 \\ m_1 & A_{11} & A_{12} \\ m - m_1 & A_{21} & A_{22} \end{matrix}, \qquad A^{-1} = \begin{pmatrix} A^{11} & A^{12} \\ A^{21} & A^{22} \end{pmatrix},$$

$$\Sigma = \begin{matrix} & m_1 & m - m_1 \\ m_1 & \Sigma_{11} & \Sigma_{12} \\ m - m_1 & \Sigma_{21} & \Sigma_{22} \end{matrix}, \qquad \Sigma^{-1} = \begin{pmatrix} \Sigma^{11} & \Sigma^{12} \\ \Sigma^{21} & \Sigma^{22} \end{pmatrix},$$

where, in each instance, the (1, 1)th submatrix is of size $m_1 \times m_1$ and the (2, 2)th submatrix is of size $m - m_1 \times m - m_1$. Further, let

$$A_{11\cdot 2}^{-1} = (A_{11} - A_{12}A_{22}^{-1}A_{21})^{-1} = A^{11}$$

and

$$\Sigma_{11\cdot 2}^{-1} = (\Sigma_{11} - \Sigma_{12}\Sigma_{22}^{-1}\Sigma_{21})^{-1} = \Sigma^{11}.$$

Then the following properties of the $W(\Sigma, \nu, m)$ pdf are known to hold:[15]

3. The joint pdf for the distinct elements of A_{11} is $W(\Sigma_{11}, \nu, m_1)$.
4. The joint pdf for the distinct elements of $A_{11\cdot 2}$ is $W[\Sigma_{11\cdot 2}, \nu - (m - m_1), m_1]$.
5. The marginal pdf for $r_{12} = a_{12}/(a_{11}a_{22})^{1/2}$ is

$$h(r_{12}|\rho_{12}, \nu) = k_1(1 - r_{12}^2)^{(\nu-3)/2}(1 - \rho_{12}^2)^{\nu/2} I_\nu(\rho_{12}r_{12}), \tag{B.59}$$

with $k_1 = [(\nu - 1)/\sqrt{2\pi}][\Gamma(\nu)/\Gamma(\nu + \frac{1}{2})]$, $\rho_{12} = \sigma_{12}/(\sigma_{11}\sigma_{22})^{1/2}$, and[16]

$$I_\nu(\rho r) = \int_0^\infty \frac{dy}{(\cosh y - \rho r)^\nu}; \tag{B.60}$$

(B.59) gives the pdf for a sample correlation coefficient based on ν pairs of observations drawn independently from a bivariate normal pdf with zero means and 2×2 PDS covariance matrix Σ.

Property 1 is a fundamental relationship between the MN pdf and the W pdf. It is established as follows.[17] The joint pdf for ν normal, mutually independent $m \times 1$ vectors, $\tilde{\mathbf{z}}_1, \tilde{\mathbf{z}}_2, \ldots, \tilde{\mathbf{z}}_\nu$, each with zero mean vector and common PDS covariance matrix, Σ, is

$$\begin{aligned} p(Z|\Sigma, \nu, m) &= \frac{|\Sigma|^{-\nu/2}}{(2\pi)^{\nu m/2}} \exp\left(-\tfrac{1}{2} \textstyle\sum_{i=1}^{\nu} \mathbf{z}_i'\Sigma^{-1}\mathbf{z}_i\right) \\ &= \frac{|\Sigma|^{-\nu/2}}{(2\pi)^{\nu m/2}} \exp\{-\tfrac{1}{2}\, tr\, \Sigma^{-1}ZZ'\}, \end{aligned} \tag{B.61}$$

where $Z = (\mathbf{z}_1, \mathbf{z}_2, \ldots, \mathbf{z}_\nu)$, an $m \times \nu$ matrix with $m \le \nu$. Now make the following transformation $Z = TK$, where K is an $m \times \nu$ matrix such that $KK' = I_m$; that is, K is a semiorthogonal matrix, and T is an $m \times m$ lower

[15] The following well known properties have been listed in S. Geisser, "Bayesian Estimation in Multivariate Analysis," *Ann. Math. Statistics*, **36**, 150–159 (1965).
[16] The "cosh" function is defined by $\cosh u = (e^u + e^{-u})/2$.
[17] The derivation follows that presented in S. N. Roy, *Some Aspects of Multivariate Analysis*. New York: Wiley, 1957, p. 33.

triangular matrix:

$$\text{(B.62)} \qquad T = \begin{pmatrix} t_{11} & 0 & \cdots & 0 \\ t_{21} & t_{22} & \cdots & 0 \\ \vdots & \vdots & \cdots & \vdots \\ t_{m1} & t_{m2} & \cdots & t_{mm} \end{pmatrix},$$

with $|T| = \prod_{i=1}^{m} t_{ii} > 0$. Note that $KK' = I_m$ places $m(m+1)/2$ constraints on the elements of K. Thus there are really only $\nu m - m(m+1)/2$ independent elements in K. Choose an independent set of elements from those in K, say $(k_{11}, k_{12}, \ldots, k_{1,\nu-1}), (k_{21}, k_{22}, \ldots, k_{2,\nu-2}), \ldots, (k_{m1}, k_{m2}, \ldots, k_{m,\nu-m})$, and call this set K_I. Thus we can regard the transformation $Z = TK$ subject to $KK' = I_m$ as equivalent to the transformation from the νm elements of Z to the $m(m+1)/2$ elements of T and the $\nu m - m(m+1)/2$ elements of K_I. Then, on substituting $Z = TK$ in (B.61), we have

$$\text{(B.63)} \qquad p(T, K_I|\Sigma, \nu, m) = \frac{J|\Sigma|^{-\nu/2}}{(2\pi)^{\nu m/2}} \exp\{-\tfrac{1}{2} tr\, \Sigma^{-1}TT'\},$$

where J denotes the Jacobian of the transformation from the elements of Z to those of T and K_I. To obtain an explicit expression for J we use the following result[18]:

If $y_i = f_i(x_1, x_2, \ldots, x_p, x_{p+1}, \ldots, x_{p+q})$ for $i = 1, 2, \ldots, p$, where the x_j's, $j = 1, 2, \ldots, p+q$, are subject to q constraints, $f_i(x_1, x_2, \ldots, x_p, x_{p+1}, \ldots, x_{p+q}) = 0$ for $i = p+1, p+2, \ldots, p+q$, then[19] the Jacobian J associated with the transformation from $x_1, x_2, \ldots, x_p$ to $y_1, y_2, \ldots, y_p$ is

$$\text{(B.64)} \quad J = \left|\frac{\partial(f_1, f_2, \ldots, f_p, f_{p+1}, \ldots, f_{p+q})}{\partial(x_1, x_2, \ldots, x_p, x_{p+1}, \ldots, x_{p+q})}\right| \div \left|\frac{\partial(f_{p+1}, \ldots, f_{p+q})}{\partial(x_{p+1}, \ldots, x_{p+q})}\right|.$$

In applying this result to the present problem, $Z = TK$ takes the place of $y_i = f_i$ and $KK' - I_m = 0$ takes the place of $f_i = 0$. Further, the elements of K_I are to be associated with $x_1, x_2, \ldots, x_p$, whereas the remaining elements of K, denoted K_D, are to be associated with $x_{p+1}, \ldots, x_{p+q}$. Then the Jacobian in (B.60) is

$$\text{(B.65)} \qquad J = \left|\frac{\partial(Z, KK')}{\partial(T, K)}\right|_{T, K_I} \div \left|\frac{\partial(KK' - I_m)}{\partial(K_D)}\right|_{K_I}.$$

The explicit expression for the numerator of (B.65) is[20]

$$\text{(B.66)} \qquad \left|\frac{\partial(Z, KK')}{\partial(T, K)}\right|_{T, K_I} = 2^m \prod_{i=1}^{m} t_{ii}^{\nu-1}.$$

[18] This result from the calculus is presented in S. N. Roy, *op. cit.*, p. 165.

[19] It is assumed that the usual conditions for the existence of the Jacobian, including the nonvanishing of the numerator and denominator in (B.64), are satisfied.

[20] Cf. Roy, *op. cit.*, pp. 170–174.

Thus (B.63) becomes

$$p(T, K_I|\Sigma, m, \nu) = \frac{2^m \prod_{i=1}^m t_{ii}^{\nu-i}}{(2\pi)^{\nu m/2}|\Sigma|^{\nu/2}} e^{-\frac{1}{2}\operatorname{tr}\Sigma^{-1}TT'} \div \left|\frac{\partial(KK')}{\partial(K_D)}\right|_{K_I}. \tag{B.67}$$

Since[21] $\int dK_I \div |\partial(KK')/\partial(K_D)|_{K_I}$ integrated over the region $KK' = I_m$, equals $\pi^{\nu m/2-[m(m-1)]/4}/\prod_{i=1}^m \Gamma[(\nu - i + 1)/2]$, the marginal distribution of the elements of T is

$$p(T|\Sigma, m, \nu) = c\frac{\prod_{i=1}^m t_{ii}^{\nu-i}}{|\Sigma|^{\nu/2}} \exp\{-\tfrac{1}{2} tr\, \Sigma^{-1} TT'\} \tag{B.68}$$

with

$$c^{-1} = 2^{\nu m/2 - m}\pi^{m(m-1)/4} \prod_{i=1}^m \Gamma\left(\frac{\nu - i + 1}{2}\right).$$

Now the sample covariance matrix S, with $m(m + 1)/2$ distinct elements, is given by $\nu S = ZZ' = TKK'T' = TT'$. Transform (B.68), which involves $m(m + 1)/2$ elements of T, to a pdf for the distinct elements of S. This yields[22]

$$\begin{aligned} p(S|\Sigma, \nu, m) &= c\frac{\prod_{i=1}^m t_{ii}^{\nu-i}}{|\Sigma|^{\nu/2}} \frac{\nu^{m(m+1)/2}}{2^m \prod_{i=1}^m t_{ii}^{m-i+1}} \exp\{-\tfrac{1}{2}\operatorname{tr}\nu\Sigma^{-1}S\} \\ &= \frac{c\nu^{m(m+1)/2}}{2^m} \frac{|T|^{\nu-m-1}}{|\Sigma|^{\nu/2}} \exp\{-\tfrac{1}{2}\operatorname{tr}\nu\Sigma^{-1}S\} \\ &= c_1 \frac{|S|^{(\nu-m-1)/2}}{|\Sigma|^{\nu/2}} \exp\{-\tfrac{1}{2}\operatorname{tr}\nu\Sigma^{-1}S\}, \end{aligned} \tag{B.69}$$

where

$$c_1^{-1} = \frac{\pi^{[m(m-1)]/4}}{(\nu/2)^{\nu m/2}} \prod_{i=1}^m \Gamma\left(\frac{\nu - i + 1}{2}\right).$$

The pdf in (B.69) is $W[(1/\nu)\Sigma, \nu, m]$, as was to be shown. By a simple change of variables the pdf for $A = \nu S$ can be obtained from (B.69) and is $W(\Sigma, \nu, m)$, the pdf given in (B.55).[23]

[21] Cf. Roy, *op. cit.*, p. 197.

[22] The Jacobian of the transformation from the elements of T to the distinct elements of S is $\nu^{m(m+1)/2} \div (2^m \prod_{i=1}^m t_{ii}^{m-i+1})$. Also, in the second line of (B.69), from $\nu S = TT'$, we use $|T| = |\nu S|^{1/2} = \nu^{m/2}|S|^{1/2}$.

[23] Note that we have defined $\tilde{A} = \tilde{Z}\tilde{Z}' = \nu\tilde{S}$, with the ν columns of $\tilde{Z}$ assumed to be independent. Frequently we have an $n \times m$ random matrix $\hat{U} = [I_n - \iota(\iota'\iota)^{-1}\iota']\tilde{Y}$, where $\iota' = (1\ldots1)$, a $1 \times n$ vector of ones and $\tilde{Y} = (\tilde{y}_1, \tilde{y}_2, \ldots, \tilde{y}_m)$, with each column of $\tilde{Y}$ an $n \times 1$ vector. The *rows* of $\tilde{Y}$ are assumed to be independently and normally distributed, each with a $1 \times m$ mean vector μ' and common $m \times m$ PDS covariance matrix Σ. Although the rows of $\tilde{Y}$ are independent, the rows of the residual matrix $\hat{U}$ are not. Writing $\tilde{Y} = \iota\mu' + \tilde{U}$, where $\tilde{U}$ is an $n \times m$ matrix whose rows are independently and normally distributed, each with zero mean vector and common PDS covariance matrix Σ, we have $\hat{U} = (I_n - \iota(\iota'\iota)^{-1}\iota')\tilde{U}$ and $\hat{U}'\hat{U} = \tilde{U}'[I_n - \iota(\iota'\iota)^{-1}\iota']\tilde{U}$,

The formulas for the moments, (B.56) to (B.58), have been obtained in the literature[24] from those of the elements of $\tilde{Z}\tilde{Z}'$, since $\tilde{A} = \tilde{Z}\tilde{Z}'$, with $\tilde{Z}$ having the MN pdf in (B.61):

$$E\tilde{a}_{ij} = E\sum_{\alpha=1}^{\nu} \tilde{z}_{\alpha i}\tilde{z}_{\alpha j} = \nu\sigma_{ij}$$

and

$$\begin{aligned} E\tilde{a}_{ij}\tilde{a}_{kl} &= E\left(\sum_{\alpha=1}^{\nu} \tilde{z}_{\alpha i}\tilde{z}_{\alpha j}\right)\left(\sum_{\alpha'=1}^{\nu} \tilde{z}_{\alpha' k}\tilde{z}_{\alpha' l}\right) \\ &= E\sum_{\alpha=1}^{\nu} \tilde{z}_{\alpha i}\tilde{z}_{\alpha j}\tilde{z}_{\alpha k}\tilde{z}_{\alpha l} + E\sum_{\substack{\alpha,\alpha'=1 \\ \alpha\neq\alpha'}}^{\nu} \tilde{z}_{\alpha i}\tilde{z}_{\alpha j}\tilde{z}_{\alpha' k}\tilde{z}_{\alpha' l} \\ &= \nu(\sigma_{ij}\sigma_{kl} + \sigma_{ik}\sigma_{jl} + \sigma_{il}\sigma_{jk}) + \nu(\nu - 1)\sigma_{ij}\sigma_{kl} \\ &= \nu^2\sigma_{ij}\sigma_{kl} + \nu\sigma_{ik}\sigma_{jl} + \nu\sigma_{il}\sigma_{jk}. \end{aligned}$$

Then

$$\begin{aligned} \operatorname{Cov}(\tilde{a}_{ij}, \tilde{a}_{kl}) &= E(\tilde{a}_{ij} - \nu\sigma_{ij})(\tilde{a}_{kl} - \nu\sigma_{kl}) \\ &= \nu(\sigma_{ik}\sigma_{jl} + \sigma_{il}\sigma_{jk}), \end{aligned}$$

and, for $i = k$ and $j = l$,

$$\operatorname{Var}(\tilde{a}_{ij}) = \nu(\sigma_{ii}\sigma_{jj} + \sigma_{ij}^2).$$

Property 3 is most easily shown by partitioning $\tilde{Z}' = (\tilde{Z}_1'\tilde{Z}_2')$ where $\tilde{Z}_1$ is an $m_1 \times \nu$ random matrix with columns independently and normally distributed with zero vector mean and $m_1 \times m_1$ PDS covariance matrix Σ_{11}. Then, using (1), $\tilde{A}_{11} = \tilde{Z}_1\tilde{Z}_1'$ has a $W(\Sigma_{11}, \nu, m_1)$ pdf. In the case that $m_1 = 1$, $W(\sigma_{11}, \nu, 1)$ is in the form of a univariate gamma (G) pdf, which, of course, can be transformed to a χ^2 pdf. Thus the W pdf can be viewed as a multivariate generalization of the univariate G pdf.

To prove Property 4 let us write $\tilde{A} = \tilde{V}'\tilde{V}$, where $\tilde{V}$ is an $\nu \times m (m \leq \nu)$ random matrix with *rows* independently and normally distributed, each with

---

where $I_n - \iota(\iota'\iota)^{-1}\iota'$ is idempotent, with rank $n - 1$. Now let $\tilde{U} = L\tilde{V}$, where L is an $n \times n$ nonsingular orthogonal matrix such that

$$L'[I_n - \iota(\iota'\iota)^{-1}\iota']L = \left(\begin{array}{c|c} I_{n-1} & \mathbf{0} \\ \hline \mathbf{0}' & 0 \end{array}\right).$$

Then $\tilde{U}'\tilde{U} = \tilde{V}'L'[I_n - \iota(\iota'\iota)^{-1}\iota']L\tilde{V} = \tilde{V}_1'\tilde{V}_1$, where $\tilde{V}' = (\tilde{V}_1' \vdots \tilde{\mathbf{v}}_n)$; that is, $\tilde{V}_1$ is an $(n - 1) \times m$ matrix formed from $\tilde{V}$ by deleting the last row. Since the $n - 1$ rows of $\tilde{V}_1$ are independently and normally distributed, each with zero vector mean and common covariance matrix Σ (Note: $E\tilde{V}'\tilde{V} = E\tilde{U}'L'L\tilde{U} = E\tilde{U}'\tilde{U} = \Sigma \otimes I_n$), it satisfies the conditions of Property 1. Thus $\tilde{U}'\tilde{U} = \tilde{V}_1'\tilde{V}_1$ has a $W(\Sigma, \nu, m)$ pdf with $\nu = n - 1$.

[24] See, for example, T. W. Anderson, *An Introduction to Multivariate Statistical Analysis*. New York: Wiley, 1958, p. 161. The fourth-order moments required in the derivation are given on p. 39 of Anderson's book.

zero vector mean and common $m \times m$ PDS covariance matrix Σ. Then, partitioning $\tilde{V} = (\tilde{V}_1 \vdots \tilde{V}_2)$, we have

$$\tilde{A} = \tilde{V}'\tilde{V} = \begin{pmatrix} \tilde{V}_1'\tilde{V}_1 & \tilde{V}_1'\tilde{V}_2 \\ \tilde{V}_2'\tilde{V}_1 & \tilde{V}_2'\tilde{V}_2 \end{pmatrix} = \begin{pmatrix} \tilde{A}_{11} & \tilde{A}_{12} \\ \tilde{A}_{21} & \tilde{A}_{22} \end{pmatrix},$$

where $\tilde{V}_1'\tilde{V}_1$ and $\tilde{V}_2'\tilde{V}_2$ are of size $m_1 \times m_1$ and $m_2 \times m_2$, respectively, where $m_1 + m_2 = m$. Then

$$\begin{aligned} \tilde{A}_{11\cdot 2} &= \tilde{A}_{11} - \tilde{A}_{12}\tilde{A}_{22}^{-1}\tilde{A}_{21} \\ &= \tilde{V}_1'\tilde{V}_1 - \tilde{V}_1'\tilde{V}_2(\tilde{V}_2'\tilde{V}_2)^{-1}\tilde{V}_2'\tilde{V}_1 \\ &= \tilde{V}_1'[I_\nu - \tilde{V}_2(\tilde{V}_2'\tilde{V}_2)^{-1}\tilde{V}_2']\tilde{V}_1. \end{aligned} \tag{B.70}$$

Now, given $\tilde{V}_2$, if we let $\tilde{V}_1 = L\tilde{Z}_1$, where L is a $\nu \times \nu$ orthogonal matrix such that[25]

$$L'[I_\nu - \tilde{V}_2(\tilde{V}_2'\tilde{V}_2)^{-1}\tilde{V}_2']L = \begin{bmatrix} I_{\nu-m_2} & 0 \\ 0 & 0 \end{bmatrix},$$

we have from the third line of (B.70)

$$\begin{aligned} \tilde{A}_{11\cdot 2} &= \tilde{Z}_1'L'[I_\nu - \tilde{V}_2(\tilde{V}_2'\tilde{V}_2)^{-1}\tilde{V}_2']L\tilde{Z}_1 \\ &= \tilde{Z}_{1a}'\tilde{Z}_{1a}, \end{aligned}$$

where $\tilde{Z}_{1a}$ is a $\nu - m_2 \times m_1$ submatrix of $\tilde{Z}_1$; that is, $\tilde{Z}_1' = (\tilde{Z}_{1a}'\tilde{Z}_{1b}')$. Therefore $\tilde{A}_{11\cdot 2}$ can be expressed as $\tilde{Z}_{1a}'\tilde{Z}_{1a}$, where the rows of $\tilde{Z}_{1a}$ are independently and normally distributed, each with zero mean vector and covariance matrix $\Sigma_{11\cdot 2} = \Sigma_{11} - \Sigma_{12}\Sigma_{22}^{-1}\Sigma_{21}$.[26] Thus, using Property 1, $\tilde{A}_{11\cdot 2}$ has a $W(\Sigma_{11\cdot 2}, \nu - m_2, m_1)$ pdf, where $m - m_1 = m_2$.

Property 5 is derived from the pdf for $\tilde{A}_{11} = \{\tilde{a}_{ij}\}$ $i, j = 1, 2$; that is, $p(a_{11}, a_{22}, a_{12}|\Sigma_{11}, \nu)$, which is $W(\Sigma_{11}, 2, \nu)$, by expressing it in terms of a_{11}, a_{22}, and $r = a_{12}/(a_{11}a_{22})^{1/2}$ and integrating out a_{11} and a_{22}.[27]

[25] For given $\tilde{V}_2$, $I_\nu - \tilde{V}_2(\tilde{V}_2'\tilde{V}_2)^{-1}\tilde{V}_2'$ is an idempotent matrix of rank $\nu - m_2$. Thus it has $\nu - m_2$ roots equal to 1 and m_2 equal to 0.

[26] From $\tilde{V}_1 = L\tilde{Z}_1$, $E\tilde{Z}_1 = 0$. Further, given $\tilde{V}_2$, $E\tilde{Z}_1'L'L\tilde{Z}_1 = E\tilde{Z}_1'\tilde{Z}_1 = E\tilde{V}_1'\tilde{V}_1 = \Sigma_{11\cdot 2} \otimes I_\nu$, where $\Sigma_{11\cdot 2}$ is the covariance matrix of the m_1 elements of any row of $\tilde{V}_1$, given the m_2 elements in the corresponding row of $\tilde{V}_2$.

[27] See, for example, T. W. Anderson, *op. cit.*, pp. 68–69, for details. Since the function $I_\nu(\rho r)$ in (5) can be expressed as a hypergeometric function, the pdf for r_{12} can be expressed as a rapidly converging series; that is

$$h(r_{12}|\rho_{12}, \nu) = \frac{(\nu - 1)\Gamma(\nu)}{\sqrt{2\pi}\,\Gamma(\nu + \frac{1}{2})} \frac{(1 - r_{12}^2)^{(\nu-3)/2}(1 - \rho^2)^{\nu/2}}{(1 - r_{12}\rho)^{(2\nu-1)/2}} S_\nu(\rho r_{12}),$$

where $S_\nu(\rho r_{12})$ denotes the hypergeometric function, $F(\frac{1}{2}, \frac{1}{2}; \nu + \frac{1}{2}; (1 + r_{12}\rho)/2)$, which

B.4 THE INVERTED WISHART (IW) Pdf

The $m(m+1)/2$ distinct elements of an $m \times m$ PDS random matrix $\tilde{G}$ follow the IW distribution if, and only if, they have the following pdf[28]:

$$\text{(B.71)} \quad p(G|H, \nu, m) = k \frac{|H|^{\nu/2}}{|G|^{(\nu+m+1)/2}} \exp\{-\tfrac{1}{2} \operatorname{tr} G^{-1}H\}, \qquad |G| > 0,$$

where $k^{-1} = 2^{\nu m/2}\pi^{m(m-1)/4} \prod_{i=1}^{m} \Gamma[(\nu+1-i)/2]$, $\nu \geq m$, and H is an $m \times m$ PDS matrix. The IW pdf is defined as (B.71) in the region $|G| > 0$ and zero elsewhere. If the elements of $\tilde{G}$ have the pdf (B.71), we say that they have an IW(H, ν, m) pdf. Some properties of (B.71) follow:

1. The joint pdf for the $m(m+1)/2$ distinct elements of $G^{-1} = A$ is a W(H^{-1}, ν, m) pdf.
2. Let G and H be partitioned correspondingly as

$$G = \begin{matrix} & m_1 & m_2 \\ m_1 & G_{11} & G_{12} \\ m_2 & G_{21} & G_{22} \end{matrix}, \qquad H = \begin{pmatrix} H_{11} & H_{12} \\ H_{21} & H_{22} \end{pmatrix},$$

where $m_1 + m_2 = m$. Then the joint pdf for the $m_1(m_1+1)/2$ distinct elements of G_{11} is an IW($H_{11}, \nu - m_2, m_1$) pdf.

3. When in (2) G_{11} is a scalar, say g_{11}, the pdf for g_{11} is[29]

$$\text{(B.72)} \qquad p(g_{11}|h_{11}, \nu', m_1) = \frac{k_1}{g_{11}^{(\nu'+2)/2}} e^{-h_{11}/2g_{11}}, \qquad 0 < g_{11},$$

with $k_1 = (h_{11}/2)^{\nu'/2}/\Gamma(\nu'/2)$, with $\nu' = \nu - m + 1$.

4. By virtue of (B.72) the moments of the diagonal elements of $\tilde{G}$ can be obtained from those associated with a univariate inverted gamma pdf.

Property 1 is a fundamental one connecting the W and IW pdf's. To establish it we require the Jacobian of the transformation of the $m(m+1)/2$ distinct elements of G to the $m(m+1)/2$ distinct elements of $A = G^{-1}$. The

---

is given by

$$S_\nu(\rho r_{12}) = \frac{\Gamma(\nu + \frac{1}{2})}{[\Gamma(\frac{1}{2})]^2} \sum_{i=0}^{\infty} \frac{[\Gamma(\frac{1}{2}+i)]^2}{i!\,\Gamma(\nu+\frac{1}{2}+i)} \left(\frac{1+r_{12}\rho}{2}\right)^i$$
$$= 1 + \frac{1}{\nu+\frac{1}{2}} \frac{1+r_{12}\rho}{8} + \frac{1^2 \cdot 3^2}{2!\,(\nu+\frac{1}{2})(\nu+\frac{3}{2})} \left(\frac{1+r_{12}\rho}{8}\right)^2 + \cdots.$$

See, for example, H. Jeffreys, *Theory of Probability* (3rd ed.), Oxford: Clarendon, 1961, p. 175, for the details of expressing $I_\nu(\rho r)$ in terms of a hypergeometric function.

[28] In the analysis of the multivariate regression model with a diffuse prior pdf in Chapter 8 we found that the posterior pdf for the disturbance covariance matrix is given by $p(\Sigma|\mathbf{y}) \propto |\Sigma|^{-\nu'/2} \exp(-\frac{1}{2} \operatorname{tr} \Sigma^{-1}S)$. This is in the form of (B.71), as can be seen by letting $G = \Sigma$, $H = S$, and $\nu' = \nu + m + 1$. Thus $p(\Sigma|\mathbf{y})$ is an IW pdf.

[29] Equation B.72 can be obtained from (A.37b) by letting $g_{11} = \sigma^2$ and $h_{11} = \nu s^2$. Of course, the positive square root of g_{11} will have a pdf precisely in the form of (A.37b).

Jacobian of the transformation is $|A|^{-(m+1)}$[30] and thus (B.71) can be expressed in terms of $A = G^{-1}$:

$$\text{(B.73)}\quad \begin{aligned} p(A|H, \nu, m) &= k\frac{|H|^{\nu/2}|A|^{-(m+1)}}{|A|^{-(\nu+m+1)/2}}\exp\{-\tfrac{1}{2}\operatorname{tr} AH\}, \qquad |A| > 0. \\ &= k|H|^{\nu/2}|A|^{(\nu-m-1)/2}\exp\{-\tfrac{1}{2}\operatorname{tr} AH\}, \qquad |A| > 0, \end{aligned}$$

with k given in connection with (B.71). If, in (B.73), we define $\Sigma^{-1} = H$, it is seen that (B.73) is in precisely the form of the $W(\Sigma, \nu, m)$ pdf shown in (B.55).

Property 2 can be established by noting that $G_{11} = (A_{11} - A_{12}A_{22}^{-1}A_{21})^{-1} = A_{11\cdot 2}^{-1}$. As shown in the preceding section, if A has a W pdf, $A_{11\cdot 2}$ also has a W pdf. Then Property 1 of the IW pdf can be employed to obtain the pdf for $G_{11} = A_{11\cdot 2}^{-1}$ from that for $A_{11\cdot 2}$. Given this result, the special case in which G_{11} is a scalar, g_{11}, leads to the result in (B.72) which is Property 3. Property 4 is an immediate consequence of Property 3.

B.5 THE GENERALIZED STUDENT *t* (GSt) Pdf[31]

The pq elements of a $p \times q$ random matrix, $\tilde{T} = \{\tilde{t}_{ij}\}$, have the GSt distribution if, and only if, they have the following pdf[32]:

$$\text{(B.74)}\qquad p(T|P, Q, n) = k\frac{|Q|^{(n-p)/2}|P|^{q/2}}{|Q + T'PT|^{n/2}}, \qquad -\infty < t_{ij} < \infty,$$

where $k^{-1} = \pi^{pq/2}\prod_{i=1}^{q}\Gamma[(n-p-i+1)/2]/\prod_{i=1}^{q}\Gamma[(n-i+1)/2]$, $n > p + q - 1$, and P and Q are PDS matrices of sizes $p \times p$ and $q \times q$, respectively. For convenience we denote the pdf in (B.74) as $T(P, Q, 0, n)$, where

[30] To show that the Jacobian is $|A|^{-(m+1)}$ write $AG = I$. Then $(\partial A/\partial\theta)G + A(\partial G/\partial\theta) = 0$ or $\partial G/\partial\theta = -G(\partial A/\partial\theta)G$. If $\theta = a_{ij}$, we have $\partial g_{\alpha\beta}/\partial a_{ij} = -g_{\alpha i}g_{\beta j}$ for α, β, i, and $j = 1, 2, \ldots, m$, with $\beta \le \alpha$ and $j \le i$, since G and A are symmetric matrices and the transformation from the elements of G to those of A involves just $m(m+1)/2$ distinct elements of G. On forming the Jacobian matrix and taking its determinant, we have $|G|^{m+1} = |A|^{-(m+1)}$. See T. W. Anderson, *op. cit.*, p. 162, for a derivation of this Jacobian which relies on properties of the W pdf. Also on pp. 348 to 349 Anderson provides the Jacobian of the transformation $G \to A = G^{-1}$ for the general case in which G and A are not symmetric, a result not applicable to the present case in which G and A are symmetric.

[31] Some call this pdf the matrix *t* pdf. See J. M. Dickey, "Matricvariate Generalizations of the Multivariate *t* Distribution and the Inverted Multivariate *t* Distribution," *Ann. Math. Statistics*, **38**, 511–518 (1967); S. Geisser, "Bayesian Estimation in Multivariate Regression," *cit. supra*; G. C. Tiao and A. Zellner, "On the Bayesian Estimation of Multivariate Regression," *cit. supra*; and the references cited in these works for further analysis of this pdf.

[32] The pdf in (B.74) was encountered in Chapter 8 in connection with the analysis of the multivariate regression model. With a diffuse prior pdf the posterior pdf for the regression coefficients was found to be $p(B|Y) \propto |S + (B - \hat{B})'X'X(B - \hat{B})|^{-n/2}$. If we let $S = Q$, $P = X'X$, and $T = B - \hat{B}$, $p(B|Y)$ is exactly in the form of (B.74).

0 appears to denote that the mean of (B.74), by symmetry, is a zero matrix. Some properties of (B.74) follow:

1. The pdf in (B.74) can be obtained as the marginal distribution of $p(G, T) = p_1(G)\,p_2(T|G)$, where $p_1(G)$ denotes an inverted Wishart pdf and $p_2(T|G)$ denotes a multivariate normal pdf.

Let

$$Q = \begin{matrix} & q_1 & q_2 \\ q_1 & Q_{11} & Q_{12} \\ q_2 & Q_{21} & Q_{22} \end{matrix} \quad \text{and} \quad P = \begin{matrix} & p_1 & p_2 \\ p_1 & P_{11} & P_{12} \\ p_2 & P_{21} & P_{22} \end{matrix},$$

with $q_1 + q_2 = q$ and $p_1 + p_2 = p$ and

$$Q_{11\cdot 2} = Q_{11} - Q_{12}Q_{22}^{-1}Q_{21}, \qquad P_{22\cdot 1} = P_{22} - P_{21}P_{11}^{-1}P_{12}.$$

These quantities appear in the properties of (B.74).[33]

2. If $T = (T_1, T_2)$, the conditional pdf for T_1, given T_2, is a GSt pdf with parameters $(P^{-1} + T_2Q_{22}^{-1}T_2')^{-1}$, $Q_{11\cdot 2}$, $T_2Q_{22}^{-1}Q_{21}$, n. The mean is $T_2Q_{22}^{-1}Q_{21}$.

3. If $T' = (X_1, X_2)$, the conditional pdf for X_1, given X_2, is a GSt pdf with parameters P_{11}, $Q + X_2'P_{22\cdot 1}X_2$, $P_{11}^{-1}P_{12}X_2$, n. The mean is $P_{11}^{-1}P_{12}X_2$.

4. If $T = (T_1, T_2)$, with $T_1 p \times q_1$ and $T_2 p \times q_2$, the marginal pdf for T_2 is GSt with parameters P, Q_{22}, 0, $n - q_1$.

5. If $T' = (X_1, X_2)$, with $X_1 p_1 \times q$ and $X_2 p_2 \times q$, the marginal pdf for X_2 is GSt, with parameters $P_{22\cdot 1}$, Q, 0, $m - p_1$.

6. If in (2) T_1 is a $p \times 1$ vector, the conditional pdf for T_1, given T_2, is in the multivariate Student t form. Similarly, if in (4) T_2 is a $p \times 1$ vector, it has a marginal pdf in the multivariate Student t form.

7. With $T = (\mathbf{t}_1, \mathbf{t}_2, \ldots, \mathbf{t}_q)$,

$$p(T) = p(\mathbf{t}_1)\,p(\mathbf{t}_2|\mathbf{t}_1)\,p(\mathbf{t}_3|\mathbf{t}_1, \mathbf{t}_2)\cdots p(\mathbf{t}_q|\mathbf{t}_1, \ldots, \mathbf{t}_{q-1}) \tag{B.75}$$

and each of the pdf's on the rhs of (B.75) is in the form of a multivariate Student t pdf.

To establish Property 1 write the IW pdf for the $q(q + 1)/2$ distinct elements of G as

$$\text{IW}(G|Q, \nu, p) = k_1 \frac{|Q|^{\nu/2}}{|G|^{(\nu+q+1)/2}} \exp\{-\tfrac{1}{2}\operatorname{tr} G^{-1}Q\}, \tag{B.76}$$

[33] Some of the following properties established in Chapter 8 draw on the results in papers by J. M. Dickey, S. Geisser, and G. C. Tiao and A. Zellner, cited in the preceding footnote.

where Q is a $q \times q$ PDS matrix and k_1 is the normalizing constant, and the multivariate normal pdf for the elements of T, a $p \times q$ matrix, given G, as[34]

$$\text{(B.77)} \qquad \text{MN}(T|G, P) = k_2|P|^{q/2}|G|^{-p/2} \exp\{-\tfrac{1}{2} \operatorname{tr} T'PTG^{-1}\},$$

where P is a $p \times p$ PDS matrix and k_2 is a normalizing constant. Then the joint pdf for the distinct elements of G and those of T, $p(G, T)$, is the product of (B.76) and (B.77); that is,

$$\text{(B.78)} \qquad p(G, T) = k_1 k_2 \frac{|Q|^{\nu/2}|P|^{q/2}}{|G|^{(\nu+p+q+1)/2}} \exp\{-\tfrac{1}{2} \operatorname{tr}(Q + T'PT)G^{-1}\}.$$

Now note from the properties of the IW pdf that

$$\text{(B.79)} \quad \int_{|G|>0} \frac{|Q + T'PT|^{(\nu+p)/2}}{|G|^{(\nu+p+q+1)/2}} \exp\{-\tfrac{1}{2} \operatorname{tr}(Q + T'PT)G^{-1}\}\, dG = \frac{1}{k_3},$$

where k_3 is the normalizing constant of the IW pdf, $p(G|Q + T'PT, \nu + p, q)$. Using (B.79), we find that the integration of (B.78) with respect to the elements of G over the region $|G| > 0$ yields

$$\text{(B.80)} \qquad p(T) = \frac{k_1 k_2}{k_3} \frac{|Q|^{\nu/2}|P|^{q/2}}{|Q + T'PT|^{(\nu+p)/2}}, \qquad -\infty < t_{ij} < \infty.$$

If we let $\nu = n - p$, we see that (B.80) is in precisely the form of (B.74), a GSt pdf.[35]

Property 2 is most easily established if we note that the GSt pdf can be written in the following alternative form[36]:

$$\text{(B.81)} \qquad p(T|P, Q, n) = k \frac{|P|^{-[(n-q)/2]}|Q|^{-p/2}}{|P^{-1} + TQ^{-1}T'|^{n/2}}, \qquad -\infty < t_{ij} < \infty.$$

Then

$$\begin{aligned}
P^{-1} + TQ^{-1}T' &= P^{-1} + (T_1 T_2)\begin{pmatrix} Q^{11} & Q^{12} \\ Q^{21} & Q^{22} \end{pmatrix}\begin{pmatrix} T_1' \\ T_2' \end{pmatrix} \\
&= P^{-1} + T_1 Q^{11} T_1' + T_2 Q^{21} T_1' + T_1 Q^{12} T_2' + T_2 Q^{22} T_2' \\
&= P^{-1} + T_2[Q^{22} - Q^{21}(Q^{11})^{-1}Q^{12}]T_2' \\
&\quad + [T_1 + T_2 Q^{21}(Q^{11})^{-1}]Q^{11}[T_1 + T_2 Q^{21}(Q^{11})^{-1}]' \\
&= P^{-1} + T_2 Q_{22}^{-1} T_2' + (T_1 - T_2 Q_{22}^{-1} Q_{21}) Q_{11\cdot 2}^{-1} \\
&\quad \times (T_1 - T_2 Q_{22}^{-1} Q_{21})',
\end{aligned}$$

[34] Note that with $T = (\mathbf{t}_1 \cdots \mathbf{t}_q)$, we can write the MN pdf for the elements of T, given G as $\text{MN}(T|G, P) = k_2|G^{-1} \otimes P|^{1/2} \exp[-\tfrac{1}{2}(\mathbf{t}_1', \mathbf{t}_2', \ldots, \mathbf{t}_q')G^{-1} \otimes P(\mathbf{t}_1', \mathbf{t}_2', \cdots, \mathbf{t}_q')']$ and that $|G^{-1} \otimes P|^{1/2} = |P|^{q/2}|G|^{-p/2}$; see, for example, T. W. Anderson, *op. cit.*, p. 348.

[35] We could transform (B.76) to a Wishart pdf for $A = G^{-1}$ and obtain (B.80) as the marginal pdf of a Wishart pdf times a conditional MN pdf for T given A.

[36] See J. M. Dickey, *op. cit.*, p. 512.

where Q^{ij}, $i,j = 1,2$, are submatrices of Q^{-1} and $Q^{21}(Q^{11})^{-1} = -Q_{22}^{-1}Q_{21}$. On substituting this result in (B.81), we have

$$p(T_1, T_2|Q, P, n) = k \times \frac{|P|^{-(n-q)/2}|Q|^{-p/2}}{|P^{-1} + T_2Q_{22}^{-1}T_2' + (T_1 - T_2Q_{22}^{-1}Q_{21})Q_{11\cdot 2}^{-1} \times (T_1 - T_2Q_{22}^{-1}Q_{21})'|^{n/2}} \tag{B.82}$$

From (B.82) it is apparent that the conditional pdf for T_1, given T_2, is in the GSt form (B.81) with parameters $(P^{-1} + T_2Q_{22}^{-1}T_2')^{-1}$, $Q_{11\cdot 2}$, and $T_2Q_{22}^{-1}Q_{21}$, n, where the conditional mean is given by $T_2Q_{22}^{-1}Q_{21}$. Further, the marginal pdf for T_2 is obtained from (B.82) by integrating with respect to the elements of T_1. Note that (B.82) can be expressed as

$$p(T_1, T_2|Q, p, n) \propto [|P^{-1} + T_2Q_{22}^{-1}T_2'|^{(n-q_1)/2}]^{-1} \times \frac{|P^{-1} + T_2Q_{22}^{-1}T_2'|^{(n-q_1)/2}|Q_{11\cdot 2}|^{-p/2}}{|P^{-1} + T_2Q_{22}^{-1}T_2' + (T_1 - T_2Q_{22}^{-1}Q_{21})Q_{11\cdot 2}^{-1} \times (T_1 - T_2Q_{22}^{-1}Q_{21})'|^{n/2}} \tag{B.83}$$

and that on integrating with respect to the elements of T_1 the second factor integrates to a numerical factor. Thus the marginal pdf for the elements of T_2 is proportional to the first factor on the rhs of (B.83) which is in the GSt form with parameters P, Q_{22}, 0, $n - q_1$. This is Property 4.

Properties 3 and 5 can be established in the same way as Properties 2 and 4. However, in proof of the former two properties, it is convenient to use the form of the GSt pdf in (B.74). Property 6 follows from Property 4. Also, it and Property 7 have been demonstrated in the text of Chapter 8.

APPENDIX C[1]

FORTRAN Programs for Numerical Integration

Frequently in statistics and other branches of applied mathematics we encounter definite integrals which are either difficult or impossible to integrate analytically. When this happens either of two procedures may be used. First, the integrand may be approximated by a series expansion and integration done termwise or, second, the integration may be done numerically by the use of the trapezoidal rule, Simpson's rule, or gaussian quadrature.

The purpose of this writeup is to explain FORTRAN programs for numerical integration of univariate and bivariate integrals that use Simpson's rule. Simpson's rule is chosen because it combines reasonable accuracy with simplicity.

C.1 UNIVARIATE INTEGRATION

General

Let $f(x)$ be a function of x defined, continuous, and non-negative on the closed interval $a \leq x \leq b$. The object of discussion is

$$A = \int_a^b f(x)\, dx;$$

that is, the area under the curve $f(x)$ and above the x-axis between a and b.

We may transform $f(x)$ into a proper density function by

$$p(x) = \frac{1}{A} f(x).$$

Then

$$\int_a^b p(x)\, dx = \int_a^b \frac{1}{A} f(x)\, dx = 1;$$

[1] This appendix was prepared by Martin S. Geisel.

A is called the *normalizing constant for* f and is one of the quantities we may wish to find by numerical integration. We may also wish to know various moments of the distribution and the mathematical expectations of functions of x; for example,

$$\mu_1' = \int_a^b x\, p(x)\, dx \qquad \text{(mean)},$$

$$\mu_2 = \int_a^b (x - \mu_1')^2\, p(x)\, dx \qquad \text{(variance)}.$$

The procedure described below is applicable to all these problems.

Simpson's Rule

Divide $[a, b]$ into n equal parts, where n is an even integer. Label the end-points of the subintervals $x_0, x_1, x_2, \ldots, x_{n-1}, x_n$ and let $y_i = f(x_i)$, $i = 0, 1, 2, \ldots, n$. Consider the first three points x_0, x_1, x_2 and the corresponding points on the curve $y = f(x)$. If the points on the curve are not collinear, there is a unique parabola $g(x)$ with axis parallel to the y-axis which passes through the three points. Write its equation $g(x) = a + b(x - x_1) + c(x - x_1)^2$, where a, b, and c are chosen so that the three points lie on the parabola. Let $\Delta x = x_i - x_{i-1}$ $i = 1, 2, \ldots, n$. Then, if y_0, y_1, y_2 lie on the parabola, we have

$$\begin{aligned} y_0 &= a - b\,\Delta x + c(\Delta x)^2, \\ y_1 &= a, \\ y_2 &= a + b\,\Delta x + c(\Delta x)^2 \end{aligned}$$

or

$$\begin{aligned} b\,\Delta x + c(\Delta x)^2 &= y_2 - y_1, \\ -b\,\Delta x + c(\Delta x)^2 &= y_0 - y_1, \\ 2c(\Delta x)^2 &= y_0 - 2y_1 + y_2. \end{aligned}$$

Using the parabola as an approximation to $f(x)$ in $[x_0, x_2]$, we obtain

$$\begin{aligned} \int_{x_0}^{x_2} f(x)\, dx &\doteq \int_{x_0}^{x_2} [a + b(x - x_1) + c(x - x_1)^2]\, dx \\ &= [ax + \tfrac{1}{2}b(x - x_1)^2 + \tfrac{1}{3}c(x - x_1)^3]\Big|_{x_0}^{x_2} \\ &= 2a\,\Delta x + \tfrac{2}{3}c(\Delta x)^3 \\ &= \frac{\Delta x}{3}(y_0 + 4y_1 + y_2), \end{aligned}$$

wherein the values for a and c, shown above, have been used.

Repeat the procedure for $[x_2, x_4], [x_4, x_6], \ldots, [x_{n-2}, x_n]$ and sum the results to get

$$\int_a^b f(x)\,dx \doteq \frac{\Delta x}{3}(y_0 + 4y_1 + 2y_2 + \cdots + 2y_{n-2} + 4y_{n-1} + y_n).$$

This is Simpson's rule.

FORTRAN Programs for Simpson's Rule

1. *Simpson's Rule I*

In this program the user specifies the function to be integrated by means of a function-defining statement. He enters the limits of integration on a data card along with the value of a parameter which we call TOL. The program then computes a first approximation to the integral and then does Simpson's rule for 2, 4, 8, 16, . . . subintervals (at each step, only the functional values corresponding to the new x-values are actually computed). After a specified number of subintervals has been reached (16 in the example below) the result from the current computation of Simpson's rule is compared with the result of the preceding computation. If the absolute difference between the two is less than the value of TOL, computation stops and the answer is printed out. If not, the number of subintervals is doubled and Simpson's rule is computed again and the procedure is repeated. If the user is not certain of the limits of integration (e.g., he is approximating $(-\infty, \infty)$ by a finite interval), he simply enters another data card which specifies the new limits of integration.

The obvious advantage to Simpson's Rule I is that it provides a good idea of the accuracy of the results for most functions. Also, the ease of changing the limits of integration is desirable at times. Furthermore, it requires very little programming effort on the part of the user. On the other hand, use of the function-defining statement makes it difficult to change the function within a given run of the program (parameters of the function may, of course, be changed as shown in the example below). Also, computer time adds up rapidly as the number of subintervals increases. Finally, this procedure does not extend easily to bivariate or multivariate integration.

Simpson's Rule I is illustrated in the computation of the area under several normal densities. Program statements are shown in Table C.1.

1. After the user's ID card comes the XEQ card and the DIMENSION statement in which all arrays are declared.

2. The next statement, which must appear before any executable statements, is the function-defining statement. The following rules are pertinent to construction of function-defining statements.

(a) The name of the function must have four to seven characters, the last of which is F and the first of which is a letter other than X. The name must not be the same as any of the computer's built-in functions.
(b) There may be as many arguments as the user wishes but they may not be subscripted in this statement.
(c) The following are functions on most computers that may be useful:
SQRTF(X) = $\sqrt{x}$
LOGF(X) = $\log_e x$
EXPF(X) = e^x
NEXPF(X) = e^{-x}
ABSF(X) = $|x|$
GAMMAF(X) = $\Gamma(x)$
LGAMAF(X) = $\log_e \Gamma(x)$
FLOATF(I) converts a number without a decimal to one with a decimal.

Consult a manual for the particular computer that is being used for information regarding limitations on the arguments for these functions.

3. *Statements 1 to 4.* Statement 1 instructs the computer to print the heading in Statement 2. Statement 3 instructs the computer to read values of the variables listed from a data card. They are punched according to the FORMAT of Statement 4.

4. *Statements 5 to 8.* These statements define the values of the arguments (other than X) of FUNCF.

5. *Statement 9.* This statement calculates the width of the subinterval for the first iteration.

6. *Statements 10 to 15.* These statements perform the initial computations for Simpson's rule.

7. *Statements 16 to 20.* These statements finish the first computation of Simpson's rule.

8. *Statements 21 and 22.* Statement 21 doubles the number of subintervals and Statement 22 halves the width of the subinterval.

9. *Statement 23.* This statement instructs the computer to repeat the above if (the new value of) N is less than 16 and to go on if N is greater than or equal to 16.

10. *Statements 24 to 26.* These statements compute Simpson's rule again.

11. *Statement 27.* This statement tells the computer to continue to Statement 28 if N is less than or equal to 4000 but to go to Statement 39 if N is greater than 4000. This sets a limit on the amount of computation to be done.

12. *Statement 28.* This statement computes the current result of Simpson's rule (actually, three times the result).

13. *Statement 29.* In this statement the difference between the current result of the Simpson's rule computation and the preceding result is compared

Table C.1 SIMPSON'S RULE I: EXAMPLE FOR NORMAL DENSITIES

```
1 2 3 4 5 6 7 8 9 101 2 3 4 5 6 7 8 9 201 2 3 4 5 6 7 8 9 301 2 3 4 5 6 7 8 9 401 2 3 4 5 6 7 8 9 501 2 3 4 5 6 7 8 9 601 2 3 4 5 6 7 8 9 701 2 3 4 5

*       XEQ
        DIMENSION D(2)
        FUNCF(X,SIGMA,XBAR)=1./SQRTF(6.2832*SIGMA**2)*NEXPF((X-XBAR)**2/(2
       1.*SIGMA**2))
      1 WOT 6,2
      2 FORMAT(1H1,70H     XBAR        SIGMA         XMIN        F(XMIN)         XMAX        F(X
       1MAX)      N       AREA)
      3 RIT 5,4,XMIN,XMAX,TOL
      4 FORMAT(2F10.6,F8.7)
      5 DO 41 I=1,2
      6 SIGMA=I
      7 DO 41 J=1,2
      8 XBAR=2.*FLOATF(J-1)
      9 H=(XMAX-XMIN)/2.
     10 E=FUNCF(XMIN,SIGMA,XBAR)
     11 G=FUNCF(XMAX,SIGMA,XBAR)
     12 C=H*(E+G)
     13 D(2)=3.*C
     14 N=1
     15 S=0.
     16 DO 17 L=1,N
     17 S=S+FUNCF(XMIN+(FLOATF(2*L-1))*H,SIGMA,XBAR)
     18 D(1)=C+4.*H*S
     19 C=(C+D(1))/4.
     20 D(2)=D(1)
     21 N=2*N
     22 H=.5*H
     23 IF (N-16) 15,24,24
     24 S=0.0
     25 DO 26 L=1,N
     26 S=S+FUNCF(XMIN+(FLOATF(2*L-1))*H,SIGMA,XBAR)
     27 IF (N-4000) 28,28,39
     28 D(1)=C+4.*H*S
     29 IF(ABSF(D(1)-D(2))-3.*TOL) 35,35,30
     30 C=(C+D(1))/4.
     31 D(2)=D(1)
```

```
1 2 3 4 5 6 7 8 9 101 2 3 4 5 6 7 8 9 201 2 3 4 5 6 7 8 9 301 2 3 4 5 6 7 8 9 401 2 3 4 5 6 7 8 9 501 2 3 4 5 6 7 8 9 601 2 3 4 5 6 7 8 9 701 2 3 4 5
    32  N=2*N
    33  H=.5*H
    34  GO TO 24
    35  B=D(1)/3
    36  WOT 6,37,XBAR,SIGMA,XMIN,E,XMAX,G,N,B
    37  FORMAT(F8.2,F10.2,F8.3,F12.5,F8.3,F12.5,I6,F10.5)
    38  GO TO 41
    39  WOT 6,40,N,C,S,D(1)
    40  FORMAT(I10,3F20.5)
    41  CONTINUE
    42  GO TO 3
        END

*       DATA
-25.000000 25.000000.0001000
 25.000000 50.000000.0001000
```

Table C.2 SIMPSON'S RULE II: EXAMPLE FOR NORMAL DENSITIES

```
1 2 3 4 5 6 7 8 9 101 2 3 4 5 6 7 8 9 201 2 3 4 5 6 7 8 9 301 2 3 4 5 6 7 8 9 401 2 3 4 5 6 7 8 9 501 2 3 4 5 6 7 8 9 601 2 3 4 5 6 7 8 9 701 2 3 4 5

*         XEQ
          DIMENSION W(500), X(500), Y(500)
          COMMON W
     1    WOT 6,2
     2    FORMAT(1H1,50H         AREA      1ST MOMENT       2ND MOMENT      SIGMA    ZBAR)
     3    DO 20 L=1,2
     4    SIGMA=L
     5    DO 20 LL=1,2
     6    ZBAR=FLOATF(LL-1)*2.
     7    E=1./SQRTF(6.2832*SIGMA**2)
     8    DO 12 I=1,201
     9    Z=-30.+.3*FLOATF(I-1)
    10    W(I)=E*NEXPF((Z-ZBAR)**2/(2.*SIGMA**2))
    11    X(I)=E*NEXPF((Z-ZBAR)**2/(2.*SIGMA**2))*Z
    12    Y(I)=E*NEXPF((Z-ZBAR)**2/(2.*SIGMA**2))*Z**2
    13    AREA=FUNC1(30.,-30.,200)
    14    DO 15 I=1,201
    15    W(I)=X(I)
    16    FM=FUN C1(30.,-30.,200)
    17    DO 18I=1,201
    18    W(I)=Y(I)
    19    SM=FUN C1(30.,-30.,200)
    20    WOT 6,21,AREA,FM,SM,SIGMA,ZBAR
    21    FORMAT(F11.5,F14.5,F13.5,F7.3,F6.2)
```

```
1 2 3 4 5 6 7 8 9 101 2 3 4 5 6 7 8 9 201 2 3 4 5 6 7 8 9 301 2 3 4 5 6 7 8 9 401 2 3 4 5 6 7 8 9 501 2 3 4 5 6 7 8 9 601 2 3 4 5 6 7 8 9 701 2 3 4 5
      CALL EXIT
      END

      FUNCTION FUNC1 (UP, SL, MM)
      DIMENSION W(500)
      COMMON W
      G=FLOATF(MM)
      H=(UP-SL)/G
      N1=MM/2
      N2=N1-1
      Q1=0.0
      Q2=0.0
      MP=MM+1
      DO 27 I=1,N1
   27 Q1=Q1+W(2*I)
      DO 28 I=1,N2
   28 Q2=Q2+W(2*I+1)
      ORQ=(H/3.)*(W(1)+W(MP)+Q1*4.+Q2*2.)
      FUNC1=ORQ
      RETURN
      END
```

with TOL. If the difference is greater than TOL, the computer goes to Statement 30. If it is less than or equal to TOL, it goes to Statement 35.

14. *Statements 30 to 34.* These statements perform the initial computations for the next computation of Simpson's rule.

15. *Statements 35 to 38.* Statement 35 computes the final result and Statement 36 prints out the results according to the FORMAT of Statement 37. Statement 38 sends the computer to Statement 41, which is the last statement in the DO-loops started in Statements 5 and 7.

16. *Statements 39 to 40.* These statements print out the results (D(1) = three times the last result of Simpson's rule) in the event that N is greater than 4000.

17. *Statement 42.* This statement sends the computer back to Statement 3 which instructs the computer to read another data card and proceed. If there are no more data cards, computation stops.

18. *Data Cards.* These cards are punched according to FORMAT Statement 4. The second card provides a check on the computations done with the first data card by computing the tail area.

2. *Simpson's Rule II*

In this program the computation of Simpson's rule is performed by the subprogram FUNCTION FUNCl (UP, SL, MM). Whenever FUNCl is used, the user specifies the upper limit of integration (UP), the lower limit (SL), and the number of subintervals (MM). The user writes his own main program to compute the values of the integrand at the endpoints of all the subintervals and to store them in an array (called W in the example below). After the computation of the W array is completed the next statement is of the form ANSWER = FUNCl (., ., .). This is all that is required to perform the computation of Simpson's rule. If another integral is desired, the values of this integrand at the endpoints of its subintervals are computed and placed in the W array and FUNCl is used again. Naturally the values of UP, SL, and MM may be changed each time.

The main advantage of Simpson's Rule II is that several different integrals may be computed easily in one program. Also, as shown below, it can be extended easily to multivariate integrations. However, it requires more programming on the part of the user and within a single run provides no idea of the accuracy of the results.

Simpson's Rule II is illustrated in the computation of the area under several normal densities and its first and second moments. Program statements are shown in Table C.2.

1. After the XEQ card and the DIMENSION statement the statement COMMON W is included. It occupies a similar position in the FUNCl

subprogram. This tells the computer that W in the subprogram is the same array as in the main program.

2. *Statements 1 and 2.* Statement 1 prints out the title shown in FORMAT Statement 2.

3. *Statements 3 to 7.* Values of SIGMA and XBAR are computed here as well as the constant of the integrand.

4. *Statements 8 to 12.* A loop is set up to compute the various values of Z, the variable of integration, and to compute the various integrands. In Statement 9 we have picked the lower limit of integration Z = −30. and the width of the subinterval = .30. The upper index of the DO (Statement 8) tells us the number of subintervals (201 − 1 = 200). These facts imply that the upper limit of integration is +30.

5. *Statement 13.* FUNC1 is used to compute the area.

6. *Statements 14 to 19.* The values of the X (first moment) array are placed in the W array and FUNC1 performs the integration for the first moment (Statements 14 to 16). Similarly, the computation for the second moment is done.

7. *Statements 20 and 21.* The results are printed out. Statement 20 ends the loops started in Statements 3 and 5. After Statement 20 is completed, the computer returns to Statement 5 and the Statements 7 to 20 are executed again for the new value of ZBAR. After this is completed, it returns to Statement 3 and does Statements 5 to 20 for the new value of SIGMA. When this is done the computation stops and the program ends via CALL EXIT and END.

8. FUNCTION FUNC1 subprogram follows the main program as shown.

Suppose we wished to check the accuracy of our results by doubling the number of subintervals. The way the program is now written we have lost the original entries in the W array and would thus have to recompute them. We still have the values of X and Y, however. We can do the following: set up new arrays, XX and YY. Transfer X(I) to XX(2 * I) and Y(I) to YY(2 * I). Then we need only compute XX(3), XX(5), ... XX(401) and YY(3), YY(5), ... YY(401). (i.e., only the new values) and perform the integration by transferring XX and YY to W. The same procedure could be followed for the original values in the W array by transferring them to another array.

C.3 BIVARIATE INTEGRATION

General

Here the problem is of the form

$$V = \int_R \int f(x, y)\, dA.$$

Table C.3 LISTING OF BIVARIATE NUMERICAL INTEGRATION PROGRAM

```
1 2 3 4 5 6 7 8 9 101 2 3 4 5 6 7 8 9 201 2 3 4 5 6 7 8 9 301 2 3 4 5 6 7 8 9 401 2 3 4 5 6 7 8 9 501 2 3 4 5 6 7 8 9 601 2 3 4 5 6 7 8 9 701 2 3 4
            IDCARD
*           XEQ
*           LINES(10000)
            DIMENSION W(300),PB(300),PG(300),A(300),C(300)
            DIMENSION P(122,102),Z1(50),Z2(50),Z3(50),X(50),Y(50)
            COMMON W
a)          RIT 5, 2,(Z1(I),Z2(I),Z3(I),I=1,20)
          2 FORMAT(12X,3F7.0)
            DO 3I=1,20
            Z2(I)=Z2(I)*2.
            Z3(I)=5.+Z3(I)*4.
            X(I)=Z3(I)+Z2(I)
          3 Y(I)=2.+Z3(I)+Z1(I)
            SUM1=0.
            SUM2=0.
            DO 5 I=1,20
            SUM1=SUM1+X(I)
          5 SUM2=SUM2+Y(I)
            XM=SUM1/20.
            YM=SUM2/20.
            XX=0.
            YY=0.
            XY=0.
            DO 7 I=1,20
            XX=XX+(X(I)-XM)**2
            YY=YY+(Y(I)-YM)**2
          7 XY=XY+(X(I)-XM)*(Y(I)-YM)
            HG=.2
            HB=.025
            DO 20 I=1,121
b)          B=.5+FLOATF(I-1)*HB
            DO 10 J=2,101
c)          G=FLOATF(J-1)*HG
d)          R=28.5*LOGF(G)+57.*LOGF(1.+G*B**2)
            S=58.*LOGF(G*(YY-2.*B*XY+B**2*XX)+(1.+G*B**2)*(40.*G+160.))
            W(J)=EXPF(370.+R-S)
e)       10 P(I,J)=W(J)
f)       20 PB(I)=FUNC1(20.0,0.0,100)
g)          DO 30 K=1,121
         30 W(K)=PB(K)
```

```
1 2 3 4 5 6 7 8 9 101 2 3 4 5 6 7 8 9 201 2 3 4 5 6 7 8 9 301 2 3 4 5 6 7 8 9 401 2 3 4 5 6 7 8 9 501 2 3 4 5 6 7 8 9 601 2 3 4 5 6 7 8 9 701 2 3 4
i)            DQ 4Q J=1,121
              PB(J)=PB(J)/VOL
j)            A(J)=.5+FLOATF(J-1)*HB
         40   W(J)=PB(J)*A(J)
              BM=FUNC1(3.5,.5,120)
k)            DO 50 I=1,121
         50   W(I)=PB(I)*(A(I)-BM)**2
              BV=FUNC1(3.5,.5,120)
l)            WRITE OUTPUT TAPE 6,51
              WRITE OUTPUT TAPE 6,55, VOL, BM, BV
              DO 60 J=1,121
         60   WRITE OUTPUT TAPE 6,65, A(J), PB(J)
         51   FORMAT(10X,39HMARGINAL POSTERIOR DISTRIBUTION OF BETA)
         55   FORMAT(//20X,E14.8,5X,2F10.5/)
         65   FORMAT(20X,F7.3,10X,E14.8)
m)            DO 200 L=1,101
              DO 100 K=1,121
        100   W(K)=P(K,L)
        200   PG(L)=FUNC1(3.5,.5,120)
              DO 300 I=1,121
              DO 300 J=1,101
        300   P(I,J)=P(I,J)/VOL
              DO 400  J=1,101
              C(J)=FLOATF(J-1)/5.
              PG(J)=PG(J)/VOL
        400   W(J)=PG(J)*C(J)
              GM=FUNC1(20.,0.,100)
              DO 500 I=1,101
        500   W(I)=PG(I)*(C(I)-GM)**2
              GV=FUNC1(20.,0.,100)
              WRITE OUTPUT TAPE 6,510
              WRITE OUTPUT TAPE 6,550, VOL, GM, GV
              DO 600 J=1,101
        600   WRITE OUTPUT TAPE 6,650, C(J), PG(J)
        510   FORMAT(1H1,10X,40HMARGINAL POSTERIOR DISTRIBUTION OF GAMMA)
        550   FORMAT(//20X,E14.8,5X,2F10.5/)
        650   FORMAT(20X,F7.3,10X,E14.8)
              CALL EXIT
              END
```

Table C.3 (continued)

```
1 2 3 4 5 6 7 8 9 101 2 3 4 5 6 7 8 9 201 2 3 4 5 6 7 8 9 301 2 3 4 5 6 7 8 9 401 2 3 4 5 6 7 8 9 501 2 3 4 5 6 7 8 9 601 2 3 4 5 6 7 8 9 701 2 3 4

n)          FUNCTION  FUNC1(UP,SL,MM)
            DIMENSION  W(300)
            COMMON  W
            G=FLOATF(MM)
            H=(UP-SL)/G
            N1=MM/2
            N2=N1-1
            Q1=0.0
            Q2=0.0
            MP=MM+1
            DO 290  I=1,N1
      290   Q1=Q1+W(2*I)
            DO 295  I=1,  N2
      295   Q2=Q2+W(2*I+1)
            ORQ=(H/3.0)*(W(1)+W(MP)+Q1*4.0+Q2*2.0)
            FUNC1=ORQ
            RETURN
            END
o)  *       DATA
              1.556    0.119   -0.078   -0.856
              0.647    1.029    1.186    0.276
              0.329    0.407    1.169    0.379
              1.188    1.171    1.170    1.468
             -0.917   -0.616   -1.589   -1.805
              0.414    0.469    0.182    2.285
              0.107   -0.187    1.343   -0.602
              0.497    1.907    0.191    0.229
              0.501    0.083   -0.321    1.382
              1.382   -0.738    1.225    0.978
             -0.590    0.699   -0.162   -0.678
              1.125    1.111   -1.065   -0.366
              0.849    0.169   -0.351   -1.074
              1.233   -0.585    0.306   -0.600
              0.104   -0.468    0.185    0.918
              0.261    1.883   -0.181    0.791
             -0.007    1.280    0.568    0.598
              0.794   -0.111    0.040    0.567
              0.431   -2.300   -1.081    0.963
             -0.149    1.294    0.580    0.489
```

In calculus the above integral is evaluated by iterated integration; that is,

$$V = \int_c^d \left(\int_{x_1}^{x_2} f(x, y)\, dx \right) dy,$$

where x_1, x_2 in general depend on y. Here we use the same procedure; that is, we use Simpson's rule twice.

The objects of interest may be a proper density function $p(x, y) = (1/V)f(x, y)$, marginal density functions, and various moments.

FORTRAN Program

The program used is a straightforward extension of Simpson's Rule II. We now have a grid of points (x_i, y_j) corresponding to the endpoints of the x-subintervals and the endpoints of the y-subintervals. The procedure used to find a volume is to fix x_i and evaluate the integrand at all y_j and use Simpson's rule (FUNC1) to compute this integral. This is done for all x_i. Application of Simpson's rule to these results then yields the desired volume.

We illustrate with a program for finding marginal posterior densities of two parameters B and G, given some sample evidence. See the listing in Table C.3.

1. Computation and definition of constants used in evaluating the integrand. Here, some of the constants are computed from data and read into the computer.
2. Definition of "current" B value (indexed by I).
3. Definition of "current" G value (indexed by J).
4. Evaluation of integrand for one G value with the value of B held constant.
5. Storage of this value in a two-dimensional array P(I, J). On each "pass" an element corresponding to the current B and G values is added to this array and computer returns to "DO 10 . . ." and repeats (3), (4), and (5) until an entire row in P(I, J) has been computed, that is, does all values of G.
6. For given B, Simpson's rule is used to compute the univariate integral over G and the computer returns to "DO 20 . . .", repeating the above for the next value of B.
7. The results of the preceding integrations are put into the W array.
8. Integration over B is performed to find the volume.
9. PB (J) is transformed into a proper marginal density.
10. The first moment of this distribution is found by use of Simpson's rule on elements of the form $x_i \cdot p(x_i)$.
11. Similarly, the variance is computed.
12. The results are printed.
13. In a like manner the marginal posterior density of G is found by

transferring a column of P(I, J) to the W array and using Simpson's rule. Similarly, the mean and variance are found and the results printed.

14. FUNCTION FUNC1 (UP, SL, MM)—Simpson's rule computation.
15. Input data.

C.4 COMMENTS

The IBM 7094 computer, for example, has approximately 32,000 storage locations. This places definite limitations on the size of the P(I, J) array. If a plot of the bivariate density is not desired (no printout for it was provided in the program above), it is possible to do away with the P(I, J) array and thereby greatly increase the maximum number of intervals the computer can handle.

Bibliography

Aigner, D. J., and A. S. Goldberger, "On the Estimation of Pareto's Law," Workshop Paper 6818, Social Systems Research Institute, University of Wisconsin, Madison, 1968.

Anderson, T. W., *An Introduction to Multivariate Statistical Analysis.* New York: Wiley, 1958.

Ando, A., and G. M. Kaufman, "Bayesian Analysis of Reduced Form Systems," manuscript, MIT, 1964.

———, "Bayesian Analysis of the Independent Multinormal Process—Neither Mean nor Precision Known," *J. Am. Statist. Assoc.*, **60**, 347–353 (1965).

Anscombe, F. J., "Bayesian Statistics," *Am. Statist.*, **15**, 21–24 (1961).

Aoki, M., *Optimization of Stochastic Systems.* New York: Academic, 1967.

Arrow, K., H. Chenery, B. Minhas, and R. Solow, "Capital-Labor Substitution and Economic Efficiency," *Rev. Econ. Statist.*, **43**, 225–250 (1961).

Barlow, R., H. Brazer, and J. N. Morgan, *Economic Behavior of the Affluent.* Washington, D.C.: Brookings Institution, 1966.

Barten, A. P., "Consumer Demand Functions under Conditions of Almost Additive Preferences," *Econometrica*, **32**, 1–38 (1964).

Bartholomew, D. J., "A Comparison of Some Bayesian and Frequentist Inferences," *Biometrika*, **52**, 19–35 (1965).

Bartlett, M. S., "A Comment on D. V. Lindley's Statistical Paradox," *Biometrika*, **44**, 533–534 (1957).

Bayes, Rev. T., "An Essay Toward Solving a Problem in the Doctrine of Chances," *Phil. Trans. Roy. Soc. (London)*, **53**, 370–418 (1763); reprinted in Biometrika, **45**, 293–315 (1958), and *Facsimiles of Two Papers by Bayes* (commentary by W. Edwards Deming). New York: Hafner, 1963.

Bellman, R., *Introduction to Matrix Analysis.* New York: McGraw-Hill, 1962.

———, and R. Kalaba, "Dynamic Programming and Adaptive Processes: Mathematical Foundations," reprinted from *IRE Trans. Aut. Control*, **AC-5** (1) (January 1960), in R. Bellman and R. Kalaba (Eds.), *Selected Papers on Mathematical Trends in Control Theory.* New York: Dover, 1964, pp. 195–200.

Boot, J. C. G., and G. M. de Wit, "Investment Demand: An Empirical Contribution to the Aggregation Problem," *Intern. Econ. Rev.*, **1**, 3–30 (1960).

Box, G. E. P., and D. R. Cox, "An Analysis of Transformations," *J. Roy. Statist. Soc., Series B*, **26**, 211–243 (1964).

Box, G. E. P., and W. J. Hill, "Discrimination Among Mechanistic Models," *Technometrics*, **9**, 57–71 (1967).

Box, G. E. P., and G. C. Tiao, "A Further Look at Robustness via Bayes Theorem," *Biometrika*, **49**, 419–433 (1962).

———, "Multiparameter Problems from a Bayesian Point of View," *Ann. Math. Statist.*, **36**, 1468–1482 (1965).

Brown, P. R., *Some Aspects of Valuation in the Railroad Industry*, unpublished doctoral dissertation, University of Chicago, 1968.

Carlson, F. D., E. Sobel, and G. S. Watson, "Linear Relationships between Variables Affected by Errors," *Biometrics*, **22**, 252–267 (1966).

Chetty, V. K., "Bayesian Analysis of Haavelmo's Models," *Econometrica*, **36**, 582–602 (1968).

———, *Bayesian Analysis of Some Simultaneous Equation Models and Specification Errors*, unpublished doctoral dissertation, University of Wisconsin, Madison, 1966.

———, "Discrimination, Estimation and Aggregation of Distributed Lag Models," manuscript, Columbia University, 1968.

———, "On Pooling of Time Series and Cross-Section Data," *Econometrica*, **36**, 279–290 (1968).

Chow, G., "Multiplier, Accelerator and Liquidity Preference in the Determination of National Income in the United States," *IBM Record Rept.*, RC 1455 (1966).

Cochran, W. G., "The Planning of Observational Studies of Human Populations," *J. Roy. Statist. Soc., Series A*, Part 2, 234–255 (1965).

Copas, J. B., "Monte Carlo Results for Estimation in a Stable Markov Time Series," *J. Roy. Statist. Soc., Series A*, No. 1, 110–116 (1966).

Cook, M. B., "Bivariate κ-statistics and Cumulants of their Joint Sampling Distribution," *Biometrika*, **38**, 179–195 (1951).

Cragg, J. G., "On the Sensitivity of Simultaneous-Equations Estimators to the Stochastic Assumptions of the Models," *J. Am. Statist. Assoc.*, **61**, 136–151 (1966).

———, "Small Sample Properties of Various Simultaneous Equation Estimators: The Results of Some Monte Carlo Experiments," Research Memo 68, Econometric Research Program, Princeton University, 408 pp., 1964.

Crockett, J., "Technical Note," in I. Friend and R. Jones (Eds.), *Proceedings of the Conference on Consumption and Savings*, Vol. II. Philadelphia: University of Pennsylvania, 1960.

Cyert, R. M., and M. H. de Groot, "Bayesian Analysis and Duopoly Theory," manuscript, Carnegie-Mellon University, April 1968.

Dickey, J. M., "Matricvariate Generalizations of the Multivariate t Distribution and the Inverted Multivariate t Distribution," *Ann. Math. Statist.*, **38**, 511–518 (1967).

Drèze, J., "The Bayesian Approach to Simultaneous Equations Estimation," Research Memorandum No. 67, Technological Institute, Northwestern University, 1962.

———, "Limited Information Estimation from a Bayesian Viewpoint," CORE Discussion Paper 6816, University of Louvain, 1968.

Feldstein, M. S., "Production with Uncertain Technology: Some Econometric Implications," manuscript, Harvard University, 1969.

Fieller, E. F., "The Distribution of the Index in a Normal Bivariate Population," *Biometrika*, **24**, 428–440 (1932).

Fisher, F. M., "Dynamic Structure and Estimation in Economy-wide Econometric Models," in J. S. Duesenberry *et al.*, *The Brookings Quarterly Econometric Model of the United States*. Chicago: Rand-McNally, 1965, pp. 589–653.

Fisher, W. D., "Estimation in the Linear Decision Model," *Intern. Econ. Rev.*, **3**, 1–29 (1962).

Frazer, F. A., W. J. Duncan, and A. R. Collar, *Elementary Matrices*. Cambridge: Cambridge University Press, 1963.

Freimer, M., "A Dynamic Programming Approach to Adaptive Control Processes," *IRE Trans.*, **AC-4**, **2** (2), 10–15 (1959).

Friedman, M., and D. Meiselman, "The Relative Stability of Monetary Velocity and the Investment Multiplier in the United States, 1897–1958," in the Commission on Money and Credit Research Study, *Stabilization Policies*. Englewood Cliffs, N.J.: Prentice-Hall, 1963.

Fuller, W. A., and J. E. Martin, "The Effects of Autocorrelated Errors on the Statistical Estimation of Distributed Lag Models," *J. Farm Econ.*, **44**, 71–82 (1962).

Geary, R. C., "The Frequency Distribution of the Quotient of Two Normal Variates," *J. Roy. Statist. Soc.*, **93**, 442–446 (1930).

Geisel, M. S., "Comparing and Choosing Among Parametric Statistical Models: A Bayesian Analysis with Macroeconomic Applications," unpublished doctoral dissertation, University of Chicago, 1970.

Geisser, S., "A Bayes Approach for Combining Correlated Estimates," *J. Am. Statist. Assoc.*, **60**, 602–607 (1962).

———, "Bayesian Estimation in Multivariate Analysis," *Ann. Math. Statist.*, **36**, 150–159 (1965).

Goldberger, A. S., *Econometric Theory*. New York: Wiley, 1964.

Graybill, F. A., *An Introduction to Linear Statistical Models*. New York: McGraw-Hill, 1961.

Greville, T. N. E., "Some Applications of the Pseudoinverse of a Matrix," *SIAM Rev.*, **2**, 15–22 (1960).

———, "The Pseudoinverse of a Rectangular or Singular Matrix and Its Application to the Solution of Systems of Linear Equations," *SIAM Rev.*, **1**, 38–43 (1959).

Griliches, Z., "Distributed Lag Models: A Review Article," *Econometrica*, **35**, 16–49 (1967).

——, G. S. Maddala, R. Lucas, and N. Wallace, "Notes on Estimated Aggregate Quarterly Consumption Functions," *Econometrica*, **30**, 491–500 (1965).

Grunfeld, Y., "The Determinants of Corporate Investment," unpublished doctoral dissertation, University of Chicago, 1958.

Haavelmo, T., "Methods of Measuring the Marginal Propensity to Consume," *J. Am. Statist. Assoc.*, **42**, 105–122 (1947), reprinted in Wm. C. Hood and T. C. Koopmans (Eds.), *Studies in Econometric Methods*. New York: Wiley, 1953, pp. 75–91.

Hadamard, J., *The Psychology of Invention in the Mathematical Field*. New York: Dover, 1945.

Hanson, N. R., *Patterns of Discovery*. New York: Cambridge University Press, 1958.

Hartigan, J., "Invariant Prior Distributions," *Ann. Math. Statist.*, **35**, 836–845 (1964).

Hildreth, C., "Bayesian Statisticians and Remote Clients," *Econometrica*, **31**, 422–438 (1963).

Hildreth, C., and J. Y. Lu, "Demand Relations with Autocorrelated Disturbances," Tech. Bull. 276. East Lansing, Mich.: Michigan State University Agricultural Experiment Station, 1960.

Hobson, E. W., *The Theory of Functions of a Real Variable and the Theory of Fourier's Series*, Vol. II. New York: Dover, 1957.

Holt, C., J. F. Muth, F. Modigliani, and H. A. Simon, *Planning Production, Inventories and Work Force.* Englewood Cliffs, N.J.: Prentice-Hall, 1960.

Hood, W. C., and T. C. Koopmans (Eds.), *Studies in Econometric Method.* New York: Wiley, 1953.

James, W., and C. M. Stein, "Estimation with Quadratic Loss," in J. Neyman (Ed.), *Proceedings of the Fourth Berkeley Symposium on Mathematical Statistics and Probability*, Vol. I. Berkeley: University of California Press, 1961.

Jeffreys, H., *Scientific Inference* (2nd ed.). Cambridge: Cambridge University Press, 1957.

———, *Theory of Probability* (3rd ed.). Oxford: Clarendon, 1961 and 1966.

Johnson, R. A., "An Asymptotic Expansion for Posterior Distributions," Tech. Report No. 114, May 1967, Department of Statistics, University of Wisconsin, Madison, *Ann. Math. Statist.*, **38**, 1899–1906 (1967).

Johnston, J., *Econometric Methods.* New York: McGraw-Hill, 1963.

Kakwani, N. C., "The Unbiasedness of Zellner's Seemingly Unrelated Regression Equations Estimators," *J. Am. Statist. Assoc.*, **63**, 141–142 (1967).

Kendall, M. G., and A. Stuart, *The Advanced Theory of Statistics*, Vol. I. London: Griffen, 1958.

———, *The Advanced Theory of Statistics*, Vol. II. New York: Hafner, 1961 and 1966.

Kenney, J. F., and E. S. Keeping, *Mathematics of Statistics*: *Part Two* (2nd ed.). New York: Van Nostrand, 1951.

Kiefer, J., and J. Wolfowitz, "Consistency of the Maximum Likelihood Estimator in the Presence of Infinitely Many Incidental Parameters," *Ann. Math. Statist.*, **27**, 887–906 (1957).

Kmenta, J., and R. F. Gilbert, "Small Sample Properties of Alternative Estimates of Seemingly Unrelated Regressions," *J. Am. Statist. Assoc.*, **63**, 1180–1200 (1968).

Koyck, L., *Distributed Lags and Investment Analysis.* Amsterdam: North-Holland, 1954.

Kullback, S., *Information Theory and Statistics.* New York: Wiley, 1959.

Le Cam, L. "Les Propriétés Asymptotiques des Solutions de Bayes," *Publ. Inst. Statist.*, University of Paris, **7**, 17–35 (1958).

———, "On Some Asymptotic Properties of Maximum Likelihood and Related Bayes Estimates," *Univ. Calif. Publs. Statist.*, **1**, 277–330 (1953).

Lindley, D. V., "The Use of Prior Probability Distributions in Statistical Inference and Decisions," in J. Neyman (Ed.), *Proc. Fourth Berkeley Symp. Math. Statist. and Probab.*, Vol. I, 1961, 453–468.

———, "Regression Lines and the Linear Functional Relationship," *J. Roy. Statistical Soc.* (Supplement), **9**, 218–244 (1947).

———, *Introduction to Probability and Statistics from a Bayesian Viewpoint*, *Part 2. Inference.* Cambridge: Cambridge University Press, 1965.

———, "A Statistical Paradox," *Biometrika*, **44**, 187–192 (1957).

Lindley, D. V., and G. M. El-Sayyad, "The Bayesian Estimation of a Linear Functional Relationship," *J. Roy. Statist. Soc.*, Series B, **30**, 190–202 (1968).

Luce, R. D., and H. Raiffa, *Games and Decisions*. New York: Wiley, 1958.

Madansky, A., "The Fitting of Straight Lines When Both Variables are Subject to Error," *J. Am. Statist. Assoc.*, **54**, 173–205 (1959).

Marsaglia, G., "Ratios of Normal Variables and Ratios of Sums of Uniform Variables," *J. Am. Statist. Assoc.*, **60**, 193–204 (1965).

Miller, M. H., and F. Modigliani, "Some Estimates of the Cost of Capital to the Electric Utility Industry, 1954–57," *Am. Econ. Rev.*, **56**, 333–391 (1966).

Moore, E. H., *General Analysis*, Part I, Philadelphia: *Mem. Am. Phil. Soc.*, Vol. I, 1935.

Moore, H., "Notes of Sculpture," in B. Ghiselin (Ed.), *The Creative Process*. Mentor Book, 1952.

Nagar, A. L., "The Bias and Moment Matrix of the General k Class Estimators of the Parameters in Simultaneous Equations," *Econometrica*, **27**, 575–595 (1959).

Nerlove, M., "A Tabular Survey of Macro-Econometric Models," *Intern. Econ. Rev.*, **7**, 127–173 (1966).

Neyman, J., and E. L. Scott, "Consistent Estimates Based on Partially Consistent Observations," *Econometrica*, **16**, 1–32 (1948).

Orcutt, G. H., and H. S. Winkour, Jr., "First Order Autoregression: Inference, Estimation, and Prediction," *Econometrica*, **37**, 1–14 (1969).

Pearson, K. (Ed.), *Tables of the Incomplete Beta Function*. Cambridge: Cambridge University Press, 1948.

———, *The Grammar of Science*. London: Everyman, 1938.

Penrose, R., "A Generalized Inverse for Matrices," *Proc. Cambridge Phil. Soc.*, **51**, 406–413 (1955).

Plackett, R. L., "Current Trends in Statistical Inference," *J. Roy. Statist. Soc.*, Series A, **129**, Part 2, 249–267 (1966).

Popper, K. R., *The Logic of Scientific Discovery*. New York: Science Editions, 1961.

Prescott, E. C., *Adaptive Decision Rules for Macro Economic Planning*, unpublished doctoral dissertation, Graduate School of Industrial Administration, Carnegie-Mellon University, 1967.

Press, S. J., "On Control of Bayesian Regression Models," manuscript (undated).

———, "The t-Ratio Distribution," *J. Am. Statist. Assoc.*, **64**, 242–252 (1969).

———, and A. Zellner, "On Generalized Inverses and Prior Information in Regression Analysis," manuscript, September 1968.

Raiffa, H. A., and R. S. Schlaifer, *Applied Statistical Decision Theory*, Boston: Graduate School of Business Administration. Harvard University, 1961.

Rao, C. R., "A Note on a Generalized Inverse of a Matrix with Applications to Problems in Mathematical Statistics," *J. Roy. Statist. Soc.*, Series B, **24**, 152–158 (1962).

———, *Linear Statistical Inference and Its Applications*. New York: Wiley, 1965.

Reichenbach, H., *The Rise of Scientific Philosophy*. Berkeley: University of California Press, 1958.

Reiersol, O., "Identifiability of a Linear Relation Between Variables which are Subject to Error," *Econometrica*, **18**, 375–389 (1950).

Richardson, D. H., "The Exact Distribution of a Structural Coefficient Estimator," *J. Am. Statist. Assoc.*, **63**, 1214–1226 (1968).

Roberts, H. V., "Statistical Dogma: One Response to a Challenge," *Am. Statist.*, **20**, 25–27 (1966).

Rothenberg, T. J., "A Bayesian Analysis of Simultaneous Equation Systems," Report 6315, Econometric Institute, Netherlands School of Economics, Rotterdam, 1963.

Roy, S. N., *Some Aspects of Multivariate Analysis.* New York: Wiley, 1957.

Samuelson, P. A., "Interactions between the Multiplier Analysis and the Principle of Acceleration," *Rev. Econ. Statist.*, **21**, 75–78 (1939).

Savage, L. J., "Bayesian Statistics," in *Decision and Information Processes.* New York: Macmillan, 1962.

———, "Subjective Probability and Statistical Practice," in L. J. Savage et al., *The Foundations of Statistical Inference.* London and New York: Methuen and Wiley, 1962, pp. 9–35.

———, "The Subjective Basis of Statistical Practice," manuscript, University of Michigan, 1961.

Sawa, T., "The Exact Sampling Distribution of Ordinary Least Squares and Two-Stage Least Squares Estimators," *J. Am. Statist. Assoc.*, **64**, 923–937 (1969).

Shannon, C. E., "The Mathematical Theory of Communication," *Bell System Tech. J.* (July-October 1948), reprinted in C. E. Shannon and W. Weaver, *The Mathematical Theory of Communication.* Urbana: University of Illinois Press, 1949, pp. 3–91.

Simon, H. A., "Dynamic Programming under Uncertainty with a Quadratic Criterion Function," *Econometrica*, **24**, 74–81 (1956).

Smirnov, N. V., *Tables for the Distribution and Density Functions of t-Distribution.* New York: Pergamon, 1961.

Solow, R. M., "On a Family of Lag Distributions," *Econometrica*, **28**, 393–406 (1960).

Stein, C. M., "Confidence Sets for the Mean of a Multivariate Normal Distribution," *J. Roy. Statist. Soc.*, Series B, **24**, 165–285 (1962).

———, "Inadmissibility of the Usual Estimator for the Mean of a Multivariate Normal Distribution, in J. Neyman (Ed.) *Proceedings of the Third Berkeley Symposium on Mathematical Statistics and Probability*, Vol. I. Berkeley: University of California Press, 1956.

Stone, M., "Generalized Bayes Decision Functions, Admissibility and the Exponential Family," Technical Report 74, Department of Statistics, University of Wisconsin, Madison, *Ann. Math. Statist.*, **38**, 618–622 (1967).

Summers, R., "A Capital Intensive Approach to the Small Sample Properties of Various Simultaneous Equation Estimators," *Econometrica*, **33**, 1–41 (1965).

Swamy, P. A. V. B., *Statistical Inference in Random Coefficient Regression Models*, unpublished doctoral dissertation, University of Wisconsin, Madison, 1968.

Theil, H., "A Note on Certainty Equivalence in Dynamic Planning," *Econometrica*, **25**, 346–349 (1957).

———, "On the Use of Incomplete Prior Information in Regression Analysis," *J. Am. Statistical Assoc.*, **58**, 401–414 (1962).

———, and J. C. G. Boot, "The Final Form of Econometric Equation Systems," *Rev. Intern. Statist. Inst.*, **30**, 136–152 (1962).

Theil, H., and A. S. Goldberger, "On Pure and Mixed Statistical Estimation in Economics," *Intern. Econ. Rev.*, **2**, 65–78 (1961).

Thornber, H., "Applications of Decision Theory to Econometrics," unpublished doctoral dissertation, University of Chicago, 1966.

———, "Bayes Addendum to Technical Report 6603 'Manual for B34T—A Stepwise Regression Program'," Graduate School of Business, University of Chicago, September 1967.

———, "Finite Sample Monte Carlo Studies: An Autoregressive Illustration," *J. Am. Statist. Assoc.*, **62**, 801–818 (1967).

———, "The Elasticity of Substitution: Properties of Alternative Estimators," manuscript, University of Chicago, 1966.

Tiao, G. C., and W. Y. Tan, "Bayesian Analysis of Random-Effect Models in the Analysis of Variance. I. Posterior Distribution of Variance Components," *Biometrika*, **52**, 37–53 (1965).

Tiao, G. C., and A. Zellner, "Bayes' Theorem and the Use of Prior Knowledge in Regression Analysis," *Biometrika*, **51**, 219–230 (1964).

———, "On the Bayesian Estimation of Multivariate Regression," *J. Roy. Statist. Soc.*, Series B, **26**, 277–285 (1965).

Tocher, K. D., "Discussion on Mr. Box and Dr. Wilson's Paper," *J. Roy. Statist. Soc.*, Series B, **13**, 39–42 (1951).

Welch, B. L., and H. W. Peers, "On Formulae for Confidence Points Based on Integrals of Weighted Likelihoods," *J. Roy. Statist. Soc.*, Series B, **25**, 318–324 (1963).

Widder, D. V., *Advanced Calculus*. Englewood Cliffs, N.J.: Prentice-Hall, 1947.

Wright, R. L., "A Bayesian Analysis of Linear Functional Relations," manuscript, University of Michigan, 1969.

Zarembka, P., "Functional Form in the Demand for Money," Social Systems Research Institute Workshop Paper, University of Wisconsin, Madison, 1966, *J. Am. Statist. Assoc.*, **63**, 502–511 (1968).

Zellner, A., "An Efficient Method of Estimating Seemingly Unrelated Regressions and Tests for Aggregation Bias," *J. Am. Statist. Assoc.*, **57**, 348–368 (1962).

———, "Estimators for Seemingly Unrelated Regression Equations: Some Exact Finite Sample Results," *J. Am. Statist. Assoc.*, **58**, 977–992 (1963).

———, "Bayesian Inference and Simultaneous Equation Econometric Models," paper presented to the First World Congress of the Econometric Society, Rome, 1965.

———, "On Controlling and Learning about a Normal Regression Model," manuscript, 1966, presented to Information, Decision and Control Workshop, University of Chicago.

———, "On the Analysis of First Order Autoregressive Models with Incomplete Data," *Inter. Econ. Rev.*, **7**, 72–76 (1966).

———, "Estimation of Regression Relationships Containing Unobservable Independent Variables," Intern. Econ. Rev., **11**, 441–454 (1970).

——— (Ed.), *Readings in Economic Statistics and Econometrics*. Boston: Little, Brown, 1968.

———, and V. K. Chetty, "Prediction and Decision Problems in Regression Models from the Bayesian Point of View," *J. Am. Statist. Assoc.*, **60**, 608–616 (1965).

Zellner, A., and M. S. Geisel, "Analysis of Distributed Lag Models with Applications to Consumption Function Estimation," invited paper presented to the Econometric Society, Amsterdam, September 1968, and published in *Econometrica*, **38**, 865–888 (1970).

———, "Sensitivity of Control to Uncertainty and Form of the Criterion Function," in D. G. Watts (Ed.), *The Future of Statistics*. New York: Academic, 1968.

Zellner, A., and D. S. Huang, "Further Properties of Efficient Estimators for Seemingly Unrelated Regression Equations," *Intern. Econ. Rev.*, **3**, 300–313 (1962).

Zellner, A., J. Kmenta, and J. Drèze, "Specification and Estimation of Cobb-Douglas Production Function Models," *Econometrica*, **34**, 784–795 (1966).

Zellner, A., and C. J. Park, "Bayesian Analysis of a Class of Distributed Lag Models," *Econometric Ann. Indian Econ. J.*, **13**, 432–444 (1965).

Zellner, A., and N. S. Revanker, "Generalized Production Functions," Social Systems Research Institute Workshop Paper 6607, University of Wisconsin, Madison, 1966, *Rev. Econ. Studies*, **36**, 241–250 (1969).

Zellner, A., and U. Sankar, "Errors in the Variables," manuscript, 1967.

Zellner, A., and H. Theil, "Three-Stage Least Squares: Simultaneous Estimation of Simultaneous Equations," *Econometrica*, **30**, 54–78 (1962).

Zellner, A., and G. C. Tiao, "Bayesian Analysis of the Regression Model with Autocorrelated Errors," *J. Am. Statist. Assoc.*, **59**, 763–778 (1964).

Author Index

Subject Index